Power Electronics and Power Systems

T0190054

For further volumes:
http://www.springer.com/series/6403

Slobodan N. Vukosavic

Electrical Machines

Springer

Slobodan N. Vukosavic
Dept. of Electrical Engineering
University of Belgrade
Belgrade, Serbia

ISBN 978-1-4899-8890-4   ISBN 978-1-4614-0400-2 (eBook)
DOI 10.1007/978-1-4614-0400-2
Springer New York Heidelberg Dordrecht London

Springer is part of Springer Science+Business Media (www.springer.com)

Preface

This textbook is intended for undergraduate students of Electrical Engineering as their first course in electrical machines. It is also recommended for students preparing a capstone project, where they need to understand, model, supply, control and specify electric machines. At the same time, it can be used as a valuable reference for other engineering disciplines involved with electrical motors and generators. It is also suggested to postgraduates and engineers aspiring to electromechanical energy conversion and having to deal with electrical drives and electrical power generation. Unlike the majority of textbooks on electrical machines, this book does not require an advanced background. An effort was made to provide text approachable to students and engineers, in engineering disciplines other than electrical.

The scope of this textbook provides basic knowledge and skills in Electrical Machines that should be acquired by prospective engineers. Basic engineering considerations are used to introduce principles of electromechanical energy conversion in an intuitive manner, easy to recall and repeat. The book prepares the reader to comprehend key electrical and mechanical properties of electrical machines, to analyze their steady state and transient characteristics, to obtain basic notions on conversion losses, efficiency and cooling of electrical machines, to evaluate a safe operating area in a steady state and during transient states, to understand power supply requirements and associated static power converters, to comprehend some basic differences between DC machines, induction machines and synchronous machines, and to foresee some typical applications of electrical motors and generators.

Developing knowledge on electrical machines and acquiring requisite skills is best suited for second year engineering students. The book is self-contained and it includes questions, answers, and solutions to problems wherever the learning process requires an overview. Each Chapter is comprised of an appropriate set of exercises, problems and design tasks, arranged for recall and use of relevant knowledge. Wherever it is needed, the book includes extended reminders and explanations of the required skill and prerequisites. The approach and method used in this textbook comes from the sixteen years of author's experience in teaching Electrical Machines at the University of Belgrade.

Readership

This book is best suited for second or third year Electrical Engineering undergraduates as their first course in electrical machines. It is also suggested to postgraduates of all Engineering disciplines that plan to major in electrical drives, renewables, and other areas that involve electromechanical conversions. The book is recommended to students that prepare capstone project that involves electrical machines and electromechanical actuators. The book may also serve as a valuable reference for engineers in other engineering disciplines that are involved with electrical motors and generators.

Prerequisites

Required background includes mathematics, physics, and engineering fundamentals taught in introductory semesters of most contemporary engineering curricula. The process of developing skills and knowledge on electrical machines is best suited for second year engineering students. Prerequisites do not include spatial derivatives and field theory. This textbook is made accessible to readers without an advanced background in electromagnetics, circuit theory, mathematics and engineering materials. Necessary background includes elementary electrostatics and magnetics, DC and AC current circuits and elementary skill with complex numbers and phasors. An effort is made to bring the text closer to students and engineers in engineering disciplines other than electrical. Wherever it is needed, the book includes extended reinstatements and explanations of the required skill and prerequisites. Required fundamentals are recalled and included in the book to the extent necessary for understanding the analysis and developments.

Objectives

- Using basic engineering considerations to introduce principles of electrome-chanical energy conversion and basic types and applications of electrical machines.
- Providing basic knowledge and skills in electrical machines that should be acquired by prospective engineers. Comprehending key electrical and mechani-cal properties of electrical machines.
- Providing and easy to use reference for engineers in general.
- Acquiring skills in analyzing steady state and transient characteristics of electri-cal machines, as well as acquiring basic notions on conversion losses, efficiency and heat removal in electrical machines.

- Mastering mechanical characteristics and steady state equivalent circuits for principal types of electrical machines.
- Comprehending basic differences between DC machines, induction machines and synchronous machines, studying and comparing their steady state operating area and transient operating area.
- Studying and apprehending characteristics of mains supplied and variable frequency supplied AC machines, comparing their characteristics and considering their typical applications.
- Understanding power supply requirements and studying basic topologies and characteristics of associated static power converters.
- Studying field weakening operation and analyzing characteristics of DC and AC machines in constant flux region and in the constant power region.
- Acquiring skills in calculating conversion losses, temperature increase and cooling methods. Basic information on thermal models and intermittent loading.
- Introducing and explaining the rated and nominal currents, voltages, flux linkages, torque, power and speed.

Teaching approach

- The emphasis is on the system overview - explaining external characteristics of electrical machines - their electrical and mechanical access. Design and construction aspects are of secondary importance or out of the scope of this book.
- Where needed, introductory parts of teaching units comprise repetition of the required background which is applied through solved problems.
- Mathematics is reduced to a necessary minimum. Spatial derivatives and differential form of Maxwell equations are not required.
- The goal of developing and using mathematical models of electrical machines, their equivalent circuits and mechanical characteristics persists through the book. At the same time, the focus is kept on physical insight of electromechanical conversion process. The later is required for proper understanding of conversion losses and perceiving the basic notions on specific power, specific torque, and torque-per-Ampere ratio of typical machines.
- Although machine design is out of the overall scope, some most relevant concepts and skills in estimating the machine size, torque, power, inertia and losses are introduced and explained. The book also explains some secondary losses and secondary effects, indicating the cases and conditions where the secondary phenomena cannot be neglected.

Field of application

Equivalent circuits, dynamic models and mechanical characteristics are given for DC machines, induction machines and synchronous machines. The book outlines the basic information on the machine construction, including the magnetic circuits and windings. Thorough approach to designing electric machines is left out of the book. Within the book, machine applications are divided in two groups; (i) Constant voltage, constant frequency supplied machines, and (ii) Variable voltage, variable frequency machines fed from static power converters. A number of most important details on designing electric machines for constant frequency and variable frequency operation are included. The book outlines basic static power converter topologies used in electrical drives with DC and AC machines. The book also provides basic information on loses, heating and cooling methods, on rated and nominal quantities, and on continuous and intermittent loading. For most common machines, the book provides and explains the steady state operating area and the transient operating area, the area in constant flux and field weakening range.

Acknowledgment

The author is indebted to Professors Miloš Petrović, Dragutin Salamon, Jožef Varga, and Aleksandar Stanković who read through the first edition of the book and made suggestion for improvements.

The author is grateful to his young colleagues, teaching assistants, postgraduate students, Ph.D. students and young professors who provided technical assistance, helped prepare solutions to some problems and questions, read through the chapters, commented, and suggested index terms.

Valuable technical assistance in preparing the manuscript, drawings, and tables were provided by research assistants Nikola Popov and Dragan Mihic.

The author would also like to thank Ivan Pejcic, Ljiljana Peric, Nikota Vukosavic, Darko Marcetic, Petar Matic, Branko Blanusa, Dragomir Zivanovic, Mladen Terzic, Milos Stojadinovic, Nikola Lepojevic, Aleksandar Latinovic, and Milan Lukic.

Contents

List of Figures

Chapter 1
Introduction

This chapter provides introduction to electromechanical energy conversion and rotating power converters. This chapter explains the role of electrical machines in electrical power systems, industry applications, and commercial and residential area and supports the need to study electrical machines and acquire skills in their modeling, supplying, and control. This chapter also discusses notation and system of units used throughout this book, specifies target knowledge and skills to be acquired, and explains prerequisites. This chapter concludes with remarks on further studies.

1.1 Power Converters and Electrical Machines

Electrical machines are power converters, devices that convert energy from one form into another. They convert mechanical work into electrical energy or vice versa. There are also power converters that convert electrical energy of one form into electrical energy of another form. They are called static power converters. Some sample power converters are listed below:

- Power converters that generate mechanical work by using electrical energy are called electrical motors. Electrical motors are electrical machines.
- Power converters that use the electrical energy of direct currents and voltages and convert this energy into electrical energy of AC currents and voltages are called inverters. Inverters belong to static power converters, and they make use of semiconductor power switches.
- Electrical generators convert mechanical work into electrical energy. They belong to electrical machines.
- Power transformers convert (transform) electrical energy from one system of AC voltages into electrical energy of another system of AC voltages, wherein the two AC systems have the same frequency.

S.N. Vukosavic, *Electrical Machines*, Power Electronics and Power Systems,
DOI 10.1007/978-1-4614-0400-2_1, © Springer Science+Business Media New York 2013

1.1.1 Rotating Power Converters

Electrical machines converting electrical energy to mechanical work are called *electrical motors*. Electrical machines converting mechanical work to electrical energy are called *electrical generators*. Mechanical energy usually appears in the form of a rotational movement; thus, electrical motors and generators are called *rotational power converters* or rotating electrical machines. The process of converting electrical energy to mechanical work is called *electromechanical conversion*. Different from rotational converters, *power transformers* are electrical machines which have no moving parts and convert electrical power of one system of AC voltages and currents into another AC system. The two AC systems have the same frequency, but their voltage levels are different due to transformation. This book deals with the rotating electrical machines, electrical generators and motors, whereas power transformers are dealt with by other textbooks.

Electrical machines comprise current circuits made of insulated conductors and magnetic circuits made of ferromagnetic materials. The machines produce mechanical work due to the action of electromagnetic forces on conductors and ferromagnetics coupled by a magnetic field. Conductors and ferromagnetic elements belong either to the moving part of the machine (*rotor*) or to the stationary part (*stator*). Rotation of the machine moving part contributes to variation of the magnetic field. In turn, an electromotive force is induced in the conductors, which allows generation of electrical energy. Similarly, electrical current in the machine conductors, called *windings*, interacts with the magnetic field and produces the forces that excite the rotor motion. Unlike electrical machines, the power transformers do not involve moving parts. Their operation is based on electromagnetic coupling between the primary and secondary windings encircling the same magnetic circuit.

1.1.2 Static Power Converters

In addition to electrical machines and power transformers, there are power converters whose operation is not based on electromagnetic coupling of current circuits and magnetic circuit. The converters containing semiconductor power switches are known as *static power converters* or *power electronics devices*. One such example is a diode rectifier, containing four power diodes connected into a bridge. Supplied by an AC voltage, diode rectifier outputs a pulsating DC voltage. Therefore, a diode rectifier carries out conversion of AC electrical energy into DC electrical energy. Conversion of DC electrical energy into AC electrical energy is carried out by inverters, static power converters containing semiconductor power switches like power transistors or power thyristors. Static power converters are frequently used in conjunction with electrical machines, but they are not studied within this book.

1.1.3 The Role of Electromechanical Power Conversion

Electromechanical conversion has a key role in production and uses of electrical energy. Electrical generators produce electrical energy, whereas motors are the consumers converting a considerable portion of electrical energy into mechanical work, required by production processes, transportation, lighting, and other industrial, residential, and household applications. Thanks to electromechanical conversion, energy is transported and delivered to remote consumers by means of electrical conductors. Electrical transmission is very reliable, it is not accompanied by emissions of gasses or other harmful substances, and it is carried out with low energy losses.

In electrical power plants, steam and water turbines produce mechanical work which is delivered to electrical generators. Through the processes taking place within a generator, the mechanical work is converted into electrical energy, which is available at generator terminals in the form of AC currents and voltages. High-voltage power lines transmit electrical energy to industrial centers and communities where power cables and lines of the distribution network provide the power supply to various consumers situated in production halls, transportation units, offices, and households. In the course of transmission and distribution, the voltage is transformed several times by using power transformers. Electrical generators, electrical motors, and power transformers are vital components of an *electrical power system*.

1.1.4 Principles of Operation

Electromechanical energy conversion can be accomplished by applying various principles of physics. Operation of electrical machines is usually based on the magnetic field which couples current-carrying circuits and moving parts of the machine. The conductors and ferromagnetic parts in the coupling magnetic field are subjected to electromagnetic forces. Conductors form contours and circuits carrying electrical currents. Flux linkage in a contour (called *flux*) can change due to changes in electrical current or due to motion. Flux change induces electromotive force in contours. The basic laws of physics determining electromechanical energy conversion in electrical machines with magnetic coupling field are:

- Faraday law of electromagnetic induction, which defines the relationship between a changing magnetic flux and induced electromotive force
- Ampère law, which describes magnetic field of conductors carrying electrical current
- Lorentz law, which determines the force acting on moving charges in magnetic and electrical fields
- Kirchhoff laws, which give relations between voltages and currents in current circuits and also between fluxes and magnetomotive forces in magnetic circuits

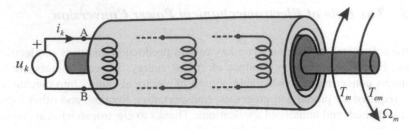

Fig. 1.1 Rotating electrical machine has cylindrical rotor, accessible via shaft. Stator has the form of a hollow cylinder, coaxial with the rotor

1.1.5 Magnetic and Current Circuits

The process of electromechanical energy conversion in electrical machines is based on interaction between the magnetic coupling field and conductors carrying electrical currents. Magnetic flux is channeled through magnetic circuits made of ferromagnetic materials. Electrical currents are directed through current conductors. Magnetic circuits are formed by stacking iron sheets separated by thin insulation layers, while current circuits are made of insulated copper conductors. The three most important types of electrical machines, DC current machines, asynchronous machines, and synchronous machines are of different constructions, and they use different ways of establishing magnetic fields and currents. Rotating electrical machines have a nonmoving part, *stator*, and a moving part, *rotor*, which can rotate around machine axis. The magnetic and current circuits could be mounted on both stator and rotor. In addition to the magnetic and current circuits, electrical machines also have other parts, like housing, shaft, bearings, and terminals of current circuits.

1.1.6 Rotating Electrical Machines

Mechanical work of electrical machines can be related to rotation or translation. Majority of electrical machines is made of rotating electromechanical converters producing rotational movement and having cylindrical rotors, like the one shown in Fig. 1.1. Electrical machines creating linear movement are called *linear motors*. Linear motors are rather rare.

Current circuits of a machine are called *windings*. They can be connected to external electrical sources or to electrical energy consumers. The ends of the winding are accessible as electrical terminals. In Fig. 1.1, terminals of kth winding are denoted by letters A and B. The electrical terminals permit *electrical access* to the machine. Since electrical machines perform electromechanical conversion, they have both electrical and mechanical accesses. Via electrical terminals, the machine can receive electrical energy from external sources or supply electrical energy to consumers in the circuits which are external to the machine. When an electrical machine has N windings, the power at electrical access of the machine is given by (1.1):

$$p_e = \sum_{k=1}^{N} u_k i_k. \tag{1.1}$$

Rotor is positioned inside a hollow cylindrical stator. Along rotor axis is a steel shaft, accessible at machine ends. Angular frequency of revolution of the rotor is also called *the rotor speed*, and it is denoted by Ω_m. At one end of the shaft, shown on the right of Fig. 1.1, the electrical machine can deliver or receive mechanical work. The shaft makes *mechanical terminal* of the machine. It transmits *rotational torque* or simply *torque* of the machine to external sources or consumers of mechanical work. The torque T_{em} in Fig. 1.1 is created by the interaction of the magnetic field and electrical current. Therefore, it is also called *electromagnetic torque*. In cases when the torque contributes to motion and acts toward the speed increase, it is called *driving torque*.

An electrical motor converts electrical energy to mechanical work. The later is delivered via shaft to a machine operating as a mechanical load, also called *work machine*. The motor acts on the work machine through the torque T_{em}, while the work machine opposes the rotation by the load torque T_m. In the case when the driving and load torques are equal, angular frequency of the rotation Ω_m does not change. Power delivered to a work machine by the electrical motor is determined by the product of the torque and speed:

$$p_m = \Omega_m T_m. \tag{1.2}$$

An electrical generator converts mechanical work to electrical energy. It receives the mechanical work from a water or steam turbine; thus, power p_m has a negative value. Rotational torque of the turbine T_m tends to set the rotor into motion, whereas the torque T_{em}, generated by the electrical machine, opposes this movement. By adopting reference directions shown in the right-hand side of Fig. 1.1, both T_{em} and T_m have negative values. Variable p_e, given by relation (1.1), defines the electrical power taken by the machine from external electrical circuits, i.e., the power taken from a supply network. Since electrical generator converts mechanical work to electrical energy and delivers it to a supply network, the generator power p_e has a negative value. The sign of these variables has to do with reference directions. Changing the reference directions for torques and currents in Fig. 1.1 would result in positive generator torques and positive generator power.

1.1.7 Reversible Machines

Electrical machines are mainly reversible. A reversible electrical machine may operate either as a generator converting mechanical work to electrical energy or as a motor converting electrical energy to mechanical work. Transition from the

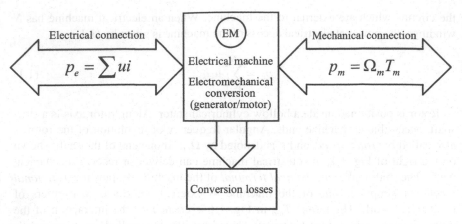

Fig. 1.2 Block diagram of a reversible electromechanical converter

generator to motor operating mode is accompanied by changes in the electrical and mechanical variables such as voltage, current, torque, and speed. The operating mode can be changed without modifications in the machine construction, with no changes in the current circuits, and without variations in the shaft coupling between the electrical machine and the work machine. An example of a reversible electrical machine is asynchronous motor. At angular rotor speeds lower than the *synchronous speed*, an asynchronous machine operates in the motor mode. If the speed is increased above the synchronous speed, the electromagnetic torque is opposed to motion while asynchronous machine converts mechanical work to electrical energy, thus operating in the generator mode.

Reversible energy conversion is shown in Fig. 1.2. Direction from left to right is taken as the reference direction for the power and energy flow. Power p_e at the electrical and p_m at the mechanical terminal of the machine have positive values in the motor mode, whereas in the generator mode these values are negative. Energy conversion is accompanied by energy losses in the current circuits, magnetic circuits, and also mechanical energy losses as a consequence of various forms of rotational friction. Due to losses, the power values at the electrical and mechanical terminals are not equal. In the motor mode, the obtained mechanical power p_m is somewhat lower than the invested electrical power p_e due to conversion losses. In the generator mode, the obtained electrical power $(-p_e)$ is somewhat lower than the invested mechanical power $(-p_m)$ because of the losses.

1.2 Significance and Typical Applications

Electrical energy is produced by operation of electrical generators. The produced energy is transmitted and distributed to energy consumers, mainly consisting of electrical motors which create controlled movement in work machines, whether household appliances, industry automation machines, robots, electrical vehicles, or machines in transportation systems.

The role of electrical machines in the processes and phases of production, transmission, distribution, and application of electrical energy is shown in Fig. 1.3. A brief description is given for each individual phase:

(a) Electrical energy can be obtained by using the potential energy of water accumulated in lakes; by using energy of coal, natural gas, or other fossil fuels; by using wind and tidal energy; by using nuclear fission, heat of underground waters, and energy of the sun; and by other means. These resources are the *primary energy sources*.

(b) In electrical power plants, the primary energy is at first converted to mechanical work. By burning fossil fuels, or using thermal springs, or in a nuclear reactor, generated heat is used to evaporate water and produce *overheated steam*. The steam acts on the blades of a steam turbine which rotates at speed Ω_m, creating rotational torque T_m. In hydroelectrical power plants, flow of water is directed to the blades of a hydroturbine. A turbine is also called *primer mover*.

(c) The obtained mechanical power $P_m = \Omega_m T_m$ is delivered to electrical generator, the electrical machine which converts mechanical work to electrical energy.

(d) Synchronous machines from 0.5 to 1,000 MW are predominantly used as generators in electrical power plants. Stator of the generator has three stationary *phase windings*. The rotor accommodates an *excitation winding* which determines the rotor flux. This flux does not move with respect to the rotor. Since the rotor revolves, the magnetic field of the rotor rotates with respect to the stator windings. Therefore, the rotor motion causes variation of the flux in the stator phase windings. Due to variation of the flux, an electromotive force is induced in the stator phase windings. Consequently, an AC voltage $u(t)$ is obtained at the stator winding terminals. When these terminals are connected to an external electrical circuit, AC currents $i(t)$ are established in the stator phase windings. The machine is connected to a transmission network which takes the role of an electrical consumer. The AC currents in the phase windings are dependent on the electrical load connected to the generator via transmission network. Electrical power obtained at machine terminals is $p_e = \sum ui$. The interaction of phase currents in the stator windings and magnetic field within the generator produces electromagnetic forces acting on the rotor which results in an electromagnetic torque T_{em}. This electromagnetic torque is a measure of mechanical interaction between the stator and the rotor. The electromagnetic torque acts on both the rotor and the stator. The stator is fixed and cannot move. The rotor speed depends on the torque T_m, acting toward the speed increase, and the generator torque T_{em}, acting toward the speed decrease. In an electrical machine operating as a generator, the torque T_m is obtained by operation of the steam or a hydroturbine. This torque tends to start and accelerate the rotor. The electromagnetic torque T_{em} opposes the rotor movement. Mechanical power input is higher than the obtained electrical power due to power losses in the electrical machine. In addition to the losses within the generator itself,

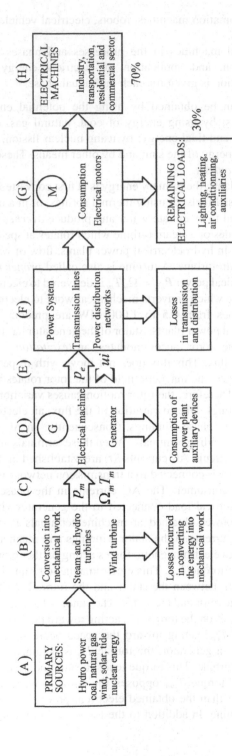

Fig. 1.3 The role of electrical machines in the production, distribution, and consumption of electrical energy

a part of the produced electrical energy is spent on covering consumption of the power plant itself. Contemporary electrical power plants are equipped with three-phase synchronous generators.[1]

(e) The electrical power obtained at generator terminals is determined by the voltages and currents. The system of AC voltages at the stator terminals has the rms[2] value from 6 to 25 kV, and frequency 50 Hz (60 Hz in some countries). A generator is connected to a *block transformer*, receiving the stator voltages on the primary side and providing the secondary side voltages compatible to the voltage level of the transmission power line.[3] At present, these voltages range up to 750 kV, and introduction of voltages of 1,000 kV is being considered.

(f) High-voltage power lines transmit electrical energy from power plants to the cities, communities, industrial zones, and transportation nodes, wherever the consumers of electrical energy are grouped. Distribution of electrical energy is carried out by the lower voltage power lines. Factories and residential blocks are usually supplied by 10–20 kV power lines or cables. Power transformers 10 kV/0.4 kV reduce the line voltages to 400 V and phase voltage to 231 V, supplied to the majority of consumers. Transmission and distribution of electrical energy are accompanied by losses in power lines and also in the transmission and distribution power transformers.

(g) Industrial consumers are using electrical motors for operating lathes, presses, rolling mills, milling machines, industrial robots, manipulators, conveyers,

---

[1] At the end of nineteenth century, the production, transmission, distribution, and application of electrical energy were dealing with DC currents and voltages. *Electrical power plants* were built in the centers of communities or close to industrial consumers, and they operated DC generators. Electrical motors were also DC machines. Both the production and application of electrical energy were relying on DC machines, either as generators or as consumers – DC motors. At the time, there were no power converters that would convert low DC voltage of the generator to a higher voltage which is more suitable for transmission. For this reason, energy transmission was carried out with high currents and considerable energy losses, proportional to the generator-load distance. Contemporary transmission networks apply three-phase system of AC voltages. The voltage level is changed by means of power transformers. A *block transformer* transforms the generator voltages to the voltage level encountered at the transmission lines. A sequence of transmission and distribution transformers reduces the three-phase line voltage to the level of 400 V which is supplied to majority of three-phase consumers. The phase voltage of single-phase consumers is 231 V. By using the three-phase system of AC currents, it is possible to achieve transmission of electrical energy to distances of several hundreds of kilometers. Therefore, contemporary power plants could be distant from consumers.

[2] "Root mean square," the thermal equivalent of an AC current, the square root of the mean (average) current squared

[3] In transmission of electrical energy by power lines over very large distances, greater than 1,000 km, transmission by AC system of voltages and currents could be replaced by DC transmission, that is, by power lines operating with DC currents and voltages. At the beginning of a very long transmission line, static power converters are applied to transform energy of the AC system into the energy of the DC system. At the end of the line, there is a similar converter which converts the energy of the DC system into energy of the AC system. In this way, voltage drops across series impedances of the power line are reduced, and the power that could be transmitted is considerably increased.

mills, pumps for various fluids, ventilators, elevators, drills, forklifts, and other equipment and devices involved in production systems and processes. In electrical home appliances, motors are applied in air conditioners, refrigerators, washing and dishwashing machines, freezers, mixers, mills, blenders, record players, CD/DVD players, and computers and computer peripherals. Approximately 8% of the total electrical energy consumption is spent for supplying motors in transportation units like railway, city transport, and electrical cars.

(h) In industrialized countries, 60–70% of the electrical energy production is consumed by electrical machines providing mechanical power and controlling motion in industry, traffic, offices, and homes. Therefore, it can be concluded that most of the electrical energy produced by generators is consumed by electrical motors which convert this energy to mechanical work. Electrical motors are coupled mechanically to machines handling mechanical work and power, carrying and transporting the goods, pumping fluids, or performing other useful operations. Most electrical motors draw the electric power from the three-phase distribution *grid* with line voltages of 400 V and *line frequency* of 50(60) Hz. If the speed of electrical motor has to be varied, it is necessary to use a *static power converter* between the motor terminals and the distribution grid. The static power converters are *power electronics* devices comprising semiconductor power switches. Their role is to convert the energy of line-frequency voltages and currents and to provide the motor supply voltages and currents with adjustable amplitude and frequency, suited to the motor needs.

1.3 Variables and Relations of Rotational Movement

Electrical machines are mainly rotating devices comprising a motionless stator which accommodates a cylindrical rotor. The rotor revolves around the axis which is common to both rotor and stator cylinders. Along this axis, the machine has a steel shaft serving for transmission of the produced mechanical work to an external work machine. There are also linear electrical machines wherein the moving part performs translation and is subject to forces instead of torques. Their use is restricted to solving particular problems in transportation and a relatively small number of applications in robotics.

Position of rotor is denoted by θ_m, and this angle is expressed in radians. The first derivative of the angle is mechanical speed of rotation, $d\theta_m/dt = \Omega_m$, expressed in radians per second. Sign of Ω_m depends on the adopted reference direction. It is adopted that positive direction of rotation is counterclockwise (CCW). Besides the rotor mechanical speed, this book also studies rotation of the magnetic field and rotation of other relevant electrical and magnetic quantities.

Speed of rotation of each of the considered variables will be denoted by the upper case Greek letter Ω, whereas the lower case Greek letter ω will be reserved

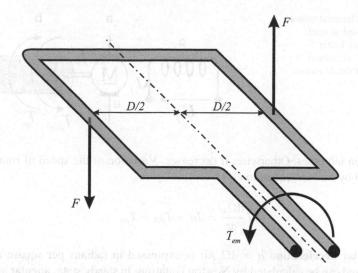

Fig. 1.4 Conductive contour acted upon by two coupled forces producing a torque

for denoting electrical angular frequency. Numerical values of the speed and frequency are usually expressed in terms of the SI system units, s^{-1}, i.e., in radians per second. The value $f = \omega/2\pi$ determines the frequency expressed as the number of cycles per second, Hz. The rotor speed can also be expressed as the number of revolutions per minute (rpm),

$$n = \frac{60}{2\pi}\Omega_m \approx 9.54\,\Omega_m. \tag{1.3}$$

Torque T_{em} generated in an interaction of the magnetic field and the winding currents is called *electromagnetic torque*. The electromagnetic torque produced by electric motors is also called *moving torque* or *driving torque*. Torque T_{em} is a measure of mechanical interaction between stator and rotor. A positive value of T_{em} turns the rotor counterclockwise. The torque contribution of individual conductors is shown in Fig. 1.4. It is assumed in Fig. 1.4 that the field of magnetic induction B extends in horizontal direction. Electrical current in conductors interacts with the field and creates the force F which acts on the conductors in vertical direction. The force depends on the field strength, on the current amplitude, and on the length of conductors. The torque exerted on single conductor is determined by the product of the force F, acting on the conductor, and perpendicular distance of the vector F from the axis of rotation, also called *the force arm*. Figure 1.4 shows a contour subjected to the action of two coupled forces producing the torque $T_{em} = FD$, where $R = D/2$ is the arm of the forces acting on conductors which are symmetrically positioned with respect to the axis of rotation.

The electromagnetic torque T_{em} is counteracted by the load torque T_m, representing the resistance of the mechanical load or work machine to the movement. In the case when the electromagnetic torque prevails, i.e., $T_{em} > T_m$, the speed

Fig. 1.5 Electrical motor (a) is coupled to work machine (b). Letter (c) denotes excitation winding of the dc motor

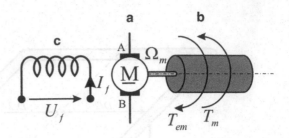

of rotation increases. Otherwise, it decreases. Variation of the speed of rotation is governed by Newton equation (1.4):

$$J\frac{d\Omega_m}{dt} = Ja = T_{em} - T_m. \tag{1.4}$$

Angular acceleration $a = d\Omega_m/dt$ is expressed in radians per square second [rad/s^2] and can be calculated by Newton equation. In steady state, angular acceleration is equal to zero. Then, the electromagnetic torque T_{em} is in equilibrium with the load torque T_m.

Moment of inertia J depends on the masses and shapes of all rotating parts. In case of stiff coupling of the shaft of an electrical machine to rotational masses of a work machine, total moment of inertia $J = J_R + J_{RM}$ is the sum of the rotor inertia J_R and inertia of rotating masses of the work machine J_{RM}. Figure 1.5 shows a work machine coupled to an electrical motor. It is assumed that rotational masses of the machine have cylindrical shapes of radius R and mass m. Moment of inertia of a solid cylinder is determined by expression $J = \frac{1}{2}mR^2$.

Question (1.1): A work machine has rotational mass of the shape of a very thin ring of radius R and mass m. Determine moment of inertia of the work machine.

Answer (1.1): $J_{RM} = mR^2$.

1.3.1   Notation and System of Units

Throughout this book, instantaneous values of the considered variables are denoted by lower case letters (u_p, i_p, $p_p = u_p i_p$), whereas steady state values, DC values, and root mean square values are denoted by upper case letters (U_p, I_p, $P_p = U_p I_p$), in accordance with recommendations of the International Electrotechnical Commission (IEC). Exceptions to these recommendations are only notations of the force F, torque T, and speed of rotation Ω. Upper case letter T is used for denoting both instantaneous and steady state values of the torque since lower case letter t is often used to denote other relevant variables of power converters. Speed of rotation is denoted by the upper case Greek letter Ω, whereas lower case letter ω denotes angular frequency. Both variables are expressed in radians per second, i.e., s^{-1}.

Relevant vectors of rotating machines are represented in the cylindrical coordinate system. These are usually planar, since their z components are equal to zero. By introducing complex plane, each vector can be represented by a complex number with real and imaginary parts corresponding to the projections of the plane vector to the axes of the coordinate system.[4] For example, voltage and current vectors are denoted by $\underline{u}_s$ and $\underline{i}_s$. It should be noted that the stated quantities are not constants; thus, $\underline{u}_s$ and $\underline{i}_s$ are not voltage and current phasors. Namely, their real and imaginary parts can vary independently during transient processes. In steady state, quantities $\underline{u}_s$ and $\underline{i}_s$, like other vectors represented in complex numbers, become complex constants and should be treated as phasors. In steady state, notation changes and becomes $\underline{U}_s$ or $\underline{I}_s$. Stationary matrices can be denoted by [A] or $\underline{A}$. All variables which can assume different numerical values are denoted by *italic*. Operators sin, cos, rot, div, mod, differentiating operator d, and others, as well as the measurement units, cannot be denoted by italic.

Notation of vectors such as magnetic induction, magnetic field, force, and other vectors in equations is usual, $\vec{B}, \vec{H}, \vec{F}$, etc. When these vectors are mentioned in the text, the upper arrow is not used and the magnetic induction vector is denoted by B, magnetic field vector by H, force vector by F, etc.

Within this book, coupled magnetic forces are called *electromagnetic torque*. In the introductory subjects of electrical engineering, forces acting on conductors in a magnetic field are called *magnetic forces*. The term *magnetic torque* is quite adequate, but the literature concerning electrical machines usually makes use of the term *electromagnetic torque*, so this term has been adopted in this book.

International System of Units (SI), introduced in 1954, has been used in this book. The system has been introduced in most countries. A merit of the SI system is that it allows calculations with no need for using special scaling factors. Therefore, by applying the SI system, remembering and using dedicated multiplication coefficients are no longer needed. For example, calculation of work W made by force F, acting along path l, is obtained by multiplying the force and path.[5] In the case of SI system, the work is determined by expression $W[\text{J}] = F[\text{N}]l[\text{m}]$, without any need for introducing additional scaling coefficients because $1\ \text{J} = 1\ \text{N} \times 1\ \text{m}$. In the case when the force is expressed in kilograms, distance in inches or feet, the result Fl would have to be multiplied by a scaling factor in order to obtain the work in joules or calories. In the analysis of electrical machines, one should check whether the results are expressed in correct units. In doing so, it is useful to know relations between the basic and derived units. Some of the useful relations are

---

[4] Representation of a plane vector by complex number is widely used in the technical literature concerning electrical machines. The vectors related to machine voltages and currents are also called *space vector*. Term *polyphasor* is also met. The complex notation is used here without any specific qualifier. A "vector $\underline{V}$," mentioned in the text, refers to the planar vector and implies that the unit vectors of the Cartesian coordinate system are formally replaced by the real and imaginary units.

[5] Work of the force is $W = Fl$ provided that the force is constant, that it moves along straight line, and that the course and direction of the force coincide with the path.

[Vs] = [Wb], [Nm rad/s] = [W], [Nm rad] = [J], [AΩ] = [V], [AH] = [Wb] = [Vs], etc.

Within this book, the rated values are denoted a subscript n, such as in U_n or I_n.

1.4   Target Knowledge and Skills

Knowledge of electrical machines is a basis for successful activity of an electrical engineer. A large number of applications and systems, designed or used by an electrical engineer, contain one or more electrical machines. Characteristics of these applications and systems are usually determined by the performances of electrical machines, their dimensions, mass, efficiency, peak torque capability, and speed range, as well as control characteristics and dynamic response. For this reason, it is necessary to acquire the knowledge and skills to understand basic operation principles of electrical machines. Basic understanding of mechanical and electrical characteristics of electrical machines is required to specify and design their power supply and controls. The knowledge concerning the origin and nature of energy losses in electrical machines is required to specify and design their cooling and conceive their loss-minimized use.

The most significant challenges in developing novel solutions are design of the magnetic circuit and windings, resolving power supply problems, and devising control laws. Machines should be designed to have the smallest possible dimensions and mass, and to operate with low energy losses. At the same time, machines should be as cheap and robust as possible. At present, the power supply and controls of electrical machines are carried out by using static power converters and digital signal controllers. Some of the goals of generator control are reduction of losses, reduction of electromagnetic and mechanical stress of materials, as well as increasing the power-to-mass ratio, also called *specific power*. Motor control aims at achieving as high as possible accuracy and speed of reaching the torque and speed targets required for performing desired movement of a work machine.

This book contains the basic knowledge concerning electrical machines necessary for future electrical engineers. The approach starts from the basic role and function of the machine. The characteristics of machine electrical and mechanical accesses (*ports*) are analyzed in order to define the mathematical model, equivalent circuits, and mechanical characteristics. This book deals with the elements of machine design, problems of heating and cooling, and also with specific imperfections of magnetic circuits and windings. The depth of the study is suited for understanding the operating principles of main machine types, for acquiring basic knowledge on power supply topologies, and for comprehending essential concepts of machine controls. Rotational electromechanical converters are studied in this book, whereas the power transformers are omitted.

1.4.1 Basic Characteristics of Electrical Machines

Design, specification, and analysis of electrical machine applications require an adequate idea on the size, mass, construction, reliability, and losses. The basic knowledge of electrical machines is required for designing systems incorporating electrical machines and solving the problems of power supply and controls. Basic electrical machine concepts are also required for designing monitoring and protection systems, designing servomotor controllers in robotics, and designing controls and protections for synchronous generators in electrical power plants. The knowledge of electrical machines is required in all situations and tasks likely to be put before an electrical engineer in industry, power generation, industrial automation, and robotics.

1.4.2 Equivalent Circuits

The torque and speed control of an electrical machine is performed by establishing the winding voltages and currents by means of an appropriate power supply. Design or selection of power supply for a given electrical machine requires establishing relations between the machine flux, torque, voltages, and currents. The steady state relations are described by a steady state equivalent circuit. The equivalent circuit is an electrical circuit containing resistors, inductances, and electromotive forces. At steady state, with constant speed and with the given amplitude and frequency of the supply, the equivalent circuit allows calculation of currents in the windings. Based upon the currents determined from the equivalent circuit, it is possible to calculate the steady state flux, torque, power of electromechanical conversion, and power losses.

1.4.3 Mechanical Characteristic

For the given voltage and frequency of the power supply, the calculation of the steady state values of the machine speed and torque requires the torque-speed characteristic of the work machine which is attached to the shaft and acts as mechanical load. The relation $T_m - \Omega_m$ of the work machine is also called *mechanical characteristic* of the load, and it is expressed by the function $T_m(\Omega_m)$. In a like manner, mechanical characteristic of the electrical machine is the relation between the electromagnetic torque T_{em} and the rotor speed Ω_m in the steady state. It is possible to express the mechanical characteristic by function $T_{em}(\Omega_m)$ and present it graphically in the T_{em}-Ω_m plane. Determination of the mechanical characteristic can be carried out by using mathematical model of the machine. The steady state operating point is found at the intersection of the two mechanical characteristics, $T_{em}(\Omega_m)$ and $T_m(\Omega_m)$.

1.4.4   Transient Processes in Electrical Machines

The equivalent circuit and mechanical characteristics can be used for the steady state analysis of electrical machines, i.e., in the operating modes where the torque, flux, speed of rotation, currents, and voltages do not change their values, amplitudes, or frequencies. There are numerous applications where it is required to accomplish fast variations of the torque and speed of rotation. In these applications, it is necessary to have a mathematical representation of the machine which reflects its behavior during transients. This representation is called *mathematical model*. Deriving the mathematical model, one cannot start from the assumption that machine operates in steady state. For this reason, such model is also called *dynamic model*. Examples of electrical machine applications where the dynamic model has to be used are the controls of industrial manipulators and robots and propulsion of electrical vehicles. The motion control implies variations of speed and position along a predefined trajectory of a tool, work piece, vehicle, or an arm of an industrial robot. Whenever the controlled object falls out of the desired trajectory, it is necessary to assert a relatively fast change of the force (or torque) in order to drive the controlled object back to the desired path and annihilate the error. The task of the position (or speed) controller is to calculate the force (torque) to be applied in order to remove the detected position (or speed) discrepancy. The task of electrical motor is to deliver desired torque as fast and accurate as possible. In such *servo applications*, electrical motors are required to realize very fast changes of torque in order to remove the influence of variable motion resistances on the speed and position of the controlled objects. The analysis of operation of an electrical machine used as a servomotor in motion control applications requires thorough knowledge of transient processes within the machine.

Another case where the mathematical model is required is the analysis of transient processes in grid-connected synchronous generators operating in electrical power plants. Sudden rises and falls of electrical consumption in transmission networks are caused by switching on and off of large consumers, or quite frequently by short circuits. They affect generators as an abrupt change of their electrical load. The analysis of generator voltages and currents during transients cannot be performed by using the steady state equivalent circuit. Instead, it is required to have a mathematical model depicting the transient phenomena within the machine.

1.4.5   Mathematical Model

The mathematical model is represented by a set of algebraic and differential equations describing behavior of a machine during transients and in steady states. The *voltage balance equations* express the equilibrium of voltage in the machine windings, and they have the form $u = Ri + \mathrm{d}\Psi/\mathrm{d}t$. The change of the rotor speed is determined by Newton differential equation $J\,\mathrm{d}\Omega_m/\mathrm{d}t = T_{em} - T_m$. Quantities such as J and R are parameters, while Ω_m and Ψ are variables describing the state of the machine (*state variables*). An electrical engineer needs the model of electrical machine for performing the analysis of the energy conversion processes, for the analysis of conversion losses, for designing of the machine power supply and controls, as well as for solving the problems that may occur during machine applications. For this reason, it is necessary to have a relatively simple and intuitive model so that it can present the processes and states of the machine in a concise and clear way. *A good model* should be thorough and concise outline of the relevant phenomena within the machine, suitable for making conclusions and taking decisions as regards power supply, controls, and use of the electrical machines.

There are aspects and phenomena within the machine which are not relevant for problems under the scope because they influence operation of an electrical machine to a very limited extent. They are called *secondary* or *parasitic* effects, as they are usually neglected in order to obtain a simpler, more practical mathematical model. As an example, the energy density $w_e = \varepsilon\, E^2/2$ of electric field E within electrical machines can be neglected. It is lower than the energy density of magnetic field by several orders of magnitude. In the process of modeling, other justifiable omissions are adopted in order to obtain a simplified model which still matches the purpose. For the problem under consideration, the most appropriate model is the simplest one, yet depicting all the relevant dynamic phenomena. Justifiable omissions of secondary effects lead to mathematical models that are less complex and more intuitive. With such models, it is easier to overview the main features of the system. The problem solving and decision-making process becomes quicker and straightforward.

In electrical engineering, the model is usually a set of differential equations describing behavior of a system. Model of an electrical machine, or a transformer, can be reduced to the equivalent circuit describing its steady state operation. On the basis of the model, it is possible to determine the mechanical characteristic of the machine.

1.5   Adopted Approach and Analysis Steps

In general, the material presented in this book is intended for electrical engineering students. The basic knowledge of mathematics, physics, and electricity is practically applied in studying electrical machines, by many students met for the very first time. The approach starts with general notion and then goes to detail. It allows the beginners to perceive at first the basic purpose, appearance, and fundamental characteristics of electrical machines. Following the introductory chapters, this book investigates operating modes in typical applications and studies equivalent circuits, mechanical characteristics, power supply topologies and controls, as well as the losses and the problems in exploitation. Later on, the focus is turned to details related to the three main types of electrical machines.

The attention is directed toward DC, asynchronous, and synchronous machines. In this book, other types of electrical machines have not been studied in detail. The problems associated with design of electrical machines are briefly mentioned. The winding techniques, magnetic circuit design, analysis of secondary phenomena, the secondary and parasitic losses, and construction details have been left out for further studies. The main purpose of this book is introducing the reader to the role of electrical machines and studying their electrical and mechanical properties in order to acquire the ability to specify their electrical and mechanical characteristics, to define their power supplies and control laws, and to design systems with electrical motors and generators.

The study of electrical machines starts by an introduction to the basic principles of operation and with a survey of functions of electrical generators and motors in their most frequent applications. The analysis steps include the principles of electromechanical conversion and study the conversion process by taking the example of an electrostatic machine. Energy of the coupling electrical field is analyzed along with the process of energy exchange between the field, the electrical source, and the mechanical port. The study proceeds with the analysis of a simple electromechanical converter with magnetic coupling field. Construction of the magnetic circuit and windings of the machine are followed by specification of conversion losses. Subsequently, rotational electromechanical converters with magnetic coupling are considered, along with the rotating electrical machines which are the main subject of this study. The basic notions and definitions include the magnetic resistances and circuits, concentrated and distributed windings, methods of calculating the flux per turn and the winding flux, and the expressions for the winding self-inductances, mutual inductances, and leakage inductances. Magnetomotive forces of the winding are explained and analyzed, as well as electromotive forces induced in concentrated and distributed windings. The magnetic field in the air gap is analyzed and applied in modeling the electromechanical conversion in cylindrical machines. The electromagnetic torque and the power of electromechanical conversion are expressed in terms of flux vectors and magnetomotive forces. The concept and creation of rotating magnetic field are detailed and used to describe the difference between direct current (DC) machines and alternating current (AC) machines. The mathematical model of a cylindrical machine with the windings having N coils is derived, along with the expressions for electrical power, mechanical power, and power losses in the windings, magnetic circuit, and the mechanical subsystem. The secondary phenomena that are usually neglected in the analysis of electrical machines are specified and explained. At the same time, some sample applications and operating conditions are named where the secondary phenomena cannot be excluded from the analysis.

The introduction is followed by the chapters dealing with DC machines, asynchronous machines (AM), and synchronous machines (SM). Each chapter starts with basic description and the operating principles of the relevant machine, followed by the most significant aspects of its construction, description of its merits, the most frequent applications, and meaningful shortcomings. The expressions are derived for the magnetomotive force, flux, electromotive force, torque, and conversion power. Mathematical model is derived for the machine under consideration, describing its behavior during transient processes and providing the grounds for obtaining the equivalent circuits and mechanical characteristics. In the case of AC electrical machines, the modeling includes introduction of the three-phase/two-phase coordinate transformation (Clarke transform) and coordinate transformation from the stationary coordinate frame to the revolving coordinate frame (dq or Park transform).

The equivalent electrical circuits of the machine are derived from the steady state analysis. They are used to calculate the currents, voltages, flux, torque, and power at steady states, where the supply voltage, the load torque T_m, and the speed of rotation are known and unchanging. The mathematical model is also used to obtain the mechanical characteristic which gives the steady state relation of the machine speed and torque delivered to the shaft. For each electrical machine, the operating regimes sustainable in the steady state are analyzed and formulated as the *steady state operating area* in the T_{em}-Ω_m plane. Likewise, the *transient operating area* of the T_{em}-Ω_m plane is defined, representing the transient operating regimes attainable in short time intervals.

Particular attention is paid to conversion losses. The losses in the windings and magnetic circuits are analyzed in depth, along with heating of electrical machines and the methods of their cooling. The highest sustainable values of the current, power, and torque are defined and explored. These values and the highest sustainable values of other relevant quantities are called *the rated values*.[6] The need for operation in the region of field weakening is emphasized, and the relevant relations and characteristics are derived.

The transient operating area is analyzed for DC and AC machines. It is derived from the short-term overload capabilities of mechanical and electrical ports of electrical machines. The analysis takes into account the impact of peak current and peak voltage capabilities of the electrical power supply on the machine transient performance. Basic information concerning the supply, controls, and typical power converter topologies used in conjunction with the electrical machine is given for DC, asynchronous, and synchronous machines.

---

[6] The *rated* values are the highest permissible values of the machine currents, voltages, power, speed, and torque in a continuous service. Permanent operation with higher values will damage the machine's vital parts due to phenomena such as overheating. They are usually the result of engineering calculation, and they also represent an important property of the machine. The rated values are usually related to specified ambient temperature. Typically, the rated power is the maximum power the electrical machine can deliver continuously at 40°C ambient temperature. The values written on the machine plate or in manufacturer's specifications are called *nameplate* or *nominal* values. The nominal and rated values are usually equal. In rare cases, manufacturer may have the reason to declare the nominal values lower than the rated values. Within this book, it is assumed that the nominal values correspond to the rated. They are denoted by a lowercase subscript n, such as in U_n or I_n.

1.5.1   Prerequisites

Precondition to understanding analysis and considerations in this book, accepting the knowledge, acquiring the target skills, and solving the problems is the knowledge of mathematics, physics, and basic electrical engineering which is normally taught at the first year of undergraduate studies of engineering. It is required to know the basic laws of motion and practical relations concerning rotation and translation. Required background includes the steady state electrical and magnetic fields, the basic characteristics of dielectric and ferromagnetic materials, and elementary boundary conditions for electrostatic and magnetic fields. Chapter 2, *Electromagnetic Energy Conversion*, deals with the analysis of the energy and forces associated with electrostatic and magnetic fields in dielectrics, ferromagnetics, and air. Further on, analysis includes solving simple electric circuits with DC or AC currents. In addition, the analysis extends on the magnetic circuits involving magnetomotive forces (*magnetic voltages*), flux linkage, and magnetic resistances. The basic laws of electrical engineering should be known, like Faraday law on electromagnetic induction, Ampere law, Lorentz law, and Kirchhoff laws and similar. The study includes spatial distribution of the current, field, and energy. Therefore, coordinates in the Cartesian or in the cylindrical coordinate systems will be used along with the corresponding unit vectors. A consistent effort is sustained throughout this book to make the developments material accessible to readers not familiar with spatial derivatives, such as rotor (**curl**, **rot**) or divergence (**div**). Therefore, familiarity with Maxwell differential equations is not inevitable. Instead of differential form of Maxwell laws, it is sufficient to know their integral form, such as Ampere law. The skill in handling complex numbers and phasors is required, as well as dealing with scalar and vector products of vectors. For determining direction of a vector product, one should be familiar with the right-hand rule. Also required are the abilities of representing and perceiving relations between three-dimensional objects, of identifying closed surfaces defining a domain, and of contours defining a surface and surface normals. Within this book, the problem solving involves relatively simple line and surface integrals and solution of first-order linear differential equations. An experience in reducing differential equations to algebraic equations by applying Laplace transform and the ability of performing basic operations with matrices and vectors are also useful.

1.6   Notes on Converter Fed Variable Speed Machines

This book has not been written with an intention to prepare a reader for designing electrical machines. The main goal is studying the electrical and mechanical characteristics of electrical machines from the user's point of view, with an intention to prepare a reader for selecting an adequate machine, for solving the problems associated with the power supply and controls, and for handling the problems that may appear during operation of electrical motors and generators. The specific knowledge required for designing electrical machines is left out for further studies. The prerequisite for exploring further is a thorough acceptance of the knowledge and skills comprised within this book. The need for skilled designers of electrical machines is higher than before. Some of the reasons for this are the following:

- During the last century, electrical machines were designed to operate from the grid, with constant voltages and with the line frequency. Development of static power converters, providing three-phase voltages of variable frequency and amplitude, permits the power supply of an electrical motor to be adjusted to the speed and torque. Most new designs with electrical motors include static power converters that convert the energy received from the grid into the form best suited to the actual speed and torque. The voltage and frequency can be adjusted to reduce the power losses while delivering the reference torque at given speed. Therefore, there is an emerging quest for electrical machines designed to operate in conjunction with static power converters and variable frequency power supply.
- Applications of electrical motors for propelling electrical vehicles or driving industrial robots often require the rotor speed exceeding several hundreds of revolutions per second, which requires the power supply frequencies of the order of $f > 500$ Hz. Therefore, within contemporary servomotors and traction motors, electrical currents and magnetic induction pulsate at the same frequency. Fast variation of magnetic field requires application of new magnetic materials and novel design solutions for magnetic circuits. The increased frequencies of electric currents demand new solutions for making the windings.
- Propelling industrial robots requires electrical motors having a fast response and low inertia. Therefore, it is required to design synchronous motors having permanent magnets in their rotors, with the rotor shape and size resulting in a low inertia and fast acceleration, such as a disc or a hollow cylinder with double air gap.
- An increased interest in alternative and renewable power sources requires design of novel synchronous generators, suitable for the operation in conjunction with wind turbines, tidal turbines, and similar. The speed and the operating frequency are variable, while in some cases, generators operate at a very low speed. At the same time, the inertia and weight of generators should be low, with the lowest possible power losses.
- Construction of thermal electric power plants with supercritical steam pressure enables design of a single block in excess of 1 GW. Mechanical power of the block, obtained from a steam turbine, is converted to electrical energy by means of a synchronous generator operating at the line frequency of 50 Hz. Designing generators of this high power demands application of new design solutions, new insulating and ferromagnetic materials, and new cooling methods and systems.

The need to increase the production of electrical energy and the need to reduce the heat released to the environment can be alleviated by reducing the losses and increasing the *energy efficiency* η of electrical machines. The efficiency of generators and motors can be increased by adequate control, but also by designing novel electrical machines and applying new materials in their construction.

1.7 Remarks on High Efficiency Machines

Reduction in power losses increases the energy efficiency of electrical machines and relieves the problem of their cooling. As all the machine losses eventually turn into heat, the loss reduction diminishes the heat emitted to the environment. The heat released by electrical machines is a form of environmental pollution, and it should be kept low. Considering the fact that industrial countries use more than 2/3 of their electrical energy in electrical motors, the loss reduction in electrical machines has the greatest potential of energy saving. In addition, an efficient heat removal (cooling) often requires specific engineering solutions, increasing in this way the cost and complexity of the design. For these reasons, there is an increasing need for designing new, more efficient electrical machines and to devise their controls that would reduce the losses. Designing new solutions for electrical generators and motors requires a thorough basic knowledge on their operating principles, and it is bound to use novel ferromagnetic materials and new design concepts. One example is the use of permanent magnet excitation which eliminates the excitation winding and cuts down the rotor losses of synchronous machines. Besides, an efficient electromechanical conversion requires as well new solutions for the machine supply. Most of contemporary machines do not have a direct connection to the grid and do not operate with line-frequency voltages and currents. Instead, they are fed from static power converters which transform the grid supply to the form which is consistent with an efficient operation of the machine. Supply from a static power converter allows for the flux changes and selection of the flux level which results in the lowest power losses. Successful design of electrical machines supplied from static power converters requires a thorough knowledge on electrical and magnetic fields within the machine, as well as the knowledge on the energy conversion processes taking place within switching power converters.

1.8 Remarks on Iron and Copper Usage

On a wider scale, the energy efficiency of electrical machines includes as well the amount of energy consumed in the course of the machine production. Manufacturing of electrolytic copper and aluminum and making of laminated steel sheets require large amounts of energy. For this reason, the machine that uses fewer raw materials is likely to be the more efficient one. Construction of electrical machines has certain similarities with the construction of power transformers. In both cases, the appliance has a magnetic circuit and some electrical current circuits. Traditionally, magnetic circuits are made of laminated steel sheets, whereas electric circuits (windings) are made of insulated copper conductors. Both transformers and electrical machines are used within systems comprising energy converters, semiconductor switches, sensors, microprocessor control systems, and the associated software. The decisive factor which governs the price of the whole system is the iron and copper weight involved. Namely, production of semiconductor devices requires relatively small quantities of raw materials, such as the silicon ingot, some donor and acceptor impurities, and relatively small quantities of ceramic or plastic materials for the casing. Moreover, development, design, and software production costs have an insignificant contribution to the cost in series production. Therefore, it is significant to design and manufacture units and systems with reduced consumption of iron and copper. Reducing the quantities of raw materials can be accomplished in three ways:

In the system design phase, the operating conditions of electrical machines involved in the system can be planned so as to manufacture them with a reduced consumption of iron and copper.

In the electrical machine design phase, it is possible to make the magnetic and electric circuits in a manner that saves on raw materials. As an example, four-pole[7] machines make a better use of the magnetic circuit than two-pole machines.

During the operation of electrical machine supplied from a static power converter, it is possible to use the control methods that maximize the torque and power available from the given magnetic and electric circuits. In this way, there is an increase in the specific torque and specific power.[8] Given the torque and power requirements, it is possible to design and make the electrical machine with less iron and less copper.

Contemporary computer tools for design of electrical machines allow anticipation of their characteristics prior to making and testing a prototype. This facilitates and speeds up the design process. Moreover, it becomes possible to test several different solutions and approaches over a relatively short period of time. Most of the software packages make use of the finite element analysis (FEM) of electrical, magnetic, mechanical, and thermal processes. Designing with computer tools brings up the risk of inadvertent errors. The problems arise in cases when designer pretends that the tool performs the creative part of the job. A computer tool will give an output for each set of input data, whether the input makes sense or not. Therefore, a user has to possess certain experience in design, in order to interpret properly the obtained results and notice errors and contradictions. A conservative use of computer tools consists of using computer for quick completion of automatic tasks and calculations which the designer would have performed himself if he had sufficient time.

---

[7] The operation of electrical machines involving multiple pairs of magnetic poles will be explained in chapter on asynchronous machines.

[8] For the given electrical machine, specific torque is the ratio between the available torque and the mass (or volume) of the machine. Hence, it is the torque per unit mass (or volume). The same holds for the specific power. With higher specific torque (or power), electrical machine is smaller and/or lighter for the same task.

Chapter 2
Electromechanical Energy Conversion

Electrical machines contain stationary and moving parts coupled by an electrical or magnetic field. The field acts on the machine parts and plays key role in the process of electromechanical conversion. For this reason, it is often referred to as the *coupling field*. This chapter presents the most significant principles of creating a force or torque on the machine moving parts. In all the cases considered, the force appears due to the action of the electrostatic or magnetic field on the moving parts of the machine. Depending on the nature of the coupling field, the machines can be magnetic or electrostatic.

2.1 Lorentz Force

Electrical machines perform conversion of electrical energy to mechanical work or conversion of mechanical work to electrical energy. The basic principles involved in the process of electromechanical conversion are presented in the considerations which follow.

One of the laws of physics which is basic for electromechanical conversion is Lorentz law which determines the force acting upon a charge Q moving with a speed v in the electrical and magnetic fields:

$$\vec{F} = Q\vec{E} + Q(\vec{v} \times \vec{B}). \tag{2.1}$$

In electrical machines, the operation is most often based on the magnetic coupling field. Conductors and ferromagnetic parts in a magnetic field are subjected to the action of electromagnetic forces. Magnetic induction B in (2.1) is also called *flux density*. Electrical current existing in the conductor is a directed motion of electrical charges. Therefore, (2.1) can be used to determine the force acting upon conductors carrying electrical currents.

S.N. Vukosavic, *Electrical Machines*, Power Electronics and Power Systems,
DOI 10.1007/978-1-4614-0400-2_2, © Springer Science+Business Media New York 2013

Fig. 2.1 Force acting
on a straight conductor
in homogeneous magnetic
field

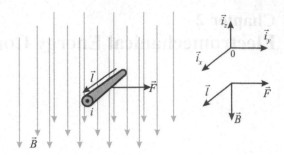

Figure 2.1 shows straight portion of a conductor of length l with electrical current i which is placed in a homogenous magnetic field with flux density B. Electromagnetic force F acting on the conductor depends on current i, conductor length l, flux density B, and angle between directions of the field and the conductor. In the example presented in Fig. 2.1, the conductor is perpendicular to the direction of the field. Applying the Cartesian coordinate system with axes x, y, and z, the vectors of magnetic induction B, conductor length, and force can be expressed in terms of the corresponding unit vectors.

$$\vec{l} = l\,\vec{i}_x,$$
$$\vec{B} = -B\,\vec{i}_z,$$
$$\vec{F} = i \cdot \left(\vec{l} \times \vec{B}\right). \tag{2.2}$$

Since the vector B is orthogonal to the conductor, the module of the force vector is equal to $F = l \cdot B \cdot i$. Direction of the force is determined by the vector product. The right-hand rule[1] can be used to determine quickly the vector product direction. If the considered part of the conductor makes a displacement Δy along the axis y, corresponding mechanical work is $\Delta W = F\Delta y$. At a constant speed of motion v_y, the mechanical power assumes the value $p_m = Fv_y$. In the case when the force acts in the direction of motion, power p_m is positive, and the system operates as a motor, delivering mechanical work and power. Otherwise, the motion and force are opposed, power p_m is negative, and the mechanical work is converted to electrical energy, while the system operates as a generator.

---

[1] The right-hand rule requires thumb and forefinger to assume right angle. The middle finger should be perpendicular to both. Now, with forefinger alligned with vector l and middle finger alligned with B, thumb determines the direction of force. Alternatively, direction of any vector product can be determined by an imaginary experiment, where the first vector of the product (l in (2.2)) is rotated toward the second vector (B). Envisaging a screw that is turned by such rotation, the screw would advance along the axis perpendicular to $l - B$ plane. The direction of the vector product is determined by the advance of the (right) screw.

In a conductor moving through a homogeneous magnetic field, induced electromotive force e depends on flux density B, speed of motion, and conductor length l. The product of the electromotive force e and current i is equal to the product of the force and speed ($p_m = Fv_y$), as shown later on. Assuming that energy losses are negligible, the motor operation can be perceived as a lossless conversion of electrical power p_e to mechanical power p_m. The relevant powers are represented by derivatives of the electrical energy and mechanical work.

2.2 Mutual Action of Parallel Conductors

In the previous example, the case is considered of a conductor in a homogenous *external* magnetic field which exists due to the action of external current circuits or permanent magnets. Force acting on conductors also exists in the case when there is no externally brought field, but there are two conductors both conducting electrical currents.

Magnetic field created by one of the conductors interacts with the current in the other conductor, according to the principle presented in Fig. 2.2. The result of this interaction is force acting on the conductor.

When currents in the conductors have the same direction, the force tends to bring the conductors closer. In the case when directions of the currents are mutually opposite, the force tends to separate the conductors.

In the case being considered, the force acting on parallel conductors is very small. If two very long and thin parallel conductors, each with current of 1A, placed at a distance $d = 0.1$ m are considered, one of the conductors will be found in the

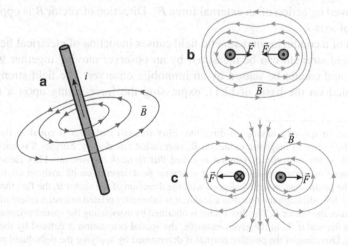

Fig. 2.2 (a) Magnetic field and magnetic induction of a straight conductor. (b) Force of attraction between two parallel conductors. (c) Force of repulsion between two parallel conductors

magnetic field created by the other conductor. The magnetic induction created by a
very long conductor is given by (2.3):

$$B = \mu_0 \frac{I}{2\pi d} = 4\pi \cdot 10^{-7} \frac{1}{2\pi \cdot 0.1} T = 2 \cdot 10^{-6} T. \tag{2.3}$$

Since the magnetic field and magnetic induction are orthogonal to the conductor,
the electromagnetic force acting on a part of the conductor 1 m long is given by (2.4):

$$F = l \cdot B \cdot I = 2 \cdot 10^{-6} N. \tag{2.4}$$

Magnetic circuits are discussed later in the book. They are made of ferromagnetic materials with high permeability μ, and they direct the magnetic flux[2] to paths
with low magnetic resistance. The lines of magnetic field are concentrated into
magnetic circuits in a way that resembles electrical current being contained in
conductors and windings. In turn, there is a considerable increase of the flux density
B, resulting in an increase in electromagnetic force and power of conversion, both
proportional to B.

2.3 Electromotive Force in a Moving Conductor

A considerable number of electrical machines convert mechanical work to electrical
energy, like synchronous generators in electrical power plants. Figure 2.3 shows the
principle where mechanical work and mechanical power are used to obtain electrical
energy and electrical power. The figure shows straight part of conductor of length l,
moving along y-axis at a speed v. The conductor placed in a homogeneous magnetic
field is moved by action of an external force F_s. Direction of vector B is opposite to
direction of axis z.

Motion of a conductor in magnetic field causes induction of electrical field E_{ind}.
Induced field strength can be measured by an observer moving together with the
conductor and cannot be sensed by an immobile observer. The field strength can
be calculated on the basis of (2.1), expressing the force acting upon a moving

---

[2] Flux is a scalar quantity having no direction. Flux through surface S is equal to the surface
integral of the vector of magnetic induction B, also called *flux density*. Surface S is encircled by
contour C; thus, the said surface integral is called *flux through the contour*. Flux through a flat
surface S placed in a homogeneous *external* magnetic field depends on its position relative to the
field. With the *positive normal* on S aligned with the direction of the vector B, the flux through S is
equal to $\Phi = BS$. Although the flux Φ is a scalar, it is inherently related to spatial orientation of the
surface S and/or the vector B. The flux vector is obtained by associating the spatial orientation (i.e.,
direction) to the scalar Φ. In the given example, the spatial orientation is defined by the *positive
normal* on S. Direction of the positive normal is determined by applying the right-hand rule to the
reference circling direction for the contour C. The external magnetic field is the one which is not
created by the electrical currents in the contour C.

Fig. 2.3 The induced
electrical field and
electromotive force in the
straight part of the conductor
moving through
homogeneous external
magnetic field

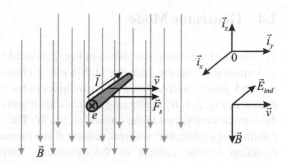

charge Q. The considered domain does not contain any electrostatic field; hence the product QE in (2.5) is equal to zero.

$$\vec{F} = Q\vec{E} + Q(\vec{v} \times \vec{B}) = Q(\vec{v} \times \vec{B}) = Q\vec{E}_{ind}. \tag{2.5}$$

The force F acts on the moving charge Q due to its motion in homogeneous magnetic field. Notice in (2.5) that the same effect can be obtained by replacing the magnetic field with the electrostatic field of the strength E_{ind}. Therefore, the induced electrical field can be determined by dividing force F by charge Q, which gives vector product of the speed and flux density B:

$$\vec{E}_{ind} = \vec{v} \times \vec{B}. \tag{2.6}$$

It is of interest to determine the electromotive force e induced in the straight part of the conductor of length l. In general, the electromotive force induced in a conductor is determined by calculating the line integral of vector E_{ind} between conductor terminals. Since the induced electrical field does not vary along the conductor, the line integral reduces to the scalar product of vectors l and E_{ind}. The electromotive force can be calculated from (2.7):

$$e = \vec{l} \cdot \vec{E}_{ind} = \vec{l} \cdot \left(\vec{v} \times \vec{B} \right). \tag{2.7}$$

In the present case, the conductor is aligned with x-axis, the magnetic field is in direction of z-axis, and the speed vector is aligned with y-axis. Therefore, the vector of the induced electrical field is collinear with the conductor (2.8); thus, the induced electromotive force is $e = lvB$. The sign of the induced electromotive force e depends on the adopted reference direction of the conductor. In the present case, it is the direction of vector l.

$$\vec{l} = -l\,\vec{i}_x, \quad \vec{B} = -B\,\vec{i}_z, \quad \vec{v} = v\,\vec{i}_y,$$
$$\vec{E}_{ind} = -vB\,\vec{i}_x, \quad e = \vec{l} \cdot \vec{E}_{ind} = lvB. \tag{2.8}$$

2.4   Generator Mode

Terminals of the conductor shown in Fig. 2.3 could be connected to the terminals of an immovable resistor, forming in this way a closed current circuit containing the induced electromotive force e, interconnections, and resistor R. This circuit is shown in Fig. 2.4. By neglecting the resistance and inductance of interconnections, the current established in the circuit is $i = e/R$. The moving conductor performs the function of a generator, whereas resistor R is a consumer of electrical energy. Since direction of the current in the conductor corresponds to the direction of electromotive force, the conductor is a source of electrical power and energy. Existence of the current in the conductor creates force F_m which opposes the movement (2.2). The external force F_s acts in the direction opposite to F_m, overcoming the resistance F_m. It is of interest to analyze the operation of the system in Fig. 2.4 with the aim of establishing the relation between the invested mechanical power $F_s v$ and obtained electrical power ei.

The electromotive force $e = lvB$, induced in the conductor, is equal to the voltage $u = Ri$, which appears across resistance R. The electromagnetic force acting on the conductor, shown in Fig. 2.4, acts from right to left and is given by (2.9):

$$\vec{F}_m = i\left(\vec{l} \times \vec{B}\right), \quad |\vec{F}_m| = ilB. \tag{2.9}$$

By maintaining the movement, the external force F_s performs the work against magnetic force F_m and delivers it to the moving conductor. Transfer of the mechanical work to electrical work is performed through electromagnetic induction. Electromotive force e, induced in the moving conductor, maintains the current $i = e/R$ in the circuit and delivers electrical energy to the resistor.

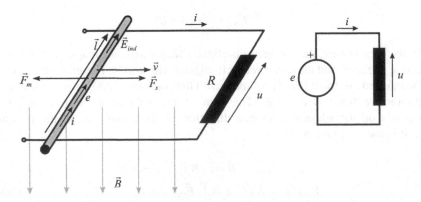

Fig. 2.4 Straight part of a conductor moves through a homogeneous external magnetic field and assumes the role of a generator which delivers electrical energy to resistor R

The sum of forces acting on the conductor is equal to zero:

$$\vec{F}_s + \vec{F}_m + m\frac{d\vec{v}}{dt} = 0.$$

In the state of dynamic equilibrium, the speed v is constant. With the acceleration dv/dt equal to zero, the inertial force $F_i = m\, dv/dt$ is equal to zero as well. Therefore,

$$\vec{F}_s + \vec{F}_m = 0; \quad \vec{F}_s = -\vec{F}_m; \quad |\vec{F}_s| = |\vec{F}_m| = ilB. \tag{2.10}$$

Mechanical power of external force F_s is equal to $P_m = F_s v = i\, l\, vB$. The induced electromotive force e develops power $P_e = ei = i\, lvB = P_m$ and delivers it to the rest of the electrical circuit. With $P_e = e^2/R > 0$, the considered system converts mechanical work to electrical energy. In the course of this analysis, energy losses have been neglected; thus, there is equality between the input (mechanical) power and output (electrical) power ($P_e = P_m$).

Question (2.1): In the case when the resistor shown in Fig. 2.4 moves together with the conductor, what will be the electrical current in the circuit?

Answer (2.1): During a parallel movement of the conductor and resistor, equal electromotive forces will be induced within each of them, thus compensating each other. Therefore, the electrical current will be equal to zero.

2.5 Reluctant Torque

Electromechanical energy conversion can be accomplished by exploiting the tendency of ferromagnetic material placed in a magnetic field to get aligned with the field and take position of minimum magnetic resistance. Figure 2.5 shows an elongated piece of ferromagnetic material of high permeability ($\mu_{Fe} \gg \mu_0$), inclined with respect to the lines of magnetic field. The electromagnetic forces tend to bring the piece in vertical position where it will be collinear with the field.

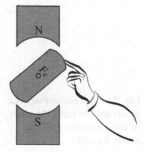

Fig. 2.5 Due to reluctant torque, a piece of ferromagnetic material tends to align with the field, thus offering a minimum magnetic resistance

In Fig. 2.5, it is assumed that the magnetic field exists owing to a permanent magnet. The moving part (rotor) of ferromagnetic material can rotate and will tend to take vertical position where magnetic resistance along the field lines (magnetic resistance along the flux path) is lower. When the rotor assumes vertical position, the flux passes from the magnet poles into the rotor whose permeability is high. The ferromagnetic rotor always tends to align with the field. The torque which appears in the considered (inclined) position tends to bring the ferromagnetic to vertical position. This torque is called *reluctant*, and the considered principle of the torque generation is called *reluctant principle*. This name stems from *reluctance*, also called magnetic resistance. Reluctant torque depends on changes in magnetic resistance due to spatial displacement of the moving part. *The reluctant torque tends to bring rotor to position where magnetic resistance is minimal.* The rotor can be connected to a work machine to deliver mechanical power.

Question (2.2): What is the value of reluctant torque acting on the rotor when it is in horizontal position?

Answer (2.2): The reluctant torque tends to bring the rotor to the position of minimal magnetic resistance. In horizontal position, magnetic resistance assumes its maximum value. A hypothetical shift of the rotor in any direction will lead to a decrease of magnetic resistance. Unless moved from horizontal position, there is no tendency to move the rotor in any direction, and the reluctant torque is equal to zero. In the considered case, there is an unstable equilibrium. Any movement of the rotor to either side would result in the reluctant torque which speeds up the initial movement.

2.6 Reluctant Force

Figure 2.6 shows a system where the *reluctant force* stimulates a translatory movement. Electromagnetic force acts on the piece of ferromagnetic material placed in a nonhomogeneous magnetic field. The force tends to bring the piece of ferromagnetic to the place where the flux density B is high.

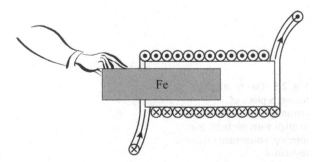

Fig. 2.6 The electromagnetic forces tend to bring the piece of ferromagnetic material inside the coil

The coil shown in Fig. 2.6 is made of circular wound conductors carrying a DC current. This system of conductors (*coil*, *winding*, or *bobbin*) creates a magnetic field that extends along the coil and has maximum intensity inside the coil. Hence, the flux path goes through the cylindrical coil. A piece of mobile ferromagnetic material can be inserted in the coil or extracted from the coil.

If the ferromagnetic piece is in the coil, the magnetic resistance (*reluctance*) along the flux path is low. When the ferromagnetic piece is outside the coil, the reluctance is high.

Taking into account that the mobile part of ferromagnetic material tends to take position where the magnetic resistance is minimal, the force will appear tempting to bring the mobile piece of ferromagnetic material in the coil.

2.7 Forces on Conductors in Electrical Field

Thanks to the action of electrical field E, one can obtain force, power, and work from the setup shown in Fig. 2.7. In the space between two parallel, charged capacitor plates, there is an electrostatic field E. In the case when the distance between the plates is small compared to their dimensions, the field can be considered *homogeneous*. Namely, the field lines are parallel, while the field strength does not change between the plates.

The charges are distributed on the interior surfaces of the plates. The field between the plates acts on the surface charges by a force tending to bring the plates closer. Force F may cause the plates to move. If one of the plates shifts by Δx, a mechanical work $F\Delta x$ is achieved.

Based on this principle, it is possible to operate electromechanical converters with electrical coupling field, also called *electrostatic machines*.

2.8 Change of Permittivity

Electromechanical conversion can be based on electrical force acting on a mobile part of dielectric material with permittivity (dielectric constant ε) different from the permittivity of the environment. Figure 2.8 shows two charged plates and a mobile

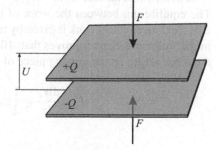

Fig. 2.7 Electrical forces act on the plates of a charged capacitor and tend to reduce distance between the plates

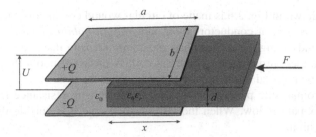

Fig. 2.8 Electrical forces tend to bring the piece of dielectric into the space between the plates. The dielectric constant of the piece is higher than that of the air

piece of dielectric material of permittivity $\varepsilon = \varepsilon_r \, \varepsilon_0$. Free space between the electrodes is filled by air of permittivity ε_0.

The piece of dielectric material of relative permittivity $\varepsilon_r > 1$ can move along a horizontal direction. By moving to the left, it comes to position $x = a$, when it fills completely the space between the plates. By moving to the right, the dielectric comes to position $x = 0$, when the space between the plates is completely filled by air. The following analysis will show that an electrical force F acts on the piece of dielectric in position $0 < x < a$, tempting to bring it into the space between the plates.

With voltage U across the plates, electrical field E in the space between the plates is $E = U/d$, where d is distance between the plates. The conductive plates represent equipotential surfaces; thus, relation $U = Ed$ applies in the air as well as in the dielectric, while the strength of the electrical field is the same in both media. Electrical induction within the dielectric is $D = \varepsilon_r \varepsilon_0 U/d$, whereas in the air, it is $D = \varepsilon_0 U/d$. Total energy of the electrical field is given by (2.11), where $S = ab$ is surface of the plates:

$$W_e = \frac{1}{2}\varepsilon_0 \left(\frac{U}{d}\right)^2 \cdot \frac{a-x}{a} Sd + \frac{1}{2}\varepsilon_r\varepsilon_0 \left(\frac{U}{d}\right)^2 \cdot \frac{x}{a} Sd$$

$$= \frac{1}{2}\varepsilon_0 \left(\frac{U}{d}\right)^2 \cdot \frac{Sd}{a}\left[(a-x) + x\varepsilon_r\right]. \tag{2.11}$$

If the plates are connected to a source of constant voltage U, a small displacement Δx will change the field energy accumulated in the space between the plates. The source U will provide an amount of electrical work, while the force F will contribute to delivered mechanical work $\Delta W_{meh} = F\Delta x$ obtained along the displacement Δx. The equilibrium between the work of the source ΔW_i, change in the field energy ΔW_e, and mechanical work is given by relation $\Delta W_i = \Delta W_e + \Delta W_{meh}$. Equation 3.8 in the following chapter proves that $\Delta W_e = \Delta W_i/2$ and $\Delta W_e = \Delta W_{meh}$. Therefore, the force acting on the moving piece of dielectric is obtained from (2.12):

$$F = \frac{dW_e}{dx} = \frac{1}{2}\varepsilon_0 \left(\frac{U}{d}\right)^2 \cdot \frac{Sd}{a}(\varepsilon_r - 1). \tag{2.12}$$

It is possible to determine electrical force F by using the equivalent pressure on the surfaces separating the media of different nature. On the basis of a conclusion from electrostatics, electrical force acting on a dividing surface that separates the spaces filled with two different dielectric materials can be determined from the equivalent pressure $p = w_{e1} - w_{e2}$. The values w_{e1} and w_{e2} are specific energies of electrostatic fields in the two separated media. They are also called the spatial energy densities of the electrostatic field. The energy of electrical field energy in the air has density of $w_0 = \frac{1}{2}\,\varepsilon_0 (U/d)^2$, whereas in the dielectric it is $w_d = \frac{1}{2}\,\varepsilon_r\varepsilon_0 (U/d)^2$. The force F can be determined from (2.13), where $S_d = bd = Sd/a$ is rectangular surface separating the two domains:

$$F = (w_d - w_0)S_d = \frac{1}{2}\varepsilon_0\left(\frac{U}{d}\right)^2 \cdot (\varepsilon_r - 1)\cdot\frac{Sd}{a}. \qquad (2.13)$$

Question (2.3): Determine the direction of force when the source is disconnected. It should be noted that total charge Q existing on the plates is then constant, whereas the voltage between the plates is variable depending on position of the dielectric.

Answer (2.3): In the space between the plates, there is a homogeneous electrical field. Conductivity $1/\rho$ of the metal plates is very high, and potential of all the points on one plate is the same. Therefore, voltage between the plates is U in the part filled by the dielectric as well as in the part filled by air. Since the field is homogeneous and orthogonal to the plates, product Ed is equal to voltage U; thus, electrical field $E = U/d$ is the same in both air and the dielectric. Since permittivity of the dielectric is higher, electrical induction D_d in the dielectric is higher than induction D_0 in the air:

$$D_0 = \varepsilon_0\frac{U}{d}, \quad D_d = \varepsilon_r\varepsilon_0\frac{U}{d}.$$

Surface charge density σ at the surface of a conductor is determined by the scalar product of the vector of electrical induction and normal to the surface at a given point:

$$\sigma = \vec{n}\cdot\vec{D}.$$

In the case being considered, the vector of electrical induction is perpendicular to the surface of the conductor and collinear with the normal n. As a consequence, the density of surface charge σ is equal to the induction D. Therefore, it will be higher in the parts of the plates which are against the dielectric. By using notation shown in Fig. 2.8, total charge Q can be expressed in terms of the shift x and values D_d and D_0,

$$Q = (a - x)b\cdot D_0 + xb\cdot D_d = (a - x)b\cdot\varepsilon_0\frac{U}{d} + xb\cdot\varepsilon_r\varepsilon_0\frac{U}{d}$$

$$= b\varepsilon_0\frac{U}{d}(a - x + x\cdot\varepsilon_r),$$

while capacitance C is determined by expression

$$C = \frac{Q}{U} = \frac{b\varepsilon_0}{d}(a - x + x \cdot \varepsilon_r).$$

Since the plates are separated from the source, the mechanical work $\Delta W_{meh} = F\Delta x$ is obtained by subtracting this amount from the field energy, $\Delta W_{meh} = -\Delta W_e$. Therefore, the electrical force can be determined according to expression $F = -dW_e/dx$. Electrical energy of the coupling field can be expressed as $W_e = \frac{1}{2}Q^2/C$ or $W_e = \frac{1}{2}CU^2$. In the present case, charge Q is constant, whereas voltage U is variable, and the electrical force can be determined according to expression

$$F = -\frac{dW_e}{dx} = -\frac{Q^2}{2}\frac{d}{dx}\left(\frac{1}{C}\right).$$

By differentiating the reciprocal value of capacitance, the following expression for the electrical force is obtained:

$$F = -\frac{Q^2 d}{2b\varepsilon_0}\frac{d}{dx}\left(\frac{1}{a - x + x \cdot \varepsilon_r}\right) = \frac{Q^2 d}{2b\varepsilon_0}(\varepsilon_r - 1)\left(\frac{1}{a - x + x \cdot \varepsilon_r}\right)^2.$$

The above expression is positive, so the direction of action of the force is the same as if the source was connected to the plates. By introducing substitution $Q = CU$ in the above expression, the electrical force is determined by

$$F = \frac{U^2}{2}(\varepsilon_r - 1)\frac{b\varepsilon_0}{d} = \frac{\varepsilon_0}{2}\frac{U^2}{d^2}(\varepsilon_r - 1)\frac{Sd}{a},$$

the expression which is fully equivalent to (2.12) and (2.13). It can be concluded that the force will not change by switching the source on or off, provided that the charge Q is the same in both cases.

Question (2.4): Consider a charged capacitor made of the plates shown in Fig. 2.8 and assume that the plates are not connected to the source. Is there any difference between E and D in the part filled by air and part filled by dielectric? Will the total energy be increased or decreased in the case that the dielectric is pushed further into the space between the plates?

Answer (2.4): In the space between the plates, the electrical field E is equal in all points, whereas the electrical induction D is ε_r times higher in the space filled by dielectric compared to induction in the space filled by air. The spatial density of the field energy in the dielectric is $w_{ed} = \frac{1}{2}\varepsilon_r\varepsilon_0 E^2$, and it is ε_r times higher than the density $w_{ea} = \frac{1}{2}\varepsilon_0 E^2$ in the air. Total field energy is $w_{ea}V_a + w_{ed}V_d$, where w_{ea} and w_{ed} are the densities of field energy in the air and in the dielectric, whereas V_a and V_d are the volumes of the interelectrode space filled by air and dielectric.

When the piece of dielectric moves toward inside of the capacitor, volume V_a decreases, whereas volume V_d increases. Since $w_{ea} < w_{ed}$, there are indications that the total field energy increases. However, filling the space between the plates by dielectric material increases the equivalent capacitance $C = Q/U$, as it is proportional to the permittivity of the dielectric material filling the space between the plates. Since the charge Q is constant, an increase in the capacitance will cause a decrease of the voltage. As a consequence, the fields E will reduce. Spatial density of the field energy depends on the square of the field strength. Therefore, it can be concluded that a deeper insertion of the dielectric reduces the total energy of the electrical field. These considerations can be verified by an analysis of the expression for field energy $W_e(x)$,

$$W_e = \frac{1}{2} \frac{Q^2}{C} = \frac{Q^2}{2} \frac{d}{(a - x + x \cdot \varepsilon_r)b\varepsilon_0},$$

which shows that in the case of a constant Q, total field energy decreases when the value of x rises, that is, when a piece of dielectric is pushed further into the space between the plates.

2.9 Piezoelectric Effect

Applying pressure on a crystal of silicon will induce charges on its surfaces and give rise to voltage between surfaces (Fig. 2.9). This phenomenon is known as *piezoelectric effect*. In a piezoelectric microphone, sound waves cause variable pressure of air against the surface of a crystal. As a consequence, variable forces act upon the crystal. A voltage which represents electrical image of the sound appears across the ends of the crystal. This voltage can be amplified and processed further.

It is possible to manufacture a crystal with linear dependence between the voltage across the crystal and the applied force. Such crystal can be used for designing precise electronic scales (weight-measuring devices).

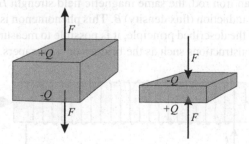

Fig. 2.9 Variation of pressure acting on sides of a crystal leads to variations of the voltage measured between the surfaces

   The inverse piezoelectric effect can be used in electromechanical conversion. If the surfaces of the crystal are covered by conducting plates connected to a variable voltage, a force varying in accordance with the variable voltage will appear. This effect can be used for creating very small displacements controlled by the applied voltage. If the connected voltage represents a record of sound, variations of the force will cause vibrations of the crystal surfaces and change the pressure of air against the surfaces, thus operating the piezoelectric loudspeakers.

   In a piezoelectric device, the crystal surface moves by a fraction of millimeter. Motors based on piezoelectric effect are used in motion control applications with very small displacements and with very high precision, such as positioning the reading heads in hard disk drives.

2.10   Magnetostriction

One of the principles applicable for electromechanical conversion is *magnetostriction*. In general, magnetization of ferromagnetic materials can change their shape and dimensions. This phenomenon is called magnetostriction. The length of the ferromagnetic rod shown in Fig. 2.10 will change with the applied magnetic field. The effect gives a rise to a force. Multiplied by mechanical displacement, the force produces mechanical work. Yet, few electromechanical converters are based on magnetostriction because of rather small displacements and a poor power-to-weight ratio. Conventional electrical machines and power transformers usually have magnetic circuits made of iron sheets, wherein magnetic field pulsates at the line frequency (50 Hz/60 Hz). The effect of magnetostriction causes magnetic circuits to vibrate. With the magnetostrictive forces proportional to the square of the magnetic field strength, the vibration frequency is twice the line frequency (100 Hz/120 Hz). These vibrations cause waves of variable air pressure and sound which are experienced as humming, frequently encountered with electrical equipment.

   The phenomenon reciprocal to magnetostriction is the change of permeability in ferromagnetic materials subjected to mechanical stress. Namely, the stress due to external forces will change magnetic properties of the material. When an external force is applied to an iron rod, the same magnetic field strength H will result in an increased magnetic induction (flux density) B. This phenomenon is called the *Villari* effect. By applying the described principle, it is possible to measure the stress in the elements of steel constructions such as the bridges or skyscrapers.

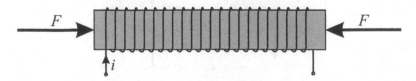

Fig. 2.10 The magnetization varies as a function of force which tends to constrict or stretch a piece of ferromagnetic material

This chapter discussed the principles of developing electromagnetic forces that act on moving parts of electromechanical converters and provide the means for the process of electromechanical conversion. The following chapter introduces some basic principles of electromechanical converters with electrical coupling field and electromechanical converters with magnetic coupling field.

Chapter 3
Magnetic and Electrical Coupling Field

Electromechanical conversion is based on forces and torques of electromagnetic origin. The force exerted upon a moving part can be the consequence of electrical or magnetic field. The field encircles and couples both moving and nonmoving parts of electromechanical converter. Therefore, the field is also called coupling field. In this chapter, some basic notions are given for electromechanical energy converters with electrical coupling field and converters with magnetic coupling field.

3.1 Converters Based on Electrostatic Field

Electromechanical conversion in electrostatic machines is based on electrical coupling field. The coupling field between moving parts is a prerequisite for electromechanical conversion. In an electrostatic machine, the field exists in the medium between mobile electrodes, and it causes electrical forces acting on the electrodes.

Preliminary insight in electromechanical energy conversion based on the electrical field can be obtained by considering the sample machine shown in Fig. 3.1, resembling the capacitor with two parallel metal plates. In the case when the plates are considerably larger compared to the distance between them ($S >> d^2$), the electrical field between the electrodes is homogeneous and equal to $E = U/d$ [V/m], where U is the voltage between the electrodes. Electrical induction vector D [As/m^2] is obtained by multiplying the vector of electric field E by the permittivity of the medium ε_0. The force acting on the plates depends on the charge stored in the capacitor. If it is possible to move one of the plates, then the product of this force and the displacement gives mechanical work. The mechanical work can be obtained at the expense of the field energy or of energy of a source connected to the plates.

S.N. Vukosavic, *Electrical Machines*, Power Electronics and Power Systems,
DOI 10.1007/978-1-4614-0400-2_3, © Springer Science+Business Media New York 2013

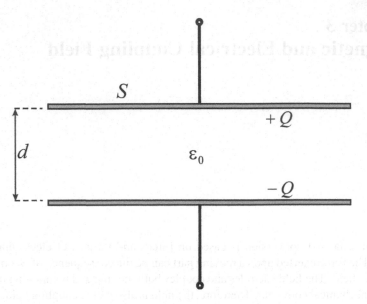

Fig. 3.1 Plate capacitor with distance between the plates much smaller compared to dimensions of the plates

3.1.1 Charge, Capacitance, and Energy

The electrical field in the interelectrode space is homogeneous. The field strength is determined by the ratio of the voltage and distance between the plates, $E = U/d$. Electrical induction D is equal to the surface charge density Q/S. At the same time, the ratio D/E is determined by permittivity (dielectric constant) ε_0.

$$E = \frac{U}{d}; D = \sigma = \frac{Q}{S} = \varepsilon_0 E; \Rightarrow Q = \varepsilon_0 ES = \varepsilon_0 S \frac{U}{d}. \qquad (3.1)$$

Capacitance C is determined by the ratio of charge Q and voltage U. The capacitance depends on the plate surface S, distance d between the plates, and permittivity of the dielectric material filling the interelectrode space:

$$C = \frac{Q}{U} = \varepsilon_0 \frac{S}{d}. \qquad (3.2)$$

Total energy of the coupling electrical field can be obtained by integrating the energy density w_e in the region where the electrical field extends. In the present case, the electrical field and the field energy exist within the interelectrode space. The energy density does not vary, and it is equal to $w_e = \frac{1}{2}\varepsilon_0 E^2$. The volume of the

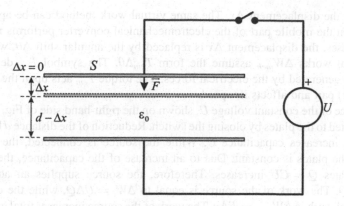

Fig. 3.2 A capacitor having mobile upper plate

region is $V = Sd$. Therefore, total energy of the coupling electrical field is $W = \frac{1}{2}CU^2 = \frac{1}{2}Q^2/C$.

$$W_e = \int_V w_e \, dV = \int_V \left(\int \vec{D} \cdot d\vec{E} \right) dV = \int_V \left(\frac{1}{2} \varepsilon_0 E^2 \right) dV$$

$$= Sd \left(\frac{1}{2} \varepsilon_0 E^2 \right) = \frac{1}{2} CU^2 = \frac{Q^2}{2C}. \tag{3.3}$$

3.1.2 Source Work, Mechanical Work, and Field Energy

Figure 3.2 shows a charged capacitor having mobile upper plate. It can be shown that by moving the upper plate downward, electrical energy is converted to mechanical work. The electric charge on the plates is of opposite polarity. Therefore, they are subjected to a force of attraction F. If the upper plate moves downward and gets closer to the lower plate by Δx, mechanical work $F\Delta x$ is obtained. During the move, there is a change in the energy W_e of the electrical coupling field. With the plates connected to the electrical source, the charge on the plates changes through an exchange of charges between the plates and the electrical source.

Electric force F acting on one plate of the capacitor can be determined by applying the method of *virtual works*, also called virtual disturbance method. It is necessary to envisage a very small displacement Δx of the mobile plate toward the opposing plate. In such case, the direction of the force F corresponds to the direction of the hypothetical displacement Δx. The method of virtual works proceeds with calculation of changes in the field energy and determines the work of the electrical source. The work $\Delta W_{meh} = F\Delta x$ is made by the electric force F during displacement Δx. The force can be calculated by dividing the work increment

ΔW_{meh} by the displacement Δx. The same virtual work method can be applied in cases when the mobile part of the electromechanical converter performs rotation. In such cases, the displacement Δx is replaced by the angular shift $\Delta \theta$, while the mechanical works ΔW_{meh} assume the form $T_{em}\Delta\theta$. The symbol T_{em} designates the torque generated by the electrical forces. The torque T_{em} acts upon the moving (revolving) part and affects its speed.

A source of the constant voltage U, shown on the right-hand side of Fig. 3.2, can be connected to the plates by closing the switch. Reduction of the distance d between the plates increases capacitance C. While the source is connected, the voltage between the plates is constant. Due to an increase of the capacitance, the charge on the plates $Q = CU$ increases. Therefore, the source supplies an additional charge ΔQ. The work of the source is equal to $\Delta W_i = U\Delta Q$, while the obtained mechanical work is $\Delta W_{meh} = F\Delta x$. The work of the source increases total energy of the system, that is, the sum of the electrical energy and mechanical work. With a constant voltage, the electrical energy is given by (3.4).

In the next considerations, it will be shown that work of the source is divided in two equal parts, that is, $\Delta W_e = \Delta W_{meh} = \frac{1}{2}\Delta W_i$.

If the switch in Fig. 3.2 is open, the source is separated from the plates, and the work of the source is equal to zero. Electrical charges on the plates cannot be changed, as well as the field D between the plates ($Q = $ const., $D = $ const.). Therefore, the density of the field energy $w_e = \frac{1}{2}D^2/\varepsilon_0$ remains unchanged.

By reducing the distance between the plates, the volume of the region comprising the electrical field is reduced as well. Therefore, the total field energy ΔW_e is also reduced. With the source separated from the system, reduction in the field energy yields the mechanical work $\Delta W_{meh} = -\Delta W_e$. In the case of a constant charge, the field electrical energy is given by (3.5):

$$W_e(\Delta x) = S(d - \Delta x)\left(\frac{1}{2}\varepsilon_0 E^2\right) = \frac{1}{2}CU^2 = \frac{U^2}{2}\frac{\varepsilon_0 S}{d - \Delta x}, \tag{3.4}$$

$$W_e(\Delta x) = S(d - \Delta x)\left(\frac{1}{2}\varepsilon_0 E^2\right) = \frac{Q^2}{2C} = \frac{Q^2}{2}\frac{d - \Delta x}{\varepsilon_0 S}. \tag{3.5}$$

3.1.3 Force Expression

The machines operating with the electrical coupling field are called electrostatic machines. Domain with the electrical field is filled with dielectric material. Dielectric is called *linear* if the vector of electrical induction D is proportional to the vector E, $D = \varepsilon E$. Electrostatic machine with linear dielectric is called *linear machine*. The structure shown in Fig. 3.2 represents a linear electrostatic machine with negligible energy losses. Therefore, in the case with $Q = $ const., the mechanical work $\Delta W_{meh} = F\Delta x$ is determined by $\Delta W_{meh} = -\Delta W_e$, whereas in the case of a constant voltage relation, $\Delta W_{meh} = +\Delta W_e = +\frac{1}{2}\Delta W_i$ applies. According to these expressions, the force can be determined as partial derivative of the coupling field energy W_e with respect to coordinate x. This coordinate represents displacement of the mobile electrode along the motion axis of the system.

In the case when the source is disconnected, the system in Fig. 3.2 has a constant charge, and the work of the source U is equal to zero. Applying the method of virtual works, the change in the field energy and the mechanical work are obtained from (3.6).

$$\Delta W_i = U\Delta Q = 0$$
$$\Delta W_i = \Delta W_{meh} + \Delta W_e \quad \Rightarrow \quad \Delta W_{meh} = -\Delta W_e. \tag{3.6}$$

When the source is disconnected, the force F acting on the mobile electrode is given by (3.7). In the case when the changes ΔW_e and Δx are very small, the ratio $\Delta W_e/\Delta x$ assumes the value of the first derivative of $W_e(x)$,

$$F = -\frac{\Delta W_e}{\Delta x},$$
$$F = -\frac{dW_e}{dx} = -\frac{d}{dx}\left\{\frac{Q^2}{2}\frac{d-x}{\varepsilon_0 S}\right\} = \frac{Q^2}{2S\varepsilon_0}. \tag{3.7}$$

If the source is connected, the considered system has a constant voltage. By applying the method of virtual works, the work of the source U, the change in the field energy, and the mechanical work are obtained in (3.8):

$$\Delta W_i = U\,\Delta Q; \quad \Delta W_e = \Delta\left(\frac{CU^2}{2}\right) = \frac{1}{2}U\Delta Q = \frac{1}{2}\Delta W_i,$$
$$\Delta W_i = \Delta W_{meh} + \Delta W_e \quad \Rightarrow \quad \Delta W_{meh} = \Delta W_i - \Delta W_e = \Delta W_e. \tag{3.8}$$

With the source connected, the force F acting on the mobile electrode is given by (3.9). With infinitesimally small changes ΔW_e and Δx, the ratio $\Delta W_e/\Delta x$ assumes the value of the first derivative $dW_e(x)/dx$,

$$F = +\frac{dW_e}{dx} = \frac{d}{dx}\left\{\frac{U^2}{2}\frac{\varepsilon_0 S}{d-x}\right\} = \frac{U^2}{2}\frac{\varepsilon_0 S}{(d-x)^2}$$
$$= \frac{E^2}{2}\varepsilon_0 S = \frac{D^2}{2\varepsilon_0}S = \frac{Q^2}{2S\varepsilon_0}. \tag{3.9}$$

Expressions for electrical force, given by (3.7) and (3.9), are applicable only when the medium is linear, that is, when the permittivity of the dielectric material does not depend on the field strength. In cases when the source U is not connected, displacement of the mobile electrode does not cause any change in charge Q.

Instead, it leads to changes in the capacitance C and the voltage across the plates. The expression for electrical force when the source is disconnected takes the following form:

$$F = -\frac{dW_e}{dx} = -\frac{d}{dx}\left\{\frac{Q^2}{2}\frac{1}{C}\right\} = -\frac{Q^2}{2}\frac{d}{dx}\left\{\frac{1}{C}\right\}.$$

If the source is connected, the voltage across the plates is constant. Therefore, the shift of the mobile electrode changes the capacitance C and the charge Q. The force expression assumes the following form:

$$F = +\frac{dW_e}{dx} = +\frac{d}{dx}\left\{\frac{U^2}{2}C\right\} = +\frac{U^2}{2}\frac{dC}{dx}.$$

Question (3.1): Equation 3.7 gives force F acting on the mobile electrode in the case when the source is disconnected, whereas (3.9) gives this force when the source is connected. Note that in both cases, the same result is obtained, proportional to Q^2. Is it possible that the force acting on the mobile electrode does depend on source U being connected or disconnected? Provide an explanation.

Answer (3.1): The electrical force acting on the mobile plate can be represented as a sum of forces acting on electrical charges distributed over the plate surface. Individual forces are dependent on the density of electrical charge and the field strength in the vicinity of the plate. It is necessary to compare the force obtained with the source U connected to the force obtained with the source detached from the plates. If the plates accommodate the same electrical charge Q in both cases, the surface charge density remains the same. The surface charge density determines the electrical induction D. Therefore, in both cases, the electrical field strength $E = D/\varepsilon_0$ is the same. From this, it can be concluded that in both cases the same force acts on the mobile plate.

3.1.4 Conversion Cycle

In the preceding section, it has been shown that the electromechanical conversion can be performed in two different modes. With the source disconnected, mechanical work ΔW_{meh} is obtained on account of the energy accumulated in the coupling field, $\Delta W_{meh} = -\Delta W_e$. If the source is connected, the work of the source ΔW_i is divided in two equal parts, that is, $\Delta W_e = \Delta W_{meh} = \frac{1}{2}\Delta W_i$. Graphical representation of electromechanical conversion is shown in Fig. 3.3. It is of interest to note that none of the two presented modes can last continuously.

If the source is connected, the electromechanical conversion is performed by turning one part of the source work into mechanical energy, whereas the rest of the

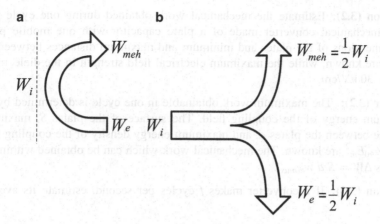

Fig. 3.3 One cycle of electromechanical conversion includes phase (**a**) when the plates of the capacitor are disconnected from the source U and phase (**b**) when the plates are connected to the source

source work increases the energy stored within the coupling field. The field energy W_e is dependent of the density $w_e = \frac{1}{2}\varepsilon_0 E^2$ and the volume of the domain where the field exists. There is an upper limit to the field energy. The maximum strength of the electrical field is limited by the dielectric strength of the material. The maximum electrical field in the air is $E_{max} \approx 30$ kV/cm. Exceeding the maximum field strength leads to dielectric breakdown, wherein the electrical current passes through the dielectric material and creates an electrical arc. The breakdown results in destruction and permanent damage. Therefore, the field strength and the field energy density w_e have to be limited. The volume of the domain is also restricted and defined by the surface of the plates and the distance between them. Therefore, there is a limit $W_{e(max)}$ to the field energy, and it cannot be exceeded. For this reason, it is not possible to withstand a permanent growth of the field energy. Hence, the operation where the source is connected cannot go on indefinitely.

In the case when the source is not connected, mechanical work is obtained on account of the field energy. This energy decreases, and the operation would eventually stop when the field energy is exhausted. Therefore, the operation where the source is disconnected cannot hold indefinitely.

When the need exists for a continuous operation of an electromechanical converter, it is necessary to use concurrently both operating modes. Namely, they should be altered in cycles by switching the source on and off. An interval of operation when the source is connected (on) is followed by another interval when the source is disconnected from the converter (off). In such way, it is possible to provide mechanical work in a continuous manner while keeping the field energy from either reaching $W_{e(max)}$ or dropping to zero. Hence, the process of electromechanical conversion is mostly performed in cycles. Cyclic exchange of the two operating modes is illustrated in Fig. 3.3. In rotating electrical machines, one conversion cycle corresponds to one revolution of the rotor (sometimes, one fraction of the rotor revolution).

Question (3.2): Estimate the mechanical work obtained during one cycle with electromechanical converter made of a plate capacitor with one mobile plate. The dimensions of the plates and minimum and maximum distances between the plates are known, while the maximum electrical field strength in the dielectric is $E_{max} = 30$ kV/cm.

Answer (3.2): The maximum work obtainable in one cycle is determined by the maximum energy of the coupling field. The surface of the plates S, maximum distance between the plates d, and maximum energy density of the coupling field $w_e = \frac{1}{2}\varepsilon_0 E_m^2$ are known. The mechanical work which can be obtained within one cycle is $\Delta W = S\, d\, w_{e(max)}$.

Question (3.3): If a converter makes f cycles per second, estimate its average power.

Answer (3.3): Average power of the converter making f cycles per second is $P_{a\,v} = f\Delta W = f S\, d\, w_{e(max)}$.

3.1.5 Energy Density of Electrical and Magnetic Field

The power of an electromechanical converter is dependent on the density of energy accumulated within the coupling field. A converter of given dimensions will have higher average power if its coupling field has a higher energy density. Given the converter power, dimensions and mass will be reduced for an increased density of energy. The power-to-size ratio is also called *specific power*. The considerations which follow show that electromechanical converters involving magnetic coupling field possess higher specific power compared to electrostatic machines.

The mechanical work obtained within one cycle of electromechanical converter is dependent on the energy stored in the coupling field. The maximum amount of the field energy is dependent on the energy density and the volume of the converter. If two electrical machines of the same size are considered, the machine with higher density of the field energy will produce higher mechanical work within each conversion cycle. If the repetition rates of conversion cycles are the same for the two machines, the machine having higher energy density will have higher average power.

The energy density of magnetic field exceeds by large the density of energy in electrical field. Permittivity (D/E) in vacuum is $\varepsilon_0 = 8.85 \cdot 10^{-12} \approx 10^{-11}$, whereas permeability (B/H) amounts $\mu_0 = 4\pi \cdot 10^{-7} \approx 10^{-6}$. Therefore, the energy density of magnetic field $w_m = \mu_0 H^2/2$ is considerably higher than the energy density of electrical field $w_e = \varepsilon_0 E^2/2$. For this reason, electrical machines are mostly operating with magnetic coupling field.

Density of energy accumulated in the coupling field depends on the square of the field strength. In air, electrical field is limited by dielectric strength, $E_{max} \approx 30$ kV/cm ≈ 3 MV/m. In electrical machines with magnetic field, the field is comprised by magnetic circuit including air gaps and ferromagnetic materials such as iron.

Magnetic inductance B in ferromagnetic materials is limited to $B_{max} = 1–2$ T, thus limiting the magnetic inductance achievable in air. Consequently, the maximum field strength H which can be met in electrical machines is close to $H_{max} \approx B_{max}/ \mu_0 \approx 1$ MA/m. With $\varepsilon_0 \approx 10^{-11}$ and $\mu_0 \approx 10^{-6}$, the achievable energy density is much higher in the case of magnetic field. Considering two electromechanical converters of the same size, the converter operating with magnetic field could accumulate much higher energy in the coupling field ($10^3–10^4$ times) and proportionally higher average power of electromechanical conversion.

3.1.6 Coupling Field and Transfer of Energy

It is of interest to note that both the electromechanical energy conversion with magnetic coupling field and the conversion with electrical field involve both field vectors, electrical field vector E and magnetic field vector H. The exchange of energy between electrical and mechanical terminals of electrical machine implies that in the space surrounding the moving part of the machine, there is transfer of energy toward the moving part (motor) or from the moving part (generator). The energy transfer through the surrounding space is measured by *Poynting vector*. Hence, the energy streams through domain if the Poynting vector has a nonzero algebraic intensity. Poynting vector is equal to the vector product of the electrical and magnetic field. It represents the surface density of power, and it is expressed in W/m^2. Surface integral of Poynting vector over a surface separating two domains represents the rate of energy transfer from one to the other domain (i.e., the power passed from one domain to another). The course and direction of Poynting vector indicate the course and direction of energy transfer. In the absence of either electrical field E or magnetic field H, Poynting vector is equal to zero; thus, no energy transfer is possible. Therefrom, the question arises on how do electrical machines with electrical coupling field acquire magnetic field H required for mandatory Poynting vector.

In an electromechanical converter involving electrical coupling field which is at the state of rest, the magnetic field will be equal to zero and so will be Poynting vector. This is an expected situation since the power of electromechanical conversion is zero in the case when mobile parts of the converter do not move. Namely, the power is equal to the product of the force and speed of motion. At rest, although the electrical force may be present, the speed is equal to zero, thus resulting in zero mechanical power. If the considered converter is in the state of motion, its mobile part moves in the electrical coupling field. This leads to variations in the field strength E and the electrical induction D within the converter. The first time derivative of D contributes to the spatial derivative (i.e., *curl*) of magnetic field H. The second Maxwell equation expresses generalized Ampere law, and it reads

$$\mathrm{rot}\, \vec{H} = \vec{J} + \frac{\partial \vec{D}}{\partial t}. \tag{3.10}$$

Since a nonzero spatial derivative of the field H exists, algebraic intensity of the vector H cannot be equal to zero at all points of the considered domain. The conclusion is that a certain magnetic field H exists in electrostatic machines in the state of motion. The field strength H is proportional to the speed of the machine moving parts. In conjunction with the field E, magnetic field H results in Poynting vector $P = E \times H$.

The same considerations can be derived for an electromechanical converter based on magnetic coupling field. At rest, the magnetic field H exists in the converter, but the electrical field E and Poynting vector P are equal to zero. With $P = 0$, there is no flow of energy toward the mobile part of the machine, and the mechanical power is equal to zero. This corresponds to the conclusion that the mechanical power at rest must be zero, as it is the product of the force and the speed. When the considered converter is in the state of motion, its mobile parts move in the magnetic coupling field. This leads to variations in the magnetic field H and the magnetic induction B within the converter. The first Maxwell equation expresses the Faraday law in differential form, and it reads

$$\text{rot } \vec{E} = -\frac{\partial \vec{B}}{\partial t}. \tag{3.11}$$

Hence, the variation of magnetic induction B results in the spatial derivative (curl) of electrical field, which causes the appearance of the electrical field E within the converter and leads to nonzero values of the Poynting vector.

3.2 Converter Involving Magnetic Coupling Field

Electromechanical conversion in converters involving magnetic coupling field is possible by means of the field acting on the mobile windings and mobile parts made of ferromagnetic materials. In such converters, magnetic field is a precondition for electromechanical conversion of energy. It exists in the space between the stationary and mobile parts of magnetic circuits and current circuits. The mobile parts can perform either linear or rotational movement.

Forces acting on mobile parts are dependent on the magnetic induction and current in conductors. Mechanical work can be obtained on account of the field energy or work of the source which is connected to the current carrying conductors.

3.2.1 Linear Converter

Figure 3.4 shows a simple electromechanical converter involving homogeneous magnetic field and a straight part of the conductor performing linear motion. The subsequent analysis is focused on motoring operation of the converter, wherein

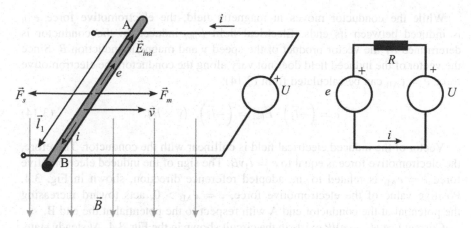

Fig. 3.4 A linear electromechanical converter with magnetic coupling field

the electrical energy, obtained from a constant voltage source U, is converted to mechanical work. Mobile conductor AB of length l_1 touches fixed parallel conductors connected to the source U. The mobile conductor AB, fixed parallel conductors, and source U make current circuit shown on the right-hand side of Fig. 3.4. The resistance of conductor AB can be neglected, whereas the sum of resistances of all remaining conductors in the current circuit is denoted by R.

The source U causes the current i in the circuit. Direction of the current corresponds to the direction of vector l_1 shown in Fig. 3.4 along conductor AB. The conductor is placed in an *external*[1] homogeneous magnetic field of induction B. The electromagnetic force F_m acting on the conductor is determined by (3.12):

$$\vec{F}_m = i\left(\vec{l}_1 \times \vec{B}\right). \tag{3.12}$$

Since vector l_1 is orthogonal to the vector of magnetic induction, algebraic intensity of the force is equal to $F_m = l_1 i B$. The electromagnetic force in Fig. 3.4 is directed from left to right. It is assumed that the force makes the conductor move in the same direction at a speed v. The conductor is subjected to an external force F_{ex}, which opposes this movement. In the state of dynamic equilibrium, acceleration of the conductor is zero, the speed of motion v is constant, and the sum of the forces acting on the conductor is equal to zero. Therefore, the algebraic intensities of the external and electromagnetic forces are equal:

$$\vec{F}_{ex} + \vec{F}_m = 0; \quad \vec{F}_{ex} = -\vec{F}_m; \quad |\vec{F}_{ex}| = |\vec{F}_m| = il_1B. \tag{3.13}$$

---

[1] Magnetic field caused by external phenomena is called *external field*. External phenomena do not make part of the system under consideration, and they are not related or caused by the considered system. External magnetic field can be created by external conductors carrying electrical current, external permanent magnets, the Earth magnetic poles, and other sources.

While the conductor moves in magnetic field, the electromotive force e_{AB} is induced between its ends. Electrical field E_{ind} induced in the conductor is determined by the vector product of the speed v and magnetic induction B. Since the vector of the induced field does not vary along the conductor, the electromotive force $e = e_{AB}$ can be calculated from (3.14):

$$e = \left(-\vec{l_1}\right) \cdot \vec{E}_{ind} = \left(-\vec{l_1}\right) \cdot \left(\vec{v} \times \vec{B}\right). \tag{3.14}$$

Vector of the induced electrical field is collinear with the conductor. Therefore, the electromotive force is equal to $e = l_1 vB$. The sign of the induced electromotive force $e = e_{AB}$ is related to the adopted reference direction, shown in Fig. 3.4. Positive value of the electromotive force, $e = e_{AB} > 0$, acts toward increasing the potential at the conductor end A with respect to the potential at the end B.

Current $i = (U - e)/R$ exists in the circuit shown in the Fig. 3.4. At steady state, time varying electrical current $i(t)$ assumes a constant value $I = (U - l_1 vB)/R$. Power of the source $P_i = Ui = ei + Ri^2$ contains the component $P_{AB} = ei = l_1 vB$ as well as the losses $P_\gamma = Ri^2$. The losses in conductors are caused by Joule effect, and they depend on the equivalent resistance and square of the current. The remaining power P_{AB} is transferred to the moving conductor. By maintaining the movement, electromagnetic force $\boldsymbol{F}_m$ performs the work against external force $\boldsymbol{F}_{ex}$ which is opposite to motion. Vectors of the force and speed of motion are collinear. Therefore, the mechanical power is equal to $P_{meh} = F_m v = l_1 ivB$. Power P_{meh} is the output power of the electromechanical converter which converts electrical energy obtained from the source U to mechanical work. Since $P_{meh} = F_m v = P_{AB} = ei = l_1 vBi$, distribution of the source power P_i can be described by expression

$$P_i = Ui = ei + Ri^2 = F_m v + Ri^2 = P_{meh} + Ri^2. \tag{3.15}$$

Therefore, power from the source is divided in the thermal losses and mechanical power, the latter being the result of electromechanical conversion. The power delivered by the induced electromotive force e is equal to $P_e = e(-i) = -ei < 0$. Consequently, the electromotive force e behaves as a receiver, taking over the electrical power $ei = l_1 vBi$ which is then converted to mechanical power $P_{meh} = F_m v = ei$. In the presented example, the mechanical power of the electromechanical converter is equal to the product of the electromotive force and current. Equation 3.16 in certain form is present in all electrical machines:

$$ei = F_m v. \tag{3.16}$$

Joule losses are determined by the power $P_\gamma = Ri^2$, and they are turned into heat. Conductors and other parts of the converter are heated. Compared to ambient temperature, their temperatures are increased. Due to elevated temperatures, these parts of the converter transfer their heat to the ambient by convection, conduction, or radiation. When the power of losses P_γ becomes equal to the heat power transferred to the ambient, the temperature increase stops and the system enters the thermal equilibrium. Since the electromechanical converters are used for

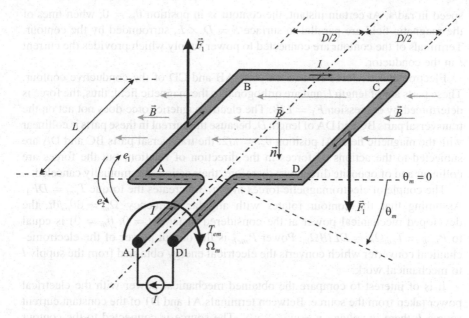

Fig. 3.5 A rotational electromechanical converter involving magnetic coupling field

converting electrical energy to mechanical work, it is necessary to keep the conversion losses as small as possible. Due to thermal losses, the coefficient of efficiency η of power converters is reduced. In addition, generated heat has to be removed so that the converter does not become overheated. It is required, therefore, to have a corresponding solution for heat transfer and cooling. The losses can be reduced by decreasing the equivalent resistance R. However, reducing resistance by increasing the cross section of conductors leads to an increased consumption of copper, increasing in this way the cost, weight, and size of converters.

The power converter shown in Fig. 3.4 can also run in generator mode. Direction of the current will be reversed and also direction of the electromagnetic force. In order to support the motion, direction of the external force $\boldsymbol{F}_{ex}$ has to be changed as well. In generator mode, mechanical power is converted to electrical energy. Generator operation is analyzed in more detail in Sect. 2.4.

3.2.2   Rotational Converter

Electromechanical conversion is most frequently performed by using rotational machines, which convert electrical energy to mechanical work of rotational movement. An example of simple rotational converter is shown in Fig. 3.5. Contour ABCD is made out of copper conductors. It has dimensions $D \times L$, and it rotates in homogeneous external magnetic field B. The contour rotates clockwise around horizontal axis, shown in Fig. 3.5. The position of the contour is determined by angle θ_m, and it varies at the rate $\Omega_m = \mathrm{d}\theta_m/\mathrm{d}t$, where Ω_m represents the angular speed in rad/s. At certain instant, the contour is in position $\theta_m = 0$, when lines of the magnetic field are parallel to surface $S = D \times L$, surrounded by the contour. Terminals of the contour are connected to power supply which provides the current I in the conductor.

Electromagnetic force F_1 acts on parts AB and CD of the conductive contour. These parts are of length L and are orthogonal to the magnetic field; thus, the force is determined by expression $F_1 = LIB$. The electromagnetic force does not act on the transversal parts BC and DA of length D, because the current in these parts is collinear with the magnetic field. At position $\theta_m = \pi/2$, the transversal parts BC and DA are subjected to the actions of forces in the direction of rotation, but the forces are collinear and of opposite directions; therefore, their actions are mutually canceled.

The couple of electromagnetic forces in Fig. 3.5 creates the torque $T_{em} = DF_1$. Assuming that the contour rotates with angular frequency $\Omega_m = d\theta_m/dt$, the developed mechanical power at the considered instant ($t = 0$, $\theta_m = 0$) is equal to $P_{meh} = T_{em}\Omega_m = DLIB\Omega_m$. Power P_{meh} is the output power of the electromechanical converter which converts the electrical energy obtained from the supply I to mechanical work.

It is of interest to compare the obtained mechanical power with the electrical power taken from the source. Between terminals A1 and D1 of the constant current source I, there is voltage $u = v_{A1} - v_{D1}$. The source is connected to the contour ABCD, and the voltage is $u = RI + d\Phi/dt$, where Φ denotes the flux through surface S encircled by the contour, while R denotes the equivalent resistance of the conductors making the contour. Reference direction of the flux is the direction of the positive normal n to surface S. This normal is in accordance with the direction of circulation along the contour ABCD, that is, the current in designated direction of circulation along the contour creates a magnetic field which is aligned with the normal n. In Figs. 3.5 and 3.6, the normal is denoted by vector n. The contour can rotate around horizontal axis; thus, the flux through the surface S depends upon the angle θ_m between the vectors of magnetic induction and the plane in which the surface S reclines.

In accordance with the notation in Fig. 3.6, the angle θ_m is equal to zero at the position where the magnetic field is parallel to the surface S. In zero position, flux Φ is equal to zero. When the contour makes an angular shift of θ_m, the normal n to surface S is shifted to position n_1. Assuming that the external field is homogeneous, Φ can be represented by the expression $\Phi(\theta_m) = \Phi_m\sin(\theta_m) = \Phi_m\sin(\Omega_m t)$, where $\Phi_m = BS$ is the maximum value of flux which is attained at position $\theta_m = \pi/2$. By using the obtained expression for the flux, the voltage across the terminals of the source is calculated as $u = RI + \Omega_m\Phi_m\cos(\Omega_m t)$. At position $\theta_m = 0$, the power delivered by the source I to the converter is given in (3.17):

$$P_i = uI = RI^2 + I\Omega_m\Phi_m = RI^2 + I\Omega_m BS$$
$$= RI^2 + DLIB\Omega_m = P_{meh} + RI^2. \tag{3.17}$$

Therefore, the power of the source I is partially converted to mechanical power, whereas the remaining part accounts for conversion losses that are turned into heat

Fig. 3.6 Variations of
the flux and electromotive
force in a rotating contour

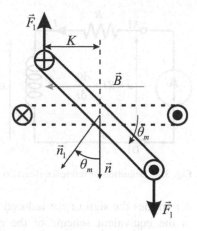

due to Joule effect. At position $\theta_m = 0$, the electromagnetic torque acting on the
contour is equal to $T_{em} = \Phi_m I = P_{meh}/\Omega_m$. After the angle is shifted to θ_m, the arm
K of force F_1 is shown in Fig. 3.6, and it is equal to $K = (D/2)\cos\theta_m$. Therefore,
torque T_{em} varies as function of angle θ_m in accordance with (3.18):

$$T_{em} = \Phi_m I \cos \theta_m. \tag{3.18}$$

Equations similar to (3.18) determine the electromagnetic torque of all rotating
electrical machines. The analysis of operation of the converter shown in Fig. 3.5
leads to the conclusion that the average value of the torque during one full revolution is zero. This can be changed by insertion of additional contours or by changing
the supply current, as will be elaborated in due course.

By changing the direction of the current or direction of rotation, the electromechanical converter shown in Fig. 3.5 will operate in the generator mode of operation,
converting mechanical work to electrical energy. Voltage and current of the current
source I will have opposite signs, while the source I will act as a receiver of electrical
energy.

3.2.3 Back Electromotive Force[2]

The arrows denoted by e_1 and e_2 in Fig. 3.5 indicate two possible reference
directions for the induced electromotive force. The choice of reference direction

---

[2] Back electromotive force (abbreviated BEMF) is also called counter-electromotive force
(abbreviated CEMF), and it refers to the induced voltage that acts in opposition to the electrical
current which induces it. BEMF is caused by changes in magnetic field, and it is described by Lenz
law. The only difference between the electromotive force (EMF) and BEMF is the reference
direction and, hence, the sign.

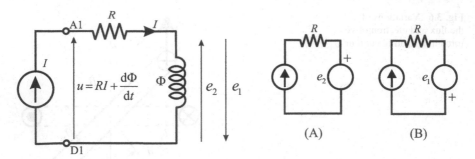

Fig. 3.7 Definition of reference direction for electromotive and back electromotive forces

determines the sign of the induced electromotive force, as well as its connection in the equivalent scheme of the electric circuit; thus, it is useful to give the corresponding explanation and expressions for the induced electromotive force in both cases.

Figure 3.7 shows the equivalent scheme of the mobile contour, fed from a constant current source via terminals A1 and D1. Character Φ denotes flux through the surface S encircled by the contour. Reference direction for the flux is determined by the normal on the surface S, denoted by n in Figs. 3.5 and 3.6.

The normal n is aligned with the magnetic field created by the current I which circulates along the contour (ABCD) in designated direction. Flux Φ depends on the angle θ_m. Rotation of the contour in the direction indicated in Fig. 3.5 leads to a growth of the flux Φ that the external magnetic field makes through the surface S.

The total magnetic flux through the surface S depends on the external magnetic field B, but it also changes with the current that circulates within the contour. Namely, the contour current creates a magnetic field of its own, and this field contributes to the total magnetic flux. The total flux can be expressed by $\Phi = LI + BS \sin(\theta_m)$. The coefficient L defines the ratio between the flux Φ and the current I in cases where the external magnetic field does not exist. The ratio $L = \Phi(I)/I$ is called *the self-inductance* of the contour.

Each change of the flux Φ induces an electromotive force in the contour. This electromotive force depends on the first time derivative of the flux. Under the action of electromotive force, a current appears in the contour. The intensity of this current depends upon the equivalent resistance of the circuit. In the case shown in Fig. 3.7, the contour is fed from a constant current source. The equivalent resistance of a constant current source is $R_{eq} = \infty$; thus, presence of an electromotive force e_1 does not cause any change of current. In the case when the contour is galvanically closed, that is, when terminals A1 and D1 are short-circuited or connected to a voltage source or a receiver of finite equivalent resistance, the presence of electromotive force e_1 will cause a change of current and a change of flux.

According to Lenz rule, electromotive forces are induced in coils due to changes in magnetic flux. Electrical currents appear as a consequence of induced electromotive forces. Induced currents oppose to the flux change and tend to maintain the initial flux value. Electrical current in a coil creates magnetic field and the flux which is proportional to the self-inductance of the coil. Direction of this *self-flux* is opposite to the original flux change. Hence, the induced electromotive force produces the current and the self-flux in direction that tends to cancel the original flux change. For that reason, induced electromotive forces are also called *counter-electromotive forces* or *back electromotive forces*.

Considering the setup in Fig. 3.5, during rotation of the contour in the direction indicated in the figure, the flux due to external magnetic field rises. Electromotive force e_1, given by (3.19), appears in the contour:

$$e_1 = -\frac{d\Phi}{dt}. \tag{3.19}$$

Since an increase of the flux results in $e_1 < 0$, a current appears opposite to the direction of circulation along the galvanically closed contour ABCD. Therefore, the induced current creates its own magnetic field and the self-flux of the contour in the direction opposite to the indicated normal n. Total flux is equal to the sum of fluxes due to external field, which is growing, and the self-flux which is of negative sign.

Electromagnetic induction opposes to changes of the flux to the degree which depends on the circuit parameters. When the equivalent resistance of the circuit is $R_{eq} = \infty$, the induced electromotive force does not cause any change in electrical current which would have opposed to changes in the flux. In cases when the equivalent resistance of the contour is zero ($R_{eq} = 0$, the case of a superconductive contour with short-circuited terminals), the phenomenon of electromagnetic induction prevents any changes of flux. Since the voltage balance equation is given in (3.20)

$$u = Ri - e_1 = Ri + \frac{d\Phi}{dt}, \tag{3.20}$$

in conditions with $u = 0$ and $R = 0$, the flux cannot change due to $d\Phi/dt = 0$. Therefore, notwithstanding eventual changes in the external magnetic field, the total flux through a short-circuited superconductive contour is constant.

For a contour fed from a constant current source, shown in Fig. 3.5, the equivalent schemes of the electrical circuit are shown in Fig. 3.7. For the reference direction of the induced electromotive force, it is possible to use e_1 or e_2, as indicated in Figs. 3.5 and 3.7. If the reference direction e_1 is chosen, the equivalent scheme (B) of Fig. 3.7 applies, and algebraic intensity of the electromotive force is determined by (3.19). Alternatively, the equivalent scheme (A) and (3.21) apply. Quantity $e_2 = +d\Phi/dt$ is called *back electromotive force* or *counter-electromotive force*:

$$e_2 = -e_1 = \frac{d\Phi}{dt}. \tag{3.21}$$

Chapter 4
Magnetic Circuit

This chapter introduces and explains magnetic circuits of electrical machines. Basic laws and skills required to analyze magnetic circuits are reinstated and illustrated on examples and solved problems. The terms such as magnetic resistance, magneto-motive force, core flux, and winding flux are recalled and applied. Dual electrical circuit is introduced, explained, and applied in solving magnetic circuits. Basic properties of ferromagnetic materials are recalled, including saturation phenomena, eddy current losses, and hysteresis losses. Laminated magnetic circuits as the means of reducing the iron losses are explained and analyzed.

One of the key operating principles of electromechanical converters based on magnetic field is creation of Lorentz force acting on a current-carrying conductor placed in the magnetic field. Magnetic field can be obtained from a permanent magnet or by using an *electromagnet*. Electromagnet is a system of windings carrying electrical currents that create magnetic field. It is useful in replacing the permanent magnets by coils carrying a relatively small electrical current. For the electromagnet currents to be moderate, it is necessary to employ magnetic circuits made of ferromagnetic material (iron), conducting the magnetic flux in a way similar to copper conductor directing electrical current. As the copper conductor provides a low-resistance path to electrical current, so does the magnetic circuit provide a path to magnetic flux that has a low magnetic resistance. An example of magnetic circuit is shown in Fig. 4.1.

Figure 4.1 shows a magnetic circuit made of iron, a ferromagnetic material with permeability $\mu = B/H$ higher than that of the vacuum (μ_0) by several orders of magnitude. This magnetic circuit has an air gap of size δ. Within the gap, it is possible to place a conductor carrying current in order to obtain Lorentz force and accomplish electromechanical conversion of energy (the conductor is not shown in the figure). Magnetic flux within the magnetic circuit is created by means of the *excitation* winding with N series-connected contours, also called *turns*. Each turn encircles the magnetic circuit. Assuming that there are no losses and that the lines of

S.N. Vukosavic, *Electrical Machines*, Power Electronics and Power Systems, DOI 10.1007/978-1-4614-0400-2_4, © Springer Science+Business Media New York 2013

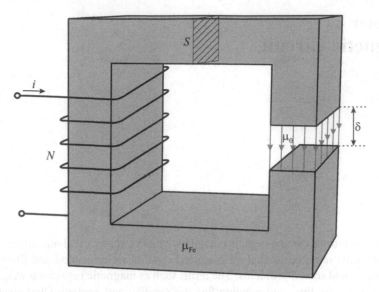

Fig. 4.1 Magnetic circuit made of an iron core and an air gap

the field are parallel, it is concluded that magnetic induction in iron (B_{Fe}) is equal to the magnetic induction in the air (B_0). The strength of the magnetic field in iron is $H_{Fe} = B_{Fe}/\mu_{Fe}$, whereas in the air gap, it is equal to $H_0 = B_0/\mu_0$. Since permeability of iron is much higher, the magnetic field in iron will be considerably lower than the field in the air gap. Ampere law thus reduces to $Ni = H_0\delta$, and the current required for obtaining magnetic field H_0 in the gap is equal to $i = H_0\delta/N$.

In order to obtain magnetic induction B in the air gap, it is necessary to establish the current $i = B\delta/(N\mu_0)$ in the excitation winding. Hence, the required excitation current is proportional to the air gap δ. An attempt to remove the iron part of the magnetic circuit can be represented as an increase of the gap δ to $\delta + l_{Fe}$, where l_{Fe} is the length of the iron part of the magnetic circuit. The required current would increase $1 + l_{Fe}/\delta$ times. Since $l_{Fe} >> \delta$, removal of the iron would result in a multiple increase of the excitation current and the associated losses. Therefore, it is concluded that the magnetic circuit is a key part of electrical machinery. It directs and concentrates the magnetic field to the region where the conductors move and the electromechanical conversion takes place. The presence of an iron magnetic circuit allows the necessary *excitation* to be accomplished with considerably smaller currents and lower losses.

In the preceding section, an analysis of a simple magnetic circuit has been done. In the analysis, certain simplifications have been made. In order to analyze more complex magnetic circuits, a list of the basic laws and usual approximations to simplify the analysis is presented within the next section.

4.1 Analysis of Magnetic Circuits

Magnetic circuit is a domain where magnetic field is created by one or several current circuits or permanent magnets. The laws applicable for analysis of magnetic circuits are:

- The flux conservation law
- Generalized form of Ampere law
- Constitutive relation $B(H)$ which describes a magnetic material

4.1.1 Flux Conservation Law

$$\oint_S \vec{B} \cdot d\vec{S} = 0. \tag{4.1}$$

Taking into account ferromagnetic properties of magnetic materials used in making magnetic circuits, a series of simplifications can be introduced in order to facilitate their analysis. One of the assumptions is that there is no leakage of magnetic lines outside magnetic circuit. Neglecting the leakage, it can be shown that the flux remains constant along the magnetic circuit. In other words, the magnetic flux in each cross-section of the magnetic circuit is the same. The flux in each cross-section is also called *the core flux* or *the flux per turn*, meaning the flux in a single turn of the winding encircling the magnetic circuit. The algebraic value of the flux in the cross-section is defined in accordance with the normal to the cross-section surface, and it is denoted by Φ. In most cases, magnetic circuit is encircled by a winding made of N series-connected turns having the orientation. Assuming that there is no flux leakage from the magnetic circuit, the flux in each turn is equal to Φ. Therefore, the flux of the winding is $\Psi = N\Phi$, with the same reference direction as for the flux in one turn.

The windings are connected in electrical circuits. The voltage across a winding is equal to $u = Ri + d\Psi/dt$, where R is resistance of the series-connected turns, i is winding current, while $d\Psi/dt$ is back electromotive force. It is of uttermost importance to match the reference direction of the electrical circuit (current) with the orientation of the magnetic circuit (flux). As a rule, the reference normal for the flux is determined from the reference direction of the current by the right-hand rule.

By applying the flux conservation law, it can be shown that the flux in one contour (turn) is equal to the flux through any other surface leaning on the same contour. This equality will be used to simplify calculation of the flux in the windings of cylindrical machines.

Commonly used assumption is that the magnetic field is homogeneous over the cross-section of a magnetic circuit and that the length of any magnetic field line is equal to the length of the average representative line of the magnetic circuit.

Fig. 4.2 The reference
normal *n* to surface *S* which
is leaning on contour *c*

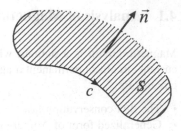

4.1.2   Generalized Form of Ampere Law

Generalized form of Ampere law for fields with stationary electrical currents is
given by (4.2). Contour *c* and surface *S* are shown in Fig. 4.2:

$$\oint_c \vec{H} \cdot d\vec{l} = \int_S \vec{J} \cdot d\vec{S}. \tag{4.2}$$

Electrical currents in electrical machines are not distributed in space, but they
exist in conductors forming the turns, windings, and current circuits. The conductors are usually made of copper wires. With a layer of insulating material wrapped
around wires, they do not have galvanic contact with other parts. Therefore, the
current is directed along wires and does not leak away. Consequently, instead of a
surface integral of current density *J*, one should use the sum of currents in the
conductors passing through a surface *S*, respecting the reference direction determined by the unit vector. Equation 4.2 thus takes the form (4.3):

$$\oint_c \vec{H} \cdot d\vec{l} = \Sigma I. \tag{4.3}$$

4.1.3   Constitutive Relation Between Magnetic
Field H and Induction B

The relation between the vector of magnetic field *H* and magnetic induction *B* in
individual parts of a magnetic circuit is determined by the properties of the
magnetic material, and it is given by (4.4):

$$\vec{B} = \vec{B}(\vec{H}). \tag{4.4}$$

In linear media, magnetic induction *B* is proportional to magnetic field *H*.
Coefficient of proportionality is a scalar quantity μ called magnetic permeability
(4.5). Magnetic permeability in vacuum is $\mu_0 = 4\pi \cdot 10^{-7}$ [H/m]. In ferromagnetic

materials like iron, the characteristic $B(H)$ is not linear. It is usually presented graphically or by the corresponding analytical approximation called *the characteristic of magnetization*. For small values of magnetic field, the magnetization characteristic of iron $B(H)$ is linear and has the slope $\Delta B/\Delta H$ which is several thousand times higher than the permeability of vacuum μ_0.

$$\vec{B} = \mu\vec{H} = \mu_0\mu_r\vec{H}. \tag{4.5}$$

4.2 The Flux Vector

Flux through the contour of Fig. 4.2 is a scalar quantity. Flux through surface S, leaning on contour c, is determined by surface integral of the vector of magnetic induction B. In the analysis of electrical machines, flux through a contour is often considered as a vector. The *flux vector* is obtained by associating the course and direction with scalar Φ. The spatial orientation is obtained from the unit normal to surface S. In cases with several contours (turns) forming a winding where all of the contours share the same orientation, it is possible to define the flux vector of the winding. This flux has algebraic intensity of $\Psi = N\Phi$ while its course and direction are determined by the unit normal to surface S. The winding can be made of series-connected contours (turns) with different spatial orientation. In such cases, the vector of the winding flux is obtained as a vector sum of flux vectors in individual contours.

4.3 Magnetizing Characteristic of Ferromagnetic Materials

Magnetic circuits of electrical machines and transformers are most frequently made of iron sheets. Iron is ferromagnetic material with magnetization characteristic B (H) shown in Fig. 4.3. The characteristic extends between the two straight lines. The line with the slope $\Delta B/\Delta H = \mu_0$ describes magnetization characteristic of vacuum, while the line with the slope μ_{Fe} corresponds to the first derivative of the function $B(H)$ at the origin. The abscissa of the B-H coordinate system is the external field H, which may be obtained by establishing a current in the excitation winding, while the ordinate is magnetic induction B existing in the ferromagnetic material.

The magnetic properties of iron originate from microscopic Ampere currents within a molecule or a group of molecules. These currents make the origin of the magnetic field of permanent magnets and other ferromagnetic materials. The said currents are the cause of forces acting on ferromagnetic parts brought in a magnetic field. The presence of microscopic currents can be taken into account by treating

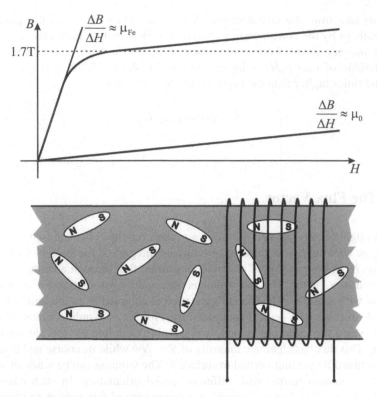

Fig. 4.3 The magnetization characteristic of iron

ferromagnetic materials as a vast collection of miniature magnetic dipoles, as
shown in Fig. 4.3. In the absence of external field H, the magnetic dipoles do not
have the same orientation. They oscillate and change directions at a speed that
depends on the temperature of the material. Therefore, in the absence of an external
magnetic field, resulting magnetic induction in the material is equal to zero.

With an excitation current giving rise to magnetic field H, magnetic dipoles turn
in an attempt to get aligned with the field. Thermal motion of dipoles prevents them
to stay aligned and makes them change the orientation. The higher the field H, the
more dipoles get aligned to the field. As a consequence, resulting magnetic induction takes the value $B = \mu_{Fe}H$ which is much higher than the corresponding value
in vacuum ($B = \mu_0 H$). In this way, ferromagnetic materials help providing the
required magnetic induction B with much smaller excitation current.

When magnetic induction reaches $B_{max} \in [1 \ldots 2]$ T, all miniature dipoles get
oriented in the same direction, aligned with the excitation field H. Any further
increase of the field strength H cannot improve the orientation of dipoles, as there
are no more disoriented dipoles. This state is called *saturation* of magnetic material.
In the region of saturation, further increase of induction is the same as it would have
been in vacuum, $\Delta B = \mu_0 \Delta H$. The saturation region is expressed in the right-hand
side of the curve in Fig. 4.3.

4.4   Magnetic Resistance of the Circuit

The role of magnetic circuit in electrical machines is to direct the lines of magnetic coupling field to the space where the electromagnetic conversion takes place. The magnetic induction and flux Φ in the magnetic circuit appear under the influence of current in the winding. The strength of the field H is dependent on the product Ni, where N is the number of turns in a winding while i is the electrical current. In a way, the value Ni tends to establish the flux Φ in the magnetic circuit. Therefore, the ratio Ni/Φ is *magnetic resistance of the circuit*. A circuit having smaller magnetic resistance will reach the given flux with smaller currents. A magnetic circuit can have several parts, which can be made of ferromagnetic material, permanent magnets, nonmagnetic materials, or air. Air-filled parts of magnetic circuits are also called *air gaps*. It is of interest to determine magnetic resistance of a magnetic circuit comprising several heterogeneous parts.

Magnetizing characteristics of ferromagnetic parts of magnetic circuit are non-linear and shown in Fig. 4.3. Operation of a magnetic circuit is usually performed in the vicinity of the origin of $B(H)$ diagram. It is therefore justifiable to linearize the magnetization characteristics and consider that the permeability μ_{Fe} of ferromagnetic (iron) parts is constant. In the linearized ferromagnetic circuits, nonlinearity of ferromagnetic material is neglected, and permeability of every part of the magnetic circuit is considered constant. In addition, it is assumed that there is no leakage of magnetic field outside magnetic circuit. The basic assumptions and steps in the analysis of linearized magnetic circuits are given by the following considerations.

On the basis of (4.3), the line integral of magnetic field H along contour c, indicated in Fig. 4.4, is equal to the product Ni. *Magnetomotive force $F = Ni$* is equal to the integral of the field H through the closed contour passing through all the parts of the magnetic circuit. Magnetomotive force F is a scalar quantity. Vector of the magnetomotive force is obtained by associating the spatial orientation to the scalar $F = Ni$. The orientation of the magnetomotive force F is determined by the vector H. Both the orientations of F and H are related to electrical currents in the winding that encircles the magnetic circuit. In Fig. 4.4, vector of the magneto-motive force F is collinear with the normal n_k related to the reference direction of the electric currents by the right-hand rule.

Surface integral of magnetic induction over surface S is denoted by Φ and is called *flux of the core* or *flux across the cross-section of the magnetic circuit* or *flux in one turn*. Assuming that there is no leakage of magnetic field outside of the magnetic circuit, the line integral of the magnetic field H along contour c is equal Ni for every and each contour passing through the magnetic circuit. Since the basic assumption is that the lengths of magnetic lines are equal to the length of the representative average line of the magnetic circuit, it can be considered that the magnetic field is homogeneous across each cross-section of the magnetic circuit. Thus, the flux through one turn is $\Phi = BS$. *Winding flux* $\Psi = N\Phi$ represents the flux through the winding with N series-connected turns.

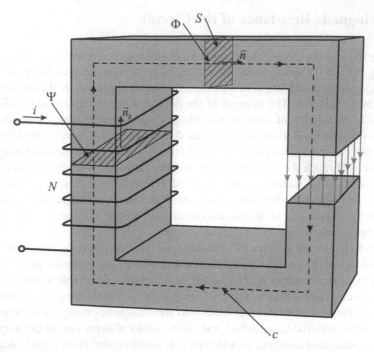

Fig. 4.4 Sample magnetic circuit with definitions of the cross-section of the core, flux of the core, flux of the winding, and representative average line of the magnetic circuit. Magnetic circuit has a large iron core with a small air gap in the right-hand side

Flux through any cross-section of the magnetic circuit is constant. Since $S = S_0 = S_{Fe}$, equality $SB_{Fe} = SB_0$ applies. Therefore, $B_{Fe} = B_0$, where B_0 is magnetic induction in the air gap while B_{Fe} is magnetic induction in the ferromagnetic material (iron). Magnetic field in the air gap is $H_0 = B_0/\mu_0$, whereas the field in the ferromagnetic material is $H_{Fe} = B_{Fe}/\mu_{Fe}$. Generalized Ampere law results in (4.6), where l is average length of the ferromagnetic circuit and l_0 is length of the air gap:

$$H_{Fe}l + H_0l_0 = Ni. \tag{4.6}$$

By inserting $H_0 = B_0/\mu_0$, $H_{Fe} = B_{Fe}/\mu_{Fe} = H_{Fe} = B_0/\mu_{Fe}$ in (4.6), one obtains (4.7), which gives magnetic induction $B_{Fe} = B_0$

$$\frac{B_0}{\mu_{Fe}}l + \frac{B_0}{\mu_0}l_0 = Ni = F. \tag{4.7}$$

Since flux of the core is $\Phi = BS$, its dependence on magnetomotive force F can be represented by (4.8):

$$\Phi = \frac{Ni}{\frac{l}{\mu_{Fe}S} + \frac{l_0}{\mu_0 S}} = \frac{F}{\frac{l}{\mu_{Fe}S} + \frac{l_0}{\mu_0 S}} = \frac{F}{R_\mu}. \tag{4.8}$$

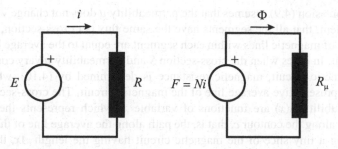

Fig. 4.5 Representation of the magnetic circuit by the equivalent electrical circuit

A dual electrical circuit can be associated with the magnetic circuit, as shown in Fig. 4.5. In this circuit, *electromotive force E* causes the current $i = E/R$ in resistance R. Electromotive force E is equal to the integral of the external electrical field in the electrical generator. In magnetic circuit, *magnetomotive force F* produces the flux $\Phi = F/R_\mu$. The flux Φ in magnetic circuit of magnetic resistance R_μ is dual to the electrical current $i = E/R$ in electrical circuit of resistance R. For this reason, the electrical circuit is an equivalent representation of the magnetic circuit and is therefore called *dual circuit*. A more detailed analysis can show that Kirchhoff laws can be applied to complex magnetic circuits in the same way they apply to electrical circuits.

Magnetomotive force F can be considered as *magnetic voltage* of the considered contour c. By analogy with electrical circuit with $i = E/R$, the flux in magnetic circuit is $\Phi = F/R_\mu$, where R_μ is *resistance of the magnetic circuit* or *reluctance*. Therefore, the flux in a magnetic circuit is obtained by dividing the magnetomotive force Ni by the magnetic resistance R_μ. This applies for linear magnetic circuits with constant permeability μ, with no magnetic leakage, and with constant core flux along the whole magnetic circuit. The last condition stems from the law of conservation of magnetic flux. Quantities Φ, F, and R_μ of a magnetic circuit are duals to quantities i, U, and R of the equivalent electrical circuit. Equation 4.8 represents "Ohm law" for magnetic circuit or Hopkins law.

Magnetic resistance of a uniform magnetic circuit of length l, constant cross-section S, and permeability μ is equal to $R_\mu = l/(S\mu)$. Magnetic circuit may consist of several segments of different dimensions and different magnetic properties. The segments of magnetic circuits are usually connected in series. The equivalent magnetic resistance of the magnetic circuit can be obtained by adding the individual resistances of series-connected segments. For a magnetic circuit with n segments, the equivalent magnetic resistance can be determined by adding resistances $R_{\mu k} = l_k/(S_k\mu_k)$, as shown in (4.9):

$$R_\mu = \sum_{k=1}^{n} \frac{l_k}{\mu_k S_k}. \tag{4.9}$$

The expression (4.9) assumes that the permeability μ does not change within the same segment, that all the segments have the same flux per cross-section, and that the lengths of magnetic lines within each segment are equal to the average length of the segment. In cases when the cross-section S and permeability μ vary continually along magnetic circuit, magnetic resistance is determined by (4.10), where c is oriented representative average line of the magnetic circuit. The cross-section $S(x)$ and permeability $\mu(x)$ are functions of variable x, which represents the path of circulation along the contour c, that is, the path along the average line of the circuit. Considering a tiny slice of the magnetic circuit having the length Δx, the cross-section $S(x)$, and permeability $\mu(x)$, it is reasonable to assume that $S(x) \approx S(x + \Delta x)$ and $\mu(x) \approx \mu(x + \Delta x)$. Therefore, magnetic resistance ΔR_μ of the considered part of magnetic circuit is equal to $\Delta x/(S\mu)$. The equivalent resistance of the magnetic circuit is obtained by adding resistances of all such parts of the magnetic circuit, resulting into integral (4.10). Equation 4.10 is in accordance with the formula for calculating resistance of a resistor with variable cross-section $S(x)$ and variable conductivity $\sigma(x)$:

$$R_\mu = \oint_c \frac{dx}{\mu(x)S(x)}.$$

(4.10)

Magnetic resistance can be used in determining the self-inductance of the winding with N turns encircling the magnetic circuit. Inductance of the winding is equal to the ratio of the flux in the winding $\Psi = N\Phi$ and the electrical current in the winding. On the basis of (4.11), inductance of the winding is equal to the ratio of the squared number of turns and magnetic resistance:

$$L = \frac{\Psi}{i} = \frac{N\Phi}{i} = \frac{N}{i}\frac{Ni}{R_\mu} = \frac{N^2}{R_\mu} = \frac{N^2}{\sum_{i=1}^{k} \frac{l_i}{\mu_i S_i}}.$$

(4.11)

4.5 Energy in a Magnetic Circuit

Energy of magnetic field is determined by integration of the spatial energy density w_m within the domain where the magnetic field exists. In a linear ferromagnetic and in air, spatial density of magnetic energy is $BH/2$. In the case of a magnetic circuit with no leakage, magnetic field is present only within the circuit. Therefore, the space V where the integration (4.12) is carried out is limited to the magnetic circuit under the scope:

$$W_e = \int_V w_m \, dV = \int_V \left(\int \vec{H} \cdot d\vec{B} \right) dV = \int_V \left(\frac{1}{2}BH \right) dV.$$

(4.12)

Magnetic circuit can be divided into elementary volumes $dV = Sdl$, where S is the cross-section of the magnetic circuit and dl is the length of the elementary volume, measured along the representative average line of the magnetic circuit (contour c). According to the flux conservation law, the flux is the same through any cross-section of the magnetic circuit. Therefore, the surface integral of magnetic induction B is equal to Φ on any cross-section of the circuit. The usual and well-founded assumption is that the magnetic field is homogeneous at every cross-section, namely, that the magnetic induction B across the cross-section does not change. Therefore, it can be concluded that magnetic induction B on each cross-section S is Φ/S. With $dV = Sdl$, the integral (4.12) can be simplified by substituting $w_m dV$ by $\frac{1}{2}\Phi Hdl$. The vector H is collinear with the oriented element of contour dl. Therefore, the scalar product of the two vectors can be replaced by the product of their algebraic intensities.

$$W_e = \frac{1}{2}\int_V (BH)dV = \frac{1}{2}\int_S dS \oint_c BH\,dl = \frac{\Phi}{2}\oint_c H\,dl = \frac{\Phi}{2}\oint_c \vec{H}\cdot d\vec{l}. \qquad (4.13)$$

According to Ampere law, line integral of the magnetic field H along contour c which represents average line of the magnetic circuit is equal to Ni. Therefore, the expression for energy of magnetic field takes the form (4.14). It should be noted that the result (4.14) cannot be applied to magnetic circuits with nonlinear magnetic materials:

$$W_e = \frac{\Phi}{2}\oint_c \vec{H}\cdot d\vec{l} = \frac{\Phi}{2}Ni = \frac{\Psi i}{2} = \frac{1}{2}Li^2. \qquad (4.14)$$

Question (4.1): The magnetic circuit shown in Fig. 4.4 is made of ferromagnetic material whose permeability can be considered infinite. Determine self-inductance of the winding.

Answer (4.1): Assuming that μ is infinite, magnetic resistance of the circuit reduces to $R_\mu = l_0/(S\mu_0)$. Inductance of the winding is $L = \mu_0 SN^2/l_0$.

4.6 Reference Direction of the Magnetic Circuit

Magnetic circuit can have more than one winding around the core. Figure 4.6 shows a magnetic circuit having two windings, N_1 and N_2. *Winding flux* arises in each of the windings. The two windings are coupled by the magnetic circuit. Therefore, the flux in each winding depends on both currents, i_1 and i_2. Reference direction of the winding flux is related by the right-hand rule to the reference direction of the

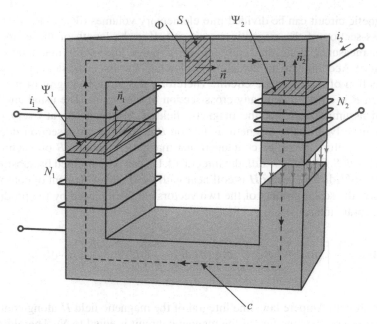

Fig. 4.6 Two coupled windings on the same core

current of the considered winding. Reference direction of flux Ψ_1 is denoted by unit vector n_1 in Fig. 4.5. This direction is in accordance with the adopted direction of circulation around contour c. The unit vector denoting this direction is the normal n. The adopted direction is called *reference direction of the magnetic circuit*. The flux intensity is determined by the product of the core flux and number of turns, thus $\Psi_1 = N_1\Phi$. The reference direction of the flux Ψ_2 in the other winding is denoted by unit vector n_2, and it is opposite to the adopted direction. Since the flux of the core Φ is defined as the flux though cross-section S in the direction of unit vector n, the flux in the second winding is negative, $\Psi_2 = -N_2\Phi$. Choice of the reference direction of magnetic circuit can be arbitrary; therefore, in the analysis of circuits having several windings, each winding should be allocated reference direction according to the right-hand rule and compared with the reference direction of the magnetic circuit.

Relations $\Psi_1 = N_1\Phi$ and $\Psi_2 = -N_2\Phi$ have been obtained under the assumption that there is no leakage of magnetic field, that is, that the flux over cross-section is maintained constant. In the absence of leakage, flux in the turns of winding N_1 is equal to the flux in the turns of winding N_2; therefore, the ratio Ψ_1/Ψ_2 is equal to N_1/N_2, the ratio of the number of turns. The same holds for the ratio e_1/e_2 between the electromotive forces induced in the windings. In real magnetic circuits, a certain amount of flux is leaking away from the magnetic circuit. A small portion of flux in winding N_1 can escape the core before arriving at winding N_2. This flux is called *stray* or *leakage flux* of the first winding. In the same manner, the leakage flux of the second winding encircles the winding N_2, but it leaks away from the core before reaching the winding N_1. In the case when the leakage flux cannot be neglected, ratio $|\Psi_1/\Psi_2|$ deflects from N_1/N_2. Strength of magnetic coupling between two windings is described by the *coefficient of inductive coupling* $k \leq 1$. In the absence of leakage, the coupling coefficient is equal to 1. With $k = 0.9$, the relative amount of leakage flux is 10%.

4.7 Losses in Magnetic Circuits

The energy accumulated in the field of electromechanical converters exhibits a cyclic change. Therefore, magnetic induction in magnetic circuits varies within conversion cycles. In AC current machines and transformers, magnetic induction has a sinusoidal variation. Variations of induction B in ferromagnetic materials cause energy losses. These can be divided into eddy current losses and hysteresis losses. Power of losses per unit mass is also called *specific* power or *loss power density*.

4.7.1 Hysteresis Losses

Variation of magnetic field in a ferromagnetic material implies setting in motion magnetic dipoles and changing their orientation. Rotation of magnetic dipoles requires a certain amount of energy. This energy can be estimated from the surface of *hysteresis curve* of the $B = f(H)$ diagram. When induction B oscillates with a cycle time (period) T, as shown in Fig. 4.7, the operating point in the $B = f(t)$ diagram runs along the trajectory called *hysteresis curve*. The energy consumed by rotation of dipoles within one cycle T is proportional to the surface encircled by the hysteresis curve swept by the (B-H) operating point. The origin of hysteresis losses

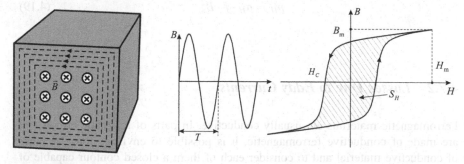

Fig. 4.7 Eddy currents in a homogeneous piece of an iron magnetic circuit (*left*). An example of the magnetization characteristic exhibiting hysteresis (*right*)

is friction between neighboring magnetic dipoles in the course of their cyclic rotation. This internal friction causes consumption of energy which is converted into heat.

Specific power losses due to hysteresis p_H are proportional to operating frequency and to surface encircled by the hysteresis curve in the B-H plane. The energy lost in each operating cycle due to hysteresis in ferromagnetic material of volume V is

$$W_H = V \oint H \, dB = V \cdot S_H, \tag{4.15}$$

where S_H is surface encircled by hysteresis curve. With the operating frequency f, power loss due to hysteresis is

$$P_H = f V \cdot S_H. \tag{4.16}$$

The specific power losses, that is, losses per unit volume, are

$$p_{H1} = \frac{P_H}{V} = f \, S_H. \tag{4.17}$$

Surface of the hysteresis curve S_H depends on the shape of the curve and peak values of the magnetic field H_m and induction B_m. The surface is proportional to the product $B_m H_m$. The peak values H_m and B_m are in mutual proportion. Therefore, the surface S_H is also proportional to B_m^2. Therefore, the losses per unit volume can be expressed as

$$p_{H1} = \sigma_{H1} \cdot f \cdot B_m^2. \tag{4.18}$$

By introducing coefficient σ_H which is equal to the ratio of the coefficient σ_{H1} and specific mass of ferromagnetic material, specific losses due to hysteresis per unit mass are

$$p_H = \sigma_H \cdot f \cdot B_m^2. \tag{4.19}$$

4.7.2   Losses Due to Eddy Currents

Ferromagnetic materials are usually conductive. In parts of magnetic circuit that are made of conductive ferromagnetic, it is possible to envisage toroidal tubes of conductive material and to consider each of them a closed contour capable of carrying electrical currents. Variation of magnetic induction $\boldsymbol{B}$ changes the flux in such contours. As a consequence, electromotive forces are induced in such contours,

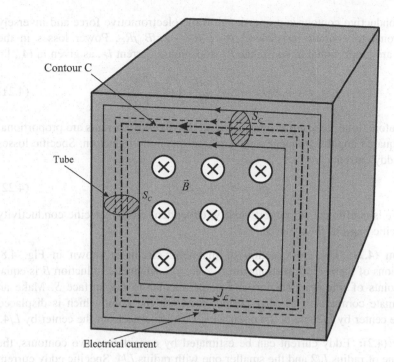

Contour C

Tube

$\vec{B}$

S_C

S_C

J_C

Electrical current

Fig. 4.8 Eddy currents cause losses in iron. The figure shows a tube containing flow of spatially distributed currents

and they produce electrical currents that oppose to the flux changes. A number of conductive contours can be identified within each piece of ferromagnetic. Therefore, the change in magnetic induction causes spatially distributed currents which contribute to losses in magnetic circuits. Such currents are also called *eddy currents*. The losses associated to such currents are called *eddy current losses*.

Figure 4.8 shows a piece of ferromagnetic material with oscillatory induction B of amplitude B_m and angular frequency ω. Lines of magnetic induction are encircled by contour C which is at the same time the average line of the tube having cross-section S_C and length l_C. Since the tube is in a ferromagnetic material of finite conductivity σ, it can be represented by a conductive contour with equivalent resistance $R_C = l_C/(S_C\sigma)$. Changes in inductance B result in flux changes. In turn, flux changes give rise to induced electromotive force in the contour

$$e = -\frac{d\Phi}{dt} = -\frac{d}{dt}(-SB_m \sin \omega t) = \omega SB_m \cos \omega t, \qquad (4.20)$$

where S is the surface encircled by the contour C in Fig. 4.8. Amplitude of the electromotive force induced in the contour is proportional to the angular frequency and magnetic induction B, hence $E \sim \omega B_m \sim 2\pi f B_m$. Electrical current established in the conductive contour is proportional to the electromotive force and inversely proportional to contour resistance,[1] $I \sim E/R_C \sim 2\pi f B_m/R_C$. Power losses in the contour are proportional to resistance R_C and square current I_C, as given in (4.21):

$$P_C \sim R_C I_C^2 \sim R_C \left(\frac{\omega B_m}{R_C}\right)^2 \sim \frac{\omega^2 B_m^2}{R_C}. \tag{4.21}$$

Therefore, total losses in magnetic circuit due to eddy currents are proportional to the squared angular frequency and squared magnetic induction. Specific losses due to eddy currents are

$$p_V = \sigma_V \cdot f^2 \cdot B_m{}^2, \tag{4.22}$$

where σ_V is coefficient of proportionality, dependent on the specific conductivity and specific mass of the material.

Question (4.2): The cross-section of magnetic circuit is shown in Fig. 4.8. Dimensions of these cross-sections are $L \times L = S$. Magnetic induction B is equal in all points of this cross-section and perpendicular to the surface S. Make an approximate comparison of eddy current losses at the point which is displaced from the center by $L/2$ and at the point which is displaced from the center by $L/4$.

Answer (4.2): Eddy current can be estimated by considering two contours, the larger one of radius $L/2$ and the smaller one with radius $L/4$. Specific eddy current losses, that is, the losses per unit volume, depend on the square of the induced electrical field E_i, $p_V \sim \sigma E_i^2$. Induced electrical field can be estimated by dividing the induced electromotive force E of the contour by the length of the contour. The contour of radius $L/2$ has four times larger surface and, therefore, four times larger flux and electromotive force E. Its length is two times larger than the length of the small contour. Therefore, induced electrical field E_i along the larger contour has twice the strength of the induced electrical field along the small contour. Finally, the eddy current losses at the point further away from the center are four times larger.

4.7.3 Total Losses in Magnetic Circuit

The sum of specific losses due to hysteresis and due to eddy currents is given by (4.23). Specific losses p_{Fe} are expressed in W/kg units. With uniform flux density B, the loss distribution in magnetic circuit is uniform as well. In this case, total magnetic field losses in a magnetic circuit of mass m are $P_{Fe} = p_{Fe}m$.

$$p_{Fe} = p_H + p_V = \sigma_H \cdot f \cdot B_m{}^2 + \sigma_V \cdot f^2 \cdot B_m{}^2. \tag{4.23}$$

---

[1] Considered contour has resistance R_C and self-inductance L_C. It has an induced electromotive force E of angular frequency ω. Electrical current in the contour should be calculated by dividing the electromotive force by the contour impedance $\underline{Z}_C = R_C + j\omega L_C$. At lower frequencies where $R_C \gg \omega L_C$, reactance ωL_C of the contour can be neglected.

In magnetic circuits with variable cross-section as well as in cases where the circuit comprises parts made of different materials and different properties, specific losses p_{Fe} are not the same in all parts of the circuit. Thus, total losses P_{Fe} are determined by integrating specific losses over the volume of the magnetic circuit.

4.7.4 The Methods of Reduction of Iron Losses

Power losses in magnetic circuits of electromechanical converters reduce their efficiency. In addition, the losses are eventually turned to heat, and they increase temperature of the magnetic circuit. Overheating can result in damage to the magnetic circuit or to other nearby parts of the machine. Therefore, it is necessary to transfer this heat to the environment. In other words, it is necessary to provide the means for proper cooling. Loss reduction simplifies the cooling system, increases conversion efficiency, and reduces the amount of heat passed to the environment.

In iron sheets and other ferromagnetic materials used for making magnetic circuits of electrical machines and transformers, iron losses due to eddy currents prevail over iron losses due to hysteresis. Eddy current losses are larger than hysteresis losses by an order of magnitude. The losses can be reduced by taking additional measures in designing and manufacturing magnetic circuits, thus increasing the efficiency of electrical machines and preventing their overheating.

By adding silicon and other materials of low specific conductivity into iron used for making magnetic circuits, specific conductivity of such an alloy is reduced. The increase of resistance R_C of the eddy current contours reduces the amplitude of such currents (4.21) and reduces eddy current losses.

Another approach to reducing eddy current losses is *lamination*, the process of assembling magnetic circuits out of sheets of ferromagnetic material. The sheets are oriented along direction of the magnetic field, in the way shown in Fig. 4.9. A laminar magnetic circuit is not made of solid iron, but of iron sheets which are electrically isolated from one another.

Since the sheets are parallel with magnetic field, contours of induced eddy currents are perpendicular to the field. Electrical insulation between neighboring layers prevents eddy currents; thus, they can be formed only within individual layers. It can be shown that this contributes to a considerable reduction of eddy current losses.

Iron sheets used for designing magnetic circuits of line-frequency transformers (50 or 60 Hz) and conventional electrical machines are 0.2–0.5 mm thick. Insulation between the sheets is made by inserting thin layers of insulating material (paper, lacquer) or by short-time exposure of iron sheets to an acid which forms a thin layer of nonconductive iron compound (salt).

In contemporary electrical machines used in electrical vehicles, hybrid cars, and alternative power sources, the operating frequency may be in excess of 1 kHz. Magnetic circuits of such machines are made of very thin iron sheets (0.05–0.1 mm) or of amorphous strips based on alloys of iron, manganese, and other metals, as well

a   b

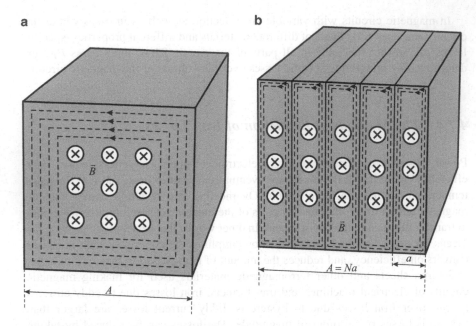

Fig. 4.9 Electrical insulation is placed between layers of magnetic circuit to prevent flow of eddy currents

as of *ferrites*. Ferrite is material obtained from molten iron alloy exposed to an increased pressure and fed to a nozzle with a very small orifice. Expanding at the mouth of the nozzle, the molten alloy is dispersed into small balls with diameter next to 50 μm. Short oxidation of these balls creates a thin layer of insulating oxide. Consequently, miniature balls fall into a cooling oil. By collecting them, one obtains a fine dust made of insulated ferromagnetic balls. Put under pressure (*sintering*), this dust becomes a hard and fragile material called ferrite. Magnetic properties of ferrites are similar to those of iron. At the same time, due to a virtual absence of eddy currents, the losses in ferrites are very low.

4.7.5   Eddy Currents in Laminated Ferromagnetics

Figure 4.10 shows one sheet of iron from the package which is used in making magnetic circuit. Thickness of the sheet is a. Magnetic induction B is directed along the sheet, and it changes in accordance with $B(t) = B_m \sin \omega t$, where B_m is amplitude and ω is angular frequency. Thickness a is very small compared to the height l of the sheet. Within the cross-section of the sheet, a contour C can be identified of width $2x$. Since $x \leq a/2$, one can assume that $x \ll l$. In Fig. 4.10, reference

Fig. 4.10 Calculation of
eddy current density within
one sheet of laminated
magnetic circuit

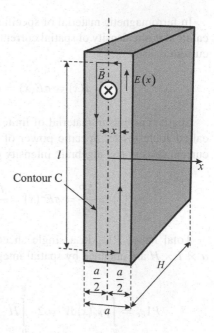

direction of contour C is opposite to the direction of the vector of magnetic
induction; thus, flux through the contour is

$$\Phi = -2 \cdot x \cdot l \cdot B_m \cdot \sin \omega t.$$

The electromotive force in the contour is determined by the first derivative of the
flux. Its amplitude is determined by the product of the frequency and amplitude of
magnetic induction. Within the contour of the width $2x$,

$$e = \oint_C \vec{E} \cdot d\vec{l} = -\frac{d\Phi}{dt} = 2 \cdot x \cdot l \cdot \omega \cdot B_m \cdot \cos \omega t.$$

The sign of the electromotive force e depends on the selected reference direction. It also changes when calculating the counter-electromotive force. When
calculating the eddy current losses, the choice of reference direction does not
influence the result of the calculation. The losses depend on the square of eddy
currents, which in turn depend on e^2. Since $x \ll l$, the part of the contour integral
along short sides of the contour C can be neglected. Therefore, the induced
electrical field E along the long sides of the contour C can be determined by (4.24):

$$e = \oint_C \vec{E} \cdot d\vec{l} = 2 \cdot l \cdot E(x) = 2 \cdot x \cdot l \cdot \omega \cdot B_m \cdot \cos \omega t,$$

$$|E(x)| = x \cdot \omega \cdot B_m \cdot \cos \omega t. \tag{4.24}$$

In ferromagnetic material of specific conductivity σ exposed to induced electrical field E, the density of spatial currents is $J = \sigma E$. In the considered sheet of iron, current density is

$$J(x) = \sigma E(x) = \sigma \cdot x \cdot \omega \cdot B_m \cdot \cos \omega t.$$

Spatial currents in material of finite conductivity give rise to power losses also called Joule losses. Specific power of these losses is equal to the product of the current density and algebraic intensity of electrical field,

$$p_{Fe}(x) = \frac{\Delta P}{\Delta V} = \sigma E^2(x) = \frac{J^2(x)}{\sigma} = \sigma \cdot (x \cdot \omega \cdot B_m \cdot \cos \omega t)^2.$$

Total losses $P1_{Fe}$ in a single sheet of ferromagnetic material of dimensions $a \times l \times H$ are obtained by spatial integration and are determined by (4.25):

$$P1_{Fe} = \int_V p_{Fe}(x) dV = 2 \cdot \int_0^{\frac{a}{2}} H \cdot l \cdot \sigma \cdot (x \cdot \omega \cdot B_m \cdot \cos \omega t)^2 dx$$

$$= \frac{a^3}{12} H \cdot l \cdot \sigma \cdot B_m^2 \cdot \omega^2 \cdot (\cos \omega t)^2 = k \cdot a^3 \cdot B_m^2 \cdot \omega^2. \qquad (4.25)$$

Coefficient k is dependent on the dimensions H and l, specific conductivity σ, and factor $\cos^2 \omega t$, whose average value is 0.5. Result (4.25) can be used in the analysis of the reduction of losses due to splitting magnetic core to layers (sheets) of thickness a.

Figure 4.9a shows a homogenous piece of ferromagnetic material which could be considered as one layer of thickness $a = A$. Starting from the assumption that thickness of the considered part is considerably smaller than the height, it is possible to apply the result (4.25) and determine losses P_{hom} by (4.26):

$$P_{hom} = k \cdot B_m^2 \cdot \omega^2 \cdot A^3. \qquad (4.26)$$

The considered part of magnetic circuit can be made to consist of N mutually insulated layers (sheets) of thickness $a = A/N$, as shown in Fig. 4.9b. If the layer of electrical insulation between the ferromagnetic sheets is considerably smaller than a, it can be assumed that the cross-section of laminated magnetic circuit is filled with iron. Therefore, magnetic resistance of laminated magnetic circuit is equal to the resistance of magnetic circuit of the same shape, made of homogenous piece of ferromagnetic material, as shown in Fig. 4.9a. Equation 4.25 gives the eddy current losses $P1_{Fe}$ in one sheet (layer), whatever the size. It has been applied (4.26) to homogeneous magnetic circuit in Fig. 4.9a, which is considered as a single sheet of iron, N times wider than the sheets shown in Fig. 4.9b, where the total number of such sheets is assumed to be N. The losses P_{lam} in laminated magnetic circuit of width $A = aN$ are determined by expression (4.27):

$$P_{lam} = N \, k \cdot B_m^{\,2} \cdot \omega^2 \cdot \left(\frac{A}{N}\right)^3 = k \cdot B_m^{\,2} \cdot \omega^2 \cdot A \cdot a^2 = \frac{P_{hom}}{N^2}. \qquad (4.27)$$

Result (4.27) indicates that losses due to eddy currents in a part of magnetic circuit of given dimensions decrease N^2 times if the ferromagnetic material is split into N insulated layers (sheets) of equal thickness oriented along the direction of magnetic field. In cases with variable magnetic field perpendicular to the iron sheets, the lamination does not reduce the eddy current losses. In addition, lamination of a magnetic circuit does not reduce the losses due to hysteresis.

In the case of a well-positioned laminar structure of magnetic circuit, losses due to eddy currents are proportional to the squared laminar thickness a, which leads to the conclusion that one should be using iron sheets as thin as possible. Consequently, the question arises, why not use the iron sheets thinner than $0.1 \div 0.2$ mm? Thinner sheets are more difficult to cut and to assemble. At the same time, a decrease in thickness would reduce the equivalent cross-section of iron, decrease the peak flux, and increase the magnetic resistance. Namely, there is an insulating layer between the sheets, made of paper or nonconductive iron compounds. It is several tens of micrometers thick, and it exists on both sides of the sheets. Any further reduction of sheet thickness would reduce the amount of iron in the cross-section of magnetic circuit below reason.

In magnetic circuits made of solid material where eddy currents are considerable, magnetic field does not have homogeneous distribution over the cross-section. An increase in operating frequency leads to significant eddy currents which, in turn, result in uneven distribution of magnetic induction B across the cross-section of the core. Namely, eddy currents create magnetic field which opposes to variations of magnetic induction in the core. Such an effect of eddy currents is more emphasized in the middle of the core, the region which is encircled by all the eddy current contours (see Fig. 4.9a). This phenomenon results in difference between magnetic induction in the center of the core and the induction at the peripheral regions. In magnetic circuits made of iron sheets, these effects are reduced considerably, and there is no significant difference in the field intensity across the cross-section of the core.

Chapter 5
Rotating Electrical Machines

This chapter provides basic information on cylindrical machine. Typical machine windings are introduced and explained, along with the basic forms of magnetic circuits with slots and teeth. This chapter introduces common notation, symbols, and conventions in representing the windings, their magnetic exes, their flux, and magnetomotive force. Typical losses and power balance charts are explained and presented for cylindrical motors and generators. Calculation of the magnetic field energy in the air gap of cylindrical machines is given at the end of this chapter, along with considerations regarding the torque per volume ratio.

Electrical machines are usually rotating devices creating electromagnetic torque due to the magnetic coupling field. The machines perform electromechanical conversion of energy; thus, they are called rotational converters. The stationary part of rotating machines is called stator. The mobile part which rotates is called rotor. The rotating movement of the rotor is accessible via shaft, which serves for rotor mechanical coupling to a work machine. The magnetic coupling field creates torque which acts on the rotor forcing it to rotation. The torque is the result of interaction of electromagnetic forces, and for this reason, it is called electromagnetic torque.

5.1 Magnetic Circuit of Rotating Machines

Electrical machines are mainly of cylindrical shape. Stationary part of the machine (stator) is mostly made in the form of a hollow cylinder which accommodates cylindrical rotor capable of rotating in its bearings with negligible friction. Both stator and rotor are made of ferromagnetic material, and between them, there is an air gap. Along the rotor axis, there is a shaft which serves for transferring the electromagnetic torque to a mechanical subsystem. The shaft protrudes out of the machine to facilitate the coupling to work machines. Both stator and rotor contain windings and/or permanent magnets which create the stator and rotor fields. By interaction of

S.N. Vukosavic, *Electrical Machines*, Power Electronics and Power Systems,   81
DOI 10.1007/978-1-4614-0400-2_5, © Springer Science+Business Media New York 2013

Fig. 5.1 Cross-section
of a cylindrical electrical
machine. (*A*) Magnetic circuit
of the stator. (*B*) Magnetic
circuit of the rotor. (*C*) Lines
of magnetic field. (*D*)
Conductors of the rotor
current circuit are subject
to actions of electromagnetic
forces F_{em}

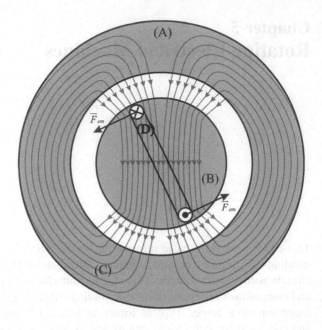

these fields, the electromagnetic torque is created, and it acts on the rotor and creates rotational movement. Figure 5.1 shows cross-section of the magnetic circuit of an electrical machine.

5.2   Mechanical Access

Rotating machines are connected via shaft to a load or a work machine. The rotor and shaft are rotating at the mechanical angular speed Ω_m. The torque T_{em} is created by a couple of electromagnetic forces F_{em} which tend to move the rotor. The product of the torque T_{em} and speed of rotation Ω_m gives the power of electromagnetic conversion $P_{em} = T_{em}\Omega_m$. In cases when the torque acts in the direction of rotation, power P_{em} is positive. Then, electrical energy is being converted to mechanical work; thus, the machine operates in the motor mode. Reference directions for the torque and speed are indicated in Fig. 5.2.

The shaft represents mechanical connection, that is, mechanical access or mechanical output of the machine. It rotates at angular speed Ω_m and does the transfer of electromagnetic torque T_{em}. In the case when the machine operates as a motor, the electromagnetic torque excites movement ($T_{em} > 0$), while mechanical load (load or work machine) resists to motion by an opposing torque, a torque of the opposite sign, denoted by T_m. In this operating mode, electromagnetic torque $T_{em} > 0$ tends to accelerate the rotor, while the opposing torque $T_m > 0$ tends to

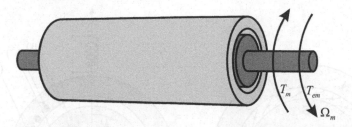

Fig. 5.2 Adopted reference directions for the speed, electromagnetic torque, and load

slow it down. The change of speed is determined by Newton law applied to a rotational movement:

$$J\frac{d\Omega_m}{dt} = T_{em} - T_m. \tag{5.1}$$

In general, electrical machine can be operated as a motor or as a generator. A motor performs electromechanical conversion of electrical energy to mechanical work; a generator performs conversion in the reverse direction. For a machine operating in motor mode, torque T_{em} is of positive sign, whereas in generator mode, the sign of the torque T_{em} is negative. Electrical generators have their rotor connected to a turbine which turns the rotor, supplying the mechanical power into the machine. Thus, the torque T_m assumes a negative value with respect to the reference direction shown in Fig. 5.2. Electromagnetic torque of the generator T_{em} opposes this movement, and it also takes a negative value with respect to the reference direction. Considering adopted reference directions, power of electromagnetic conversion P_{em} is negative, indicating that the machine converts mechanical work into electrical energy.

5.3 The Windings

In addition to magnetic circuits, which direct the magnetic field, electrical machines also have current circuits, also called windings, which conduct electrical current. The windings are made of a number of series connected, insulated copper conductors. In cylindrical machines, the conductors are positioned along the cylinder axis (coaxially). By connecting a number of conductors in series, one obtains a winding. Two conductors connected in series and positioned diametrically constitute one contour or one *turn*. A machine could have a number of windings. They may be placed on both stator and rotor. Each winding has two terminals, which could be short circuited, open, or connected to a power source feeding the machine. By connecting them to a voltage or current source, electrical current is established in the windings. Terminals of the windings are *electrical access* (connection, input) of the machine. Current i in a winding having N turns creates magnetomotive force

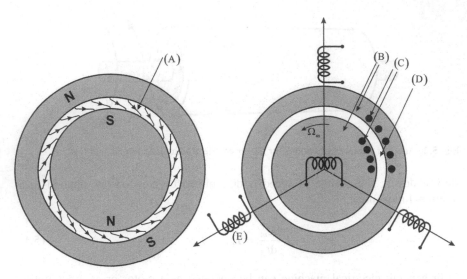

Fig. 5.3 Magnetic field in the air gap and windings of an electrical machine. (*A*) An approximate appearance of the lines of the resultant magnetic field in the air gap. (*B*) Magnetic circuits of the stator and rotor. (*C*) Coaxially positioned conductors. (*D*) Air gap. (*E*) Notation used for the windings

of $F = Ni$. By dividing the magnetomotive force with magnetic resistance R_μ, one obtains flux Φ. Stator windings create stator flux, whereas rotor windings create rotor flux. There are machines where stator or rotor does not have windings, but the flux is created by permanent magnets.

The resultant flux of the machine is obtained by joint action of the magnetomotive forces of the stator and rotor. Lines of the resulting field run through the magnetic circuit of rotor, air gap, and magnetic circuit of stator. Figure 5.3(A) shows the lines of magnetic field in the air gap. The figure shows the zone where magnetic lines leave magnetic circuit of the rotor and enter the air gap.[1] This zone is called *north*

---

[1] Figure 5.3 shows an approximate shape of the lines of magnetic field in the air gap, which does not correspond to the air gap field of real machines. Electrical machines have magnetic circuit containing slots and teeth which are described in the following subsections. The presence of slots has an influence on the shape of the air gap field, making it relatively more complicated. In the hypothetical case when the magnetic circuit is of an ideal cylindrical shape and permeability of the ferromagnetic material is considerably higher than that of the air, the lines of magnetic field are perpendicular to the surface separating the air gap and the ferromagnetic material. It is of interest to envisage the surface that separates the air gap from the ferromagnetic material. The boundary conditions relate tangential components of magnetic field H on either side of this surface with surface electrical currents J_S. With $J_S = 0$, tangential component of magnetic field H in the air is equal to tangential component of magnetic field in ferromagnetic material. Since $B_{Fe} < 1.7$ T and $H_{Fe} = B_{Fe}/\mu_{Fe} \approx 0$, tangential component of magnetic field in the air is close to zero. Thus, the lines of magnetic field in the air gap are perpendicular to the surface separating the air gap and the ferromagnetic circuit.

magnetic pole of the rotor. Diametrically opposed is the south magnetic pole. In the same way, the north and south poles of the stator can be identified. The electromagnetic torque arises from the tendency of rotor poles to take place against the opposite poles of the stator.

The right-hand side of Fig. 5.3 shows cross-section of the machine. Conductors of stator and rotor windings are marked by (C). The stator and rotor could have several windings. For clarity, individual windings are represented by the symbol marked by (E) in Fig. 5.3.

Stator flux can be represented by a vector whose course and direction are determined by positions of its poles, while its algebraic intensity (amplitude) is determined by the flux itself, namely, by the surface integral of magnetic induction *B*. Rotor flux can be represented in the same way. In the following subsections, it will be shown that the electromagnetic torque is determined by the vector product of the two fluxes, that is, by the product of the flux amplitudes and the sine of the angle between them.

5.4   Slots in Magnetic Circuit

Magnetic circuits of the stator and rotor are made of iron sheets in order to reduce power losses. The iron sheets are laid coaxially. Each individual sheet has a cross-section of the form indicated in Fig. 5.4. A number of sheets are assembled and fastened, producing is such way magnetic core. Stator usually assumes the form of a hollow cylinder, whereas rotor is cylindrical, fitting in the stator cavity. Distance between the stator and rotor (air gap) can be from one to several millimeters.

Windings of the machine consist of series connected, mutually insulated copper conductors. Conductors are insulated between each other, as well as from the magnetic circuit and other parts of the machine. The conductors are insulated

Fig. 5.4 Cylindrical magnetic circuit of a stator containing one turn composed of two conductors laid in the opposite slots

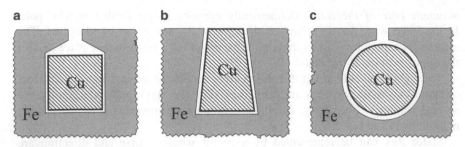

Fig. 5.5 Shapes of the slots in magnetic circuits of electrical machines. (**a**) Open slot of rectangular cross-section. (**b**) Slot of trapezoidal shape. (**c**) Semi-closed slot of circular cross-section

by lacquer, paper, silicon rubber, or some other insulating material. The insulated copper conductors are placed in *slots* which are positioned coaxially (parallel to machine axis) along the inner side of the stator magnetic circuit or outer side of the rotor magnetic circuit. Examples of some of the slots are shown in Fig. 5.5. Shape of the cross-section of a slot is determined by the need for achieving a smaller or larger leakage flux as well as by the need for mechanical tightening of the conductors placed in these slots. The slots shown in Fig. 5.5 may have one or more conductors. In most cases, the stator slots may have conductors of two different phase windings.

In Fig. 5.6, the cross-section of the stator magnetic circuit shows the axial grooves along the inner surface of the stator. These grooves are called slots. Part of the magnetic circuit between two neighboring slots is called *tooth*. The teeth are formed by cutting the same slot through all the iron sheets which are assembled when forming the magnetic circuit. After the sheets are arranged, one obtains a slot of trapezoidal, or oval, or of some other cross-section. The way the insulated conductors are placed in the slots is illustrated in Fig. 5.6, where the front side of the stator is shown as (F), size view of the stator is shown as (G), whereas the views (H) and (I) show 3D view of the stator with one *section*.

One turn can be obtained by a series connection of conductors placed in different slots (A–B in Fig. 5.6). The two conductors making one turn can be placed in diametrical slots, but there are also turns where this is not the case. The two conductors which belong to one turn pass through the slots and get out of the magnetic circuit at the rare side. At that point, they get connected by the end turns, denoted by A, C, and D in Fig. 5.6. Conductors that are placed in slots, the end turns (D), and front connections (C) between conductors are usually made of a single piece of insulated copper wire.

The slots under consideration may hold more than one copper conductor. Therefore, several turns may reside in the same pair of slots. These turns are connected in series. In such way, one obtains a *coil* or a *section* (C–E in Fig. 5.6). To connect the turns in series, one section has the end turns at the front side of the stator (detail C) as well as at the rear side of stator (detail D).

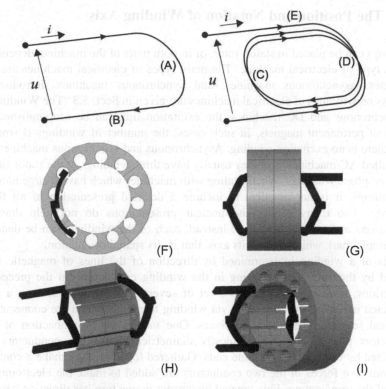

Fig. 5.6 Definitions of one turn and one section

One section can have one or more turns connected in series. One winding can have one or more sections connected in series. Terminals of the winding can be connected to electrical source or electrical load. They represent electrical access to the machine. Electrical machines can have several windings in the stator and/or rotor.

Flux of one turn is equal to the flux through the contour determined by the conductors of the turn. Flux of a turn is denoted by Φ, and it is equal to the surface integral of magnetic induction over the surface leaning on the contour. Flux of a coil having N turns is equal to $N\Phi$.

Question (5.1): Is magnetic induction in the teeth of higher or lower intensity compared to the rest of the magnetic circuit? Why?

Answer (5.1): Flux of the machine passes through magnetic circuit of the stator and through magnetic circuit of the rotor. Within the iron parts of the circuit, there are no air gaps of high magnetic resistance. Passing toward the air gap, lines of magnetic field get through the teeth. The equivalent cross-section is then reduced, since the field is directed toward teeth and not toward slots, where magnetic resistance is much higher. Since the same flux now passes through a smaller equivalent cross-section, magnetic induction in the teeth is higher than the induction in the other parts of magnetic circuit.

5.5   The Position and Notation of Winding Axis

Windings can be placed in stator, rotor, or in both parts of the machine, depending on the type of electrical machine. The main types of electrical machines are DC machines, asynchronous machines, and synchronous machines. Introductory remarks on windings of electrical machines are given in Sect. 5.3 "The Windings". In synchronous and DC machines, the excitation flux can be accomplished by means of permanent magnets. In such cases, the number of windings is smaller since there is no excitation winding. Asynchronous and synchronous machines are also called AC machines, and they usually have three windings in the stator called the three-phase windings. When dealing with machines which have a large number of windings, it is not practical to include a detailed presentation of all these windings. Too many details and unclear presentations do not help drawing conclusions and making decisions. Instead, each of the windings can be denoted by a simple mark which defines its axis, that is, its spatial orientation.

Axis of a winding is determined by direction of the lines of magnetic field created by the currents circulating in the winding conductors. In the preceding subsections, *winding* is defined as a set of several conductors placed in a slot, connected in series, and accessible via winding terminals which are connected to electrical sources or electrical receivers. One *turn* is series connection of two conductors placed in different, mostly diametrical slots. The conductors are connected by end turns at machine ends. Gathered together, they make a contour. Electromotive forces of the two conductors are added to make the electromotive force of the turn/contour. Flux created by current in one turn has direction perpendicular to the surface encircled by the contour. This normal on this surface defines spatial orientation of the turn. The turns making one winding can be distributed along machine perimeter and can be of different spatial orientation.

In cases where all the turns that constitute winding reside in the same pair of slots, the turns share the same magnetic axis. Electrical currents in these turns create magnetomotive force and flux in the same direction. Such winding is called *concentrated winding*. The winding current in concentrated winding creates magnetic field in the air gap. Lines of this field pass through the iron core, where intensity of the field H_{Fe} is very small due to high permeability of iron. In addition, lines of the field pass through the air gap twice, as shown in Fig. 5.7. Therefore, intensity of the field in the air gap can be determined from $Ni = 2H_0\delta$. Magnetic field created by the winding has two distinct zones in the air gap, one where the lines of magnetic field come out of the rotor, pass through the air gap, and enter magnetic circuit of the stator, and the other where the field is in the opposite direction. These zones are called *magnetic poles*. Positions of the poles are determined by the *direction of the field*. This direction extends along the axis of the winding. For a concentrated winding with turns made of series connected diametrically positioned conductors, the axis of the winding corresponds to the axis of each individual turn. This axis is perpendicular to the surface encircled by diametrical conductors.

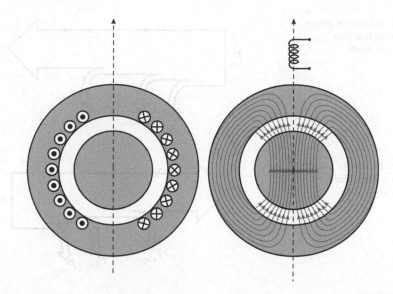

Fig. 5.7 Notation of a winding and its axis

While analyzing electrical machines, consideration of all conductors of all windings would be too complex and of little use. Therefore, the windings are represented by special marks, similar to those of the coils. Orientations of each mark should be such that it extends along the axis of the winding. Namely, direction of the mark should be aligned with the lines of magnetic field established by electrical current in the winding. The way of marking the axis of a winding is shown in Fig. 5.7.

Windings of electrical machines can be made in such way that one slot contains more than one conductor. Conductors placed in one slot do not have to belong to the same winding. In a three-phase machine, there are three separate stator windings, having a total of six terminals. One slot may contain conductors belonging to two or even three separate windings. Three parts of one stator winding are often called *phases* (three-phase windings).

5.6 Conversion Losses

Conversion process in electrical machines involves power losses in magnetic circuits, in windings, and in mechanical subsystem. Losses in magnetic circuits are a consequence of alternating magnetic induction in ferromagnetic materials, and it is divided in hysteresis losses and eddy currents losses. Losses due to eddy currents can be reduced by lamination. Laminated magnetic circuit is made of iron sheets separated by thin layers of electrical insulation. In such way, eddy currents are suppressed along with eddy current losses. Winding losses are proportional to the winding resistance and square of electrical current. Mechanical losses are

Fig. 5.8 Balance of power
of electrical machine
in motoring mode

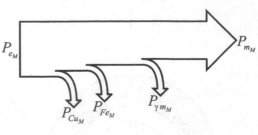

Fig. 5.9 Balance of power
of electrical machine
in generator mode

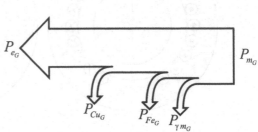

consequence of resistance to rotor motion. They are caused mainly by friction in the bearings and air resistance in the air gap. When electrical machine operates in the steady state motoring mode, it takes the power $P_{eM} = \Sigma u i_M$ from the electrical power source. In motoring mode, it is convenient to assume the reference direction for the power P_{eM} and current i_M from the source toward the machine. During the process of electromechanical conversion, certain amount of energy is lost in magnetic circuit at the rate of P_{FeM}, also called power losses in iron. In windings, energy is lost at the rate of P_{CuM}, also called power losses in copper. Internal mechanical power which is transferred to the rotor is the product of electromagnetic torque and speed of rotation, $T_{em}\Omega_m$. The motion resistance caused by friction in the bearings and friction in the air results in mechanical losses $P_{\gamma mM}$. Power $P_{mM} = T_{em}\Omega_m - P_{\gamma mM} = T_m\Omega_m$ is transferred via shaft to a work machine. In the motor mode, the source power is the machine input, power P_{mM} is the output, whereas the sum $P_{FeM} + P_{CuM} + P_{\gamma mM}$ determines the power of losses. Ratio $\eta = P_{mM}/P_{eM}$ is *the coefficient of efficiency*, and it is always less than one. The balance of power for an electrical machine operating in motoring mode is shown in Fig. 5.8.

In the case when machine operates in generator mode, it converts mechanical work to electrical energy. The balance of power for the generator mode is shown in Fig. 5.9. Generator receives mechanical power P_{mG}, obtained from a hydroturbine, an endothermic motor, or some other similar device.

In Fig. 5.8, the mechanical power is considered positive if it is directed from the turbine toward electrical machine. In this case, the turbine is the source of mechanical power. With the reference directions for power and current adopted for motoring mode (Fig. 5.8), where the power is considered positive when being supplied from the electrical machine and being delivered to the work machine, then the mechanical power P_m in generator mode has a negative value. Therefore, the reference direction in generator mode is often changed and determined so as to obtain positive values of power P_{mG}, received from the turbine, and a positive value of power P_{eG}, supplied by the generator to electrical circuits and receivers, connected to the stator winding. The same reference direction is usually taken for the electrical current i_G.

Power P_{mG} represents input to the generator, and it comes from the turbine or other source of mechanical power. Within the machine, one part of the input power is lost on overcoming motion resistances encountered by the rotor. By subtracting power $P_{\gamma mG}$ from input power P_{mG}, one obtains internal mechanical power which is converted to electrical power. One part of the obtained electrical power is lost in windings, where the copper losses are P_{CuG}, and in magnetic circuit, where the iron losses are P_{FeG}. The remaining power is at disposal to electrical consumers supplied by the generator. At the ends of stator winding, one obtains currents and voltages which determine the generated electrical power $P_{eG} = \Sigma ui_G$, which can be transferred to electrical consumers.

The coefficient of efficiency can be increased by designing the machine to have reduced losses windings (copper losses) and magnetic circuit (iron losses). By increasing the cross-section of conductors, the resistance of copper conductors is decreased which leads to reduced copper losses. By increasing the cross-section of magnetic circuit, the magnetic induction decreases for the same flux. Consequently, the iron losses are smaller. On the other hand, this approach to reducing the current density and magnetic induction leads to an increased volume and mass of the machine. The specific power, determined by the ratio of the power and mass of the machine, becomes smaller as well. Therefore, for an electrical machine of predefined power, decreased current and flux densities lead to an increase in quantities of copper and iron used to make the machine. At the same time, dimensions of the machine are increased as well.

Design policy of reducing the flux and current densities decreases the overall energy losses in copper and iron in the course of electrical machine service. Nevertheless, the overall effects of this design policy may eventually be negative. Namely, the increase in efficiency is obtained on account of an increased consumption of iron and copper. At this point, along with the energy spent during the operating lifetime of electrical machines, it is of interest to take into account the energy spent in their manufacturing. Production of the electrolytic copper, used to make the windings, requires considerable amounts of energy. The same way, production of insulated iron sheets for making magnetic circuits requires energy. Therefore, the energy savings due to reduced copper and iron losses are counteracted by increased energy expenditure in machine manufacturing.

Choice of the flux and current density in an electrical machine is made in the design phase, and it represents a compromise. For machines to be used in short time intervals, followed by prolonged periods of rest, it is beneficial to use increased flux and current density. Increased copper and iron losses are of lesser importance, as the machines are mostly at rest. On the other hand, savings in copper and iron contribute to significant reduction in energy used in machine manufacturing.

Contemporary electrical motors are fed from power converters which can adjust voltages and currents of the primary source to the requirements of machines. Among other things, the possibility of varying conditions of supply is used for the purpose of bringing a machine to the operating regime where power losses are reduced.

5.7   Magnetic Field in Air Gap

Between stator and rotor, there is a clearance, often called *air gap*. In Fig. 5.10, the air gap is denoted by (C). The clearance δ is considerably smaller than diameter of the machine and ranges from a fraction of millimeter for very small machines up to 10 mm for large machines.

The permeability of ferromagnetic materials (iron) is very large. Since the flux through the air gap is equal to the flux through magnetic circuit, similar values of magnetic induction are encountered in both the air gap and iron. Since $\mu_{Fe} \gg \mu_0$, the magnetic field H in iron is negligible. It can be assumed that the magnetic field H has a significant, nonzero value H_0 only in the air gap (Fig. 5.11). The contour integral of the field H is reduced to the value given by (5.2) which relates the magnetomotive force F to the line integral of magnetic field along the closed contour:

$$F = \sum Ni = \oint_C \vec{H} \cdot d\vec{l} \cong 2H_0\delta. \tag{5.2}$$

Doubled value of the product $H_0\delta$ in (5.2) exists since the lines of magnetic field pass through the air gap twice, as shown in Fig. 5.11.

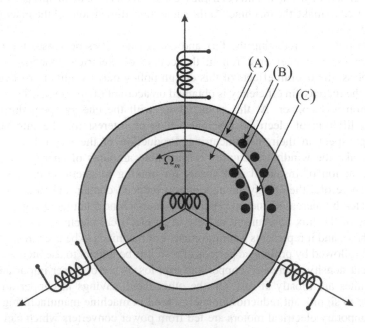

Fig. 5.10 Cross-section of an electrical machine. (*A*) Magnetic circuits of the stator and rotor. (*B*) Conductors of the stator and rotor windings. (*C*) Air gap

Fig. 5.11 The magnetic field
lines over the cross-section
of an electrical machine

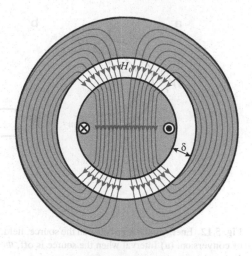

Since $Ni = 2H_0\delta$, it is of significance to have smaller gap δ. In this way, the required field H_0 can be accomplished with a smaller current in the windings and lower losses. However, there are limits to the minimum applicable air gap. The air gap must be sufficient to ensure that the stator and rotor do not touch under any circumstances. A finite precision in manufacturing mechanical parts and a finite eccentricity of the rotor as well as the existence of elastic radial deformation of the shaft in the course of operation prevent the use of air gaps inferior to 0.5–1 mm. Otherwise, there is a risk that the rotor could scratch the stator in certain operating conditions.

5.8 Field Energy, Size, and Torque

Cylindrical electrical machines based on magnetic coupling field develop the electromagnetic torque through an interaction of magnetic field with winding currents. The available electromagnetic torque can be related to the machine size. In addition, the available torque can be estimated from the energy of the magnetic field in the electrical machine.

Cylindrical electrical machines based on magnetic coupling field have an immobile stator and a revolving rotor. The rotor is turning around the axis of cylindrical machine. The axis is perpendicular to the cross-section of the machine presented in Figs. 5.10 and 5.11. The measure of mechanical interaction of the stator and rotor is the torque. When the torque is obtained by action of the magnetic coupling field, it is called the *electromagnetic torque*. The torque is created due to the interaction of the stator and rotor fields. Magnetic fields of stator and rotor can be obtained either by inserting permanent magnets into magnetic circuit or by electrical currents in the windings. Left part in Fig. 5.3 illustrates the torque which tends to align different magnetic poles of the stator and rotor. In order to determine the relation between the

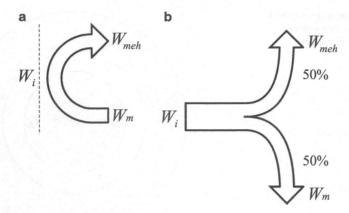

Fig. 5.12 Energy exchange between the source, field, and mechanical subsystem within one cycle of conversion. (**a**) Interval when the source is off, $\Phi = $ const. (**b**) Interval when the source is on, $I = $ const

available torque and the energy accumulated in the coupling magnetic field, it is of use to summarize the process of electromechanical conversion, taking into account the cyclic nature of the process as well as the two phases in one conversion cycle. During the first phase, electromechanical converter is connected to the electrical source. The source supplies the energy which is split in two parts. One part increases the energy accumulated in the coupling field, while the other part feeds the process of electromechanical conversion. During the second phase of the conversion cycles, the electrical source is disconnected from the electromechanical converter, and the mechanical energy is obtained from the energy stored in the coupling field. In Fig. 5.12, marks W_i, W_{meh}, and W_m denote energy of the source, mechanical energy obtained from the converter, and energy accumulated in the magnetic coupling field, respectively. The cycle of electromechanical conversion in converters with magnetic coupling field is analogous to the cycle of converters based on electrical coupling field, the later being described in Sect. 3.1.4, *Conversion Cycle*.

In the case when the source is separated from machine windings (Fig. 5.12a), the voltage across terminals of the winding is $u \approx Nd\Phi/dt = 0$. Neglecting the voltage drop Ri, the flux in a short-circuited winding is constant. In the absence of electrical source, mechanical work can be obtained only on account of the energy of the coupling field. Therefore, in such conditions, $dW_{meh} = -dW_m$. This assertion can be illustrated by example where a mobile iron piece is brought into magnetic field of a coil. A shift dx of the iron piece produces mechanical work $dW_{meh} = Fdx$, where F is the force acting on the piece. Self-inductance of the coil $L(x)$ is dependent on the position of the piece of iron, as the piece changes the magnetic resistance to the coil flux. In the case when this coil is separated from the source and short circuited, and resistance R of the coil is negligible, the first derivative of the flux is zero, and the flux Ψ is constant. Variation of the field energy due to a shift dx is given by (5.3), where i is the coil current:

$$dW_m = d\left(\frac{1}{2}Li^2\right) = d\left(\frac{\Psi^2}{2L}\right) = -\frac{\Psi^2}{2L^2}dL = -\frac{1}{2}i^2 dL. \qquad (5.3)$$

Since $dW_{meh} = -dW_m$, the force acting on mobile piece of iron in magnetic field of a short-circuited coil can be determined from (5.4):

$$dW_{meh} = F\, dx = -dW_m$$

$$\Rightarrow F = \frac{1}{2}i^2\frac{dL}{dx}. \qquad (5.4)$$

If the electrical source is connected (Fig. 5.12b), electrical current in the winding is determined by the source current I. The current is constant, while the flux changes. Upon shift dx, the source delivers energy $dW_i = u\, I\, dt$, where $u = d\Psi/dt$ is the voltage across the coil terminals. Work of the source is given by (5.5):

$$dW_i = u\, I\, dt = I\, d\, \Psi. \qquad (5.5)$$

Since the coil current is constant, the corresponding increase of energy of the magnetic field can be obtained by applying (5.6):

$$dW_m = d\left(\frac{1}{2}LI^2\right) = \frac{1}{2}I^2 dL = \frac{1}{2}I\, d\, \Psi = \frac{1}{2}dW_i. \qquad (5.6)$$

From the previous equation, it follows that work of the source is split to equal parts[2] $dW_m = dW_{meh} = dW_i/2$. Expression for the force acting on the mobile piece of iron in cases where the source is connected is $F = \frac{1}{2}I^2 dL/dt$, which corresponds to expression (5.4), developed for the case when the source is disconnected and the coil is short circuited.

Neither of the two described processes could last for a long time. If the source is disconnected permanently, the energy of the coupling field W_m is converted to

---

[2] Distribution of the work delivered by the source corresponds to expressions $dW_m = dW_{meh} = dW_i/2$ if the medium is linear, that is, if permeability μ of the medium does not depend on the field strength. Then, the coefficient of self-inductance $L(x)$ depends exclusively on the position x of the iron piece. Thus, ratio Ψ/I does not depend neither on flux nor on the electrical current. Under the described conditions, the statements (5.5) and (5.6) are correct. Consequently, relations $dW_m = dW_{meh} = dW_i/2$ hold as well. In cases when magnetic induction in iron reaches the level of magnetic saturation, the characteristic of magnetization of iron $B(H)$ becomes nonlinear. The saturation is followed by a drop in permeability μ of the medium (iron). In such cases, inductance of the coil is a function of both position and flux, $L = \Psi/i = f(x, \Psi)$. Consequently, the expression for increase of the field energy would take another form. Subsequent analysis leads to the conclusion that with nonlinear medium, the work of the source is not to be split in two equal parts.

mechanical work until it is completely exhausted. On the other hand, should the source be permanently connected, one part of the source energy is converted to mechanical work, while the other part increases energy of the field. An increase in the field energy is followed by a raise in the magnetic field H and magnetic induction B. Magnetic induction in a magnetic circuit comprising iron parts cannot increase indefinitely. Accumulation of energy W_m is limited by magnetic saturation in iron sheets, which limits magnetic induction to $B_{max} < (1.7 \div 2)$T. Since neither of the two phases in one conversion cycle cannot persist indefinitely, they have to be altered in order to keep the field energy within limits. Therefore, the electromechanical conversion is performed in cycles which include interval when the source is disconnected (left side of Fig. 5.12) and interval when the source is connected (right side of the figure). An interval when the source is connected must be followed by another interval when the source is disconnected in order to prevent an excessive increase or decrease of the energy accumulated in the coupling field.

The expressions for electrical force given by (3.7) and (3.9) can be applied in the cases when the medium is linear, that is, when permeability μ of the medium does not depend on the field strength.

All electromechanical converters are operating in cycles. In the first phase, mechanical work is obtained from the source, while in the second phase, it comes from the energy accumulated in the coupling field. The cyclic connection and disconnection of the source does not have to be made by a switch. Instead, the electrical source can be made in such way to provide a pulsating or alternating voltage which periodically changes direction or stays zero for a certain amount of time within each cycle. *In the case of an AC voltage supply, the cycle of electromechanical conversion is determined by the cycle of the supply voltage.* The mechanical work which can be obtained within one cycle is comparable to the energy of the coupling field. In the example presented in Fig. 5.12, energy of the coupling field assumes its maximum value $W_{m(max)}$ at the instant when the source is switched off. If the field energy is reduced to zero at the end of the cycle, then the mechanical work obtained during one full cycle is twice the peak energy of the field, $W_{meh(1)} = 2W_{m(max)}$. For rotating machines, one cycle is usually determined by one turn of the rotor.[3] Mechanical work obtained by action of the electromagnetic torque T_{em} during one turn is equal to the product of the torque and angular path $(2\pi T_{em})$. Therefore, the electromagnetic torque of an electrical machine can be estimated on the basis of the peak energy accumulated in the coupling field.

The relevant values of magnetic field H are exclusively those in the air gap. In ferromagnetic materials, the field H is negligible due to a rather large permeability. Therefore, energy of the coupling field is located mainly in the air gap.

Product of the torque T_{em} and angular speed of rotation Ω_m is the mechanical power which is transferred to work machine via shaft. In the case of a generator, the energy is converted in the opposite direction. Namely, mechanical work is

---

[3] Exceptions are electrical machines with more than one pair of magnetic poles, explained later on.

converted into electrical energy. Given the reference directions, the product of generator torque T_{em} and speed Ω_m is negative, indicating that the mechanical power is transferred to electrical machine via shaft.

Question (5.2): If dimensions of a machine and the peak value of magnetic induction are known, estimate the electromagnetic torque which can be developed by this machine.

Answer (5.2): Energy accumulated in the electromagnetic field is located mainly in the air gap. Volume of the air gap is $V = \pi L D \delta$, where D is diameter of the machine, L is axial length, and δ is the air gap: Magnetic induction B in the air gap depends on electrical current in the winding. Lines of the magnetic field which pass through the air gap enter the ferromagnetic material. Usually, the ferromagnetic parts are made of iron sheets which make up the stator and rotor magnetic circuits. Therefore, induction in the air gap cannot exceed value $B_{max} \approx 1.7$ T. Excessive values of B would cause magnetic saturation in the iron sheets. Therefore, the density of energy accumulated in magnetic field in the air gap cannot exceed ½ B_{max}^2/μ_0. The maximum energy of the coupling field can be estimated as $W_{m(max)} \approx$ ½ $\pi L D \delta B_{max}^2/\mu_0$, while the electromagnetic torque of the machine can be estimated by dividing the obtained energy by 2π.

Chapter 6
Modeling Electrical Machines

This chapter introduces, develops and explains generalized mathematical model of electrical machines. It explains the need for modeling, introduces and explains approximations and neglected phenomena, and formulates generalized model as a set of differential and algebraic equations.

Working with electrical machines requires mathematical representation of the process of electromechanical conversion. It is necessary to determine equations which correlate the electrical quantities of the machine, such as the voltages and currents, with mechanical quantities such as the speed and torque. These equations provide the link between the electrical access to the machine (terminals of the windings) and the mechanical access to the machine (shaft). The two accesses to the machine are shown in Fig. 6.1, which illustrates the process of electromechanical conversions and presents the principal losses and the energy accumulated in magnetic field. Equations of the mathematical model are used to calculate changes in electromagnetic torque, electromotive forces, currents, speed, and other relevant variables. Besides, the model helps calculating conversion losses in windings, in magnetic circuits, and in mechanical parts of the machine.

The set of equations describing transient processes in electrical machines contains differential and algebraic equations, and it is also called *dynamic model*. Operation of a machine in steady states is described by the steady state equivalent scheme, which gives relations between voltages and currents at the winding terminals, and by mechanical characteristic, which describes relation between the torque and speed at the mechanical access. In the following subsection, an introduction to the modeling of electrical machines is presented. Figure 6.1 presents a diagram showing power of the source P_e, mechanical power P_m, winding losses P_{Cu}, losses in the coupling field P_{Fe}, mechanical losses due to rotation $P_{\gamma m}$ (motion resistance losses), and energy of the coupling field W_m.

S.N. Vukosavic, *Electrical Machines*, Power Electronics and Power Systems,
DOI 10.1007/978-1-4614-0400-2_6, © Springer Science+Business Media New York 2013

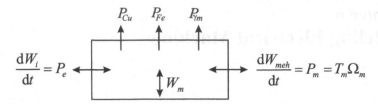

Fig. 6.1 Power flow in an electromechanical converter which is based on magnetic coupling field

6.1   The Need for Modeling

A good knowledge of electrical machines is a prerogative for successful work of electrical engineers. Knowledge of the equivalent schemes in steady states and mechanical characteristics is required for selecting a machine which would be adequate for a particular application, for designing systems containing electrical machines, as well as for solving the problems which may arise in industry and power engineering. Knowledge of the dynamic model of electrical machines is necessary for solving the control problems of generators and motors, for designing protection and monitoring systems, for determining the structures and control parameters in robotics, as well as for solving the problems in automation of production, electrical vehicles, and other similar applications.

In all cases mentioned above, one should know the basic concepts concerning the size, mass, construction, reliability, and coefficient of efficiency of electrical machines.

Further on, a general model of electrical machines is presented in this section. Along with the model, common approximations made in the course of modeling are listed, explained, and justified. The main purpose of studying the general model is to determine the dynamic model for commonly used electrical machines and to obtain their steady state equivalent schemes and mechanical characteristics.

The diagram shown in Fig. 6.1 presents power P_e which the electromechanical system receives from the source, power P_m which is transferred via shaft to the mechanical subsystem, losses in the electrical subsystem P_{Cu}, iron losses P_{Fe}, as well as the losses due to friction $P_{\gamma m}$ in the mechanical subsystem. It is necessary to develop corresponding mathematical model which describes the phenomena within the electromechanical converter shown in Fig. 6.1.

What is a *good model*? How to obtain a *good model*?

Generally speaking, a *model* is a mathematical representation of the system which is under consideration. In most cases, less significant interactions are neglected, and then, a simplified representation is obtained, yet still adequate for the purposes and uses. In electrical engineering, a model is usually a set of differential equations describing behavior of certain system. In some cases, like steady state operation, these equations can be reduced to an equivalent electrical circuit (expressions *replacement scheme*, *replacement circuit*, and *equivalent scheme* are also used).

The phenomena and systems of interest for an electrical engineer are usually complex and include some interactions which are not of uttermost importance and should not be taken into account. Considering gas pressure pushing the head of a piston in the cylinder of an endothermic motor, the action is the result of continuous collisions of a large number of gas particles with the surface of the piston. Strictly speaking, the force is not constant, but it consists of a very large number of strikes (pulses) per second. However, in the analysis of the torque which the motor transfers to the work machine, only the average value of the force is of interest. Therefore, the effects of micro strikes are neglected, and the force is considered to be proportional to the surface of the piston head and to the gas pressure in the cylinder.

In electrical engineering, passive components, such as resistors, capacitors, and inductive coils (chokes), are very often mentioned and used. Strictly speaking, models of real chokes, capacitors, and resistors are more complex compared to widely accepted models that are rather simple. At high frequencies, influences of parasitic inductance and equivalent series resistance of a capacitor become noticeable. Under similar conditions, parallel capacitance of a choke cannot be neglected at very high frequencies. Similar conclusion can be made for a resistor. In a rigorous analysis, real components would have to be modeled as networks with distributed parameter. Yet the frequencies of interest are often low. At low frequencies, parasitic effects are negligible, and the well-known elements R, L, and C are considered as lumped parameter circuit elements described by relations $Ri = u$, $u = Ldi/dt$, and $i = Cdu/dt$. Therefore, parasitic effects and distributed parameters do not have to be taken into account when solving problems and tasks at low frequencies. In cases when all the parasitic and secondary effects are modeled, the considered RLC networks become rather complex, and their analysis becomes difficult. Drawing conclusions or making design decisions based upon too complex models becomes virtually impossible.

Therefore, a *good model* is not the one that takes into account all aspects of dynamic behavior of a system, but the one which is simplified by justifiable approximations, thus facilitating and improving the process of making conclusions and taking design decisions, while retaining all relevant (significant) phenomena within the system. It is not possible to develop an analytical expression which would help in defining the *relevance*, but in making approximations, it is necessary to distinguish the essential from nonessential on the basis of a deeper knowledge of the system and material and through the use of experience. A tip to apprentices in modeling conventional electrical machines is to take into account all the phenomena up to several tens of kHz and to neglect the processes at higher frequencies.

6.1.1 Problems of Modeling

In the process of modeling, it is required to neglect insignificant phenomena in order to obtain a simple, clear-cut, and usable model. For successful modeling and use of the model, it is necessary to make correct judgment as regards phenomena that can be neglected. If important phenomena are neglected or overlooked, the result of modeling will not be usable. Here, an example is presented which shows that taking correct decisions often relies on a wider knowledge of the considered objects and phenomena which is acquired by engineering practice.

Consider a capacitor which consists of two parallel plates with a dielectric of thickness d in between. It is customary to consider that the field in dielectric is homogeneous and equal to $E = U/d$. If this capacitor is used in a system where its high-frequency characteristics are not significant, it can be considered ideal and its interelectrode field homogeneous. Consider the case when a pulse-shaped voltage having a very large slope dV/dt is connected to the plates. At low frequencies, the capacitor can still be considered ideal since the simplification made cannot be of any influence to the low-frequency response. However, it is possible that a very high dV/dt values result in breakdown of dielectric, even in cases where the steady state field strength $E = U/d$ is considerably smaller than the *dielectric strength*. The term dielectric strength corresponds to the maximum sustainable electric field in dielectric, exceeding of which results in breakdown of the dielectric. High dV/dt contributes to a transient nonuniform distribution of electric field between the plates, with $E < U/d$ in the middle and $E > U/d$ next to the plates. Namely, what actually happens in the process of feeding the voltage to the plates is, in fact, propagation of an electromagnetic wave which comes from the source and is directed by conductors (*waveguides*) toward the plates. The propagation of the electromagnetic wave continues in the dielectric; therefore, the highest intensity of the electric field is in the vicinity of the plates, whereas in the space between the plates, it is lower. Uneven initial distribution of the field is established within a very short interval of time, which is dependent on dimensions and is of the order of nanoseconds.

As a consequence of this uneven field distribution, the process of an abrupt voltage rise (very high slope dV/dt) may lead to a situation when, for a short time, the field exceeds dielectric strength of the material in close vicinity of the capacitor plates, even in cases when $E = U/d$ is very small. A breakdown results in destruction of the structure and chemical contents of the dielectric, but nevertheless, it has local character. The damaged zones of the dielectric are next to the plates, whereas toward inside the dielectric is preserved. However, if described incidents occur frequently (say 10,000 times per second), damaged zones tend to spread, and they change characteristics of the capacitor. Prolonged operation in the prescribed way eventually leads to dielectric breakdown between the plates, and it puts the capacitor out of service. Similar phenomenon occurs in insulation of AC motors fed from three-phase transistor inverters, commonly used to provide the so-called U/f frequency control. Three-phase inverters provide variable voltage by feeding a train of voltage pulses to the motor. The pulse frequency is next to 10 kHz. The width of the pulses is altered so as to obtain the desired change in the average voltage (*PWM – pulse width modulation*). The voltage pulses are of very sharp edges, with considerable values of dV/dt, and they bring up an additional stress to the insulation of the windings. Hence, certain high-frequency phenomena cannot be neglected when analyzing PWM-supplied electrical machines.

The example considered above requires a deeper understanding of the process and is founded on experience. The knowledge required for thorough understanding of given example is not a prerequisite for further reading. However, it is of interest to recognize the need to enrich the studies by laboratory work, practice, written papers, and projects, acquiring in this way the experience necessary for a successful engineering practice. A successful engineer combines the theoretical knowledge, skill in solving analytically solvable problems, but also the experience in modeling processes and phenomena. In order to make the most out of the knowledge and skill, it is necessary primarily to use the experience and deeper understanding of the process of electromechanical conversion and reduce a complex system to a mathematical model to be used in further work.

6.1.2 Conclusion

A *good model* is the simplest possible model still representing the relevant aspects of the dynamic behavior of a system – process – machine in a satisfactory way.

In the process of generating models, justifiable approximations are made in order to make a simple model suitable for recognizing relevant and significant phenomena, for making conclusions, and for taking engineering and design decisions. When introducing the approximations, care should be taken that these do not jeopardize the accuracy to the extent that makes the model useless.

This book is the first encounter with cylindrical electromechanical converters with magnetic coupling field for a number of readers. Therefore, initial steps in machine analysis and modeling are made with certain approximations. Among the four principal approximations, the losses in magnetic circuits or iron losses are also neglected. Omission of these losses makes the basic models of electrical machines easier to understand. In most electrical machines, iron losses are marginalized by lamination, and the mentioned approximation is partially justifiable. It is necessary, however, to have in mind that at higher frequencies and larger magnetic induction, the iron losses can be considerable and should be taken into account in calculating the total losses and coefficient of efficiency, as well as in designing the corresponding cooling systems.

6.2 Neglected Phenomena

In the course of developing a model, it is justifiable to neglect less significant phenomena, the omission of which does not cause significant deviations of the obtained results. The four most common approximations are:

- The system is considered as a lumped parameter network.
- Parasitic capacitances are neglected.

- The iron losses are neglected.
- Ferromagnetic materials are considered linear

6.2.1 Distributed Energy and Distributed Parameters

Electrical machines are usually considered as networks with lumped parameters and represented by circuits comprising discrete inductances and resistances. Considering actual L and C elements, the coil energy resides in spatially distributed magnetic field, while the capacitor energy resides in spatially distributed electrical field. It is well known[1] that changes in magnetic field give a rise to induced electrical field and *vice versa*. The induced field is proportional to the rate of change of the inducting field, that is, to the operating frequency. Hence, a coil with an AC current is surrounded by magnetic field, but also with an induced electrical field, the strength of which depends on the operating frequency. Similar conclusion can be drawn for a capacitor. The presence of both fields contributes to parasitic capacitance of the coil and parasitic inductance of the capacitor.

Lumped parameter approach neglects the spatial distribution of the coil and capacitor energy. It is assumed that both energies reside within discrete elements and that the amounts $\frac{1}{2}Li^2$ and $\frac{1}{2}Cu^2$ do not reside in space. The coils and capacitors are considered ideal, *lumped parameter L-C* elements. The adopted models are $u_L = Ldi/dt$ and $i_C = Cdu/dt$, neglecting the secondary effects such as capacitance of a coil or inductance of a capacitor. With the induced fields being proportional to the operating frequency, lumped parameter approach introduces a negligible error at relatively low frequencies that are in use in typical applications of electrical machines.

One of the consequences of neglecting distributed energy and distributed parameters is concealing the energy transfer. In a lumped parameter network, a pair of conductors with electrical current i and voltage u transmits the energy at a rate of $p = ui$. This expression involves macroscopic quantities like voltage and current and suggests that the energy passes through conductors. In reality, the energy is transmitted through the surrounding space with the presence of electrical and magnetic field.

6.2.2 Neglecting Parasitic Capacitances

For electrical machines operating on the basis of a magnetic coupling field, the effects of parasitic capacitances of the windings and the amounts of energy accumulated in the electrical field are negligible. Since the spatial density of

---

[1] Consider Maxwell equations, such as rot $\vec{E} = -\partial \vec{B}/\partial t$.

magnetic field is considerably higher than that of electrical field ($\mu H^2 \gg \varepsilon E^2$), it is justified to neglect the capacitances between insulated conductors and capacitances between the windings and magnetic circuit.

6.2.3 Neglecting Iron Losses

It is considered that the losses due to hysteresis and losses due to eddy currents are considerably smaller compared to the power of conversion; thus, they can be neglected. Specific losses in ferromagnetic materials (iron losses) are dependent on magnetic induction and operating frequency and they can be represented by the following expression:

$$p_{Fe} = \frac{P_{Fe}}{m} = p_H + p_V = \sigma_H \cdot f \cdot B_m^2 + \sigma_V \cdot f^2 \cdot B_m^2.$$

Since the losses due to eddy currents are dependent on squared frequency and squared magnetic induction, the iron losses are to be reconsidered in cases when electrical machine operates with elevated frequencies. In such cases, it is necessary to check whether neglecting the iron losses can be justified.

6.2.4 Neglecting Iron Nonlinearity

The characteristic of magnetization of magnetic materials is considered linear. Therefore, the effects of saturation of the ferromagnetic material (iron) are neglected. Permeability B/H is considered constant and equal to differential permeability $\Delta B/\Delta H$ at all operating points of the magnetization characteristic. In applications where induction exceeds 1.2T, it is necessary to check whether this is justified.

General model of electrical machine based on magnetic coupling field is developed hereafter relying on the four basic approximations mentioned above. It is assumed that the converter has N windings which can be either short-circuited or connected to a source. The windings are mounted on the rotor or stator.

6.3 Power of Electrical Sources

Figure 6.2 shows a converter having N magnetically coupled contours (windings) which could be either connected to a power source or separated from it and brought into short circuit. Windings of electrical machines can be fed from current or voltage sources. Real voltage sources have finite internal resistance (impedance), whereas current sources have finite internal conductance (admittance). With no loss in generality, in further text, it is assumed that electrical sources are ideal.

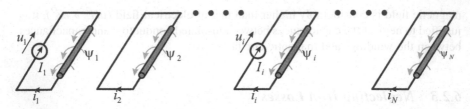

Fig. 6.2 Model of electromechanical converter based on magnetic coupling field with N contours (windings). Contours 1 and i are connected to electric sources, while contours 2 and N are short circuited thus voltages at their terminals are zero

In windings connected to a current source, winding current is constant and determined by the current of the source. If the winding is short-circuited, then the voltage balance in the winding is given by expression $u = Ri + d\Psi/dt = 0$. If resistance of the winding is negligible, then $d\Psi/dt = 0$, and flux in the short-circuited winding is constant. In the case where the winding terminals are connected to a voltage source, voltage of the source determines the change of flux ($u \approx d\Psi/dt$).

Electrical power delivered by the sources to the electromechanical converter is determined by (6.1), where $\underline{u}$ and $\underline{i}$ are vector-columns with their elements being voltages and currents of individual windings. The expression for the power the sources deliver to the machine does not depend on whether the windings are connected to current sources and voltage sources or are short-circuited.

$$
\begin{aligned}
P_e &= \sum_{j=1}^{N} u_j i_j = \underline{i}^T \cdot \underline{u}, \\
\underline{i}^T &= [i_1, \ i_2, \ \ldots \ i_i, \ \ldots \ i_{N-1}, \ i_N], \\
\underline{u}^T &= [u_1, \ u_2, \ \ldots \ u_i, \ \ldots \ u_{N-1}, \ u_N].
\end{aligned} \tag{6.1}
$$

6.4   Electromotive Force

Voltage balance in the winding is given by (6.2), where u is the voltage across winding terminals, i is the winding current, and $\Psi = N\Phi$ is the winding flux. Parameter R denotes the winding resistance.

$$
u = Ri + \frac{d\Psi}{dt} = Ri + e_{CEMF} \tag{6.2}
$$

The considered winding is shown in Fig. 6.3. Flux derivative determines the electromotive force induced in the winding. When making the equivalent scheme, the electromotive force can be represented as an ideal voltage generator attaching the sign + pointed downward, in accordance with the adopted reference direction

Fig. 6.3 The electromotive and counter-electromotive forces

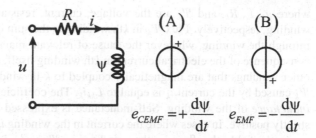

$$e_{CEMF} = +\frac{d\psi}{dt} \qquad e_{EMF} = -\frac{d\psi}{dt}$$

for the current. Then, the force is $e_{EMF} = -d\Psi/dt$. Quantity $-d\Psi/dt$ is the electromotive force induced in the winding, as shown in Fig. 6.3, in the part denoted by (B).

On the other hand, it is possible to alter the reference direction and sign ($e = +d\Psi/dt$), as shown in the part (A) of Fig. 6.3. Quantity $e_{CEMF} = +d\Psi/dt$ is the counter-electromotive force induced in the winding.

Approach (B) is used in majority of courses in Electrical Engineering Fundamentals and Electromagnetics, since it undoubtedly illustrates the circumstance that in each contour, the induced electromotive force and current are opposing the change of flux. For example in the case that intensity of the current decreases, flux through the contour decreases, and a positive value $e = -d\Psi/dt$ appears. Taking into account that sign $+$ is pointed downward, it is concluded that the induced electromotive force supports current in the circuit opposing the change of flux.

Approach (A) results in an equivalent scheme where the reference positive terminals of the voltage and electromotive force are pointed upward. Taking that $e = +d\Psi/dt$, current in the circuit can be determined as ratio $(u-e)/R$. Defined in this way, the electromotive force opposes the voltage; thus, it is called *counter-electromotive force*. Approach (A) is often applied when solving electrical circuits containing electromotive forces, as is the case of replacement schemes of electrical machines. The question of choice of the reference direction of the electromotive force is not of essential significance since the choice does not lead to essential changes in voltage balance equation in the winding. In the Anglo-Saxon, German, and Russian literatures, the approaches are different, which should confuse the reader. In practice, both approaches are accepted, provided that the adopted reference direction corresponds to the sign taken for the electromotive force ($e = +/- d\Psi/dt$).

Electromotive force and counter-electromotive force induced in a contour are discussed further on, in Chap. 10, *"Electromotive Forces Induced in the Windings."*

6.5 Voltage Balance Equation

Voltage balance in each winding is given by (6.2). For a system having N windings, equilibrium of k-th winding is given by expression

$$u_k = R_k i_k + \frac{d\Psi_k}{dt}, \tag{6.3}$$

where u_k, i_k, R_k, and Ψ_k are the voltage, current, resistance, and flux of the kth winding, respectively. Flux Ψ_k in kth winding is the sum of all the fluxes that pass through the winding, whatever the cause of relevant magnetic field. This flux is a consequence of the electrical current in kth winding itself, as well as the currents in other windings that are magnetically coupled to k-th winding. The part of the flux Ψ_k caused by the current i_k is equal to $L_{kk}i_k$. The coefficient L_{kk} is also called *self-inductance* of the winding. Self-inductance is expressed in H = Wb/A, and it is strictly positive. In cases where the current in the winding i_k is the only originator of magnetic field, the flux in the winding is $\Psi_k = L_{kk}i_k$. Electrical currents in remaining windings (Fig. 6.2.) can also contribute to the flux Ψ_k. Current i_j in the turns of the winding j changes the flux Ψ_k proportionally to the coefficient of mutual inductance between windings k and j, L_{kj}. Parameter L_{kj} can be positive, negative, or zero. Spatial orientation of the two windings may be such that a positive current in one of the windings contributes to a negative flux in the other.

Voltage balance of a system with N windings is described by a set of N differential equations. A shorter and more clear-cut record of these equations can be obtained by introducing vectors of the voltages and currents

$$\underline{i}^T = [i_1, \ i_2, \ \ldots \ i_k, \ \ldots \ i_{N-1}, \ i_N]$$
$$\underline{u}^T = [u_1, \ u_2, \ \ldots \ u_k, \ \ldots \ u_{N-1}, \ u_N], \tag{6.4}$$

by defining vectors of winding fluxes

$$\underline{\Psi}^T = [\Psi_1, \ \Psi_2, \ \ldots \ \Psi_k, \ \ldots \ \Psi_{N-1}, \ \Psi_N], \tag{6.5}$$

as well as by introducing matrix of resistances $\underline{R}$ in (6.6), which contains resistances of the windings along the main diagonal. Voltage balance equations in matrix form are given by (6.7), which represents N differential equations of the form (6.2). Voltage balance equations define dynamics of the electrical part of an electromechanical converter, that is, dynamics of the electrical subsystem.

$$\underline{R} = \begin{bmatrix} R_1 & 0 & \ldots & 0 & \ldots & 0 \\ 0 & R_2 & \ldots & 0 & \ldots & 0 \\ \ldots & \ldots & \ldots & \ldots & \ldots & \ldots \\ 0 & 0 & \ldots & R_k & \ldots & 0 \\ \ldots & \ldots & \ldots & \ldots & \ldots & \ldots \\ 0 & 0 & \ldots & 0 & \ldots & R_N \end{bmatrix} \tag{6.6}$$

$$\underline{u} = \underline{R} \cdot \underline{i} + \frac{d\underline{\Psi}}{dt} \tag{6.7}$$

Flux vector-column $\underline{\Psi}$ is determined by the winding currents, self-inductances, and mutual inductances. Flux of kth winding is determined by the coefficient of self-inductance of the winding k, as well as by the coefficients of mutual inductance

L_{kj} between the kth winding and remaining windings, as given by the following equation:

$$\Psi_k = L_{k1}i_1 + L_{k2}i_2 + \ldots + L_{kk}i_k + \ldots + L_{kN}i_N.$$

Since the above expression applies for each winding, the flux vector can be obtained by multiplying the inductance matrix $\underline{L}$ (6.8) and current vector-column $\underline{i}$, in the way shown by (6.9):

$$\underline{L} = \begin{bmatrix} L_{11} & L_{12} & \ldots & L_{1k} & \ldots & L_{1N} \\ L_{21} & L_{22} & \ldots & L_{2k} & \ldots & L_{2N} \\ \ldots & \ldots & \ldots & \ldots & \ldots & \ldots \\ L_{k1} & L_{k2} & \ldots & L_{kk} & \ldots & L_{kN} \\ \ldots & \ldots & \ldots & \ldots & \ldots & \ldots \\ L_{N1} & L_{N2} & \ldots & L_{Nk} & \ldots & L_{NN} \end{bmatrix} \tag{6.8}$$

$$\underline{\Psi} = \underline{L} \cdot \underline{i} \tag{6.9}$$

Along the main diagonal of inductance matrix, there are self-inductances of individual windings, while the remaining coefficients residing off the main diagonal are mutual inductances. Since $L_{ij} = L_{ji}$, the inductance matrix is symmetrical, that is, $\underline{L} = \underline{L}^T$.

Elements of the inductance matrix can be variable. Variations of the self-inductances and mutual inductances can be due to relative movement of the moving parts of the electromechanical converter (rotor) with respect to the immobile parts (stator). Windings may exist in both parts; thus, the movement causes changes in relative positions of individual windings. For each winding, it is possible to define the winding axis (Sect. 5.5). Considering a pair of windings, rotation of one with respect to the other changes the angle between their axes. Consequently, their mutual inductance is also changed. The rotor motion can also change self-inductances. Self-inductance of a winding depends on the magnetic resistance R_μ. Considering a stator winding, its flux passes through the stator magnetic circuit, passes through the air gap, and proceeds through the rotor magnetic circuit. There are cases where the rotor has unequal magnetic resistances in different directions. In such cases, the rotor motion changes the equivalent magnetic resistance R_μ of the stator winding and changes the self-inductance of the winding. An example where movement changes self-inductance of the winding is given in Fig. 2.6, where the magnetic resistance decreases and self-inductance increases by inserting a piece of iron in the magnetic circuit of the coil.

6.6 Leakage Flux

With the current i_k being the sole originator of magnetic field, the flux in kth winding is $\Psi_k = L_{kk} i_k$. One portion of this flux passes to other windings as well, and it is called *mutual flux*. The remaining flux encircles only the kth winding

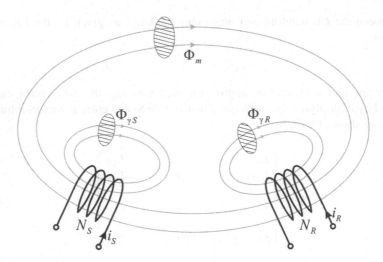

Fig. 6.4 Definitions of the leakage flux and mutual flux

and does not pass to any other winding. As this component, in a way, "misses the opportunity" to effectuate magnetic coupling, it is also called *leakage flux*. Figure 6.4 depicts the mutual and leakage fluxes of a system having two windings, one on stator and the other on rotor. The flux in one turn is denoted by Φ. Fluxes of individual windings are obtained by multiplying flux in each turn Φ by the number of turns.

Self-inductances of the stator and rotor windings are equal to the ratio of the flux established due to the winding current and the current intensity. In other words, self-inductance is the ratio of the flux and current of the winding in cases where the flux does not get affected by external magnetic fields or by currents in other windings, but it is the consequence of the electrical current in the winding itself. In a system comprising a number of coupled windings, self-inductance of the considered winding can be determined by dividing the flux and current in conditions when all the remaining windings are with zero current. Self-inductances of practical windings are *strictly positive*, whereas mutual inductances could be negative. Mutual inductance of the stator and rotor windings determines the measure of the contribution of stator current to the total flux of the rotor winding. Since $L_{SR} = L_{RS} = L_m$, the impact of stator currents on rotor flux is the same as the impact of rotor currents on stator flux. By rotation, relative positions of the two windings may become such that positive current in one winding reduces the flux in the other, thus resulting in a negative value of the mutual inductance. For the system of windings shown in Fig. 6.4, the matrix of inductances is of dimensions 2×2. Along the main diagonal of the matrix, there are positive coefficients of self-induction L_S and L_R. At the remaining places of the matrix are mutual inductances that may be positive, negative, or equal to zero, which is the case when the winding axes are displaced by the angle of $\pi/2$. The mutual inductance is dependent of the coupling coefficient

$$k = \frac{L_m}{\sqrt{L_S L_R}}.$$

The coefficient k is a measure of magnetic coupling of the stator and rotor windings. In cases where the leakage flux is negligible, all of the flux is mutual, and it encircles both windings. In such cases, $k = 1$. As the leakage flux increases, the mutual flux decreases as well as the coefficient k. In cases when the two windings do not have any mutual flux, $k = 0$. It is important to notice that k cannot exceed 1.

Relation between the mutual inductance L_m and coefficient k can be illustrated by the example where the stator and rotor windings have the same axis and the same number of turns N. If the magnetic resistance R_μ is the same, the self-inductances are equal, $L_S = L_R = N^2 / R_\mu$, while the mutual inductance is $L_m = kL_S = kL_R$. Since $k < 1$, mutual inductance is smaller than the self-inductance. With the introduced assumptions, the difference,

$$L_{\gamma S} = L_S - L_m = (1 - k)L_S$$

is called *leakage inductance* of the stator, a measure of the *leakage flux* which encircles the stator winding and does not reach to the rotor winding. A stronger magnetic coupling between the windings means the higher coupling coefficient and the smaller leakage flux and leakage inductance. The example outlined above assumes that $N_S = N_R$. It is of interest to consider the leakage flux and leakage inductance in the case when the windings have different magnetic circuits and different numbers of turns.

Fluxes through one turn of the stator and rotor windings are shown in Fig. 6.4 and given by expressions

$$\Phi_S = \Phi_{\gamma S} + \Phi_m,$$
$$\Phi_R = \Phi_{\gamma R} + \Phi_m.$$

The mutual flux has a component which is a consequence of the stator current (Φ^S) and a component which is a consequence of the rotor current (Φ^R),

$$\Phi_m = \Phi_m^S + \Phi_m^R.$$

Fluxes of the windings are obtained by multiplying flux through one turn by the number of turns:

$$\Psi_S = N_S \Phi_S = N_S \Phi_m + N_S \Phi_{\gamma S} = N_S \Phi_m + \Psi_{\gamma S},$$
$$\Psi_R = N_R \Phi_R = N_R \Phi_m + N_R \Phi_{\gamma R} = N_R \Phi_m + \Psi_{\gamma R}.$$

Flux $\Psi_{\gamma S}$ is the *leakage flux* of the stator winding, while $\Psi_{\gamma R}$ is the leakage flux of the rotor winding. Leakage flux in each winding is proportional to the current. The coefficient of proportionality is *the leakage inductance* of the winding. For the windings shown in Fig. 6.4, the leakage inductances are given by expressions

$$L_{\gamma S} = \frac{\Psi_{\gamma S}}{i_S}, \quad L_{\gamma R} = \frac{\Psi_{\gamma R}}{i_R}.$$

The mutual inductance is determined by expression

$$L_m = L_{SR} = L_{RS} = \frac{N_S \Phi_m^R}{i_R} = \frac{N_R \Phi_m^S}{i_S}.$$

In order to define the winding self-inductance, it is necessary to identify the component of the winding flux which is caused by the electrical currents of the same winding. Self-inductance is the quotient of this flux component ($L_S i_S$) and the current intensity (i_S). One part of the flux component ($L_S i_S$) is partially mutual (that is, encircling both windings) and partially leakage (encircling only the stator winding). Self-inductances of the stator and rotor are

$$L_S = \frac{N_S \Phi_m^S + N_S \Phi_{\gamma S}}{i_S} = \frac{N_S \Phi_m^S + \Psi_{\gamma S}}{i_S}$$

$$= \frac{N_S}{N_R} L_{RS} + L_{\gamma S} = \frac{N_S}{N_R} L_m + L_{\gamma S},$$

$$L_R = \frac{N_R \Phi_m^R + N_R \Phi_{\gamma R}}{i_R} = \frac{N_R \Phi_m^R + \Psi_{\gamma R}}{i_R}$$

$$= \frac{N_R}{N_S} L_{SR} + L_{\gamma R} = \frac{N_R}{N_S} L_m + L_{\gamma R}.$$

Therefore, the leakage inductance is a part of the self-inductance of the winding. The leakage inductance is higher when the magnetic coupling between the coupled windings is weaker. In the case when the numbers of turns of the stator and rotor are equal, as well as in the case when the rotor quantities are scaled (transformed) to the stator side, previous equations take the following form:

$$L_S = L_m + L_{\gamma S},$$

$$L_R = L_m + L_{\gamma R}.$$

6.7   Energy of the Coupling Field

The coupling field has a key role in the process of electromechanical conversion. The energy obtained from the source can be accumulated in the coupling field and then taken from the field and converted to mechanical work. It is of significance to determine the relation between the energy of this field, winding currents, and parameters such as the self-inductances and mutual inductances. Spatial density of energy accumulated in the coupling magnetic field is

$$w_m = \int \vec{H} \cdot d\vec{B}.$$

In linear media, permeability $\mu = B/H$ is constant. Therefore, the energy density in the coupling field is equal to $w_m = \frac{1}{2}BH = \frac{1}{2}\mu H^2$. Total energy can be obtained by integrating the density w_m over the domain where the field exists. The field energy can be expressed as function of electrical currents and winding inductances such as L_S, L_R, L_m, and similar. Mutual inductance between coils L_1 and L_2 is denoted by L_m or by $L_{12} = L_m$. For a system with two coupled windings, the field energy is equal to $f(i_1, i_2, L_1, L_2, L_{12}) = \frac{1}{2}L_1i_1^2 + \frac{1}{2}L_2i_2^2 + L_{12}\,i_1\,i_2$. A rigorous proof of this statement is omitted at this point. Instead, it is supported by considerations which indicate that the spatial integral of energy density w_m corresponds to $f(i_1, i_2, L_1, L_2, L_{12})$. Spatial integration of the energy density involves the sum of minute energy portions $\frac{1}{2}BH/dV$ comprised in infinitesimal volumes dV. Taking into account that $dV = dS/dx$, the problem can be reduced to calculating integral $(\frac{1}{2}\,BH)\,dS\,dx = \frac{1}{2}\,(B\,dS)\,(H\,dx)$. In general, integration of (BdS) results in a *flux*, whereas integration of (Hdx) results in a magnetomotive force Ni, that is, in electrical current (ampere-turns). Therefore, the formula for the field energy contains members of the form Φi or Li^2.

For a system containing N coupled windings, energy of the coupling field is

$$W_m = \int_V w_m dV = \int_V \left(\int \vec{H} \cdot d\vec{B} \right) dV = \frac{1}{2}\sum_{i=1}^{N}\sum_{j=1}^{N} L_{ij}i_i i_j.$$

In the above expression, elements L_{ii} correspond to self-inductances of the windings, and they are strictly positive. Elements L_{ij} are mutual inductances, and they can be positive or negative. A more illustrative expression for the coupling field energy is obtained by introducing the flux and current vectors

$$\underline{i}^T = [i_1, i_2, \ldots i_k, \ldots i_{N-1}, i_N],$$
$$\underline{\Psi}^T = [\Psi_1, \Psi_2, \ldots \Psi_k, \ldots \Psi_{N-1}, \Psi_N],$$

resulting in (6.10), where $\underline{L}$ is matrix of inductances of the considered system of windings:

$$W_m = \frac{1}{2}\underline{i}^T \underline{L}\underline{i} \qquad (6.10)$$

Question (6.1): Consider two windings having self-inductances L_1 and L_2. Is it possible for the coefficient of mutual inductance to exceed $(L_1 \cdot L_2)^{0.5}$?

Answer (6.1): Mutual inductance of the two windings is $L_{12} = k \cdot (L_1 \cdot L_2)^{0.5}$, where k is coupling coefficient. Maximum value of k is 1, and it exists in cases without any leakage, when the total flux of one winding goes through the other winding as well. Since the coupling coefficient cannot be greater than 1, mutual inductance cannot be greater than $(L_1 \cdot L_2)^{0.5}$.

Question (6.2): Is it possible that expression for the field energy

$$2W_m = \sum_j \sum_k L_{jk} i_j i_k$$

gives a negative result? Derive the proof taking the example of a system having two coupled windings.

Answer (6.2): The above expression gives magnetic field energy, and therefore, it cannot have a negative value. In the case of two windings, the expression takes the form

$$W_m = \tfrac{1}{2}L_1 i_1{}^2 + \tfrac{1}{2}L_2 i_2{}^2 + L_{12} i_1 i_2 = \tfrac{1}{2}\left[L_1 i_1{}^2 + L_2 i_2{}^2 + 2k \cdot (L_1 \cdot L_2)^{0.5} i_1 i_2\right].$$

By introducing notation $a = (L_1)^{0.5} i_1$ and $b = (L_2)^{0.5} i_2$, the expression takes the form $2W_m = a^2 + b^2 + 2k \cdot a \cdot b$. It is required to prove that this expression cannot take a negative value, whatever the current intensities i_1 and i_2 might be. Since only the third member of the sum may assume a negative value, and this happens in the event when current intensities are of opposing signs, it is necessary to prove that $2W_m \geq 0$ for $k = 1$. If so, the statement holds for any $k < 1$. With $k = 1$, $2W_m = (a + b)^2$, which completes the proof.

6.8   Power of Electromechanical Conversion

For the considered system of N windings coupled in a magnetic field, it is required to determine the power at the electrical and mechanical accesses, power losses, and power of the electromechanical conversion. Power of the source is supplied through the electrical access of the machine, and it is determined by the sum of powers $p_k = u_k i_k$ supplied to each individual winding.

$$p_e = \sum_k u_k i_k = \underline{i}^T \underline{u} = \underline{u}^T \underline{i} \tag{6.11}$$

Since the voltage vector is expressed by the voltage balance equations (6.7), given in matrix form, power of the source can be expressed as function of the current vector, resistance matrix, and inductance matrix:

$$p_e = \underline{i}^T \left(\underline{R}\,\underline{i} + \frac{d\underline{\Psi}}{dt} \right) = \underline{i}^T \left(\underline{R}\,\underline{i} + \frac{d}{dt}(\underline{L}\,\underline{i}) \right)$$

$$= \underline{i}^T \underline{R}\,\underline{i} + \underline{i}^T \frac{d\underline{L}}{dt}\underline{i} + \underline{i}^T \underline{L}\frac{d\underline{i}}{dt}. \tag{6.12}$$

Power losses in the coupling field are neglected at this point. The losses in the windings can be expressed in matrix form as well. In a winding of resistance R_k, with electrical current i_k, the losses are determined by expression $R_k i_k^2$. Total winding losses of a system containing N windings are given by expression

$$p_{Cu} = \sum_k R_k i_k^2 = i^T \underline{R}\, i \tag{6.13}$$

where $\underline{R}$ is a square matrix of dimensions $N \times N$ having winding resistances along the main diagonal, while the remaining elements are zeros.

One part of work supplied from the source is accumulated in the coupling field. Since the energy of the coupling field is given by (6.10), the power p_{wm} depicting the rate of change of energy accumulated in the field is given by (6.14):

$$p_{wm} = \frac{\mathrm{d}W_m}{\mathrm{d}t} = \frac{\mathrm{d}}{\mathrm{d}t}\left(\frac{1}{2} i^T \underline{L} i\right)$$

$$= \frac{1}{2}\left(\frac{\mathrm{d}i^T}{\mathrm{d}t}\right)\underline{L}i + \frac{1}{2} i^T\left(\frac{\mathrm{d}\underline{L}}{\mathrm{d}t}\right)i + \frac{1}{2} i^T \underline{L}\left(\frac{\mathrm{d}i}{\mathrm{d}t}\right). \tag{6.14}$$

Expression for power p_{wm} can be written in a more convenient form. It should be noted that expression (6.14) represents a sum of three scalar quantities, each obtained by multiplying the vector of electrical currents and the matrix of system inductances. It can be shown that values of the first and third member are equal. For any scalar quantity $\underline{s} = s$ (i.e., for matrices of dimensions 1×1), it can be written that $\underline{s} = \underline{s}^T$. At the same time, the inductance matrix is symmetric ($L_{jk} = L_{kj}$, $\underline{L} = \underline{L}^T$). Therefore, it can be shown that

$$\left(\frac{\mathrm{d}i^T}{\mathrm{d}t}\right)\underline{L}i = \left[\left(\frac{\mathrm{d}i^T}{\mathrm{d}t}\right)\underline{L}i\right]^T = i^T \underline{L}^T\left(\frac{\mathrm{d}i}{\mathrm{d}t}\right) = i^T \underline{L}\left(\frac{\mathrm{d}i}{\mathrm{d}t}\right).$$

By introducing this substitution to (6.14), one obtains (6.15):

$$p_{wm} = \frac{1}{2} i^T\left(\frac{\mathrm{d}\underline{L}}{\mathrm{d}t}\right)i + i^T \underline{L}\left(\frac{\mathrm{d}i}{\mathrm{d}t}\right). \tag{6.15}$$

Equation 6.15 contains first time derivatives of the current i and inductance $\underline{L}$. Variations in the matrix occur due to the relative motion of the rotor with respect to the stator. This motion leads to variation of mutual inductance between the rotor and stator windings. In certain conditions, rotor movement may cause variation of self-inductances of individual windings. Derivative of the current vector i in (6.15) is a vector whose elements are derivatives of the currents of individual windings. In cases where a winding is connected to an ideal current source which provides constant current, derivative of the winding current is zero. Derivative of a winding current can take nonzero values if the winding is short-circuited, or connected to an ideal voltage source, or connected to real current or voltage sources.

The part p_{wm} of the power p_e determines the increase in energy W_m accumulated in the coupling field. The part p_{Cu} is lost in winding conductors due to Joule effect. What remains of the source power is $p_e - p_{Cu} - p_{wm} = p_c$. The remaining power p_c is converted to mechanical through electromagnetic processes involving conductors, magnetic circuit and coupling magnetic field. An integral of p_c represents the mechanical work. The power p_c is also called *power of electromechanical conversion* or *conversion* power. In the motor mode (Fig. 5.8), reference direction for power is such that the power p_c is positive. A positive power of electromechanical conversion means that electrical energy is being converted into mechanical work. In the generator mode (Fig. 5.9), direction of the converter power is reversed, and the power p_c, as defined above, assumes a negative value.

Since

$$p_C = p_e - p_{Cu} - p_{wm},$$

and considering (6.12), (6.13), and (6.15), p_C is expressed as

$$p_C = \left(\underline{i}^T \underline{R}\, \underline{i} + \underline{i}^T \frac{d\underline{L}}{dt} \underline{i} + \underline{i}^T \underline{L} \frac{d\underline{i}}{dt} \right) - (\underline{i}^T \underline{R}\, \underline{i})$$
$$- \left(\frac{1}{2} \underline{i}^T \left(\frac{d\underline{L}}{dt} \right) \underline{i} + \underline{i}^T \underline{L} \left(\frac{d\underline{i}}{dt} \right) \right).$$

By simple rearrangement of this expression, it is obtained that

$$p_C = \frac{1}{2} \underline{i}^T \frac{d\underline{L}}{dt} \underline{i}. \tag{6.16}$$

According to the later expression, electromechanical conversion is possible only in cases where at least one element of the inductance matrix $\underline{L}$ changes. Variation of the self-inductance or mutual inductance is generally a consequence of changing the rotor position relative to the stator. In rotating machines, the rotor displacement θ_m is tied to mechanical speed of rotation $\Omega_m = d\theta_m/dt$, and the expression for conversion power takes the form (6.17)

$$p_C = \frac{1}{2} \underline{i}^T \frac{d\underline{L}}{dt} \underline{i} = \frac{1}{2} \underline{i}^T \frac{d\underline{L}}{d\theta_m} \underline{i} \cdot \frac{d\theta_m}{dt} = \frac{\Omega_m}{2} \underline{i}^T \frac{d\underline{L}}{d\theta_m} \underline{i}. \tag{6.17}$$

Equation 6.17 shows that electromechanical conversion in rotating machines relies on variation of one or more elements of the inductance matrix in terms of the rotor movement θ_m.

In the case when a converter operates in the motor mode, power of electromechanical conversion is transferred to the mechanical subsystem. Within the mechanical subsystem, a small part of mechanical power p_c is dissipated on covering mechanical losses such as friction, while the remaining power is, via shaft, transferred to a work machine (mechanical load). In the generator mode, mechanical power of the driving turbine is, via shaft, transferred to electromechanical converter, where one part of turbine power is dissipated on covering mechanical losses. The remaining power is converted to electrical power and, reduced by the losses in the electrical subsystem, transferred to electrical receivers connected to the winding terminals.

6.9 Torque Expression

Rotating electrical machine consists of a still stator and a moving rotor, both of cylindrical shapes having a common axis. The rotor is rotating relative to the stator with speed Ω_m, and its position with respect to the stator is determined by angle θ_m at each instant. Electrical machine has N windings, and some of them are positioned on the stator while the remaining ones are on the rotor. Since self-inductances and mutual inductances depend on relative position between the stator and the rotor, elements of the inductance matrix $\underline{L}$ are also functions of the same angle.

Speed of rotation is

$$\Omega_m = \frac{d\theta_m}{dt},$$

and the inductance matrix can be represented by expression

$$\underline{L} = f_1(t) = f_2(\theta_m).$$

The magnetic coupling field acts on both stator and rotor and creates electromagnetic forces. Coupled forces create electromagnetic torque. Torque T_{em} acts upon rotor, while torque $-T_{em}$, of equal amplitude and opposite direction, acts upon stator. Since the stator is not mobile, it is only the rotor which can move. Torque T_{em} is, via rotor and shaft, transferred to work machine or driving turbine.

In motor mode, torque acts in the direction of movement. It tempts to increase the speed of rotation Ω_m. Therefore, the power $p_C = T_{em}\Omega_m$ is positive. The torque T_{em} acts in the direction of motion. It is transferred to a work machine via shaft, and it tends to start up or to accelerate its movement.

In generator mode, torque T_{em} is acting in the direction opposite to the movement; thus, the power $p_C = T_{em}\Omega_m$ is negative. Negative value of the power of electromechanical conversion indicates that the direction of conversion is changed, that is, mechanical work of the driving turbine is converted to electrical energy. Acting in the direction opposite to the movement, the torque T_{em} is, via shaft, transferred to the driving turbine and resists its rotation, tending to lower the speed of rotation.

It is of interest to determine the expression for the electromagnetic torque. Since the power is equal to the product of the torque and speed, the conversion power of expression (2.84) can be represented by equation

$$p_C = \Omega_m \left(\frac{1}{2} \underline{i}^T \frac{d\underline{L}}{d\theta_m} \underline{i} \right) = T_{em}\Omega_m,$$

where T_{em} is the electromagnetic torque determined by (6.18)

$$T_{em} = \frac{1}{2} \underline{i}^T \frac{d\underline{L}}{d\theta_m} \underline{i}, \tag{6.18}$$

This result can be verified by using the example of a rotational converter with N windings connected to ideal current sources. The winding currents are determined by the source currents and therefore constant. On the basis of (5.6), which applies in cases where magnetic field exists in linear medium, work of the source is evenly distributed between the mechanical work and the increase of the field energy; thus,

$$dW_m = dW_{meh} = T_{em}\, d\theta_m,$$

which results in the torque expression

$$T_{em} = \frac{dW_m}{d\theta_m}.$$

The energy of magnetic field is determined by (6.10). Therefore, the torque expression assumes the following form:

$$T_{em} = \frac{d}{d\theta_m} \left(\frac{1}{2} \underline{i}^T \underline{L} \underline{i} \right).$$

In accordance with the above assumptions, the winding currents are fed from ideal current sources. Therefore, the current intensities are constant. For this reason, the inductance matrix is the only factor in the above expression that may change as a function of angle θ_m. Therefore, the torque expression takes the form

$$T_{em} = \frac{1}{2} \underline{i}^T \frac{d\underline{L}}{d\theta_m} \underline{i}.$$

Electromagnetic torque can exist if at least one element of the induction matrix varies as function of angle θ_m. This could be variation of the winding self-inductance or variation of the mutual inductance between two windings. Variable inductances change as the rotor changes its position relative to the immobile stator. For a system of N windings, the electromagnetic torque is given by (6.19):

$$T_{em} = \frac{1}{2} \sum_{k=1}^{N} \sum_{j=1}^{N} \left(i_k i_j \frac{dL_{jk}}{d\theta_m} \right). \tag{6.19}$$

Question (6.3): Determine the course of change of the mutual inductance between two windings, one of them being on the stator and the other on the rotor.

Answer (6.3): Since the rotor revolves, angle θ_m between the stator and rotor reference axes varies. Without the lack of generality, it can be assumed that the case $\theta_m = 0$ corresponds to the rotor position where the magnetic axes of the two windings are collinear. In such case, the lines of the magnetic field created by the stator winding are perpendicular to the surface delineated by the turns of the rotor winding. The part of the stator flux passing through rotor winding is at a maximum. Mutual inductance between the two windings is the highest for $\theta_m = 0$. When the rotor moves, angle between the field lines and the rotor surface is no longer $\pi/2$ but takes value of $\pi/2-\theta_m$. Since the flux is determined by the sine of this angle, variation of the mutual inductance is determined by function $\cos \theta_m$. Therefore, $L_{SR} = L_m \cos \theta_m$.

Question (6.4): Give an example of a cylindrical machine with one of the stator windings having the self-inductance that varies with the rotor position.

Answer (6.4): Self-inductance of stator winding depends on the number of turns and magnetic resistance. Resistance of the magnetic circuit consists of the resistance of the iron core of the stator, resistance of the magnetic circuit of the rotor, and magnetic resistance of the air gap, where flux from the stator magnetic circuit passes to the rotor magnetic circuit and *vice versa*. Magnetic circuit of the rotor is mainly cylindrical, and has a circular cross section. By removing some iron on the rotor sides, the cross section becomes elongated and resembles an ellipse. The elliptical rotor and cylindrical stator produce a variable air gap. Therefore, the flux extending along the larger rotor diameter will encounter magnetic resistance much smaller than the flux oriented along the shorter diameter of the elliptical rotor. For this reason, the stator flux meets a variable magnetic resistance as the rotor revolves. In turn, the self-inductance $L_S = N_S^2/R_\mu$ is variable as well.

6.10 Mechanical Subsystem

Moving parts of a rotating electrical machine are magnetic and current circuits of the rotor, shaft, and bearings. Bearings are usually mounted on both shaft ends. They hold the rotor shaft firm and collinear with the axis of the stator cylinder. There are electrical machines having special fans built in the rotor for the purpose of enhancing the air flow and facilitate the cooling (*self-cooling machines*). In addition, rotor often has built-in sensors for performing measurements of the speed of rotation, position, and temperature of the rotor. In some cases, rotor may contain permanent magnets or semiconductor diodes. When modeling mechanical subsystem of an electrical machine, these details are not taken into account, and the rotor is modeled as a homogeneous cylinder of known mass and dimensions. Owing to the action of electromagnetic forces, torque T_{em} acts upon the rotor. The rotor is

Fig. 6.5 Balance of power
in mechanical subsystem
of rotating electrical machine.
Obtained mechanical power
p_c covers the losses in
mechanical subsystem and
the increase of kinetic energy
and provides the output
mechanical power $T_{em}\Omega_m$

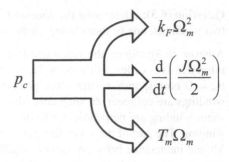

connected via shaft to a work machine or a driving turbine. The shaft transfers the
mechanical torque T_{em}.

In the mechanical subsystem, there are losses due to friction and ventilation.
A certain amount of energy is accumulated as kinetic energy in the rotating parts of
the machine. For this reason, mechanical torque T_m existing at the shaft output is not
equal to the electromagnetic torque T_{em} acting on the rotor. The power of electromechanical conversion p_c is divided into the losses of mechanical subsystem,
accumulation, and output power $p_m = T_m\Omega_m$, as shown in Fig. 6.5.

6.11 Losses in Mechanical Subsystem

Losses in mechanical subsystem consist of the energy required to overcome the
resistance due to air friction experienced by the rotor and to overcome the friction in
bearings, as well as the motion resistances of other nature and of secondary
importance. Power losses in mechanical subsystem vary as a function of speed
(Fig. 6.5). This variation may be a complex function of speed. Since the losses in
mechanical subsystem are usually small, it is not of interest to introduce a complex
model, but most often, the assumption is introduced that the friction torque is
proportional to the speed and can be modeled by expression $k_F\Omega_m$, resulting in
the expression for power losses in the mechanical subsystem

$$p_{\gamma m} = k_F\Omega_m^2. \tag{6.20}$$

This model of losses appears in a number of books and articles dealing with
electrical machines. There are, however, electrical machines and applications
where this model of losses in the mechanical subsystem is inadequate.

Model of the losses in a mechanical subsystem, shown in Fig. 6.5, has been
developed at the time when majority of electrical machines were mainly DC
machines, which will be discussed in more detail in the subsequent chapters. Stator
of these machines creates a still magnetic field, and it accommodates revolving rotor.

Variation of magnetic induction in the rotor magnetic circuit is determined by the speed of rotation Ω_m. Losses due to eddy currents in the rotor magnetic circuit are

$$p_{Fe} = k_V \Omega_m^2.$$

Dividing the losses p_{Fe} by the speed Ω_m, the torque $T_{Fe} = p_{Fe}/\Omega_m$ is obtained which resists the rotor motion and tends to diminish the speed. This torque is equal to

$$T_{Fe} = \frac{p_{Fe}}{\Omega_m} = k_V \Omega_m.$$

The obtained motion resistance T_{Fe} corresponds to the model $p_{\gamma m} = k_F \Omega_m^2$, often used in the literature.[2]

In electrical machines operating at speeds above 1,000 rad/s per second, there is a substantial air resistance. Forces resisting movement through the air are proportional to the square speed; thus, power of losses due to air friction should be modeled by expression

$$p_{\gamma m} = k_F \Omega_m^3.$$

6.12 Kinetic Energy

Accumulation of energy in the rotating masses is dependent on rotor inertia J. Rotor can be represented as a homogeneous cylinder of uniform specific mass in all its parts. With radius R and mass m, resulting moment of inertia is $J = \frac{1}{2}mR^2$. Kinetic energy W_k of a rotor with moment of inertia J and speed Ω_m is $W_k = \frac{1}{2}J\Omega_m^2$. In order to increase kinetic energy, it is necessary to supply the power $d(W_k)/dt = J\Omega_m\,d\Omega_m/dt$. Therefore, increasing the speed of rotation involves adding an amount of energy into revolving masses, while in order to slow down a speed of rotation, the energy should be taken away (by supplying a negative power).

---

[2] Losses p_{Fe} in the magnetic field of the rotor of DC machines p_{Fe} belong to the losses in magnetic circuit, that is, to iron losses. Nevertheless, the motion resistance torque T_{Fe} arises due to losses p_{Fe}. It is of interest to emphasize that in AC machines having permanent magnets, motion resistance T_{Fe} also appears even in cases with no electric current in the windings. Motion resistance T_{Fe} does not belong to mechanical losses, since it is not caused by friction in the bearings nor friction with the air, but it is a specific *electrical friction*. There is, therefore, dilemma whether power p_{Fe} should be classified as motion resistance losses or losses in the magnetic field. If all the losses that oppose to motion and diminish the speed are classified as motion resistance losses, whether their cause is mechanical friction or not, then losses in the rotor magnetic circuit of DC machines should be classified as motion resistance losses as well. Similar dilemma arises in the classification of the stator iron losses of synchronous machines having permanent magnets in their rotors.

As a consequence of this, the torque T_m obtained at the shaft of the machine is smaller than electromagnetic torque T_{em} during acceleration intervals because one part of power p_C and one part of the torque $T_{em} = p_C/\Omega_m$ are used to increase kinetic energy of revolving masses. For the same reason, the torque T_m might exceed the electromagnetic torque T_{em} during deceleration intervals.

Balance of power shown in Fig. 6.5 can be represented in analytical form. Considering a most common electromechanical converter with only one mechanical access (that is, only one shaft), the mechanical power is given by expression

$$p_m = T_m\Omega_m.$$

Kinetic energy is given by

$$W_k = \frac{1}{2}J\Omega_m^2,$$

and the rate of change of the kinetic energy is

$$\frac{dW_k}{dt} = J\Omega_m\frac{d\Omega_m}{dt}. \tag{6.21}$$

Starting from the power of electromagnetic conversion p_C, one part of this power is dissipated on the losses in the mechanical subsystem (6.20), and the other part changes kinetic energy and alters the speed of rotation (6.21); thus, the mechanical power available at the shaft (the output power) is given by (6.22):

$$p_c = \Omega_m T_{em} = \frac{dW_k}{dt} + p_{\gamma m} + p_m$$

$$= J\Omega_m\frac{d\Omega_m}{dt} + k_F\Omega_m^2 + T_m\Omega_m.$$

$$p_m = p_c - J\Omega_m\frac{d\Omega_m}{dt} - k_F\Omega_m^2. \tag{6.22}$$

6.13   Model of Mechanical Subsystem

Equation 6.22 can be divided by the rotor angular speed Ω_m to obtain (6.23) which determines the torque T_m. This torque is transferred to work machine via shaft.

$$T_m = T_{em} - J\frac{d\Omega_m}{dt} - k_F\Omega_m \tag{6.23}$$

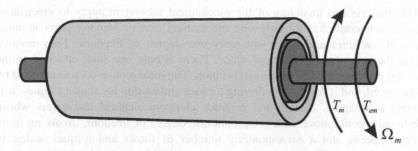

Fig. 6.6 Reference directions for electromagnetic torque and speed of rotation

In motor mode, the electrical machine acts on the work machine by the torque T_m in order to support its motion. At the same time, the work machine reacts by the torque $-T_m$ that opposes to rotor motion. Action and reaction torques are of the same magnitude, and they have opposite directions. Reference directions of the torque and speed are presented in Fig. 6.6. Former equation can be presented in the form

$$J \frac{d\Omega_m}{dt} = T_{em} - T_m - k_F \Omega_m, \tag{6.24}$$

which represents Newton equation for rotational motion. This equation is the model of mechanical subsystem of an electrical machine. In this equation, torque T_m is opposed to electromagnetic torque T_{em} as well as the friction torque. In the case when electromagnetic torque prevails ($T_{em} > T_m + k_F \Omega_m$), the speed Ω_m increases. If $T_{em} < T_m + k_F \Omega_m$, the speed decreases. In steady state, electromagnetic torque T_{em} is equal to the sum of all the torques that oppose to motion. Steady state is described by the equations

$$\frac{d\Omega_m}{dt} = 0, \quad T_{em} = T_m + k_F \Omega_m.$$

Figure 6.6 shows reference directions of the electromagnetic torque T_{em} and torque T_m of the work machine which opposes the motion. The torque with positive sign with respect to assigned reference directions corresponds to the motor mode. In the case of the *generator* mode, when mechanical work is converted to electrical energy, the meaning and signs of the above two torques are reversed. Namely, in the generator mode, the torque T_{em} has a negative sign, and it resists the motion, while the torque T_m tends to support the motion. In such cases, torque T_m is obtained from a turbine and supplied via shaft to the generator, making the rotor turn. In practice, reference directions of the torques T_{em} and T_m can be different than those shown in Fig. 6.6. Within this book, theoretical considerations and problem solving are written in accordance with directions presented in Fig. 6.6. Therefore, in the motor mode $T_{em} > 0$, $T_m > 0$, while in the generator mode $T_{em} < 0$, $T_m < 0$. As an exception, it is possible to define *generator* torque $T_G = -T_{em}$, which assumes a positive value in the generator mode.

The analysis and modeling of the mechanical subsystem apply to electrome-chanical converters having only one mechanical access. Moving parts in most electrical machines are rotors with only one degree of freedom. They revolve around the axis of the cylindrical stator. There is only one shaft attached to the rotor and positioned along the axis of rotation. The rotor motion is characterized by unique speed, and it depends on driving torques and motion resistance torques. It is possible to imagine, design, and produce electromechanical converters whose mobile parts could move with more than one degree of freedom, involving more different speeds and a corresponding number of forces and torques acting in different directions. These converters are not studied in this book.

6.14 Balance of Power in Electromechanical Converters

Diagram presented in Fig. 6.7 shows the power flow in a rotational electromechanical converter having N windings located on immobile cylindrical stator and on revolving cylindrical rotor. It is assumed that converter operates with magnetic coupling field. The relevant powers presented in the figure are explained in the following sequence.

Power at electrical access of the machine, also called input[3] power, or electrical power transferred by the source to the converter is

$$p_e = \underline{i}^T \underline{u}.$$

The power lost in the windings due to Joule effect represents losses in the electrical subsystem. This power is called *power of losses in copper*, and it is equal to

$$p_{Cu} = \underline{i}^T \underline{R}\,\underline{i} = \sum_k R_k i_k^2.$$

Power which determines the increase of energy of the coupling field is

$$p_{wm} = \frac{dW_m}{dt} = \frac{1}{2}\frac{d}{dt}\left(\underline{i}^T \underline{L}\,\underline{i}\right).$$

Power of losses in the magnetic circuit, also called power of iron losses, amounts

$$p_{Fe} = \sigma_H B^2 f + \sigma_V B^2 f^2,$$

and it is neglected in preliminary considerations.

---

[3] For electrical motors, electrical power is supplied to the motor, and it is considered an input. Mechanical power is obtained on the shaft, and it represents an output. In the case that machine operates in the generator mode, mechanical power is considered an input, while electrical power is output.

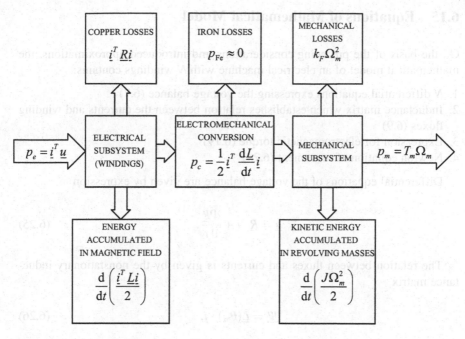

Fig. 6.7 Block diagram of the electromechanical conversion process

Power of electromechanical conversion is

$$P_c = \frac{1}{2}\, i^T \frac{\mathrm{d}L}{\mathrm{d}t}\, i.$$

Power which determines the increase in kinetic energy of revolving parts represents the accumulation in the mechanical subsystem, and it is equal to

$$p_{wk} = \frac{\mathrm{d}W_k}{\mathrm{d}t} = \frac{1}{2}\frac{\mathrm{d}}{\mathrm{d}t}\left(J\Omega_m^2\right).$$

Power of losses in the mechanical subsystem (power of losses due to rotation or motion resistance losses) is equal to

$$p_{\gamma m} = k_F \Omega_m^2.$$

Power at the mechanical access of the machine is also called output power or shaft power, and it is equal to

$$p_m = T_m \Omega_m.$$

6.15 Equations of Mathematical Model

On the basis of the preceding considerations and introduced approximations, the mathematical model of an electrical machine with N windings contains:

1. N differential equations expressing the voltage balance (6.7)
2. Inductance matrix which establishes relation between the currents and winding fluxes (6.9)
3. Expression for electromagnetic torque (6.19)
4. Newton equation of movement (6.24)

Differential equations of the voltage balance are given by expression

$$\underline{u} = \underline{R} \cdot \underline{i} + \frac{d\underline{\Psi}}{dt}. \tag{6.25}$$

The relation between fluxes and currents is given by the nonstationary inductance matrix

$$\underline{\Psi} = \underline{L}(\theta_m) \cdot \underline{i}. \tag{6.26}$$

The electromagnetic torque is determined by the following equation:

$$T_{em} = \frac{1}{2} \underline{i}^T \frac{d\underline{L}}{d\theta_m} \underline{i} = \frac{1}{2} \sum_{k=1}^{N} \sum_{j=1}^{N} \left(i_k i_j \frac{dL_{jk}}{d\theta_m} \right). \tag{6.27}$$

According to (6.24), transient phenomena in the mechanical subsystem are determined by Newton differential equation of motion. The change of the speed of rotation is determined by expression

$$J \frac{d\Omega_m}{dt} = T_{em} - T_m - k_F \Omega_m. \tag{6.28}$$

The four above expressions define general model of a rotational electromechanical converter based on the magnetic coupling field. The model is derived including the four previously mentioned approximations. Among the approximations are the assumptions that ferromagnetic materials are linear and that iron losses are negligible.

The inductance matrix is a nonstationary matrix. In general, elements of the matrix may be functions of the angle, time, as well as of the flux and current, which could change the self-inductances and mutual inductances due to nonlinearities in ferromagnetic materials and due to magnetic saturation. Within the following considerations, it is considered that the ferromagnetic material is linear and that the inductance matrix and its elements are dependent only on the angle θ_m. This approximation is justified in the majority of cases and will not present an obstacle in understanding the operation of electrical machines and deriving their characteristics.

It should be noted in 6.27 that the electromechanical conversion can be accomplished only in cases where at least one element of the inductance matrix changes with the angle θ_m, either self-inductance of a winding or mutual inductance between two windings.

In cases where an electrical machine has N windings, expression (6.25) contains N differential equations of voltage balance, expression (6.26) gives relation between the winding currents and corresponding fluxes, expression (6.27) gives electromagnetic torque, and expression (6.28) is Newton differential equation defining the speed change. Therefore, the model contains $N + 1$ differential equations and the same number of *state variables*.

Chapter 7
Single-Fed and Double-Fed Converters

In this chapter, examples of single-fed and double electromechanical converters are analyzed and explained. In both cases, the torque changes are analyzed in cases where the windings have DC currents and AC currents of adjustable frequency. Revolving magnetic field created by AC currents in the windings is introduced and explained. Using the previous considerations, some basic operating principles are given for DC current machines, induction machines, and synchronous machines.

Electrical machines where one or more self-inductances L_{kk} vary as a function of angle θ_m are usually called *single-side supplied converters* or *single-fed machines*. It is shown later that these machines may operate and perform electromechanical conversion in conditions where only the stator windings are fed from electrical source. It is also possible to envisage a single-fed electrical machine where only rotor windings are connected to the source, but this is rarely the case. There exist single-fed electrical machines having windings on the stator only, while the rotor contains no windings and has magnetic circuit with magnetic resistance which depends on the flux direction.

Machines where one or more mutual inductances L_{ij} change with angle θ_m are called *double-side supplied converters* or *double-fed* machines. They have windings on both stator and rotor. Exception to this rule is synchronous machines with permanent magnets on the rotor and DC machines with permanent magnets on the stator, which will be considered later. The effect of permanent magnets on building the flux is equivalent to effects of windings with direct current mounted instead of magnets. Therefore, electrical machines with stator windings fed from electrical source and with permanent magnet on the rotor are classified as double-fed machines. The same holds for permanent magnet DC machines.

In most cases, the windings of double-fed machine are fed from two different electrical sources; thus, there are electrical sources for the stator and the rotor. This two-sided power supply is the reason to call this type of machines *double-fed machines*.

S.N. Vukosavic, *Electrical Machines*, Power Electronics and Power Systems, 129
DOI 10.1007/978-1-4614-0400-2_7, © Springer Science+Business Media New York 2013

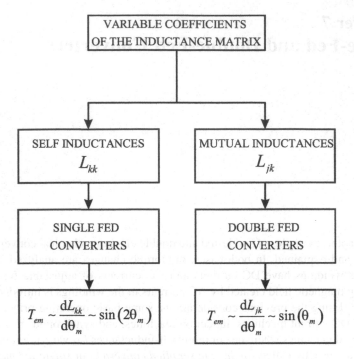

Fig. 7.1 Properties of single-fed and double-fed machines

There are machines whose classification as single- or double-fed is not immediate. An example is induction machine, which is considered later in this book. Induction machine has windings on both stator and rotor. Neglecting the secondary and parasitic effects, it can be stated that self-inductances of the stator and rotor windings of an induction motor are constant, while mutual inductances vary as functions of angle, which may be the basis for classifying induction machines as double-fed machines. Nevertheless, only stator of an induction motor is fed from electrical sources, while rotor winding is short-circuited (*squirrel cage*), and it is not connected to any source. Since an induction motor is fed from the stator side only, it cannot be classified as double-fed machine. On the other hand, there exist induction machines with rotor winding which is separately fed, and these machines truly belong to double-fed machines. Similar dilemma appears when classifying synchronous machine to single-fed or double-fed group. Synchronous machine with permanent magnet on the rotor does not have any rotor windings. Therefore, it is difficult to determine a variable mutual inductance between the stator and rotor winding, as the rotor does not have any windings. On the other hand, a permanent magnet can be represented by a sheet of electrical currents or by a winding with direct current excitation. Thus, there is a basis for classifying permanent magnet synchronous machines as double-fed machines (Fig. 7.1).

7.1 Analysis of Single-Fed Converter

Figure 7.2 shows an elementary single-fed machine. The stator winding has N_1 turns with equivalent resistance R_1, fed from a current source $i(t)$. Depending on variations of the flux and current, there is a voltage $u_1(t)$ across terminals of the winding. Magnetic circuit consists of the immobile stator part with magnetic resistance that is constant. It is considered that induction B_1 in the stator is homogeneous on the magnetic circuit cross-section. Therefore, the flux Φ_1 in one turn can be determined as $B_1 S_1$, where S_1 is the cross-section area of the stator.

Rotor is revolving and its angular displacement from horizontal position is denoted by θ_m. Magnetic circuit of the rotor is made in such way that the magnetic resistance is dependent upon direction of the flux. The rotor is not cylindrical. Instead, it has salient poles. In the case when the rotor is in horizontal position, the stator flux passes through a relatively large air gap of very low permeability (denoted by A in Fig. 7.2). After passing through the air, the flux arrives in the rotor magnetic circuit which is made of high-permeability ferromagnetic material. Then, the flux leaves the rotor magnetic circuit (denoted by B in Fig. 7.2), passing again through the air and entering the stator magnetic circuit. The resulting resistance of the magnetic circuit is then relatively high. For vertically positioned rotor, resulting magnetic resistance is much lower. The field lines pass through a very small air gap, and the magnetic resistance is relatively low. Self-inductance of the stator winding is $L_1 = N_1^2/R_\mu$, where R_μ denotes magnetic resistance across the path of the stator flux. Since the magnetic resistance varies as function of angle, the self-inductance is also variable, fulfilling the requirements of electromechanical conversion.

For the considered machine, the magnetic resistance is variable. Magnetic resistance is also called *reluctance*. For this reason, this type of machine is called reluctant machine, and the torque developed in this machine is called *reluctant torque*.

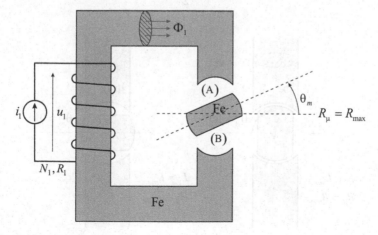

Fig. 7.2 Single-fed converter having variable magnetic resistance

7.2 Variation of Self-inductance

Figure 7.2 shows single-fed converter with variable magnetic resistance. This converter has only one winding; thus, the inductance matrix has only one element, self-inductance of the stator winding $L_1(\theta_m)$. Modeling the process of conversion requires the following function to be known:

$$L_1(\theta_m) = \frac{N_1^2}{R_\mu(\theta_m)}$$

Magnetic resistance R_μ is the ratio of the magnetomotive force $F_1 = N_1 i_1$ and the flux in a single turn $\Phi_1 = B_1 S_1$. Flux of the stator winding is $\Psi_1 = N_1 \Phi_1$. In accordance with the adopted notation, flux of the core is also flux through the contour representing one turn, and it is denoted by Φ. On the other hand, *winding flux* is denoted by Ψ. In a magnetic circuit of cross-section S_{Fe} having an air gap δ and iron core where the intensity H of magnetic field is rather small, magnetic resistance is $R_\mu = \delta/(\mu_0 S_{Fe})$.

Magnetic resistance $R_\mu(\theta_m)$ of the converter given in Fig. 7.2 has its minimum when the rotor is in vertical position. This occurs in the case when $\theta_m = \pi/2$ or $\theta_m = 3\pi/2$. The magnetic resistance is at its maximum when the rotor is in horizontal position. These are the cases with $\theta_m = 0$ or $\theta_m = \pi$, as shown in Fig. 7.3. During the rotor revolution, magnetic circuit in Fig. 7.3 changes in a way that can be modeled assuming that the air gap is variable. It can be concluded that function $R_\mu(\theta_m)$ is periodic with the period of π. For this reason, function $L_1(\theta_m)$ is also periodic and it has the same period. Actual variation of the self-inductance is dependent on the shape of the stator and rotor magnetic circuits.

In order to facilitate the analysis and get to conclusions, function $L_1(\theta_m)$ is approximated by the following trigonometric function:

$$L_1(\theta_m) \approx L_{min} + \frac{L_{max} - L_{min}}{2}(1 - \cos 2\theta_m).$$

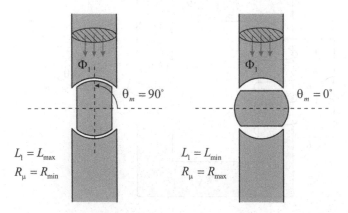

Fig. 7.3 Modeling variations of the magnetic resistance and self-inductance of the winding

which satisfies conditions $L_1(\pi/2) = L_{max}$, $L_1(3\pi/2) = L_{max}$, $L_1(0) = L_{min}$, and $L_1(\pi) = L_{min}$. Therefore, during one turn of the rotor, the inductance has two minima and two maxima. Function $L_1(\theta_m)$ can be represented in the form given by (7.1):

$$L_1(\theta_m) = \frac{1}{2}(L_{min} + L_{max}) - \frac{1}{2}(L_{max} - L_{min})\cos 2\theta_m. \tag{7.1}$$

7.3 The Expressions for Power and Torque

It is of interest to determine the electromagnetic torque and power of electromechanical conversion of the single-fed converter presented in Fig. 7.2. In the case when the stator winding is connected to a current source $i_1(t)$, the energy accumulated in the magnetic coupling field is

$$W_m = \frac{1}{2}L_1(\theta_m) \cdot i_1^2.$$

Since the winding is connected to the current source, the torque can be determined as the first derivative of the field energy W_m. By using expression (7.1) for self-inductance $L_1(\theta_m)$, the electromagnetic torque is determined by (7.2):

$$T_{em} = \frac{dW_m}{d\theta_m} = \frac{1}{2}i_1^2 \frac{dL_1}{d\theta_m} = \frac{1}{2}i_1^2(L_{max} - L_{min})\sin 2\theta_m. \tag{7.2}$$

The obtained torque is proportional to the difference between the maximum and minimum inductance, current squared, and sine of doubled angle. Since the flux is proportional to the current, the electromagnetic torque can be also expressed as a function of the flux squared,

$$T_{em} \sim i^2 \sim \Phi^2. \tag{7.3}$$

The highest value of the electromagnetic torque is obtained for $\theta_m = \pi/4$ and $\theta_m = 5\pi/4$, while in positions $\theta_m = \pi/2$, $\theta_m = 3\pi/2$, $\theta_m = 0$, and $\theta_m = \pi$, the torque is equal to zero. In the case when stator current is constant, $i_1(t) = I_1$, and the rotor is moving at a constant speed Ω_m, the torque is proportional to function $\sin(2\Omega_m t + \theta_0)$ and its average value is equal to zero. For this reason, the average value of power of electromagnetic conversion is also equal to zero. In other words, the converter shown in Fig. 7.2 with constant (DC) current in the winding cannot perform electromechanical conversion since the average torque and power in one revolution are both equal to zero.

The torque and power with an average values different than zero can be obtained where the winding has an alternating current. For current with angular frequency ω_s, with amplitude I_m, and with initial phase $-\varphi$, squared instantaneous current value is

$$i_1^2 = [I_{max} \sin(\omega_s t - \varphi)]^2 = \frac{I_{max}^2}{2}(1 - \cos(2\omega_s t - 2\varphi)).$$

On the basis of (7.2), the electromagnetic torque of single-fed converter with alternating current in its winding is

$$T_{em} = \frac{I_{max}^2}{4}(1 - \cos(2\omega_s t - 2\varphi)) \cdot (L_{max} - L_{min}) \sin 2\theta_m.$$

By introducing constant

$$K_m = \frac{I_{max}^2}{8}(L_{max} - L_{min}),$$

expression for the electromagnetic torque obtains the following form:

$$\begin{aligned}
T_{em} &= 2K_m \sin 2\theta_m - 2K_m \cos(2\omega_s t - 2\varphi) \sin 2\theta_m \\
&= 2K_m \sin 2\theta_m - K_m \sin(2\theta_m + 2\omega_s t - 2\varphi) \\
&\quad - K_m \sin(2\theta_m - 2\omega_s t + 2\varphi).
\end{aligned}$$

Since the rotor revolves at a constant speed, position of the rotor is determined by expression $\theta_m(t) = \Omega_m t + \theta_0$. Taking into account that $\theta_m(0) = \theta_0 = 0$, position of the rotor takes the value $\theta_m(t) = \Omega_m t$; thus, the torque is equal to

$$\begin{aligned}
T_{em} = \ &2K_m \sin 2\Omega_m t - K_m \sin(2\Omega_m t + 2\omega_s t - 2\varphi) \\
&- K_m \sin(2\Omega_m t - 2\omega_s t + 2\varphi) .
\end{aligned} \tag{7.4}$$

The first member in the above expression is a harmonic function with average value equal to zero. The same conclusion applies to the second member, except in cases where $\Omega_m + \omega_s = 0$. The third member has a nonzero average value if $\Omega_m = \omega_s$. Therefore, a nonzero average value of the torque can be obtained if the angular frequency of stator current is equal to the angular frequency of rotation. The torque is also dependent on the initial phase of the current. By selecting the corresponding phase, one may accomplish either positive or negative average torque. In the case when $\Omega_m = \omega_s$ and $\varphi = 3\pi/4$, average value of the electromagnetic torque is

$$T_{av} = \frac{I_{max}^2}{8}(L_{max} - L_{min}).$$

For initial phase $\varphi = \pi/4$, average value of the electromagnetic torque is

$$T_{av} = -\frac{I_{max}^2}{8}(L_{max} - L_{min}).$$

It should be noted that operation of the considered machine is based on simultaneous variation of current and rotor position. Namely, the alternating current needs to have the angular frequency ω_s equal to the rotor speed Ω_m. In other words, mechanical and electrical phenomena are to be *synchronous*.

Question (7.1): In the case when current i_1 is constant, prove that in cases with the rotor stopped at position $\theta_m = \pi/2$, the rotor stays in stable equilibrium, while at position $\theta_m = 0$, the rotor is in unstable equilibrium.

Answer (7.1): Electromagnetic torque given by (7.2) is proportional to $\sin(2\theta_m)$. If the rotor is stopped at position $\theta_m = \pi/2$, electromagnetic torque is equal to zero. A hypothetically small displacement $\Delta\theta$ in positive direction places the rotor in a new position where $2\theta_m = \pi + 2\Delta\theta$, where $\sin(2\theta_m) < 0$. A negative torque arises, tending to drive the rotor back to the previous position. In the case when the rotor makes a small move $\Delta\theta$ to negative direction, a positive torque arises, tending to return the rotor to the previous position. In the case when the rotor is stopped at position $\theta_m = 0$, the equilibrium is unstable. A hypothetically small movement $\Delta\theta$ in positive direction leads to creation of a positive torque, proportional to factor $\sin(\Delta\theta_m)$. Positive torque tends to increase the initial displacement and drive the rotor away from the initial position. The deviation is also cumulatively increased if a hypothetically small movement $\Delta\theta$ is made in the negative direction.

Question (7.2): Is it possible to accomplish a nonzero average value of the torque using an alternating current of angular frequency $\omega_s \neq \Omega_m$?

Answer (7.2): On the basis of (7.4), nonzero average value of the torque can be obtained also in the case when angular frequency of the current is $\omega_s = -\Omega_m$.

7.4 Analysis of Double-Fed Converter

A double-fed machine shown in Fig. 7.4 has windings on both moving and still parts. Stator has the magnetic circuit and winding with N_1 turns having resistance R_1. The current in the stator winding is $i_1(t)$, while the voltage $u_1(t)$ across terminals of the winding depends on variations of the flux and current. The rotor has cylindrical magnetic circuit and built-in rotor winding with resistance R_2 and with electrical current $i_2(t)$. Depending on variations of the flux and rotor current, the voltage across terminals of the rotor winding is $u_2(t)$. Electromagnetic coupling between the stator and rotor is accomplished by variable mutual inductance.

One part of the field lines representing the flux in the stator winding passes through magnetic circuit of the rotor and through rotor winding, and this part is

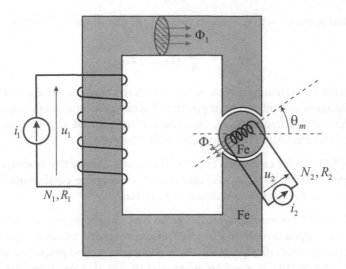

Fig. 7.4 Double-fed electromechanical converter with magnetic coupling field

called mutual flux. Since the rotor is cylindrical, it gives the same magnetic resistance in all directions. It is therefore called *isotropic*. The air gap is constant; thus, rotation of the rotor does not cause any variation of magnetic resistance along the stator flux path. Therefore, the self-inductance of the stator winding is constant.

For the particular form of the stator magnetic circuit shown in Fig. 7.4, it appears that the rotor self-inductance L_2 would change in the course of rotor revolution. The variation of self-inductance is not the key property of double-fed converters. Nevertheless, the variation of L_2 will be briefly explained for clarity. Direction of the rotor flux is determined by the position of the magnetic axis of the rotor winding, that is, by the angle θ_m. This angle determines the displacement between the rotor magnetic axis and the horizontal axis. As the rotor turns, the rotor flux is facing magnetic resistance which is dependent on the rotor position θ_m. Namely, for $\theta_m = \pi/2$, the rotor flux is passing through a relatively small air gap and it enters the magnetic circuit of the stator. When $\theta_m = 0$, the rotor flux passes from the rotor magnetic circuit into the surrounding airspace with permeability and high magnetic resistance. With $\theta_m = \pi/2$, the path of the rotor flux through the air is shorter compared to the rotor flux path through the air for $\theta_m = 0$. Therefore, the magnetic resistance and self-inductance of the rotor are both dependent on position θ_m. Variation of L_2 is dependent upon the shape of magnetic circuit. Assuming that stator magnetic circuit is modified in such way that it firmly embraces the rotor cylinder, variation of inductance L_2 would be smaller. In cases where both stator and rotor magnetic circuits are cylindrical (see Fig. 5.10.), the rotor self-inductance L_2 remains constant and does not depend on the rotor position.

In the subsequent analysis of the operation of a double-fed converter, variation of the rotor self-inductance as function of the shift θ_m is neglected, and it is assumed that $L_2 = $ const.

7.5 Variation of Mutual Inductance

Mutual inductance between the stator and rotor windings is dependent on the rotor position θ_m. When the rotor is in position where the rotor magnetic axis is horizontal, magnetic axes of stator and rotor windings are perpendicular. The lines of the stator flux do not affect the flux through the rotor turns, nor do the lines of the rotor flux contribute to the flux in the stator turns. Therefore, with the rotor axis in horizontal position, the mutual inductance L_{12} is equal to zero. On the other hand, in positions $\theta_m = \pi/2$ and $\theta_m = 3\pi/2$, magnetic coupling between the windings is strong, magnetic axes of the two windings reside on the same line, and mutual inductance L_{12} reaches its maximum absolute value L_m. The sign of the mutual inductance depends on relative position between magnetic axes of the two windings. Physically, the question is whether the fluxes add or subtract. When magnetic axes are oriented in the same direction, a positive current in one winding tends to increase the flux in the other winding. Therefore, the mutual inductance is positive. In cases where magnetic axes of stator and rotor are in opposite directions, a positive current in one winding tends to decrease the flux in the other winding and the mutual inductance is negative. Variation of the mutual inductance with the rotor angle θ_m depends on the shape of magnetic circuit and also on the distribution of conductors making up the windings. In majority of cases, this inductance can be approximated by

$$L_{12}(\theta_m) = L_m \sin \theta_m.$$

The inductance matrix expresses the total flux of the stator Ψ_1 and total flux of the rotor Ψ_2 in terms of the winding currents i_1 and i_2. Self-inductances $L_{11} = L_1$ and $L_{22} = L_2$ are positioned along the main diagonal of the matrix, while the remaining matrix elements are equal to the mutual inductance between the two windings $L_{12} = L_{21} = L_m \sin \theta_m$, as illustrated in Fig. 7.5.

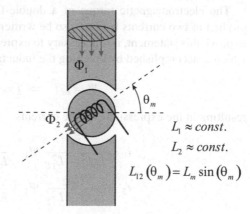

$$L_1 \approx const.$$
$$L_2 \approx const.$$
$$L_{12}(\theta_m) = L_m \sin(\theta_m)$$

Fig. 7.5 Calculation of the self-inductances and mutual inductance of a double-fed converter with magnetic coupling field

$$\underline{\Psi} = \begin{bmatrix} \Psi_1 \\ \Psi_2 \end{bmatrix} = \begin{bmatrix} L_{11} & L_{12} \\ L_{21} & L_{22} \end{bmatrix} \begin{bmatrix} i_1 \\ i_2 \end{bmatrix} = \begin{bmatrix} L_1 & L_{12} \\ L_{12} & L_2 \end{bmatrix} \begin{bmatrix} i_1 \\ i_2 \end{bmatrix} = \underline{L}\,\underline{i}. \qquad (7.5)$$

The energy of magnetic field can be expressed as function of currents and elements in the inductance matrix. Expression for the energy can be given in the form of a sum, in the matrix form, or as a scalar expression

$$W_m = \frac{1}{2} \sum_{j=1}^{2} \sum_{k=1}^{2} L_{jk} i_j i_k$$

$$= \frac{1}{2} L_1 i_1^2 + \frac{1}{2} L_2 i_2^2 + L_{12} i_1 i_2 = \frac{1}{2} \underline{i}^T \underline{L}\,\underline{i}. \qquad (7.6)$$

7.6 Torque Expression

Since the mutual inductance is variable and it changes with the rotor position, one of the three members in the expression for the field energy (7.6) varies with angle θ_m in the following manner:

$$L_{12} i_1 i_2 = i_1 i_2 L_m \sin \theta_m. \qquad (7.7)$$

The electromagnetic torque can be determined as the first derivative of the field energy W_m. Expression for the electromagnetic torque, given by (7.8), shows that the torque is proportional to the product of the currents in the stator and rotor windings and that it is dependent on the mutual inductance L_m as the angle θ_m. Namely, it changes with $\cos\theta_m$.

$$T_{em} = \frac{dW_m}{d\theta_m} = \frac{d}{d\theta_m} \left[\frac{1}{2} L_1 i_1^2 + \frac{1}{2} L_2 i_2^2 + L_{12} i_1 i_2 \right] = i_1 i_2 L_m \cos \theta_m. \qquad (7.8)$$

The electromagnetic torque of a double-fed machine can be expressed as a product of two currents but can also be written as a product of two fluxes. In order to prove this statement, it is necessary to express the currents in terms of the fluxes, which is accomplished by inverting the inductance matrix,

$$\underline{\Psi} = \underline{L}\,\underline{i} \quad \Rightarrow \quad \underline{i} = \underline{L}^{-1} \underline{\Psi},$$

resulting in the expressions for the currents

$$i_1 = \frac{L_2}{L_1 L_2 - L_{12}^2} \Psi_1 - \frac{L_{12}}{L_1 L_2 - L_{12}^2} \Psi_2,$$

$$i_2 = \frac{-L_{12}}{L_1 L_2 - L_{12}^2} \Psi_1 + \frac{L_1}{L_1 L_2 - L_{12}^2} \Psi_2.$$

By multiplying the above expressions for i_1 and i_2, one obtains the torque expression which comprises the factor $\Psi_1\Psi_2$. On the other hand, since $\Psi_1 = N_1\Phi_1$ and $\Psi_2 = N_2\Phi_2$, the torque can be expressed as function of the product of the fluxes in individual turns,

$$T_{em} \sim i_1 i_2 \sim \Phi_1 i_2 \sim \Phi_2 i_1 \sim \Phi_1\Phi_2.$$

On the basis of the obtained expression, it is possible to conclude:

- In single-fed machine, the electromagnetic torque is proportional to the current squared, $i_1{}^2$. It can be expressed in terms of the total flux squared, $\Psi_1{}^2$, or the flux in one turn squared, $\Phi_1{}^2$.
- In double-fed machine, the electromagnetic torque is proportional to the product of the two winding currents, $i_1 i_2$, or to the product of the two winding fluxes, $\Psi_1\Psi_2$, or to the product of the fluxes in one stator and rotor turn, $\Phi_1\Phi_2$.

7.6.1 Average Torque

Electromagnetic torque in a double-fed machine is given in (7.8). When the rotor revolves at a constant angular speed Ω_m, position of the rotor is $\theta_m = \Omega_m t + \theta_0$. It can be assumed that at the instant $t = 0$, angle $\theta_m(0) = \theta_0$ gets equal to zero; thus, position of the rotor is $\theta_m = \Omega_m t$. If electrical currents in the windings are constant ($i_1 = I_1$ $i_2 = I_2$), it can be concluded that the torque will changed according to cos $\Omega_m t$, with the average value equal to zero. Therefore, a double-fed machine with constant (DC) currents in the stator and rotor windings produces electromagnetic torque with average value equal to zero. Therefore, the average value of the conversion power is equal to zero as well. If one of the currents is variable, it is possible to synchronize its changes with the rotor revolution and obtain a nonzero average torque and power.

Question (7.3): Assuming that the current of the other winding is constant, $i_2 = I_2$, and that $\theta_m = \Omega_m t$, determine the variation of current i_1 which will give the torque with nonzero average value.

Answer (7.3): According to expression (7.8), the electromagnetic torque is determined by the product of functions cos $\Omega_m t$ and $i_1(t)$. Product $i_1(t)$ cos $\Omega_m t$ can have a nonzero average if i_1 is an alternating current with the ω_1 equal to the rotor angular speed Ω_m.

7.6.2 Conditions for Generating Nonzero Torque

The subsequent analysis proves that the electromagnetic torque of a double-fed machine with alternating currents in the stator and rotor windings may assume a nonzero average value, provided that the frequencies of currents and the rotor speed meet certain conditions. The currents can be expressed in terms of their amplitude I, angular frequency ω, and initial phase, $i = I \cos(\omega t - \varphi)$. When both the stator and rotor currents have nonzero angular frequencies, they change periodically, as well as the stator and rotor fluxes. It is also possible to distinguish the case where one of the currents has its angular frequency equal to zero, $\omega = 0$. This actually means that such current does not change, maintaining the value of $i = I \cos(\varphi)$. As a matter of fact, $\omega = 0$ results in a DC current. The currents of the stator and rotor may have different amplitudes, frequencies, and initial phases. Let the angular frequency of the stator current be ω_1, the angular frequency of the rotor current ω_2, the relevant amplitudes I_{1m} and I_{2m}, and the initial phases $-\varphi_1$ and $-\varphi_2$. Instantaneous values of the winding currents are given by (7.9):

$$i_1 = I_{1m} \cos(\omega_1 t - \varphi_1),$$
$$i_2 = I_{2m} \cos(\omega_2 t - \varphi_2). \tag{7.9}$$

By introducing these expressions into (7.8), one obtains the electromagnetic torque as

$$T_{em} = i_1 i_2 L_m \cos \theta_m$$
$$= I_{1m} \cos(\omega_1 t - \varphi_1) \cdot I_{2m} \cos(\omega_2 t - \varphi_2) \cdot L_m \cos \theta_m .$$

With $\theta_m = \Omega_m t$, the torque T_{em} is a product of three periodic functions. By introducing coefficient $K_n = L_m I_{1m} I_{2m}/4$, this equation assumes the form

$$T_{em} = K_n \cos(\omega_1 t - \varphi_1 + \omega_2 t - \varphi_2 + \Omega_m t)$$
$$+ K_n \cos(\omega_1 t - \varphi_1 + \omega_2 t - \varphi_2 - \Omega_m t)$$
$$+ K_n \cos(\omega_1 t - \varphi_1 - \omega_2 t + \varphi_2 + \Omega_m t)$$
$$+ K_n \cos(\omega_1 t - \varphi_1 - \omega_2 t + \varphi_2 - \Omega_m t). \tag{7.10}$$

The electromagnetic torque has the amplitude proportional to the mutual inductance and to the product of the amplitudes of the winding currents ($L_m I_{1m} I_{2m} = 4K_n$). Variation of the torque is determined by four cosine functions having different frequencies. Their frequencies can be expressed by $\omega_1 \pm \omega_2 \pm \Omega_m$. For the function $\cos(\omega t - \varphi)$ to assume a nonzero average value, it is necessary that the angular frequency ω is equal to zero. Hence, for the expression (7.10) to have a nonzero average value, one of frequencies $\omega_1 \pm \omega_2 \pm \Omega_m$ has to be equal to zero. Therefore, conclusion is reached that a nonzero average value of the torque T_{em} is obtained in cases where the angular frequencies (ω_1 and ω_2) of electrical currents in the windings and the rotor speed Ω_m meet one out of four conditions given in expression (7.11):

$$\Omega_m = \omega_1 + \omega_2,$$
$$\Omega_m = \omega_1 - \omega_2,$$
$$\Omega_m = -\omega_1 + \omega_2,$$
$$\Omega_m = -\omega_1 - \omega_2. \tag{7.11}$$

7.7   Magnetic Poles

Double-fed electrical machine has magnetic circuit where it is possible to observe two magnetic poles of the stator and two magnetic poles of the rotor. Position of the *north* magnetic pole of the rotor can be determined as a zone where the lines of magnetic field, created by electrical currents in the rotor windings, come out of the rotor magnetic circuit and enter the air gap. Similarly, one can define the *south* magnetic pole of the rotor, as well as the magnetic poles of the stator. Double-fed machine under the scope has two stator poles and two rotor poles. Since $L_{12} = L_m$ $\sin\theta_m$, it can be concluded that one cycle of variation of the mutual inductance corresponds to one full mechanical rotation of the rotor.

In due course, *multipole* machines will be defined and explained. The matter concerns electrical machines made to have more than one pair of magnetic poles. In most cases, the number of poles on the stator is equal to the number of rotor poles. A four pole machine has two north and two south poles on the stator and the same number of poles on the rotor. Such machine is said to have $p = 2$ pairs of poles. In a four pole double-fed machine, mutual inductance varies as $L_{12} = L_m \sin (p\theta_m) = L_m \sin(2\theta_m)$, thus making two cycles during one revolution of the rotor. A nonzero average value of the torque is obtained in the case when $\pm \omega_1 \pm \omega_2 \pm p\Omega_m = 0$, where Ω_m denotes the *mechanical* angular frequency of the rotor motion.

In this book, letter ω denotes the angular frequencies of voltages and currents, while letter Ω_m denotes the speed of rotor motion, also called mechanical angular speed. The former is often referred to as the *electrical* frequency ω, while the later is called *mechanical* speed Ω. Therefrom, mechanical speed Ω_m may have its electrical counterpart $\omega_m = p\Omega_m$.

Later on, revolving vectors are defined representing the spatial distribution of the magnetic induction B, magnetic field H, but also the voltages and currents in multiphase winding. The speed of rotation of such vectors in space is also denoted by letters Ω.

The expressions *electrical* frequency ω and *mechanical* speed Ω will be better defined in the course of presentation, as well as the relation $\omega = p\Omega$. For the time being, it is understood that two-pole machines are considered, resulting in $p = 1$ and $\omega = \Omega$, unless otherwise stated.

7.8   Direct Current and Alternating Current Machines

The analysis of double-fed machines can be used for demonstration of the basic operating principles of DC machines, induction machines, and synchronous machines. The latter two are also called AC machines. These machines will be studied in the remaining part of the book. All three types of machines have windings on both stator and rotor. Rotation of the rotor changes mutual inductance between the stator and rotor windings.

It has been shown that development of electromagnetic torque with nonzero average value requires the electrical stator frequency ω_1, electrical rotor frequency ω_2, and the rotor speed Ω_m[1] to meet the condition $\omega_1 \pm \omega_2 \pm \omega_m = 0$.

DC machines have a DC current in the stator windings ($\omega_1 = 0$), while in the rotor windings they have an AC current. The angular frequency of the rotor currents is determined by the speed of rotation, $\omega_2 = p\Omega_m$[2].

Induction machines have alternating currents in stator windings and alternating currents in rotor windings. According to (7.11), the sum of $p\Omega_m$[3] and rotor frequency ω_2 has to be equal to the stator frequency ω_1. Therefore, $p\Omega_m = \omega_1 - \omega_2$. The rotor mechanical speed Ω_m lags behind ω_1/p by ω_2/p. The rotor frequency $\omega_2 = \omega_1 - p\Omega_m$ of induction machines is also called *slip* frequency, as it defines the slip of the rotor speed behind the value of ω_1/p, determined by the stator frequency and called *synchronous speed*.

Synchronous machines have alternating currents in stator windings, while the rotor conductors carry DC current. Since $\omega_2 = 0$, condition (7.11) reduces to $\omega_1 = p\Omega_m$. Therefore, the rotor speed Ω_m is uniquely determined by the stator electrical frequency, $\Omega_m = \omega_1/p$. Hence, all the two-pole ($p = 1$) synchronous machines connected to the three-phase grid with the line frequency of $f_s = 50$ Hz make 50 turns per second, or $50 \cdot 60 = 3{,}000$ revolutions per minute (rpm). A four pole ($p = 2$) synchronous machine supplied by $f_s = 60$ Hz runs at $60 \cdot 60/p = 1{,}800$ rpm. Hence, these machines run *synchronously* with the supply frequency and therefore their name.

7.9 Torque as a Vector Product

The principles of operation of DC machines, induction, and synchronous machines as well as the main differences between them are more obvious when the stator and rotor fluxes are represented by corresponding vectors. Electromagnetic torque can

---

[1] In cases where machine has p pairs of poles, the condition for torque development is $\omega_1 \pm \omega_2 \pm p\Omega_m = 0$. Notation Ω_m is angular speed of rotor motion, hence mechanical speed. Angular frequency $\omega_m = p\Omega_m$ is *electrical* representation of the rotor speed. It defines the period $T_{\omega m} = 2\pi/\omega_m$ which marks passing of north magnetic poles of the rotor against north magnetic poles of the stator. With $p > 1$, this happens more than once per each mechanical revolution. In a machine with $p > 1$ pole pairs, angular distance between the two neighboring north poles is $\Omega_m T_{\omega m} = 2\pi/p$. A four pole machine ($p = 2$) has two north and two south poles. Two north poles are at angular distance of $\Omega_m T_{\omega m} = 2\pi/2 = \pi$. Therefore, any north magnetic pole of the rotor passes against stator north pole twice per turn. In a two-pole machine ($p = 1$), starting from the north magnetic pole, one should pass angular distance of $\Omega_m T_{\omega m} = 2\pi/1 = 2\pi$ in order to arrive at the next north pole, the very same pole from where one started. Namely, a two-pole machine has only one north magnetic pole and one south magnetic pole.

[2] In a two-pole DC machine, the number of pole pairs is $p = 1$. Therefore, $\omega_2 = p\Omega_m = \Omega_m$. With $p > 1$, the condition reads $\omega_2 = \omega_m = p\Omega_m$.

[3] In a two-pole induction motor, $p = 1$.

be expressed as a vector product of the stator and rotor flux vectors. In other words, the torque is obtained by multiplying the amplitude of the stator flux vector, the amplitude of the rotor flux vector, and the sine of the angle between the two vectors. A proof of this statement will be presented later on for all the machines studied in this book. Moreover, the electromagnetic torque developed by an electrical machine can be determined by calculating the vector product of:

- Stator flux and rotor flux vectors
- The stator and rotor magnetomotive force vectors (current vectors)
- The stator flux vector and the rotor magnetomotive force vector (current vector)
- The rotor flux vector and the stator magnetomotive force vector (current vector)

Obtaining electromagnetic torque as vector product of the flux and current can be demonstrated by taking the example of a contour placed in an external, homogeneous magnetic field, as shown in Fig. 7.6. The contour is made of a conductor carrying electrical current I. The conductor is shaped in the form of a flat rectangle of width D and length L, encircling the surface $S = DL$. In the considered position, angle between the normal n_1 on the surface plane and vector of magnetic induction is α. Angle α determines the electromagnetic torque acting on the contour.

Magnetic momentum of the contour is a vector collinear with the normal n_1 on the surface S surrounded by the contour. The orientation of the normal is determined by the direction of electrical current in the contour and the right-hand rule. The amplitude of the magnetic momentum m is determined by the product of the contour current I and the surface S,

$$\vec{m} = I \cdot S \cdot \vec{n}_1 \tag{7.12}$$

The electromagnetic torque acting on the contour is equal to the vector product of the magnetic momentum m and the magnetic induction B. The torque can be determined from (7.13). In Fig. 7.6, the torque vector extends in the axis of rotation of the contour, and its direction is determined from the coupled forces by the right-hand rule. Maximum value of the torque $T_m = D \cdot L \cdot I \cdot B$ is obtained at position $\alpha = \pi/2$.

$$\vec{T}_{em} = \vec{m} \times \vec{B}, \quad |\vec{T}_{em}| = S \cdot I \cdot B \cdot \sin \alpha = D \cdot L \cdot I \cdot B \cdot \sin \alpha. \tag{7.13}$$

Result (7.13) can be checked by analyzing the forces acting on parts of the rectangular contour. For contour parts of the length L, orthogonal to the lines of magnetic field, the electromagnetic force is determined by expression $F = L \cdot I \cdot B$. On parts of the contour of the length D, the forces are acting in the direction of rotation, but they are collinear and of opposite directions. Therefore, their opposing actions are canceled. Force arm K is equal to

$$K = \frac{D}{2} \sin \alpha,$$

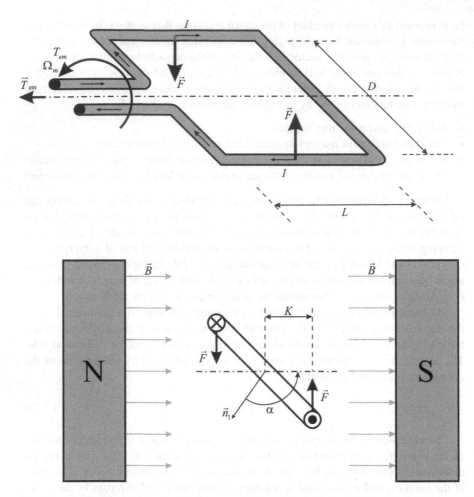

Fig. 7.6 Torque acting on a contour in homogenous, external magnetic field is equal to the vector product of the vector of magnetic induction **B** and the vector of magnetic momentum of the contour. Algebraic intensity of the torque is equal to the product of the contour current I, surface $S = L \cdot D$, intensity of magnetic induction B, and $\sin(\alpha)$. Its course and direction are determined by the normal n_1 oriented in accordance with the reference direction of the current and the right-hand rule

thus, the electromagnetic torque acting on the contour of Fig. 7.6 is

$$T_{em} = 2 \cdot F \cdot K = 2(L \cdot I \cdot B)\frac{D}{2}\sin\alpha = D \cdot L \cdot I \cdot B \cdot \sin\alpha.$$

The preceding expression obtained for the torque can be represented as function of the flux and magnetomotive force. Maximum value of the flux through the contour is $\Phi_m = SB = DLB$, and it is obtained in position $\alpha = 0$. Since the contour has one turn ($N = 1$), current I in the contour is equal to the magnetomotive force $F_m = NI = I$. Starting from expression (7.13), the electromagnetic torque can be expressed as

$$T_{em} = F_m \cdot \Phi_m \cdot \sin \alpha \qquad (7.14)$$

Flux through the contour is a scalar quantity. By associating the course and direction of magnetic induction B to the flux Φ, it is possible to conceive the flux vector. Magnetomotive force of the contour is a vector whose orientation is determined by the normal n_1, which is collinear with the vector of magnetic momentum of the contour. Therefore, the value of expression (7.14) is determined by the vector product of the magnetomotive force vector and the flux vector. In a like manner, it can be shown that the electromagnetic torque of a cylindrical rotating machine is determined by the vector product of the magnetomotive force of the stator and the rotor flux. By rearranging the expressions, it is possible to express the torque as the vector product of the stator and rotor fluxes. It is also possible to express the torque in terms of stator and rotor magnetomotive forces or in terms of the stator flux and the rotor magnetomotive force.

7.10 Position of the Flux Vector in Rotating Machines

The stator flux vector and the rotor flux vector of an electrical machine have the spatial orientation which depends on electrical currents they originate from. A DC current in stator windings creates stator flux which does not move relative to the stator. A DC current in rotor windings creates rotor flux which does not move with respect to the rotor. In cases where rotor turns, such rotor flux revolves with respect to the stator at the rotor speed. It will be shown later that a set of stator windings with AC currents may produce stator flux vector which revolves with respect to the stator at a speed determined by the angular frequency of AC currents. More detailed definition of the flux per turn, flux per winding, and the method of representing flux as a vector are given in Chap. 4.

The analysis which shows that the electromagnetic torque of a machine can be determined from the vector product of fluxes and magnetomotive forces is a part of the chapters dealing with DC and AC machines. Induction, synchronous, and DC machines differ inasmuch as they have DC or AC currents in stator and rotor windings.

The electromagnetic torque of DC and AC machines can be determined on the basis of the vector product between the stator and rotor flux vectors. Provided with the stator flux per turn (Φ_S), rotor flux per turn (Φ_R), and with the angle $\Delta\theta$ between the stator and rotor flux vectors, the electromagnetic torque can be calculated from the expression $|\Phi_S \times \Phi_R| = \Phi_S \Phi_R \sin(\Delta\theta)$.

In cases when relative position of the two flux vectors varies according to the law $\Delta\theta = \omega t$, the electromagnetic torque will, according to (7.13), exhibit oscillations

Fig. 7.7 Change of angular
displacement between stator
and rotor flux vectors in the
case when the stator and rotor
windings carry DC currents

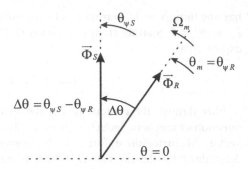

and change as ($\sin \omega t$). Average value of such torque is equal to zero. In order to create an electromagnetic torque with nonzero average, it is necessary that relative position between the stator and rotor flux vectors does not change. A constant displacement $\Delta\theta$ is obtained in cases where both flux vectors are stationary with respect to the stator but also in cases where the two vectors rotate at the same speed and in the same direction, keeping their relative displacement $\Delta\theta$ constant.

A constant displacement $\Delta\theta$ cannot be achieved in electrical machines that have DC currents in both stator and rotor windings. Namely, windings carrying DC current create a magnetomotive force and flux along the winding axis. Therefore, the flux caused by DC currents cannot move relative to the winding. Therefore, DC currents in stator windings create a stationary stator flux. DC currents in rotor windings create a rotor flux that does not move with respect to the rotor. With the rotor in motion, the rotor flux revolves at the rotor speed, moving in such a way relative to the stator flux. In these conditions, the angle $\Delta\theta$ changes while the electromagnetic torque oscillates and has the average value equal to zero.

In the considered case, the flux vectors are shown in Fig. 7.7. Stator flux Φ_S does not move, while rotor flux Φ_R revolves at rotor speed Ω_m. With $\theta_{\psi S} = 0$, the angle $\Delta\theta$ between the two vectors is function of the speed of rotor rotation $\Delta\theta = -\Omega_m \cdot t$, while variation of the torque is determined by function $\sin(-\Omega_m \cdot t)$; thus, its average value is zero. In order to accomplish a constant value of the angle between stator and rotor fluxes, both vectors have to be still or moving at the same speed. In any case, one of the windings, stator or rotor, has to create a magnetic field that revolves with respect of the originating winding. Although the principles of operation of the DC machines and induction and synchronous machines are yet to be explained and analyzed in detail, it is of interest to indicate the position of the stator and rotor flux vectors in these machines.

A **DC machine** is shown in the part A of Fig. 7.8. Stator flux is represented by vector Φ_S. Flux Φ_S is immobile, created by DC currents in the stator windings. Rotor flux is represented by vector Φ_R. Flux Φ_R is created by alternating currents in the rotor conductors. Usually, rotor winding has a large number of turns, but in Fig. 7.8a, it is represented by conductors P1 and P2. In these conductors, there is an alternating current with angular frequency of $\omega_2 = \Omega_m$. During one turn of the rotor, currents in conductors P1 and P2 make one full cycle of their periodical change, being positive during one half period and negative during another half period.

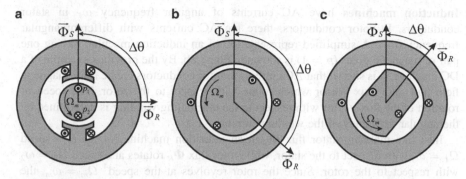

Fig. 7.8 Position of stator and rotor flux vectors in DC machines (**a**), induction machines (**b**), and synchronous machines (**c**)

It is assumed that the rotor revolves at the speed Ω_m. Since the current in rotor conductors changes sign synchronously with rotor revolutions, the current in rotor conductor passing by the south magnetic pole of the stator will always be directed toward the spectator ($\odot$). In Fig. 7.8a, the rotor is in position where the conductor P1 passes under the south magnetic pole of the stator.

The preceding statement can be supported by the following discussion. In position of the rotor shown in Fig. 7.8a, conductor P2 is below the north magnetic pole of the stator and carries the current directed away from the spectator ($\otimes$). Having passed one half of the rotor turn, conductor P2 comes in place of the conductor P1, below the south magnetic pole of the stator. At the same time, direction of rotor current changes. Hence, in conductor P2, direction ($\otimes$) changes into ($\odot$). Therefore, direction of the current in the rotor conductor below the south magnetic pole of the stator remains toward the spectator. It can be shown in a like manner that the rotor conductor passing by the north stator pole keeps the direction away from the reader ($\otimes$).

Distribution of rotor currents described above does not move with respect to the stator. Rotor currents create magnetomotive force and flux which are immobile with respect to the stator. What remains unclear at this point is the way of supplying the rotor winding with alternating currents having an angular frequency equal to the rotor speed. This will be explained in more detail in the chapter dealing with DC machines.

Under considerations, the AC currents in rotor conductors create rotor flux vector Φ_R which revolves with respect to the rotor itself. The magnetic field which rotates with respect to the originating windings is called *rotating* or *revolving magnetic field*. Conditions to be met for AC currents to create rotating magnetic field will be explained in more detail in the chapter dealing with induction machines.

In the course of rotation of the rotor in Fig. 7.8a, the rotor flux vector is still with respect to the stator and orthogonal to the stator flux vector, regardless of the speed and direction of rotation. For this reason, rotor magnetic field in a DC machine can be called *halted rotating field*.

Induction machines have AC currents of angular frequency ω_1 in stator conductors. In rotor conductors, there are AC currents with different angular frequency (ω_2). A simplified representation of an induction machine having one pair of magnetic poles ($p = 1$) is shown in Fig. 7.8b. By the previous example of a DC machine, it is shown that AC currents in rotor conductors create rotor magnetic field and rotor flux vector which rotate with respect to the rotor. The speed of rotation of the flux vector with respect to the originating winding is determined by the angular frequency of the winding currents.

In a like manner, stator flux $\boldsymbol{\Phi}_S$ in an induction machine rotates at a speed $\Omega_1 = \omega_1$ with respect to the stator, while rotor flux $\boldsymbol{\Phi}_R$ rotates at a speed $\Omega_2 = \omega_2$ with respect to the rotor. Since the rotor revolves at the speed[4] $\Omega_m = \omega_m$, the speed of rotation of rotor flux is $\Omega_m + \omega_2$. Vector of the stator flux rotates at the speed of ω_1; thus, the speed difference between stator flux vector and rotor flux vector is $\omega_1 - \Omega_m - \omega_2$. According to (7.11) which gives the condition for delivering the power and torque with nonzero average values, the sum $\omega_1 - \Omega_m - \omega_2$ must be equal to zero.

On the basis of previous considerations regarding the operation of induction machines, the following conclusions can be drawn:

- The stator and rotor flux vectors rotate at the same speed. The speed of rotation of the magnetic field in induction machine is Ω_1, and it is determined by the angular frequency ω_1 of the stator currents. In a two-pole machine, this speed is $\Omega_1 = \omega_1$.
- Angle $\Delta\theta$ between the stator and rotor flux vectors is constant in a steady state. Machine provides electromagnetic torque proportional to $\sin(\Delta\theta)$, and it is constant in a steady state.
- Rotor of a two-pole machine revolves at the speed which is different than the speed $\Omega_1 = \omega_1$ of the magnetic field. The speed difference $\omega_2 = \omega_1 - \Omega_m$ is called *slip*. Slip of a two-pole induction machine ($p = 1$) is equal to the angular frequency of the rotor currents (ω_2).

Synchronous machines have AC currents of frequency ω_1 in the stator windings and a DC current in the rotor windings.[5] The stator flux vector $\boldsymbol{\Phi}_S$ of a two-pole ($p = 1$) machine rotates at the speed $\Omega_1 = \omega_1$ with respect to the stator, while the rotor flux vector $\boldsymbol{\Phi}_R$ rotates at the same speed as the rotor, $\Omega_m = \omega_m$. A simplified representation of a two-pole synchronous machine is shown in Fig. 7.8c. Generation of the electromagnetic torque with a nonzero average value requires that relative position of the two flux vectors does not change. In other words, the angle $\Delta\theta$ has to remain constant. For this reason, the rotor speed and the speed of

---

[4] Example in Fig. 7.8b considers a two-pole machine having $p = 1$ pair of magnetic poles. Due to $\omega = p\Omega$ and $p = 1$, mechanical speed (angular frequency) Ω corresponds to electrical speed (angular frequency) ω.

[5] There exist synchronous machines that have permanent magnets in place of DC excited rotor windings.

revolving stator flux vector have to be the same. Therefore, the stator and rotor flux vectors of a synchronous machine rotate synchronously with the rotor. In a two-pole synchronous machine, angular frequency of stator currents has to be equal to the rotor speed. In machines having several pole pairs ($p > 1$), this condition takes the form $\omega_1 = p\Omega_m$.

7.11 Rotating Field

The analysis carried out in the preceding subsection shows that the condition for developing an electromagnetic torque with a nonzero average value is that relative position $\Delta\theta$ between the stator and rotor flux vectors remains constant. In DC machines, both fluxes are still with respect to the stator, while in AC machines, induction and synchronous, the two fluxes revolve at the same speed.

With the rotor revolving at a speed Ω_m, the angle $\Delta\theta$ can remain constant provided that at least one of the two fluxes (Φ_S or Φ_R) revolves with respect to the winding whose magnetomotive force originates the flux. The magnetic field which rotates with respect to the originating winding is called *rotating magnetic field*. It will be shown later that creation of a rotating field in induction and synchronous machines requires at least two separate windings on the stator, also called *phases* or *phase windings*. With two-phase windings on the stator, the spatial displacement between the winding axes has to be $\pi/2$. The alternating currents in two-phase windings have to be of the same amplitude and the same angular frequency. The difference of their initial phases has to be $\pi/2$, the same as the spatial displacement between the phase windings. In this case, stator currents result in a rotating magnetic field. The amplitude of the stator flux and its speed of rotation can be changed by varying the amplitude and frequency of the stator currents. In this chapter, an introductory example is given, illustrating the generation of a rotating field by the stator with two-phase windings.

Figure 7.9 shows two stator windings with their magnetic axes spatially displaced by $\pi/2$. Axes of the windings are denoted by α and β. The winding in axis α has the same number of turns as the winding in axis β. Both windings carry alternating currents of the same amplitude I_m and frequency ω_S,

$$i_\alpha(t) = I_m \cos(\omega_S t),$$
$$i_\beta(t) = I_m \cos\left(\omega_S t - \frac{\pi}{2}\right) = I_m \sin(\omega_S t),$$

but their initial phases differ by $\pi/2$. Each winding creates a magnetomotive force along its own axis. Magnetomotive force amplitude depends on the current and the number of turns. The winding flux is proportional to the magnetomotive force and inversely proportional to magnetic resistance. If magnetic circuits of the stator and rotor are of cylindrical shape, magnetic resistance R_μ incurred along the flux path

Fig. 7.9 Two stator phase
windings with mutually
orthogonal axes and
alternating currents with
the same amplitude and
frequency create rotating
magnetic field, described by
a revolving flux vector of
constant amplitude. It is
required that initial phases
of the currents differ by $\pi/2$

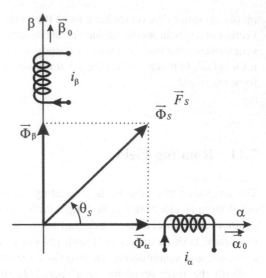

does not depend on the flux spatial orientation. For this reason, the magnetic
resistance to the flux $\boldsymbol{\Phi}_\alpha$ is equal to the magnetic resistance to the flux $\boldsymbol{\Phi}_\beta$. With
both windings having the same number of turns and the same magnetic resistances,
the fluxes $\boldsymbol{\Phi}_\alpha$ and $\boldsymbol{\Phi}_\beta$ are obtained by multiplying the number of turns N by
electrical currents i_α and i_β, respectively, and dividing the product by the magnetic
resistance R_μ. Maximum values of the fluxes $\boldsymbol{\Phi}_\alpha$ and $\boldsymbol{\Phi}_\beta$ are

$$\Phi_{\alpha\max} = \frac{N_\alpha I_m}{R_\mu} = \Phi_{\beta\max} = \frac{N_\beta I_m}{R_\mu} = \Phi_m$$

The instantaneous values of the fluxes are

$$\Phi_\alpha(t) = \Phi_m \cos(\omega_S t),$$
$$\Phi_\beta(t) = \Phi_m \sin(\omega_S t).$$

The two fluxes contribute to the resulting flux Φ in the electrical machine, which
can be represented by a vector in $\alpha - \beta$ coordinate frame. Functions $\Phi_\alpha(t)$ and $\Phi_\beta(t)$
represent projections of such flux vector on α-axis and β-axis. The amplitude
of the resulting flux is Φ_m. With the assumed electrical currents, the resultant
magnetic field created by the pair of windings in Fig. 7.9 rotates at the speed
$\Omega_S = \omega_S$. During rotation, algebraic intensity of the flux vector does not change and
it remains Φ_m. This example demonstrates the possibility for a system of two windings
to create magnetic field which rotates with respect to the windings. It is important to
notice that the windings must carry alternating currents and that the angular frequency
of electrical currents ω_S determines the speed of magnetic field rotation Ω_S.

Rotating magnetic field is a prerequisite for DC, induction, and synchronous
machines, analyzed within this book. In each of the three machine types, windings
exist with AC currents creating magnetic field that revolves with respect to the
winding itself, also called rotating magnetic field.

7.12 Types of Electrical Machines

7.12.1 Direct Current Machines

Electrical machines where the stator winding carries a DC current, while the rotor winding carries AC currents, and where the stator flux vector and the rotor flux vector do not move with respect to the stator are called *DC current machines*. Stator windings of DC machines are fed by DC, *direct current*. Rotor conductors in such machines carry AC currents with the frequency determined by the speed of rotation. The power source feeding a DC machine does not provide AC currents and voltages, but instead it gives DC currents and voltages. The method of directing DC current from the power source into the rotor conductors involves *commutator*, mechanical device explained further on. The action of commutator is such it receives DC source current and feeds the rotor winding with AC currents, the frequency of which is determined by the rotor speed.

Induction and synchronous machines have AC currents in their stator windings. The angular frequency ω_1 of these currents provides a rotating magnetic field. Therefore, these machines belong to the group of *AC machines*. The speed of rotation of the magnetic field is determined by the angular frequency ω_1. It is shown by the analysis of the structure in Fig. 7.9 that a system of two orthogonal stator windings could create magnetic field that revolves at the speed determined by the angular frequency of AC currents. Practical AC machines usually have a system of stator windings consisting of three parts, *three phases*, that is, three-*phase windings*. Magnetic axes of three-phase windings are spatially shifted by $2\pi/3$. The initial phases of AC currents carried by the windings should be displaced by $2\pi/3$ in order to provide rotating field. Amplitude I_m of AC currents determines the algebraic intensity of the flux vector, while the angular frequency $\omega_1 = \omega_S$ determines the speed of rotation Ω_S of the magnetic field.

7.12.2 Induction Machines

In addition to AC currents carried by the stator windings, induction machines also have AC currents in the rotor conductors. Magnetic field created by the stator currents rotates at the speed $\Omega_1 = \omega_1$, while the rotor field revolves at the speed $\Omega_2 = \omega_2$ with respect to the rotor. The speeds of rotation of the stator and rotor flux vectors have been discussed in the previous section, where it is shown that the angular frequency of rotor currents, also called the slip frequency, corresponds to the difference between the angular frequency of stator currents and the rotor speed.

7.12.3   Synchronous Machines

Like induction machines, synchronous machines have a system of stator windings with AC currents creating magnetic field which revolves at the speed determined by the angular frequency of stator currents. Currents of the rotor winding of a synchronous machine are constant. They are supplied from a separate DC current source. Rotor current creates the rotor flux which does not move with respect to the rotor. Therefore, the rotor flux rotates together with the rotor and has the same speed Ω_m. There are synchronous machines which do not have the rotor winding. Instead, the rotor flux is obtained by placing permanent magnets within the rotor magnetic circuit. It has been shown before that the torque generation within an electrical machine requires the angle between the stator and rotor flux vectors to be constant. Therefore, the stator flux vector of a synchronous machine has to rotate at the same speed as the rotor. In other words, the stator flux has to move *synchronously* with the rotor.

Among these machines, each type has its merits, limitations, and specific field of application.

Further analysis of electrical machines requires some basic knowledge on the machine windings, skills in analyzing the magnetic field in the air gap, and understanding the principles of rotating magnetic field.

Chapter 8
Magnetic Field in the Air Gap

This chapter presents an analysis of the magnetic and electrical fields in the air gap of a cylindrical machine. It is assumed that the fields come as a consequence of electrical current in the windings. The magnetic field in the air gap is created by the currents in both stator and rotor, which generate the corresponding stator and rotor magnetomotive forces.

Conductors of the stator winding are placed in the grooves made on the inner surface of the stator magnetic circuit, while conductors of the rotor winding are placed in the grooves made on the outer surface of the rotor magnetic circuit. The grooves are called slots, and they are opened toward the air gap (Fig. 8.1). Thus, the conductors are placed near the air gap.

It is also assumed that conductors that make up a winding are many and that they are series connected. They are not located in the same slot. Instead, the conductors are distributed along the circumference of the air gap. Conductor density can be determined by counting the number of conductors distributed along one unit length of the circumference. To begin with, it is assumed that the windings are formed with sinusoidal distribution of conductor density. Namely, the number of conductors placed in the fragment $R \cdot \Delta\theta$ of the circumference (Fig. 8.2) is determined by the function $\cos\theta$, where the angle θ determines the position of the observed fragment. When electrical currents are fed into the winding, they create a sinusoidal distributed current sheet, also called *sinusoidally distributed current sheet*. With these assumptions, the subsequent analysis determines expressions for radial and tangential components of the magnetic field in the air gap, for magnetomotive forces of the stator and rotor windings, and for fluxes per turn and the winding fluxes. The subsequent passages also introduce the notation aimed to simplify the presentation of the windings, magnetomotive forces, and fluxes. At the same time, the energy of the magnetic field in the air gap and electromagnetic torque are calculates as well, the torque being a measure of mechanical interaction between the stator and rotor. Further on, relation between the torque and machine dimensions is analyzed. Eventually, conditions for creating rotating magnetic field in the air gap are studied and specified.

S.N. Vukosavic, *Electrical Machines*, Power Electronics and Power Systems,
DOI 10.1007/978-1-4614-0400-2_8, © Springer Science+Business Media New York 2013

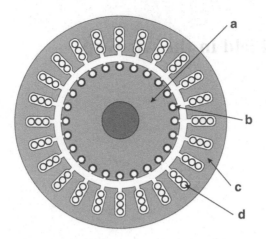

Fig. 8.1 Cross section of the magnetic circuit of an electrical machine. Rotor magnetic circuit (*a*), conductors in the rotor slots (*b*), stator magnetic circuit (*c*), and conductors in the stator slots (*d*)

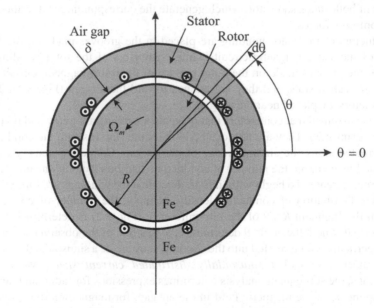

Fig. 8.2 Simplified representation of an electrical machine with cylindrical magnetic circuits made of ferromagnetic material with very large permeability. It is assumed that the conductors are positioned on the surface separating ferromagnetic material and the air gap

On the basis of the analysis of magnetomotive forces, the merits of sinusoidal spatial distribution of conductors are given a rationale. The analysis of electromotive forces in the concentrated windings and windings having periodic, non-sinusoidal spatial distribution is carried out in Chapter 10, *Electromotive Forces*.

8.1   Stator Winding with Distributed Conductors

Electrical machines are usually of cylindrical shape. An example of the cross section of a cylindrical machine is shown in Fig. 8.1. The magnetic circuit is made of iron sheets in order to reduce iron losses. The sheets forming magnetic circuits of the stator and rotor are coaxially placed, and they have shapes shown in Fig. 8.1. Stator has a form of a hollow cylinder. Rotor is a cylinder with slightly smaller diameter than the internal diameter of the stator. Distance δ between the stator and rotor is of the order of one millimeter and is called air gap. The air gap is considerably smaller than radius of the rotor cylinder R, $\delta \ll R$. The sheets are made of iron, ferromagnetic material with permeability much higher than μ_0; thus, the intensity H_{Fe} of magnetic field in iron is up to thousand times lower compared to the intensity H_0 of magnetic field in the air gap. Therefore, H_{Fe} can be neglected in most cases. Due to $\delta \ll R$, the changes of H_0 along the air gap δ can be neglected. For this reason, the value of the contour integral of magnetic field H in an electrical machine is reduced to the sum of products $H_0\delta$, also called *magnetic voltage drop* across the air gap.

Conductors of the stator and rotor are laid along the axis of the cylinder and placed next to the surface which separates the magnetic circuit and the air gap. They can be on both stator and rotor sides. Figure 8.2 shows conductors of the stator. The sign $\otimes$ represents a conductor carrying current away from the reader, while the sign $\otimes$ represents a conductor carrying current toward the reader. One pair of conductors connected in series makes up one *contour* or one *turn*. Conductors making one turn are usually positioned on the opposite sides of the cylinder, at an angular displacement of π (*diametrically positioned conductors*).

The conductors are positioned along circumference of the cylinder so that their line density (number of conductors per unit length $R \cdot \Delta\theta$) varies sinusoidally as function of angular displacement θ (i.e., $\cos\theta$). In cases where the function $\cos\theta$ suggests a negative number, it is understood that the number of actual conductors is positive, but direction of the current in these conductors is changed (diametrically positioned conductors are denoted by $\otimes$ and $\odot$).

Line density of conductors in the stator winding, shown in Fig. 8.3, changes sinusoidally, and it can be modeled by function

$$N_S'(\theta) = N_{S\,max}' \cdot \cos\theta \tag{8.1}$$

Function $N_S'(\theta)$ gives the number of conductors per unit length along the internal circumference of the stator magnetic circuit. If a very small segment $d\theta$ is considered, the corresponding fraction of the circumference length is $dl = R\,d\theta$, while the number of conductors within this fraction is

$$dN_S = N_S'(\theta)\,dl = N_S'(\theta)R\,d\theta = N_{S\,max}' \cdot \cos\theta \cdot R\,d\theta$$

In the example given in the figure, the density of conductors carrying current of direction $\otimes$ is the highest at $\theta = 0$, and it amounts $N_S'(0) = N_{S\,max}'$. The highest

Fig. 8.3 Sinusoidal spatial
distribution of conductors
of the stator winding

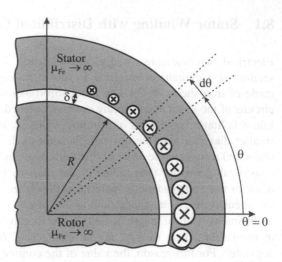

density of conductors carrying current in the opposite direction (⊙) corresponds to
position $\theta = \pi$. According to Fig. 8.2, over the interval from $\theta = -\pi/2$ up to
$\theta = \pi/2$, there are conductors with reference direction ⊗, while from $\theta = \pi/2$ up
to $\theta = 3\pi/2$, there are conductors with reference direction ⊙.

One pair of diametrically placed conductors (⊗ and ⊙) forms one turn or one
contour. The considered winding is obtained by connecting several turns in series.
The total number of *turns* N_T can be determined by counting conductors having
reference direction ⊗, that is, by integrating the function $N_S'(\theta)$ over the span
extending from $\theta = -\pi/2$ up to $\theta = \pi/2$:

$$N_T = \int_{-\frac{\pi}{2}}^{+\frac{\pi}{2}} N_S'(\theta)\, R\, d\theta = \int_{-\frac{\pi}{2}}^{+\frac{\pi}{2}} N_{S\,max}' \cos\theta\, R\, d\theta$$

$$= N_{S\,max}' R \cdot \sin\theta \Big|_{-\frac{\pi}{2}}^{+\frac{\pi}{2}} = 2R \cdot N_{S\,max}' . \tag{8.2}$$

Total number of *conductors* of the considered winding N_C is twice the number of
turns; thus, $N_C = 2N_T = 4R\, N_{S\,max}'$.

The number of conductors can be obtained by calculating the integral of the
function $|N_S'(\theta)|$ over the whole circumference of the machine, that is, over the
interval starting with $\theta = 0$ and ending at $\theta = 2\pi$. This calculation implies counting
all conductors, irrespective of their reference direction. Integration of the absolute
value of density of conductors takes into account the conductors having reference
direction from the reader ⊗ and also the conductors having reference direction
toward the reader ⊙:

$$N_C = \int_0^{2\pi} |N_S'(\theta)|\, R\, d\theta = RN_{S\,max}' \int_0^{2\pi} |\cos\theta|\, d\theta = 4RN_{S\,max}' \tag{8.3}$$

8.2   Sinusoidal Current Sheet

Electrical current in series-connected, spatially distributed stator conductors forms a current sheet on the inner surface of the stator cylinder. Current direction from the reader $\otimes$ extends in the interval $-\pi/2 < \theta < \pi/2$, while the direction toward the reader $\odot$ extends over the interval $\pi/2 < \theta < 3\pi/2$. Distribution of current over this surface is shown in Fig. 8.2.

The considered current sheet has the line density of surface currents dependent on the line density of conductors. The line density of the current sheet over the inner surface of the stator cylinder is denoted by $J_S(\theta)$, and it is function of the angular displacement θ. It is determined by the density of conductors $N'_S(\theta)$ and the current strength in a single conductor. Since the stator winding is formed by connecting the conductors in series, all the conductors carry the same current $i_1(t)$, also called the stator current. Current through conductors is determined by the reference direction, shown in Fig. 8.2, and algebraic intensity $i_1(t)$ of the current supplied to the winding at the two winding ends, also called *terminals*. Line density of the surface currents is determined by (8.4):

$$J_S(\theta) = N'_S(\theta) \cdot i_1 = (N'_{S\,max} \cdot i_1) \cos \theta \qquad (8.4)$$

If the maximum line current density is denoted by

$$J_{S0} = N'_{S\,max} \cdot i_1$$

one obtains

$$J_S(\theta) = J_{S0} \cos \theta \qquad (8.5)$$

Considering a small segment $d\theta$, the corresponding part of the circumference is $dl = R\,d\theta$, and the total current within this segment is

$$di = J_S(\theta)R\,d\theta$$

Electrical currents in axially placed conductors create magnetic field within the machine. By considering the boundary surface between the air gap and magnetic circuit made of iron (ferromagnetic), it can be noted that the magnetic flux entering ferromagnetic material from the air gap does not change its value; thus, the orthogonal components of magnetic induction B in the air (B_0) and the ferromagnetic material (B_{Fe}) are equal. Since permeability μ_{Fe} of the ferromagnetic material is considerably higher than permeability μ_0 of the air, it is justifiable to neglect the field H_{Fe} in the ferromagnetic material and consider that field H exists only in the air gap.

Question (8.1): In cases where current sheet density is zero, is it possible that the tangential component of the field H exists in the air next to the inner surface of the magnetic circuit of the stator?

Answer (8.1): It is necessary to consider magnetic field in the immediate vicinity of the surface separating the air gap and magnetic circuit of the stator. In the absence of electrical currents, the tangential component of the magnetic field in the air must be equal to the tangential component of the magnetic field in iron. Since permeability of iron is so high that intensity of the field H in iron can be neglected, the tangential component of the field H in iron is considered to be zero. Therefore, the tangential component of the magnetic field in the air is zero as well.

8.3 Components of Stator Magnetic Field

It is required to determine the components of the magnetic field H created in the air gap by the sheet of stator currents. The air gap is of cylindrical shape; therefore, it is convenient to adopt the cylindrical coordinate system. The unit vectors of this system, indicating the radial (r), axial (z), and tangential (θ) directions, are presented in Fig. 8.4. Axis (z) is directed toward the reader ($\odot$). For the purpose of denoting individual components of the magnetic field, magnetic induction, and induced electrical field in the air gap, the following rules are adopted:

- Components of the field originated by the stator currents are denoted by superscript "S" (H^S), while components of the field created by the rotor currents are denoted by superscript "R" (H^R).
- Radial components of the field are denoted by subscript "r" (H_r), tangential by subscript "θ" (H_θ), and axial by subscript "z" (H_z).

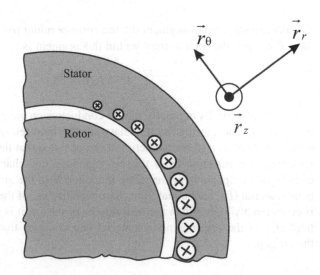

Fig. 8.4 Unit vectors of cylindrical coordinate system. Unit vectors r_r, r_z and r_θ determine the course and direction of the radial, axial, and tangential components of magnetic field

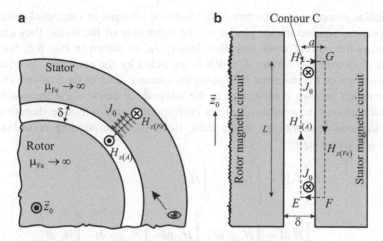

Fig. 8.5 Cross section (a) and longitudinal cross section (b) of a narrow rectangular contour C positioned along axis z. Width a of the contour EFGH is considerably smaller than its length L. Signs $\odot$ and $\otimes$ in the left-hand part of the figure indicate reference direction of the contour and do not indicate direction of the magnetic field. Reference directions of the magnetic field are indicated in Fig. 8.2.

Thus, the radial component of the magnetic field created by the stator winding is denoted by H_r^S, while the axial component of the magnetic field created by the rotor winding is denoted by H_z^R.

8.3.1   Axial Component of the Field

In electrical machines having magnetic circuits of cylindrical shape and with conductors positioned in parallel with the cylinder axis, that is, axis z of the cylindrical coordinate system, axial component of magnetic field is equal to zero. This statement can be confirmed by considering Fig. 8.5.

Figure 8.5 shows the front and side views of closed rectangular contour C. It has the length L and the width a. The two longer sides of the contour are positioned along the axis z. The longer sides of the contour are denoted by $\odot$ and $\otimes$ in the cross section of the machine, shown on the left side in Fig. 8.5. One of the two sides ($\otimes$) passes through magnetic circuit of the stator which is made of iron. The axial component of magnetic field in iron is denoted by $H_{z(Fe)}$. The other side of the rectangular contour ($\odot$) passes through the air gap. The axial component of magnetic field in the air gap is denoted by $H_{z(A)}$.

In most general case, electrical machine may have electrical currents in all the three directions: radial, tangential, and axial. Tangential current would be represented by a circular path on the left side of the figure, while on the right side of the figure, their direction is from the observer into the drawing. Assuming that the machine comprises conductors with electrical currents in tangential direction and that these conductors are placed on the inner side of the stator, they can be modeled as the current sheet with line density J_θ, as shown in Fig. 8.5. Surface integral of J_θ over the surface S which is encircled by the contour C is equal to the line integral of the magnetic field along the contour. Each of the four sides of the contour makes its own contribution to the integral. In cases where the course of circulation around the contour does not correspond to the reference direction for radial and axial components of the field, then the corresponding contributions assume a negative sign:

$$\int_S \vec{J}\,d\vec{S} = \int_S J_\theta\,dS = \int_C \vec{H}\,d\vec{l}$$

$$\int_C \vec{H}\,d\vec{l} = \int_E^H H_{z(A)}\,dl + \int_H^G H_r\,dl - \int_G^F H_{z(Fe)}\,dl - \int_F^E H_r\,dl.$$

It is assumed that the contour is very long and narrow; hence, $a \ll L$. Longer sides are positioned close to the surface which separates the air gap from the stator magnetic circuit. The other two sides of the rectangle are much shorter. Therefore, the integral of the radial component of the magnetic field along sides FE and HG can be neglected; thus, the line integral along contour C is reduced to the integral along sides GF and EH:

$$\int_S J_\theta\,dS = -\int_G^F H_{z(Fe)}\,dl + \int_E^H H_{z(A)}dl.$$

Since permeability of iron is very high and the magnetic induction in iron B_{Fe} has finite value, the magnetic field strength $H_{Fe} = B_{Fe}/\mu_{Fe}$ in iron is very low. It can be considered equal to zero. Therefore, line integral along the contour shown in Fig. 8.5 is reduced to the integral of magnetic field along side EH:

$$\int_S J_\theta\,dS = \int_E^H H_{z(A)}\,dl \qquad (8.6)$$

Electrical currents in rotating electrical machines exist in insulated copper conductors. These conductors are placed in slots, carved on the inner surface of the stator magnetic circuit and along the rotor cylinder. The slots extend axially, they are parallel to the axis of the cylinder and also parallel to z axis. Hence, in cylindrical electrical machines, only z component of electrical currents can exist. Thus, the density of tangential currents J_θ is equal to zero. Therefore, the value of the integral of the axial component of the magnetic field along the side EH is also zero. Under assumption that $H_{z(A)}$ remains constant, $J_\theta = 0$ proves that $H_{z(A)} = 0$. Yet, there is no proof at this point that $H_{z(A)}$ remains constant along the machine length.

The contour C can be chosen in such way that its length L is considerably smaller than the overall axial length of the machine. In such case, there are no significant variations of the field $H_{z(A)}$ along the side EH, and the expression (8.6) assumes the value:

$$\int_S J_\theta \, dS = 0 = \int_E^H H_{z(A)} \, dl \approx H_{z(A)}L \tag{8.7}$$

which leads to conclusion that $H_{z(A)} = 0$. There is also another way to prove that the axial component of the field is equal to zero. Statement $H_{z(A)} = 0$ can be proved even if the contour length L is longer and becomes comparable to the axial length of the machine. The integral in (8.6) is equal to zero for an arbitrary choice of points H and E, and this is possible only if the axial component of magnetic field in the air gap $H_{z(A)}$ is equal to zero at all points along the axis z. This statement can be supported by the following consideration.

The contour C (EHGF) can be slightly extended by moving the side FE into position F_1E_1, wherein the points E and E_1 are very close. In such way, the contour C_1 is formed, defined by the points E_1HGF_1. In the absence of electrical currents in tangential direction (J_θ), the line integral of the field H along the contour C is equal to zero. The same holds for the contour C_1. For the reasons given above, the line integral along the contour C reduces to the integral along the side EH, while the line integral along the contour C_1 reduces to the integral along the side E_1H. Both integrals are equal to zero. Therefore, the line integral of the field H along the side EE_1 has to be equal to zero as well. The point E_1 can be placed next to the point E, so that the changes in the field strength H from E to E_1 become negligible. At this point, the line integral along the side EE_1 reduces to the product of the path length EE_1 and the field strength $H_{z(A)}$ at the point E, leading to $H_{z(A)} = 0$. This statement applies for arbitrary choice of points E and E_1. This proves that the axial component of the magnetic field in the air gap is equal to zero. Notice that all the above considerations start with the assumption that the machine cylinder is very long and that the field changes at the ends of the cylinder are negligible.

Magnetic circuit of electrical machines has the stator hollow cylinder and the rotor cylinder, both made of iron sheets. At both ends of the cylinder, the air gap opens toward the outer space. Considering the windings, each turn has two diametrical conductors. The ends of these conductors have to be tied by the end turns, denoted by D in Fig. 5.6. The end turns are found at both the front and the rare side of the cylinder. Electrical current in end turns extends in tangential direction. Due to the air gap opening toward the outer space and due to end turns, there is local dispersion of the flux at both ends of the machine in the vicinity of the air gap opening. Therefore, a relatively small z component of the magnetic field may be established toward the ends of cylindrical machines. Above-described *end effects* and parasitic axial field are neglected throughout this book. It should be mentioned that the above-mentioned effects should be considered in the analysis of machines with an unusually small axial length L and with diameter $2R$ considerably larger than the axial length L.

8.3.2 Tangential Component of the Field

The analysis carried out in this subsection determines the tangential component of the magnetic field H_θ^S in the air gap, produced by electrical currents in the stator winding. Tangential component of the field is calculated in the air gap, next to the inner side of the stator. Namely, the observed region is close to the boundary surface separating the magnetic circuit of the stator and the air gap.

Boundary conditions for the magnetic field at the surface separating two different media are studied by electromagnetic. In the case with no electrical currents over the surface, tangential components of vector H are equal at both sides of the surface. By considering the surface separating the stator magnetic circuit and the air gap (Fig. 8.6), it can be stated that tangential component of the magnetic field in iron is equal to zero ($H_{Fe} = B_{Fe}/\mu_{Fe}$). This is due to magnetic induction B_{Fe} in iron being finite and permeability μ_{Fe} of iron being very high. Therefore, it is possible to conclude that the tangential component of magnetic field H_θ^S in the air, next to the inner stator surface, is equal to zero in all cases where the stator winding does not carry electrical currents.

In the example considered above, the magnetic field in the air gap is analyzed as a consequence of the stator currents. Besides these currents, the machine can also have electrical currents in rotor conductors. With the stator currents equal to zero ($J_S = 0$), the field H_θ^S against the inner stator surface is equal to zero, notwithstanding the rotor currents. Hence, the rotor currents do not have any influence on tangential component of the magnetic field in the air gap region next to the stator surface. Moreover, tangential components of magnetic field in the air gap are not the same against the inner surface of the stator and against the outer surface of the rotor.

It is known that in close vicinity of a plane which carries a uniform sheet of surface currents with line density σ, there is magnetic field of the strength $H = \sigma/2$, wherein the field is parallel to the plane and orthogonal to the current, while the plane resides in air or vacuum. In cases where the surface currents exist in the plane separating high-permeability ferromagnetic material and the air, the field in the air is $H = \sigma$. This statement can be proved with the help of Fig. 8.6. The figure shows the plane separating a space filled with air (left) from a space filled by ferromagnetic material (right). The boundary plane carries a uniform current sheet of line density σ. Closed contour EFGH is of the length L and width a, considerably smaller than the length. Line integral of the magnetic field along the closed rectangular contour is equal to $L\sigma$, and it sums all the currents passing through the contour. Since magnetic field in the ferromagnetic material is very low, the integral along side FG can be neglected. Because $a \ll L$, integral of the magnetic field along the closed contour is reduced to the product of side HE length and the field strength H_A. Since $L\sigma = LH_A$, it is shown that the magnetic field strength in the air is equal to the line current density σ. In the same way, it can be concluded that tangential component of the magnetic field H_θ in the air gap of a cylindrical machine in the vicinity of the inner side of the stator will be equal to the line density of stator currents, while the field H_θ

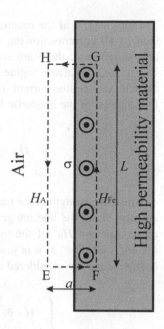

Fig. 8.6 Magnetic field strength in the vicinity of the boundary surface between the ferromagnetic material and air is equal to the line density of the surface currents

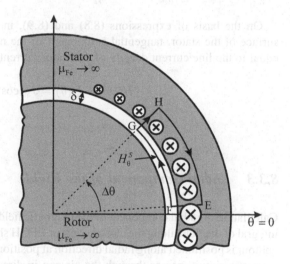

Fig. 8.7 Calculation of the tangential component of magnetic field in the air gap region next to the boundary surface between the air gap and the stator magnetic circuit

close to the rotor will be equal to the line density of rotor currents. First of the two statements will be proved by using Fig. 8.6.

It is of interest to consider the closed contour EFGH, having very short sides EF and GH, with circular arcs FG and HE having roughly the same lengths $R\Delta\theta$, where R is internal radius of the stator. Circular arc HE passes through ferromagnetic stator core, while circular arc FG passes through the air in the immediate vicinity of the stator inner surface.

Line integral of the magnetic field along closed contour EFGH is equal to the sum of all currents flowing through the surface leaning on the contour. In the considered case, there are surface currents of the stator with line density $J_S(\theta)$. If a relatively narrow segment is considered, such that $\Delta\theta \ll \pi$, it is justified to assume that the line current density $J_S(\theta)$ does not change over the arc FG, and the line integral of the magnetic field along the contour becomes

$$\int\limits_{EFGHE} \vec{H} \cdot d\vec{l} = \int\limits_{\theta_{FE}}^{\theta_{GH}} J_S(\theta)R\,d\theta \approx J_S(\theta) \cdot R \cdot \Delta\theta. \qquad (8.8)$$

Since the strength of the magnetic field in iron is very small and sides EF and GH are very short, the line integral along the closed contour reduces to the integral of the component H_θ^S of the magnetic field in the air along the arc FG. With the assumption $\Delta\theta \ll \pi$, it is justified to consider that the field strength H_θ^S does not change along the considered circular arc and that the integral is

$$\int\limits_{EFGHE} \vec{H} \cdot d\vec{l} = \int\limits_{\theta_{FE}}^{\theta_{GH}} H_\theta^S(\theta)R\,d\theta \approx H_\theta^S(\theta) \cdot R \cdot \Delta\theta. \qquad (8.9)$$

On the basis of expressions (8.8) and (8.9), in the region close to the inner surface of the stator, tangential component of the magnetic field in the air gap is equal to the line current density of the stator current sheet:

$$H_\theta^S(\theta) = J_S(\theta) = J_{S0} \cos\,\theta. \qquad (8.10)$$

8.3.3 Radial Component of the Field

Calculation of radial component of the magnetic field in the air gap relies on the line integral of the field along the closed contour EFGH shown in Fig. 8.8. Side EF of the contour is positioned along radial direction at position $\theta = 0$. It starts from the stator magnetic circuit, passes through the air gap in direction opposite to the reference direction of the radial component of the field (inside-out), and ends up in the rotor magnetic circuit. Side GH is positioned radially at $\theta = \theta_1$. It starts from the rotor magnetic circuit, passes through the air gap in the reference direction of the radial component of the field, and comes back into the stator magnetic circuit. The contour has two circular arcs, FG and HE. They have approximately equal length $R\theta_1$, and they pass through magnetic circuits of the rotor (FG) and stator (HE).

Due to a very high permeability of iron, the strength $H_{Fe} = B_{Fe}/\mu_{Fe}$ of the magnetic field is negligible in these segments of the contour which pass through

Fig. 8.8 Calculation of the
radial component of magnetic
field in the air gap

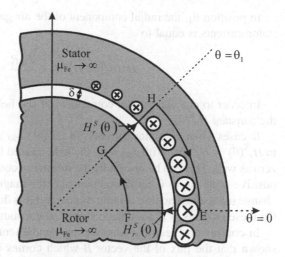

iron. Therefore, it can be considered that the magnetic field exists only along
segments EF and GH passing through the air gap. These segments are of length
δ, considerably smaller compared to the radius of the machine ($\delta \ll R$). It is thus
justified to assume that intensity of the radial component of the magnetic field along
sides EF and GH in the air gap does not change along this short path δ through the
air gap. At position $\theta = 0$, the field strength is $H_r^S(0)$, while at position $\theta = \theta_1$,
the field strength is $H_r^S(\theta_1)$. With these assumptions, line integral of the magnetic
field along the contour (circulation) becomes

$$\int_C \vec{H} \cdot d\vec{l} = +\delta \cdot H_r^S(\theta_1) - \delta \cdot H_r^S(0). \tag{8.11}$$

Negative sign in front of $H_r^S(0)$ in the preceding expression indicates that the
direction along the side EF of the contour is opposite to the reference direction for
the radial component of the magnetic field, as defined in the cylindrical coordinate
system.

Circulation of vector $\boldsymbol{H}$ along the closed contour EFGH is equal to the sum of all
currents passing through the surface leaning on the contour, that is, to the integral of
the surface currents of line density $J_S(\theta)$ between the limits $\theta = 0$ and $\theta = \theta_1$
(8.12). By comparing Fig. 8.8 to Figs. 8.2 and 8.3, it can be concluded that the
highest line density of the stator surface currents takes place at $\theta = 0$. The line
density of stator currents is determined by (8.1):

$$\int_0^{\theta_1} J_S(\theta) R \, d\theta = \int_0^{\theta_1} J_{S0} \cos \theta \cdot R \cdot d\theta = R \cdot J_{S0} \cdot \sin \theta_1 \tag{8.12}$$

In position θ_1, the radial component of the air gap magnetic field caused by the stator currents is equal to

$$H_r^S(\theta_1) = H_r^S(0) + \frac{J_{S0}R}{\delta} \sin \theta_1 \qquad (8.13)$$

In order to calculate radial component of the field, it is necessary to determine the constant $H_r^S(0)$.

In cases when the stator currents are absent ($J_{S0} = 0$), expression (8.13) reduces to $H_r^S(\theta) = H_r^S(0)$. With $J_{S0} = 0$, the field caused by the stator currents should be zero as well. This can be proved by the following consideration. If constant $H_r^S(0)$ is positive while $J_{S0} = 0$, radial component of the magnetic field in the air gap does not change along the machine circumference, and it is directed from rotor toward stator. On these grounds, it is possible to show that constant $H_r^S(0)$ has to be equal to zero.

In courses on Electrical Engineering Fundamentals and Electromagnetics, it is shown that the flux of the vector $\boldsymbol{B}$ which comes out of a closed surface S must be equal to zero. An example of the closed surface S can be the one enveloping the rotor of an electrical machine. This surface has three parts, cylindrical surface passing through the air gap and the two flat, round parts at both machine ends, representing the bases of the cylinder. The flux of the vector $\boldsymbol{B}$ through the surface S is called *the output flux,* and it is calculated according to

$$\oint_S \vec{B} \cdot d\vec{S} = 0$$

Differential form of the preceding statement is

$$\operatorname{div} \vec{B} = 0,$$

and it represents one out of four Maxwell equations. Divergence is a spatial derivative of a vector which can be used for establishing the relation between the surface integral (2D) of the vector over a closed surface S and the space integral (3D) of the spatial derivative of the same vector within the domain encircled by the closed surface S. Therefore, the information on the divergence of vector $\boldsymbol{B}$ in domain V, encircled by surface S, can be used to calculate the output flux of the vector $\boldsymbol{B}$:

$$\oint_S \vec{B} \cdot d\vec{S} = \int_V \operatorname{div} \vec{B} \ dV.$$

As a consequence of div $\boldsymbol{B} = 0$, the surface integral of vector $\boldsymbol{B}$ over the close surface S is equal to zero:

$$\oint_S \vec{B} \cdot d\vec{S} = 0. \qquad (8.14)$$

The law given by (8.14) can be used to prove that the constant $H_r^S(0)$ equals zero. It is necessary to note a closed surface S of cylindrical form, enveloping the rotor in the way that the cylindrical part S_1 passes through the air gap while the two flat round parts (basis) stay in front and at the rare of the rotor.

Equation 8.7 shows that axial component of the magnetic field H_z in electrical machines is zero. Due to $B_z = \mu_0 H_z$ in the air, the same holds for the magnetic induction; hence, $B_z = 0$. As a consequence, the flux of the vector B through the front and rear basis of the closed cylindrical surface S is equal to zero. In accordance with the law (8.14), the flux through the cylindrical surface S_1 passing through the air gap must be equal to zero as well.

Relation $B = \mu_0 H$ connects the magnetic field strength H and the magnetic induction B in the air. Since the permeability μ_0 does not vary, flux of the vector H through the cylindrical surface S_1 residing in the air gap can be obtained by dividing the flux of vector B through the same surface by the permeability μ_0. Therefore, the flux of the vector H through the same surface must be equal to zero as well as the flux of the vector B. In the case when $J_{S0} = 0$ and $H_r^S(\theta) = H_r^S(0)$, the flux of the magnetic field H through the cylindrical surface S_1 is equal to $2\pi R L \, H_r^S(0)$, where R is the radius and L is the length of the machine, which completes the proof that constant $H_r^S(0)$ in (8.13) has to be equal to zero. Having proved that $H_r^S(0) = 0$, one can obtain the expression for the radial component of the magnetic field in the air gap.

In Fig. 8.8, position θ_1 of side GH of the contour EFGH is arbitrarily chosen. Therefore, all previous considerations are applicable at any position θ_1. Thus, it can be concluded that radial component of the magnetic field created in the air gap by the stator currents is equal to

$$H_r^S(\theta) = \frac{J_{S0}R}{\delta} \sin\theta, \qquad\qquad (8.15)$$

where the above expression defines the strength of the stator magnetic field H_r at the position θ within the air gap. The expression is applicable in cases where only the stator windings carry electrical currents and when these currents can be represented by surface currents with sinusoidal distribution around the machine circumference.

Question (8.2): Consider a closed surface which partially passes through the air and partially through ferromagnetic material such as iron. Is it possible to prove that the output flux of the field H though this closed surface is equal to zero? Is it possible to prove that the output flux of induction B through this closed surface is equal to zero?

Answer (8.2): According to (8.14), the output flux of the vector of magnetic induction through any closed surface S is equal to zero. This law is applicable in homogeneous media, where permeability does not change, but also in the media with variable permeability, as well as the media comprising parts of different permeability. Therefore, the output flux of magnetic induction is also equal to zero through the closed surface passing through the air in one part and through iron in the other part.

Equation 8.14 deals with magnetic induction B. It is applicable to magnetic field H only in cases where the permeability $\mu = B/H$ does not change over the integration domain. Therefore, if surface S passes through media of different permeability, it cannot be stated that output flux of the vector H through a closed surface is equal to zero.

8.4   Review of Stator Magnetic Field

The subject of the preceding analysis is cylindrical electrical machine of the length L, with the rotor outer diameter $2R$. The rotor is placed in hollow, cylindrical stator magnetic circuit so that an air gap $\delta \ll R$ exists between the stator and rotor cores.

The magnetic field is created in the air gap by electrical currents in the stator winding. The stator windings have a sinusoidal distribution of their conductors along the circumference. Therefore, the stator currents can be replaced by a sheet of surface currents extending in axial direction, with a sinusoidal change of their density around the machine circumference. This current sheet is located on the inner side of the stator magnetic circuit, facing the air gap. The line density of the surface currents (8.4) is determined by the conductor density (8.1) and the electrical current i_1 in stator winding. As a consequence of the stator magnetomotive force, magnetic field is established in the air gap, with its axial, radial, and tangential components discussed above. Due to a very high permeability of iron, it is correct to assume that the magnetic field strength in iron is negligible.

In cylindrical coordinate system, the axial component of the field H in the air gap is equal to zero, while the tangential and radial components are given by expressions (8.17) and (8.18):

$$H_z^S(\theta) = 0 \tag{8.16}$$

$$H_\theta^S(\theta) = J_{S0}R \cdot \cos\theta \tag{8.17}$$

$$H_r^S(\theta) = \frac{J_{S0}R}{\delta} \sin\theta \tag{8.18}$$

Since $\delta \ll R$, the radial component is considerably higher compared to the tangential component. Difference in intensities between the radial and tangential components is up to two orders of magnitude.

Question (8.3): Consider a cylindrical machine of known dimensions having the stator winding with only one turn made out of conductors A1 and A2. Conductor A1 carries electrical current in direction away from the reader ($\otimes$), and its position is at $\theta = 0$. The other conductor (A2) is at position $\theta = \pi$, and it carries current in direction toward the reader ($\odot$). Conductors A1 and A2 are connected in series, and they are fed from a current source of constant current I_0. Determine the radial component of magnetic field $H_r^S(\theta)$ in an arbitrary position θ. If the rotor revolves, what is the form of the electromotive force that would be induced in a single rotor conductor axially positioned on the surface of the rotor cylinder? What is the form of this electromotive force in cases where radial component of the stator field changes according to 8.18?

Answer (8.3): It is necessary to envisage a contour which passes through both stator and rotor magnetic circuits. This contour has to pass through the turn A1–A2, encircling one of the conductors. Such contour is passing across the air gap two times, both passages extending in radial direction. The circulation of the vector H (i.e., the line integral of H around the closed contour) is equal to the current strength I_0. Thus, the radial component of the magnetic field in the air gap is $H_m = I_0/(2\delta)$. Direction of the radial field depends on the position along the circumference. Along the first half of the circumference, starting from the conductor A1 and moving clockwise toward the conductor A2, direction of the magnetic field is from the stator toward the rotor, while in the remaining half of the circumference, direction of the field is from the rotor toward the stator. Therefore, variation of the magnetic field in the air gap can be described by the function $H_r^S(\theta) = H_m \operatorname{sgn}(\sin \theta)$. In the case when the rotor revolves at a speed Ω, position of the rotor conductor changes as $\theta = \theta_0 + \Omega t$, where θ_0 denotes the position of the rotor conductor at $t = 0$. The electromotive force induced in the conductor is $e = LvB$, where L is the length of the conductor and $v = R\Omega$ is the peripheral velocity, while $B = \mu_0 H_r^S$ is algebraic intensity of the vector B around the conductor. Therefore, the change of the electromotive force is determined by the function $H_r^S(\theta) = H_r^S(\theta_0 + \Omega t)$. In the example given above, the electromotive force would change as $\operatorname{sgn}(\sin(\theta_0 + \Omega t))$. In cases where the field $H_r^S(\theta)$ changes in a sinusoidal manner, the electromotive force induced in rotor conductors would be sinusoidal as well.

8.5 Representing Magnetic Field by Vector

The subject of the previous analysis was the magnetic field created by the stator winding. Figure 8.10 shows the lines of the radial field. The stator conductors are not shown in this figure, neither is the detailed representation of sinusoidally distributed sheet of stator currents. Instead, direction of electrical currents and position of the maximum current density are denoted by placing symbols $\otimes$ and $\odot$. The field lines shown in the figure correspond to sinusoidal change of the magnetic field H and magnetic induction B along the machine circumference, in accordance with (8.18). The regions on the inner surface of the stator magnetic circuit with the highest density of the fields B and H are denoted as the north (N) and south (S) magnetic pole. In the region of the north pole of the stator magnetic circuit, the field lines come out of the stator core and enter the air gap, while in the zone of the south magnetic pole, the field lines from the air gap enter the ferromagnetic core (Fig. 8.9).

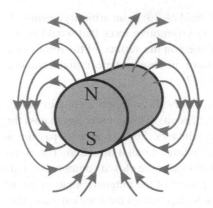

Fig. 8.9 Closed cylindrical surface S envelops the rotor. The lines of the magnetic field come out of the rotor (surface S) in the region called north magnetic pole of the rotor, and they reenter in the region called south magnetic pole

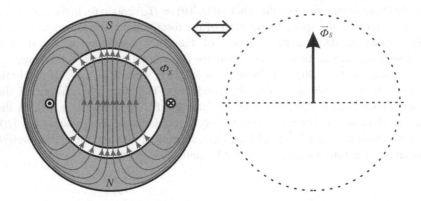

Fig. 8.10 Convention of vector representation of the magnetic field and flux

The previous analysis and Fig. 8.10 represent the magnetic field produced by only one stator winding. An electrical machine has a number of stator and rotor windings. The resulting magnetic field comes as a consequence of several magneto-motive forces. The magnetomotive force of each winding creates the field represented by the field lines similar to those in Fig. 8.10. An effort of presenting several such fields in a single drawing would be rather difficult to follow, let alone getting useful in making conclusions and design decisions.

In further analyses, the magnetic field produced by single winding can be represented in a concise way by introducing the *flux vector* of the winding. Magnetic flux is an integral of the vector B over the given surface S. The result of such integration is a scalar. Yet, the flux in an electrical machine is tied to the normal n on the surface S, and it depends on spatially oriented field of B. Therefore, the flux is also called *directed scalar*. Considering the magnetic field created by a single turn, it is possible to calculate the flux as a scalar quantity and to define the *flux vector* by associating the course and direction to the scalar value. In Sect. 4.4, the flux in a single turn is represented by the flux vector, wherein the spatial orientation and reference direction are determined from the normal to the surface S defined by the contour C, made out by the single-turn conductors.

In most cases, a winding consists of a number of turns connected in series. All the turns may not share the same spatial orientation. Therefore, the normals on the surfaces, leaning on individual turns, may not be collinear. Hence, there is a need to clarify the course and direction of the winding flux. In cases where the winding is concentrated, all the conductors reside on only two diametrical slots, and all the turns have the same orientation. Therefore, their normals coincide and define the spatial orientation of the winding flux vector. Yet, the same approach cannot be applied in cases where the winding conductors and its turns are distributed along the machine circumference.

The flux shown in Fig. 8.10 is created by the currents in conductors that are sinusoidally distributed along the inner surface of the stator. A pair of diametrical conductors constitutes one contour, that is, a single turn. The normals on individual turns are obviously not collinear. Yet, the winding flux can be represented by a vector[1] collinear with the *winding axis*. Determination of the windings axes is

---

[1] Interpretation of magnetic flux as a vector can be understood as a convention and a very suitable engineering tool in the analysis of complex electromagnetic processes taking place in electrical machines. Nevertheless, magnetic flux is a scalar by definition. It may be called *directed scalar*, as it is closely related to the spatial orientation of relevant turn or winding, and it depends on the course and direction of the vector of magnetic induction. Magnetic flux Ψ can be compared to the strength I of spatially distributed electrical currents, which describe the phenomenon of moving electrical charges. The following illustration shows spatial currents passing through the surface S which is leaning on the contour c:

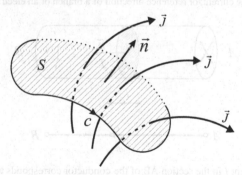

The vector of current density J gives direction of the current I through the contour. Its integral over surface S (the flux of spatial currents) gives the current intensity I. In the case when the vector of spatial currents J is of the same orientation at all points of surface S (homogeneous), the current intensity can be determined by the following expression:

$$I = \int_S \vec{J} \cdot \mathrm{d}\vec{S} = \int_S J \cos(\vec{J}, \vec{n})\, \mathrm{d}S = J \cos(\vec{J}, \vec{n})\, S$$

described in Sect. 5.5. A more elaborated definition of the winding axis in cases with spatially distributed conductors is presented further on.

The flux vector is determined by its course, direction, and amplitude. The vector presented in the right-hand side of Fig. 8.10 represents the field of magnetic induction B, distributed sinusoidally over the air gap and shown in the left-hand side of the figure. Direction of the flux is determined by the course of the field lines, which start from the north magnetic pole (N) of the stator, pass through the air gap, enter into the rotor magnetic circuit, then pass for the second time through the air gap, and enter into the stator magnetic circuit in the region of the south pole (S). Direction of the flux is determined by direction of the magnetic field H and induction B.

In linear ferromagnetic and in the air gap, the vectors B and H have the same course and direction due to $B = \mu H$. Spatial distribution of the field lines representing magnetic induction B can be represented by the flux vector $\boldsymbol{\Phi}$. The flux amplitude Φ and the magnetomotive force F are related by $F = R_\mu \Phi$, where R_μ is magnetic resistance encountered along the flux path, that is, magnetic resistance of the magnetic circuit. It is of interest to notice that the magnetomotive force F can be represented by vector F, which represents the spatial distribution of the field H. Due to $B = \mu H$, such vector is collinear with $\boldsymbol{\Phi}$, while its amplitude is $F = R_\mu \Phi$, and it is equal to the circulation of the vector H along the flux path.

The amplitude of the flux vector $\boldsymbol{\Phi}_S$ is the surface integral of the vector B over the surface leaning on one turn of the stator winding. It is possible to define the

---

For the line conductor shown in the next figure, the unit vector of normal n on surface S represents *reference direction* of the current, or reference direction of a branch of an electrical circuit:

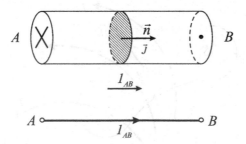

The sign of the current I in the section AB of the conductor corresponds to the direction of the vector J. For this reason, the current intensity I can be called *directed scalar*. By replacing the spatial current density J and the current intensity (strength) I by the magnetic induction B and magnetic flux Φ, the previous considerations can be used to establish the magnetic flux as a directed scalar. Flux vector through a contour c has direction of the normal on surface S and its algebraic intensity, determined by the integral of magnetic induction over the surface S.

vector of the total winding flux Ψ as the sum of flux vectors representing the flux in individual turns.[2]

Question (8.4): Consider Fig. 8.10, where symbols $\otimes$ and $\odot$ denote direction of current in conductors of the stator winding. There are $2N_T$ conductors, sinusoidally distributed along the machine circumference, all of them carrying electrical current I. Derive the expression for the maximum value of the radial component H_r^S of the air gap field which is achieved in the regions of the north and south magnetic poles (use the previously obtained expressions and the relation between the maximum line density of conductors N_{Smax} and the total number of conductors, $N_T = 2 RN_{Smax}$, $H = N_T I/(2\delta)$). Determine the amplitude of the stator magnetomotive force F.

Answer (8.4): It is necessary to determine the line integral of the field H along the closed contour starting from the north pole of the stator, going vertically toward the south magnetic pole, and closing through the stator magnetic circuit. Circulation of the vector H is $N_T I = 2\delta H$. Intensity of the magnetic field is $H_{max} = N_T I/(2\delta)$. Magnetomotive force F is equal to the circulation of the vector H, $F = N_T I$.

Question (8.5): Assume now that the number of stator conductors does not change and that stator current is the same, but the conductors are grouped at the places designated by $\otimes$ and $\odot$ in Fig. 8.10. Instead of being distributed, the conductors are concentrated in diametrical slots. Such winding is called *concentrated winding*. What is, in this case, the value of the line integral of the magnetic field H? Are there any changes in the maximum intensity of the field H below the north and south poles? What is the amplitude of the stator magnetomotive force F?

Answer (8.5): The magnetic field strength H_{max} and the magnetomotive force F are equal as in the preceding case, $H = N_T I/(2\delta)$, $F = N_T I$.

Question (8.6): Compare the field distribution $H(\theta)$ for concentrated and distributed winding.

Answer (8.6): On the basis of the previous expressions, magnetic field of the winding with sinusoidally distributed conductors has a sinusoidal distribution of the magnetic field in the air gap. In the case when the conductors are concentrated, radial component of magnetic field $H_r^S(\theta) = H_m \, \mathrm{sgn}(\sin \theta)$ has a constant amplitude along the circumference, and its direction is positive over one half and negative over the other half of the circumference. In both cases, maximum intensity of the field is $H = N_T I/(2\delta)$.

Question (8.7): Determine the flux through a contour made of two conductors denoted by $\otimes$ and $\odot$ in Fig. 8.10 in the case when the winding is concentrated and has N_T conductors. All conductors of the considered winding directed toward the

---

[2] Total flux Ψ of the stator winding with N turns, with sinusoidal distribution of conductors along circumference of the stator, and with flux Φ_S in one of the turns is not equal to $N\Phi_S$ because the fluxes of individual turns are not equal. Flux Φ_S in a single turn (contour) is function of position θ.

reader are in position denoted by $\odot$. The remaining conductors of the opposite direction are in position denoted by $\otimes$.

Answer (8.7): It is necessary to note that the magnetic field strength in the air gap is $H = +N_T I/(2\delta)$ over the interval $\theta \in [0 \ldots \pi]$ and $H = -N_T I/(2\delta)$ over interval $\theta \in [\pi \ldots 2\pi]$. The flux through the contour is obtained by calculating the integral of the magnetic induction B over the surface leaning on the contour. Since the surface integral of magnetic induction over a closed surface is equal to zero (div $B = 0$), the surface integral of B through all the surfaces leaning on the same contour is the same. Therefore, there is a possibility of selecting the proper surface that would facilitate the calculation. For the surface in the air gap, the expression for magnetic induction B is known. Over the interval $[0 \ldots \pi]$, the magnetic induction in the air gap has radial direction and intensity $B = +\mu_0 N_T I/(2\delta)$. The surface leaning on the contour can be specified by the semicircular banded rectangle which leans on conductor $\otimes$, passes through the air gap over the arc interval $[0 \ldots \pi]$, and leans on conductor $\odot$, which is positioned at $\theta = \pi$ in Fig. 8.10. The considered surface has the length L, width πR, and surface area $S = L\pi R$. In all parts, the vector of magnetic induction is vertical to the surface; thus, the flux through the surface, that is, the flux through the contour, is equal to $\Phi = BS = \mu_0 \, \pi LR \, N_T \, I \, /(2\delta)$.

Question (8.8): Determine the flux of a contour consisting of two conductors denoted by $\otimes$ and $\odot$ in Fig. 8.8 in the case when the winding has a sinusoidal distribution of conductors.

Answer (8.8): It is necessary to note that in the zones of magnetic poles, at positions $\theta = \pi/2$ and $\theta = 3\pi/2$, the magnetic induction in the air gap is equal to the one in the preceding case ($B_{max} = +\mu_0 N_T I/(2\delta)$), but the field changes along the circumference. As in the preceding case of Question 8.7, the flux through the contour can be obtained by calculating the surface integral of the magnetic induction over the semicircular banded rectangle of the length L and width πR, which passes through the air gap and leans on conductors $\otimes$ and $\odot$. The area of the considered surface is $S = L\pi R$. The flux cannot be calculated as $B_{max}S$, as the magnetic induction exhibits sinusoidal changes over the surface. The flux through the contour is equal to the product $B_{av}S$, where B_{av} is the average value of the magnetic induction in the air gap over the interval $\theta \in [0 \ldots \pi]$. It is well known that the function $\sin(\theta)$ has an average value of $2/\pi$ on the interval $\theta \in [0 \ldots \pi]$. Therefore, $B_{av} = 2/\pi B_{max}$. The flux through the contour is $\Phi = B_{av}S = \mu_0 \, LR \, N_T \, I \, /\delta$.

Question (8.9): By using the results obtained in previous two questions, specify how do the magnetomotive force of the winding and the flux in one contour change by converting a concentrated winding into winding with sinusoidal distribution of conductors. Are there any reasons in favor of using distributed windings?

Answer (8.9): If the two windings have the same current in their conductors and the same number of conductors, the maximum strength H_{max} of the magnetic field in the air gap of the machine and the magnetomotive force $F = 2\delta H_{max}$ are the same.

For the concentrated winding, the field strength retains the same value along the circumference, while for the distributed winding, the field varies in accordance with $\sin(\theta)$. For this reason, the flux in one turn is smaller for the distributed winding. The ratio of the fluxes in one turn obtained in two considered cases is $2/\pi$. Even though the flux of the distributed winding is smaller, there are reasons in favor of using the windings with sinusoidally distributed conductors. It has to do with the harmonics of the induced electromotive force. With sinusoidal distribution of the conductors along the circumference, the electromotive force induced in the winding is sinusoidal, unspoiled with harmonics, and with no distortion even in cases where the change of the magnetic field along the circumference is non-sinusoidal and when the function $B(\theta)$ comprises significant amount of harmonics. In the later case, a concentrated winding will have an electromotive force waveform which resembles $B(\theta)$. Therefore, a winding with sinusoidal distribution of conductors has the properties of a filter. A proof of this statement will be presented in Chap. 10.

8.6 Components of Rotor Magnetic Field

In addition to stator windings, electrical machines usually have windings on the rotor as well. Rotor could have several windings. The following analysis will consider magnetic field produced by one rotor winding. Conductors of the considered winding are placed on the surface of the rotor magnetic circuit in the close vicinity of the air gap, in the way shown in Fig. 8.11. In this figure, the conductors directed away from the reader are denoted by $\otimes$, while the conductors directed toward the reader are denoted by $\odot$. One pair of diametrically positioned conductors creates one turn of the rotor winding. These turns are connected in series and constitute a winding.

The rotor conductors are positioned along the rotor circumference in the manner that their line density varies as a sinusoidal function of the angular displacement θ. The function $N'_R(\theta)$ determines the number of conductors per unit length $R \cdot \Delta\theta$. The argument of the function is the angle θ, measured from the reference axis of the stator, denoted by (A) in Fig. 8.11, to the place on the rotor circumference where the conductor density $N'_R(\theta)$ is observed. The angle θ_m is also marked in the figure, and it defines the rotor displacement from the reference axis of the stator. When the rotor revolves at a constant speed Ω_m, the rotor position changes as $\theta_m = \theta_0 + \Omega_m t$, where θ_0 is the initial position. The reference axis of the rotor is denoted by (B). On the rotor reference axis, the angle θ is equal to θ_m. An arbitrary position (C) is shifted by $\theta - \theta_m$ with respect to the rotor reference axis. Since the highest line density of the rotor conductors $N'_{R\max}$ is at position $\theta = \theta_m$, the sinusoidal distribution of conductors can be described by function

$$N'_R(\theta) = N'_{R\max} \cdot \cos(\theta - \theta_m). \tag{8.19}$$

Fig. 8.11 Rotor current
sheet is shifted with respect
to the stator by θ_m. Maximum
density of the rotor
conductors is at position
$\theta = \theta_m$

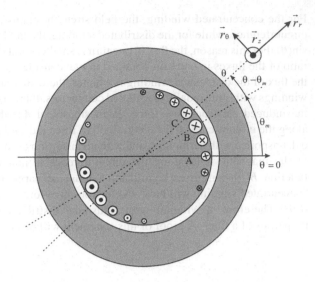

If the rotor winding carries electrical current i_2, the magnitude of sinusoidally distributed sheet of the rotor currents is $J_{R0} = N'_{R\max} i_2$.

In the case when a constant current $i_2 = I_2$ exists in the rotor conductors, the sheet of the rotor currents will create magnetic field in the air gap that would not move with respect to the rotor. The spatial orientation of such field is determined by the rotor position. By analogy with the stator field shown in Fig. 8.10, the north magnetic pole of the rotor is at position $\theta = \theta_m + \pi/2$, the radial component of the rotor field at $\theta = \theta_m$ is equal to zero, and the south magnetic pole of the rotor is at $\theta = \theta_m - \pi/2$. If the rotor does not move, position of the rotor magnetic poles does not change. When the rotor revolves, the field created by the DC current in the rotor conductors rotates with respect to the stator. The speed of the field rotation is equal to the rotor speed. In this case, position of the north magnetic pole of the rotor is $\theta = \theta_m + \pi/2 = \theta_0 + \Omega_m t + \pi/2$, where θ_0 is the rotor position at $t = 0$.

The line density of the rotor currents is given by function

$$J_R(\theta) = N'_R(\theta) \cdot i_2 = (N'_{R\max} \cdot i_2)\cos(\theta - \theta_m) = J_{R0}\cos(\theta - \theta_m) \qquad (8.20)$$

where $J_{R0} = N'_{R\max} i_2$ denotes the maximum line density of the rotor currents.

The components of the air gap magnetic field created by distributed stator winding have been analyzed in Sect. 8.3. In a like manner, it is necessary to determine the axial, tangential, and radial component of the magnetic field created in the air gap by distributed rotor winding. The air gap is cylindrical in shape; thus, it is convenient to adopt the unit vectors of the cylindrical coordinate system, the same system used in calculating the stator field. The axis (z) is oriented toward the reader ($\odot$), while the radial and tangential directions in position θ are shown in Fig. 8.11. On the basis of previously adopted notation rules, the axial, tangential, and radial components created by the rotor currents are denoted by H_z^R, H_θ^R, and H_r^R.

Question (8.10): Conductors of the stator and rotor are placed in close vicinity of the air gap. What are the negative effects of positioning the rotor conductors deeper in the rotor magnetic circuit, further away from the air gap?

Answer (8.10): The lines of the magnetic field of a single conductor placed deeper into the rotor magnetic circuit would close through the ferromagnetic material, where magnetic resistance is lower, instead of passing through the air gap and encircling the stator conductors. In cases where the rotor conductor is placed deep into the iron magnetic circuit, far away from the air gap, the rotor magnetic field and flux exist mainly in the rotor magnetic circuit and they do not extend neither to the air gap nor to the stator winding. For this reason, there is significant reduction of magnetic coupling between the rotor and stator windings. In such cases, most of the rotor flux is the leakage flux, the part of the rotor flux which does not encircle the stator windings. With the electromechanical conversion process being based on the magnetic coupling, an increased rotor leakage greatly reduces the electromagnetic torque and the conversion power. On the other hand, the rotor leakage is reduced by placing the rotor conductors in rotor slots, next to the air gap. Magnetic field of such conductors passes through the air gap and encircles conductors of the stator, contributing to the magnetic coupling between stator and rotor windings.

8.6.1 Axial Component of the Rotor Field

It is proved in Sect. 8.3 that the axial component of the magnetic field is equal to zero in cylindrical machines with axially placed conductors. Since electrical currents exist in the conductors placed along z axis of the cylindrical coordinate system, there are no currents in tangential direction. As a consequence, the axial component of the magnetic field in the air gap is equal to zero. The analysis of circulation of the field along the contour shown in Fig. 8.5 shows that the axial component of the field in the air gap is equal to zero, notwithstanding the stator and rotor currents.

8.6.2 Tangential Component of the Rotor Field

Tangential component of the rotor magnetic field H_θ^R is calculated in the air gap, next to the rotor magnetic circuit. The point of interest is in the air, and it resides on the boundary surface separating the rotor magnetic circuit and the air gap.

The line integral of the magnetic field along the contour shown in Fig. 8.6 helps calculating the magnetic field in the vicinity of the boundary surface between the ferromagnetic material and the air gap. The tangential component of the field is determined by the line density of the surface currents in the boundary plane. Conclusions drawn from Fig. 8.6 can be applied to determining the tangential

Fig. 8.12 Calculation of
the tangential component
of the magnetic field in the air
gap due to the rotor currents,
next to the rotor surface

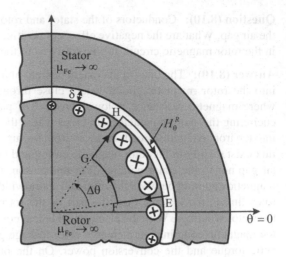

field caused by the rotor currents. The field strength H_θ^R is determined by the line
current density $J_R(\theta)$ of the current sheet representing the rotor currents. This
statement will be proved by using the example presented in Fig. 8.12.

One should consider closed contour EFGH whose radial sides EF and GH are
very short, while circular arcs FG and HE have approximately the same length $R\Delta\theta$,
where R is diameter of the rotor. Circular arc FG passes through the iron part of the
magnetic circuit, while circular arc HE passes through the air next to the rotor
surface. Circulation of the vector of magnetic field along the closed contour EFGH
is equal to the sum of all the currents passing through the surface leaning on the
contour. In the considered case, there are rotor surface currents with line density
$J_R(\theta)$. With $\Delta\theta \ll \pi$, it is justified to consider that the line current density does not
change along the circular arc HE; thus, the line integral of the magnetic field along
the closed contour is equal to the product of $J_R(\theta)$ and the length of the arc HE:

$$\oint_{EFGHE} \vec{H} \cdot \mathrm{d}\vec{l} = \int_{\theta_{EF}}^{\theta_{GH}} J_R(\theta) R \, \mathrm{d}\theta \approx J_R(\theta) \cdot R \cdot \Delta\theta. \tag{8.21}$$

Since the sides EF and GH are very short, while the magnetic field in iron, along
the arc FG, is very low, the line integral along the closed contour is reduced to the
integral of the component H_θ^R of the magnetic field in the air gap along the circular
arc HE. With $\Delta\theta \ll \pi$, it is justified to assume that the field intensity H_θ^R does not
change along the considered arc, and the integral is reduced to

$$\oint_{EFGHE} \vec{H} \cdot \mathrm{d}\vec{l} = \int_H^E H_\theta^R(\theta)\mathrm{d}l = \int_{\theta_{GH}}^{\theta_{FE}} H_\theta^R(\theta) R(-\mathrm{d}\theta) \approx -H_\theta^R(0) \cdot R \cdot \Delta\theta \tag{8.22}$$

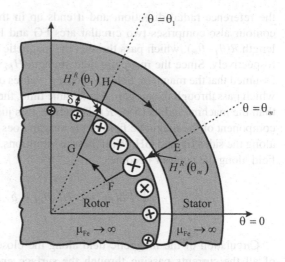

Fig. 8.13 Calculation of the radial component of the magnetic field caused by the rotor currents. Position θ_m corresponds to the rotor reference axis, while position θ_1 represents an arbitrary position where the radial component of the magnetic field is observed

Direction of the tangential component of magnetic field H_θ^R in the air gap, in close vicinity of the rotor surface, is opposite to the reference direction for tangential components in cylindrical coordinate system, and it is also opposite to the direction of the tangential component of the stator field. For this reason, there is a minus sign in (8.22).

On the basis of expressions (8.21) and (8.22), the component of magnetic field H_θ^R next to rotor surface is equal to the line density of the rotor currents:

$$H_\theta^R(\theta) = -J_R(\theta) = -J_{R0}\cos(\theta - \theta_m) \qquad (8.23)$$

8.6.3 Radial Component of the Rotor Field

Radial component of the magnetic field in the air gap due to the rotor currents can be determined by calculating the line integral along the closed contour EFGH shown in Fig. 8.13. The side EF of the contour extends in radial direction, at position $\theta = \theta_m$, in the region with the maximum density of the rotor conductors directed toward the reader. Position θ_m represents the angular displacement of the rotor, and it is measured with respect to the stator reference axis. The side EF of the contour starts from the stator magnetic circuit, it passes through the air gap in direction opposite to the reference direction, and it ends up in the rotor magnetic circuit. The side GH is directed radially at position $\theta = \theta_1$. It starts from the magnetic circuit of the rotor, passes through the air gap in direction aligned with the reference radial direction, and it ends up in the stator magnetic circuit. The contour also comprises two circular arcs FG and HE of approximately the same length $R(\theta_1 - \theta_m)$, which pass through the magnetic circuits of the rotor and stator, respectively. Since the magnetic field strength H_{Fe} in iron is very small, it can be assumed that the magnetic field has nonzero values only along the sides EF and GH, which pass through the air gap. At the same time, the air gap length is much smaller than the machine radius ($\delta \ll R$). Therefore, it is justified to assume that the radial component of the magnetic field in the air gap does not exhibit significant changes along the sides EF and GH. With these assumptions, the circulation of the magnetic field along the contour becomes

$$\oint_C \vec{H} \cdot d\vec{l} = +\delta \cdot H_r^R(\theta_1) - \delta \cdot H_r^R(\theta_m) \tag{8.24}$$

Circulation of the magnetic field along the closed contour is equal to the sum of all the currents passing through the surface encircled by the contour. In the case of the contour shown in Fig. 8.13, the sum of the currents passing through the contour is determined by calculating the integral of the line density $J_R(\theta)$ of surface currents from $\theta = \theta_m$ up to $\theta = \theta_1$:

$$\int_{\theta_m}^{\theta_1} J_R(\theta) R d\theta = \int_{\theta_m}^{\theta_1} J_{R0} \cos(\theta - \theta_m) \cdot R \cdot d\theta = R \cdot J_{R0} \cdot \sin(\theta_1 - \theta_m). \tag{8.25}$$

At position θ_1, the radial component of the magnetic field in the air gap caused by the rotor currents is

$$H_r^R(\theta_1) = H_r^R(\theta_m) + \frac{J_{R0}R}{\delta} \sin(\theta_1 - \theta_m) \tag{8.26}$$

For the purpose of deriving the radial component $H_r^R(\theta_1)$, it is necessary to determine the constant $H_r^R(\theta_m)$. In Sect. 8.3, where the calculation of the radial component of the stator magnetic field is carried out, it is shown that the average value of the radial component $H(\theta)$ in the air gap must be equal to zero. The proof was based on the fact that the field of the vector of magnetic induction B cannot have a nonzero flux through a closed surface, such as the cylinder enveloping the rotor. Namely, div $B = 0$. Under circumstances, the same holds for the flux of the vector H through the cylindrical surface passing through the air gap and enveloping the rotor. Therefore, the constant $H_r^R(\theta_m)$ in (8.26) must be equal to zero. Since the position θ_1 can be arbitrarily chosen, the final expression for the radial component of the rotor magnetic field takes the form

$$H_r^R(\theta) = \frac{J_{R0}R}{\delta} \sin(\theta - \theta_m). \tag{8.27}$$

8.6.4 Survey of Components of the Rotor Magnetic Field

In the preceding section, the air gap magnetic field caused by the rotor currents is analyzed, assuming that the rotor winding has axially placed conductors, wherein the conductor density changes along the circumference as a sinusoidal function, reaching the highest density at position θ_m, also called the reference axis of the rotor. With the electrical current i_2 fed into the rotor conductors, the sheet of currents is formed on the rotor surface. The line density $J_R(\theta)$ of the surface currents exhibits the same sinusoidal change along the circumference as the density of the rotor conductors. The magnetic field is established in the air gap, while in iron, due to a very high-permeability μ_{Fe}, the magnetic field H_{Fe} is negligible. In the cylindrical coordinate system, the axial component of the field H is zero, while the tangential and radial components are determined by the expressions (8.29) and (8.30):

$$H_z^R(\theta) = 0 \tag{8.28}$$

$$H_\theta^R(\theta) = -J_R(\theta) = -J_{R0}\cos(\theta - \theta_m) \tag{8.29}$$

$$H_r^R(\theta) = \frac{J_{R0}R}{\delta}\sin(\theta - \theta_m) \tag{8.30}$$

The air gap δ is considerably smaller than radius R of the machine; thus, the radial component of the field is much higher than the tangential component.

Question (8.11): Consider a cylindrical machine of known dimensions, having the same number of conductors on the stator and the rotor. It is known that each conductor of the stator has electrical current in direction $\otimes$, while the rotor currents across the air gap have the current of the same strength but in the opposite direction ($\odot$). Determine the magnetic field in the air gap.

Answer (8.11): Since the air gap δ is very small ($\delta \ll R$), the opposite conductors of the stator and rotor are very close. Each stator conductor carrying the current in direction $\otimes$ has its counterpart across the air gap, the rotor conductor carrying the current in the opposite direction $\odot$. The distance between the two is rather small, $\delta \ll R$. For this reason, circulation of the magnetic field along the contour EFGH, shown in Fig. 8.13, gets equal to zero, as the sum of electrical currents passing through the integration contour gets zero. Therefore, the radial component of the magnetic field is equal to zero across the air gap. Regarding tangential component, it should be noted that the opposite directions of the currents in stator and rotor conductors contribute to tangential components of vector H. This component is equal to the line density of the stator (or the rotor) sheet of surface currents.

8.7   Convention of Representing Magnetic Field by Vector

The subject of analysis in the preceding section was the magnetic field created by the rotor winding made out of series-connected conductors distributed sinusoidally along the rotor circumference. The left-hand part of Fig. 8.14 shows the lines of the radial field created by the rotor winding. The symbols $\otimes$ and $\odot$ indicate positions where the density of rotor conductors reaches its maximum. They also determine the reference axis of the rotor, which is perpendicular to the line $\otimes - \odot$ and which is determined by the angle θ_m. The symbols $\otimes$ and $\odot$ also indicate positions where the line density of the rotor current sheet has its maximum. The magnetic field lines shown in the figure correspond to sinusoidal change of the magnetic field H along the air gap circumference. The area of the rotor surface where the field lines exit the rotor and enter the air gap is denoted as the north (N) magnetic pole. In a like manner, the south (S) magnetic pole is defined and marked as the area where the field gets from the air gap into the rotor. In central parts of magnetic poles, the field strength H and the magnetic induction B assume their maximum values.

Magnetic field of the rotor winding can be represented in a concise way by introducing *the vector of the rotor flux*. Even though the flux is a directed scalar, it is possible to represent it as a vector by adding the course and direction to the scalar value.

A flux vector is determined by its course, direction, and amplitude. Vector $\boldsymbol{\Phi}_R$, shown in the right-hand side of Fig. 8.14, represents a sinusoidal distribution of the magnetic induction B, the field lines of which are shown in the left-hand side of the figure. Direction of the flux vector is in accordance with direction of the field lines of H and $B = \mu H$. The amplitude of the flux vector $\boldsymbol{\Phi}_R$ is equal to the surface integral of the vector B over the surface leaning on one turn of the rotor winding. Therefore, the flux vector $\boldsymbol{\Phi}_R$ represents the flux in one turn. Alternatively, one can define the winding flux vector $\boldsymbol{\Psi}_R$ as the vector sum of all the fluxes in individual turns.

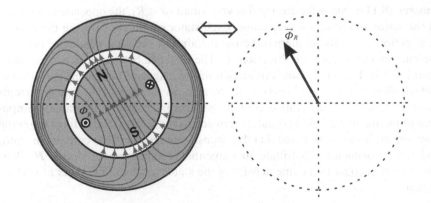

Fig. 8.14 Convention of vector representation of rotor magnetic field and flux

Question (8.12): Behold the left side of Fig. 8.14 and the two rotor conductors forming one rotor turn. Assume that these conductors are displaced several millimeters toward the stator and positioned across the air gap, on the inner surface of the stator magnetic circuit, while the electrical currents in these conductors remain the same. In the prescribed way, what used to be a rotor turn becomes a stator turn. Since the conductors denoted by $\otimes$ and $\odot$ are now on the surface of the stator magnetic circuit, the field created by the currents through these conductors becomes now the stator field. Sketch the field lines and compare them with the lines presented in the left-hand side of the figure. Denote positions of the north and south magnetic poles of the stator flux created by these conductors.

Answer (8.12): Radial component of the magnetic field in the air gap will not change by shifting the conductors. Direction of the tangential component of the field will change. Since radial component prevails over tangential component by an order of magnitude, it can be concluded that shifting the conductors will have no influence on the shape of the field lines. It is of interest to note that the north pole corresponds to the region where the magnetic field is directed from the magnetic circuit toward the air gap. In Fig. 8.14, the north pole of the stator is opposite to the south pole of the rotor.

Chapter 9
Energy, Flux, and Torque

Magnetic field in the air gap is obtained from electrical currents in stator and rotor windings. Another source of the air gap field can be permanent magnets that may be placed within magnetic circuits of either stator or rotor. The stator and rotor fields in the air gap are calculated in the previous chapter. Interaction of the two fields incites the process of electromechanical conversion.

In this chapter, expressions for the magnetic field in the air gap are used to calculate the field energy, to derive the energy accumulated in the magnetic field, and to calculate the electromagnetic torque caused by the interaction between the stator and rotor fields. In order to simplify the analysis, the flux linkages in one turns and the winding fluxes are represented by flux *vectors*. The concept of flux *vector* is introduced and explained along with magnetic axes of turns and windings. The torque expression is rewritten and expressed as the *vector* product of stator and rotor flux *vectors*. It is pointed out that continuous torque generation requires either stator or rotor windings to create the revolving magnetic field. This chapter ends with the analysis of two-phase windings systems and three-phase winding systems that create revolving magnetic field.

9.1 Interaction of the Stator and Rotor Fields

Electrical machines usually have windings on both stator and rotor. Currents through the windings create stator and rotor fluxes. There are machines which have permanent magnets instead of the stator or rotor winding. Magnetic field in the air gap has its radial and tangential components. The radial component is R/δ times larger than the tangential. It determines the spatial distribution of the magnetic energy of the field, as well as the course and direction of the field lines.

The stator and rotor fields exist in the same air gap and the same magnetic circuit. They add up and make the resulting magnetic field and the resulting flux. Assuming that magnetic circuit is linear ($\mu = $ const.), the resulting field is obtained by superposition of stator and rotor fields. Namely, the strength of the resulting

S.N. Vukosavic, *Electrical Machines*, Power Electronics and Power Systems, 185
DOI 10.1007/978-1-4614-0400-2_9, © Springer Science+Business Media New York 2013

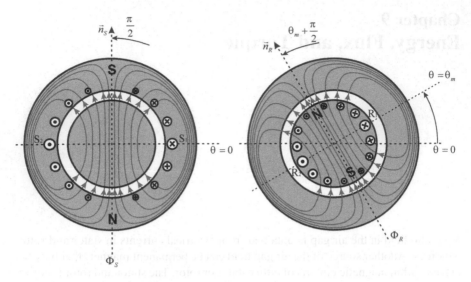

Fig. 9.1 Magnetic fields of stator and rotor

field is obtained by adding the two fields. The lines of the stator and rotor fields are shown in Fig. 9.1, which presents the course and direction of relevant flux *vectors* and magnetic axes. It is assumed that both the stator and rotor have a number of sinusoidally distributed conductors. For clarity, Fig. 9.1 shows just a few conductors which denote distributed windings. Conductors S1 and S2 of the stator winding are placed in positions with the maximum density of the stator conductors. The normal of the stator turn S1–S2 is, at the same time, the magnetic axis of the stator winding. In the same way, the axis of the rotor winding is determined by the normal of the rotor turn R1–R2.

In Fig. 9.1, direction of the stator field and flux is determined by normal n_S of the stator turn S1–S2. This normal extends in direction shifted by $\pi/2$ with respect to position $\theta = 0$, where the density of the stator conductors reaches its maximum. Direction of the rotor field and flux is determined by normal n_R to the rotor turn R1–R2, which extends in direction shifted by $\pi/2$ with respect to position $\theta = \theta_m$, where the density of the rotor conductors reaches its maximum. For this reason, direction of the rotor flux is shifted by $\theta_m + \pi/2$ with respect to position $\theta = 0$. When the stator and rotor conductors have constant currents (DC currents), the stator flux *vector* remains in its position (vertical position in Fig. 9.1), while the rotor flux *vector* revolves along with the rotor. In this case, the angle between the two flux *vectors* is $\Delta\theta = \theta_S - \theta_R = -\theta_m$, due to the rotor displacement of θ_m. In cases where the stator and/or the rotor has two or more windings with alternating currents, the angle between the two flux *vectors* can be different than θ_m. Namely, a set of stator (rotor) windings with the proper orientation of their magnetic axes creates the magnetic field and the flux *vector* which revolve with respect to their originator.

Fig. 9.2 Mutual position
of the stator and rotor fluxes

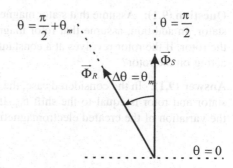

For that to be achieved, the winding currents must have the appropriate frequency and the initial phases. Figure 7.9 shows an example where the two orthogonal stator windings with alternating currents create magnetic field which revolves with respect to the stator. In this case, position of the flux *vector* and the angular difference $\Delta\theta$ between the two fluxes depend not only on the rotor position bus also on the supply frequency and the initial phase of the winding currents.

Description and further analysis are facilitated by *vector* representation of the stator and rotor fields in the manner shown in Fig. 9.2. The field of the magnetic induction **B** can be represented by the flux *vector*, in accordance with conclusions presented in Sects. 4.4 and 5.5, as well as in Sect. 8.5, formulating the convention of *vector* representation of magnetic fields. *Flux vector* of one turn is obtained by associating the course and direction with scalar Φ. This course and direction is obtained from the unit *vector* of the normal to the surface encircled by the relevant turn. Figure 9.2 shows the flux *vector* of the turn S1–S2 and the flux *vector* of the turn R1–R2. These *vectors* represent the magnetic fields of the stator and rotor shown in Fig. 9.1. Scalar value Φ_S represents the flux of the turn determined by the stator conductors S1–S2, placed in the region with maximum density of stator conductors. *Vector* Φ_S has the course and direction obtained from the normal to the surface encircled by the turn S1–S2. The same way, scalar value Φ_R represents the flux of the turn determined by the rotor conductors R1–R2, placed in the region with maximum density of rotor conductors. *Vector* Φ_R has the course and direction obtained from the normal to the surface encircled by the turn R1–R2.

By interaction of the stator and rotor magnetic fields, electromagnetic torque is created as a mechanical interaction between the stator and the rotor. Since the rotor can revolve, this torque can bring the rotor into rotation or change the speed of the rotor revolutions. The torque is created due to an interaction of the stator and rotor magnetic fields. Therefore, it is also called electromagnetic torque, T_{em}. Considering the force of attraction between different magnetic poles, it can be concluded that the electromagnetic torque tends to move the rotor in a way that brings closer the north magnetic pole of the rotor and the south magnetic pole of the stator. The torque acts toward bringing the two opposite poles one against the other. In terms of the flux *vectors*, the electromagnetic torque tends to align the stator and rotor flux *vectors*. It will be shown further on that the electromagnetic torque can be expressed as the *vector* product of the stator and rotor flux *vectors*.

Question (9.1): Assume that stator magnetic poles do not move with respect to the stator. In addition, assume that rotor magnetic poles do not move with respect to the rotor. If the rotor revolves at a constant speed, what is the change of the torque acting on the rotor?

Answer (9.1): In the considered case, the angle $\Delta\theta$ between the flux *vectors* of the stator and rotor is equal to the shift θ_m. If the rotor revolves at a constant speed, the variation of the created electromagnetic torque will be sinusoidal.

9.2   Energy of Air Gap Magnetic Field

It is of interest to determine the electromagnetic torque acting on the rotor and stator of a cylindrical machine. This torque can be determined as the first derivative of the energy accumulated in the magnetic (coupling) field in terms of the rotor displacement θ_m. On the basis of the equations given in Sect. 6.9, the increment of mechanical work $dW_{meh} = T_{em}d\theta_m$ is equal to the increment of energy of the magnetic field; thus, the torque can be determined as the first derivative of the magnetic field energy in terms of the rotor shift θ_m, $dW_m/d\theta_m$. Therefore, it is necessary to determine the energy of the magnetic field in terms of the rotor position, $W_m(\theta_m)$.

The energy of the magnetic field can be calculated by integrating the density of the field energy w_m over the entire domain where the magnetic field exists. The energy density w_m is expressed in J/m^3, and it represents the amount of the field energy comprised within unit volume; thus, $w_m = \Delta W_m/\Delta V = dW_m/dV$. Expression $w_m = \frac{1}{2}\,\mu H^2$ determines the density of the field energy in a linear medium, where the magnetic permeability does not change. Therefore, the density of the field energy in the air gap is $w_m = \frac{1}{2}\mu_0 H^2$.

Magnetic field exists in the magnetic circuits of the stator and rotor which are made of iron, as well as in the air gap. Since the same magnetic flux which passes through the air gap gets into the stator and rotor magnetic circuits, magnetic induction in iron B_{Fe} and in the air gap B_0 is roughly the same. The permeability of iron μ_{Fe} is several orders of magnitude higher than the permeability of the air μ_0. Therefore, the magnetic field in iron $H_{Fe} = B_{Fe}/\mu_{Fe}$ is negligible compared to the field H_0 in the air gap. The same way, the density of the field energy in iron $(\frac{1}{2}B^2/\mu_{Fe})$ is negligible when compared to the density of the field energy in the air gap $(\frac{1}{2}B^2/\mu_0)$. For this reason, the overall energy of the magnetic field can be determined by integrating the density of the field energy (specific energy) over the whole domain of the air gap.

In the expression for the field energy density $w_m = \frac{1}{2}\mu_0 H^2$, the symbol H represents the strength of the resultant magnetic field in the air gap, namely, the sum of the stator and rotor fields. Since the tangential components of the magnetic field are negligible ($\delta \ll R$), the strength H of the resulting magnetic field in the air gap is equal to the sum of radial components of the stator and rotor fields.

The expression for the density of the resulting magnetic field takes the form $w_m = \frac{1}{2}\,\mu_0(H_r^S + H_r^R)^2$.

By using (8.18) and (8.30), which give the radial components of the stator and rotor fields in the air gap at position θ, one obtains the function which determines the density of the magnetic field energy as a function of angle θ,

$$w_m(\theta) = \frac{\mu_0}{2}\left(\frac{R}{\delta}\right)^2 [J_{R0}\sin(\theta - \theta_m) + J_{S0}\sin\theta]^2. \tag{9.1}$$

The total energy accumulated in magnetic field is given by expression

$$W_m = \int_V w_m(\theta)\,dV,$$

where V is the total volume of the air gap. Since the elementary volume is obtained as

$$dV = L\,\delta\,R\,d\theta,$$

the total magnetic field energy in a cylindrical electrical machine of the length L, radius R, and the air gap δ becomes

$$W_m = L\delta R \int_0^{2\pi} w_m(\theta)\,d\theta. \tag{9.2}$$

By introducing (9.1) into (9.2), one obtains the expression

$$
\begin{aligned}
W_m &= \frac{\mu_0 R^3 L}{2\delta}\left[\int_0^{2\pi} J_{R0}^2 \sin^2(\theta - \theta_m)\,d\theta\right.\\
&\quad + \left.\int_0^{2\pi} J_{S0}^2 \sin^2(\theta)\,d\theta + \int_0^{2\pi} 2J_{R0}J_{S0}\sin(\theta - \theta_m)\sin(\theta)\,d\theta\right]\\
&= \frac{\mu_0 R^3 L}{2\delta}\left[J_{R0}^2 I_1 + J_{S0}^2 I_2 + 2J_{R0}J_{S0}I_3\right],
\end{aligned}
\tag{9.3}
$$

where J_{R0} represents the maximum value of the line density of the rotor currents while J_{S0} represents the corresponding value for the stator currents. Evaluation of the expression (9.3) requires finding the three integrals of trigonometric integrand functions, I_1, I_2, and I_3. Since

$$\sin^2\theta = \frac{1}{2}[1 - \cos(2\theta)], \quad \sin^2(\theta - \theta_m) = \frac{1}{2}[1 - \cos(2\theta - 2\theta_m)],$$

the integrals I_1 and I_2 take values

$$I_1 = \int_0^{2\pi} \sin^2\theta d\theta = \int_0^{2\pi} \frac{1}{2}[1 - \cos(2\theta)]d\theta = \pi,$$

$$I_2 = \int_0^{2\pi} \sin^2(\theta - \theta_m)d\theta = \int_0^{2\pi} \frac{1}{2}[1 - \cos(2\theta - 2\theta_m)]d\theta = \pi.$$

By using equation

$$\sin(\alpha)\sin(\beta) = \frac{1}{2}[\cos(\alpha - \beta) - \cos(\alpha + \beta)],$$

the integrand function of the third integral becomes

$$\sin(\theta - \theta_m)\sin(\theta) = \frac{1}{2}[\cos(-\theta_m) - \cos(2\theta - \theta_m)] .$$

Considering the integral boundaries 0 and 2π,

$$I_3 = \int_0^{2\pi} \sin(\theta - \theta_m)\sin(\theta)d\theta = \int_0^{2\pi} \frac{1}{2}[\cos(-\theta_m) - \cos(2\theta - \theta_m)]d\theta$$

$$= \int_0^{2\pi} \frac{1}{2}\cos\theta_m d\theta + \int_0^{2\pi} \frac{1}{2}\cos(2\theta - \theta_m)d\theta = \pi\cos\theta_m.$$

Finally, the expression for the energy of the magnetic field becomes

$$W_m = \frac{\mu_0 R^3 L\pi}{2\delta} \left[J_{R0}^2 + J_{S0}^2 + 2J_{R0}J_{S0}\cos\theta_m\right]. \tag{9.4}$$

It is important to recall that all previous considerations start with the assumptions that both stator and rotor windings carry DC currents; thus, the angle $\Delta\theta$ between the stator and rotor flux *vector* is equal to $-\theta_m$. On the basis of (9.4), the energy of magnetic field has its maximum value in the case when the *vector* of the stator flux is collinear with the *vector* of the rotor flux, that is, when $\Delta\theta = -\theta_m = 0$.

As already mentioned, the stator and/or rotor may have several windings with their magnetic axes shifted in space. With sinusoidal currents of the corresponding amplitudes, frequencies, and initial phases, it is possible to achieve the resultant magnetomotive force which keeps the amplitude constant while rotating at the speed determined by the frequency of the winding currents. Revolving magnetomotive force creates the revolving magnetic field and flux in the air gap which can be represented by rotating flux *vector*. One way of creating revolving magnetic field is shown in Fig. 7.9. In machines with alternating currents on the stator and/or rotor, the angle between the stator and rotor flux *vectors* depends on the rotor position, but it also depends on instantaneous values of the winding currents. For this reason, relation $\Delta\theta = -\theta_m$ is not valid unless the windings have DC currents, such as in the case shown in Fig. 9.1.

In general, expression for the total energy of magnetic field takes the form

$$W_m = \frac{\mu_0 R^3 L \pi}{2\delta} \left[J_{RO}^2 + J_{SO}^2 + 2J_{RO}J_{SO}\cos(\Delta\theta) \right] \qquad (9.5)$$

where $\Delta\theta$ is the angle between the stator and rotor flux *vectors*.

9.3 Electromagnetic Torque

The energy of the magnetic field in electrical machine shown in Fig. 9.1 is given by expression (9.4). Machine under consideration has one distributed winding on the stator and one distributed winding on the rotor. The windings carry constant (DC) currents. By using the expression for the field energy, it is possible to determine the electromagnetic torque.

The electromagnetic torque is a measure of mechanical interaction between the stator and rotor. The torque of the same amplitude acts on both stator and rotor in different directions. Under conditions when the stator is fixed and does not move, the torque cannot make the stator turn. On the other hand, the rotor has the freedom to turn. Therefore, the torque can make the rotor revolve and/or it can alter the rotor speed. The angle θ_m denotes shift of the rotor with respect to the stator. Under circumstances, the angle θ_m also determines the shift between the two windings as well as the angle $\Delta\theta$ between the stator and rotor flux *vectors*. Expression for the torque is $T_{em} = +dW_m/d\theta_m$. It has positive sign due to the assumption that the windings are connected to corresponding electrical power sources. Hence, considered electrical machine acts as an electromechanical converter connected to the power source, hence the expression $T_{em} = +dW_m/d\theta_m$. Moreover, it is assumed that the stator and rotor windings are supplied from controllable current sources. Therefore, electrical currents in the windings do not depend on the rotor position θ_m. For this reason, the line densities of electrical currents J_{RO} and J_{SO} do not depend on the rotor position θ_m, and their first derivatives $dJ_{RO}/d\theta_m$ and $dJ_{SO}/d\theta_m$ are equal to zero. Under the circumstances, electrical currents do not change as the rotor moves by $d\theta_m$. For the purpose of calculating $+dW_m/d\theta_m$, electrical currents can be considered constant, resulting in $\Delta\theta = -\theta_m$ and $\cos(\Delta\theta) = \cos(\theta_m)$. Therefore, expression for the electromagnetic torque becomes

$$T = +\frac{dW_m}{d\theta_m} = \frac{d}{d\theta_m} \left\{ \frac{\mu_0 R^3 L \pi}{2\delta} \left[J_{RO}^2 + J_{SO}^2 + 2J_{RO}J_{SO}\cos\theta_m \right] \right\}$$

or

$$T = \frac{\mathrm{d}}{\mathrm{d}\theta_m} \left\{ \frac{\mu_0 R^3 L J_{R0} J_{S0}}{\delta} \pi \cos \theta_m \right\}. \tag{9.6}$$

The torque is given by expression (9.7), and it is proportional to the fourth power of machine dimensions and inversely proportional to the air gap δ:

$$T = -\frac{\mu_0 \pi R^3 L}{\delta} J_{R0} J_{S0} \sin \theta_m. \tag{9.7}$$

The sign of the obtained torque is negative. This means that the torque acts in direction which is opposite to the reference counterclockwise direction. In the preceding sections, electrical machine is presented in cylindrical coordinate system where z-axis is directed toward the reader ($\odot$). From the reader's viewpoint, the reference direction of rotation around this axis is counterclockwise. The torque which supports the motion in counterclockwise direction can be represented as a *vector* collinear with z-axis. This association can be supported by the right-hand rule. The counterclockwise direction is adopted as the reference direction for the angular speed and torque. With that in mind, positive torque excites and supports the motion in counterclockwise (positive) direction. While the rotor revolves at a positive angular speed, a positive torque tends to increase the speed. On the other hand, torque of negative value excites and supports the motion in clockwise (negative) direction. While the rotor revolves at a positive angular speed, a negative torque tends to decrease the speed. The system in Fig. 9.1 tends to draw the north pole of the rotor toward the south pole of the stator and, hence, generates a negative torque, acting in clockwise direction.

The torque in (9.7) is proportional to the product of the stator currents, the rotor currents, and the sine of the displacement θ_m. In the case under consideration, the stator and rotor currents are constant, DC currents. Therefore, position of the stator flux Φ_S is determined by position of the stator. In other words, the stator flux does not move. At the same time, the position of the rotor flux Φ_R is determined by the position of the rotor itself. Therefore, the stator and rotor flux *vectors* are displaced by θ_m. Hence, the torque is proportional to the sine of the angle between the two fluxes. With that in mind, there are good grounds for expressing the torque *vector* in terms of the *vector* product of the stator and rotor flux *vectors*. This statement will be proved in the subsequent sections.

Question (9.2): Assume that the rotor is turning at a constant speed. What is the average value of the torque in the case where the stator and rotor windings both have DC currents?

Answer (9.2): The electromagnetic torque is a sinusoidal function of the angle between the stator and rotor flux *vectors*. In cases with no change in the relative position of the two fluxes, this angle does not change, neither does the sine of the angle. Therefore, there are conditions for generating a constant, nonzero torque. If the angle between the two fluxes keeps changing at a constant rate, the electromagnetic torque is sinusoidal function of time, and it has an average value equal to zero. In the given case, both windings have DC currents, and they generate the flux *vectors* which stay aligned with magnetic axes of corresponding windings. Since the rotor is turning, the rotor flux revolves with respect to the stator flux. Therefore, the average value of the torque will be equal to zero.

9.3.1 The Torque Expression

Equation (9.7) gives the electromagnetic torque of the electrical machine shown in Fig. 9.1, whose windings carry DC currents. In all the cases where the windings have constant (DC) currents, position of the stator flux *vector* is determined by the position of the stator itself, while position of the rotor flux *vector* tracks the position of the rotor. Therefore, the angle $\Delta\theta$ between the two *vectors* is equal to $-\theta_m$.

In cases where the stator (or rotor) has a set of windings with alternating (AC) currents, position of the flux *vector* is not uniquely determined by position of the stator (rotor); it also depends on electrical currents in the windings. Under proper conditions, AC currents create rotating magnetic field, that is, the field which revolves with respect to the windings. Creation of rotating magnetic field is analyzed in detail in Section 9.9, *Rotating magnetic field*. It is of interest to calculate the electromagnetic torque in cases where the stator and/or rotor windings have AC currents and create rotating magnetic field.

Starting from Figs. 9.1 and 9.2 and assuming that the windings carry DC currents, position of the stator flux *vector* $\theta_{\Psi S}$ and position of the rotor flux *vector* $\theta_{\Psi R}$ are

$$\theta_{\Psi S} = \frac{\pi}{2}, \quad \theta_{\Psi R} = \theta_m + \frac{\pi}{2}.$$

In the case when the stator has at least two spatially shifted stator windings with AC currents, and provided that conditions detailed in Section 9.9 are met, the stator flux *vector* rotates with respect to the very stator, and its position is

$$\theta_{\Psi S} = \frac{\pi}{2} + \theta_{is},$$

where the angle θ_{is} depends on instantaneous values of stator currents. If the rotor as well has a system of windings creating a rotating magnetic field, then angle of the rotor flux *vector* is

$$\theta_{\Psi R} = \frac{\pi}{2} + \theta_m + \theta_{iR},$$

where the angle θ_{iR} is determined by instantaneous values of the rotor currents. The angle between the stator flux **vector** and the rotor flux **vector** is equal to

$$\Delta\theta = \theta_{\Psi S} - \theta_{\Psi R} = -\theta_m + \theta_{iS} - \theta_{iR}.$$

The electromagnetic torque is calculated as the first derivative (9.8) of the energy accumulated in magnetic field. Magnetic field energy is defined by (9.5). When determining the first derivative of the magnetic field energy in terms of the coordinate θ_m, it is assumed that the electrical currents in the windings do not depend on θ_m. Validity of such an assumption is obvious in cases where the windings are supplied from external current sources. Therefore, the first derivative of the sum $-\theta_m + \theta_{\Psi S} + \theta_{\Psi R}$ in terms of θ_m is equal to -1, while the torque expression becomes

$$
\begin{aligned}
T_{em} &= +\frac{dW_m}{d\theta_m} = \frac{d}{d\theta_m}\left\{\frac{\mu_0 R^3 L\pi}{2\delta}\left[J_{R0}^2 + J_{S0}^2 + 2J_{R0}J_{S0}\cos(\Delta\theta)\right]\right\} \\
&= \frac{\mu_0 R^3 L\pi}{\delta}J_{R0}J_{S0}\frac{d}{d\theta_m}\left[\cos(-\theta_m + \theta_{iS} - \theta_{iR})\right] \\
&= \frac{\mu_0 R^3 L\pi}{\delta}J_{R0}J_{S0}\sin(-\theta_m + \theta_{iS} - \theta_{iR}) \\
&= \frac{\mu_0 R^3 L\pi}{\delta}J_{R0}J_{S0}\sin\Delta\theta.
\end{aligned}
\tag{9.8}
$$

The obtained expression shows that the torque is proportional to the product of amplitudes of the stator and rotor currents and to the sine of the angle between the stator and rotor flux **vectors**. The torque expression (9.8) holds notwithstanding the AC or DC currents in the machine windings. In order to show that the electromagnetic torque depends on the **vector** product of the stator and rotor flux **vectors**, it is necessary to probe further and clarify the relations between the single turn flux, the winding flux, and the amplitude of the flux **vector**.

9.4   Turn Flux and Winding Flux

In this section, some more detailed considerations concerning the winding flux and **vector** of the resultant flux are given. Algebraic intensity of the flux **vector** is calculated by relating the flux **vector** to the flux in one turn and the flux in the winding. The goal of these efforts is to represent the electromagnetic torque as the **vector** product of the stator and rotor flux **vectors**.

For the purpose of facilitating the analysis of electrical machines, directed scalars, such as magnetomotive forces and fluxes, can be represented by corresponding **vectors**. In Sect. 4.4, it is shown that the field of the **vector** of magnetic induction B can be represented by **vector**, thus defining the flux **vector** in a single turn (contour). In Sect. 5.5, the winding magnetic axis is introduced and defined, while Sect. 8.5 gives the convention of representing the magnetic field by *vector*. These results are used here to express the winding flux and the resultant flux.

Magnetic field in electrical machines appears as a consequence of magnetomotive forces established by stator and rotor currents. An example of a machine having one stator and one rotor winding is given in Fig. 9.1. In this example, it is assumed that electrical currents in both windings are constant and that the magnetic circuit is linear. Not all of the conductors are shown in the Fig. 9.1. It is understood that a number of stator and rotor conductors are distributed along the machine circumference and that their line density changes in a sinusoidal manner. The stator magnetic field is caused by the stator currents and shown in the left of Fig. 9.1. The rotor magnetic field is caused by the rotor currents, and it is shown in the right. The resultant flux is obtained by superposition, that is, by adding the stator and rotor fields and fluxes. At any point in the air gap, it is possible to identify the *vector* of the magnetic induction B^S created by stator currents and the *vector* of the magnetic induction B^R created by rotor currents. Resultant magnetic induction B^{Res} is equal to the *vector* sum $B^S + B^R$. The stator flux is calculated as a surface integral of the *vector B^S*, while the rotor flux is the surface integral of the *vector B^R*. The resultant flux is the surface integral of the *vector $B^S + B^R$*. Therefore, the resultant flux *vector* is the *vector* sum of the rotor flux *vector* and the stator flux *vector*.

At this point, it is of interest to clarify the terms *stator flux* and *rotor flux*. Within further developments, the references to *stator flux* imply the flux created by magnetomotive forces of stator currents. In cases where the rotor does not have any electrical currents in its windings nor does it comprise permanent magnets, the only flux in electrical machine is the stator flux. In absence of rotor currents, magnetic inductance B^R, created in the air gap by means of the rotor currents, is equal to zero. In such conditions, the resultant magnetic induction is equal to B^S. The stator flux in one turn is determined by the surface integral of the *vector B^S* over the surface S encircled by the turn. In the same way, all the developments within this book consider the *rotor flux* as the surface integral of the magnetic induction B^R, wherein the induction B^R is created by the rotor currents and corresponds to the resultant induction in cases where the stator currents are equal to zero. The resultant magnetic induction $B^{Res} = B^S + B^R$ exists in the machine with both the stator and rotor currents. The resultant flux is the surface integral of the *vector B^{Res}*.

One can consider the term *stator flux* to be the flux in the stator winding, whatever the magnetomotive force incites the flux. Adopting this viewpoint, the stator flux can be created by the stator currents, by the rotor currents, or by the contemporary action of both currents. This meaning of the term is better explained by citing *resulting stator flux*, implying the resultant flux in the stator winding, caused by any magnetomotive force and whatever magnetic inductance. The same holds for the term *rotor flux*.

Flux in one turn (contour) is determined as the surface integral of the *vector* of magnetic induction B over the surface encircled by the contour. The reference direction to be respected in the course of integration is determined by the right-hand rule. Placing the right hand so that the four fingers point to direction $\odot$ (Fig. 9.3),

Fig. 9.3 Calculation of the
flux in one turn. While the
expression for magnetic
induction B_{Fe} on the diameter
S_1S_2 is not available, the
expression $B(\theta)$ for magnetic
induction in the air gap is
known

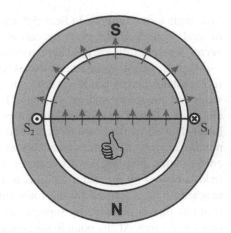

while the base of the hand is turned toward $\otimes$, the thumb will indicate the reference direction of the contour flux. A positive current in the contour will create positive flux. The field lines would extend in the reference course and direction.

In Fig. 9.3, one stator turn with conductors S1 and S2 has electrical current that creates magnetic field in the air gap. The arrows indicate the course and direction of the magnetic induction **B** in the air gap. The flux in the turn is determined by calculating surface integral of the **vector B** induction over the surface encircled by the turn S1–S2. There is a multitude of different surfaces that are all surrounded by the turn S1–S2. Due to div**B** = 0, the flux of the **vector B** on all such surfaces has the same value. Therefore, the flux calculation can be performed by selecting the surface that leads to less difficulty in calculation of the surface integral. This surface may be a rectangle $D \times L$, with one side being the diameter S1–S2 and the other side being the axial length L of the machine. However, analytical expression for the magnetic induction B is unknown along the diameter S1–S2 and within the rotor magnetic circuit. On the other hand, magnetic induction $B(\theta)$ in the air gap is known. For this reason, the integration is carried out over the surface which is passing through the air gap, residing at the same time on the considered contour.

9.4.1   Flux in One Stator Turn

It is of interest to determine the flux in the turn S1–S2 of the stator winding, created by the electrical currents in stator conductors. It is assumed that the stator has sinusoidally distributed conductors creating the stator current sheet. Considered turn S1–S2 is a part of the stator winding. It is connected in series with a multitude of other turns, displaced along the circumference. The turn S1–S2 resides in the position where the density of stator conductors is at the maximum.

The stator conductors with sinusoidal distribution and with DC current in the direction shown in Figs. 9.1 and 9.3 create the stator magnetic field in the air gap. Prevailing radial component of the magnetic field H^S is determined by (8.18),

$$H_r^S(\theta) = \frac{J_{S0}R}{\delta} \sin \theta.$$

The air gap permeability is constant (μ_0). Therefore, the corresponding magnetic induction in the air gap is

$$B_r^S(\theta) = \mu_0 \frac{J_{S0}R}{\delta} \sin \theta.$$

Magnetic induction $B_r^S(\theta)$ is created by action of the stator currents. In cases where the rotor currents are equal to zero, B_r^S determines the resultant magnetic induction in the air gap. The maximum intensity of magnetic induction is $B_m = \mu_0(R/\delta)J_{S0}$, and it is reached in the region of magnetic poles, such as the upper part of the figure, where the field lines leave the air gap and enter into magnetic circuit of the stator. In order to determine the flux in the turn S1–S2, it is necessary to select the surface convenient for the calculation of the surface integral. Since the expression $B_r^S(\theta)$ for the magnetic induction in the air gap is readily available, it is most suitable to adopt the surface which passes through the air gap. Hence, the choice is semicylinder of diameter R and length L. It looks like a rectangle of dimensions L × (πR), folded to make a semicylinder which starts from S1, passes through the air gap, and gets to S2. In Fig. 9.3, the cross section of such semicylinder corresponds to the upper semicircle where the field lines leave the air gap and enter into stator magnetic circuit. The surface S is

$$S = \pi \cdot R \cdot L.$$

The flux Φ_{S1} in the turn S1–S2 is obtained by calculating the surface integral of the magnetic induction over the surface S. The subscript "S1" intends that the symbol Φ_{S1} stands for the flux in one (1) turn of the stator (S). With,

$$\Phi_{S1} = \int_S B_r^S(\theta) \, dS,$$

where

$$dS = L \cdot R \cdot d\theta.$$

The flux in one turn is obtained as

$$\Phi_{S1} = \int_0^\pi B_r^S(\theta) \, L \cdot R \, d\theta = \frac{\mu_0 L R^2}{\delta} J_{S0} \int_0^\pi \sin(\theta) d\theta$$

$$= \frac{\mu_0 L R^2}{\delta} J_{S0}(-\cos \theta)\big|_0^\pi = \frac{2\mu_0 L R^2}{\delta} J_{S0}. \tag{9.9}$$

9.4.2   Flux in One Rotor Turn

Preceding analysis calculates the flux in one stator turn. In a similar way, it is possible to obtain the flux in one rotor turn, namely, the flux in the contour R1–R2 of the rotor winding (Fig. 9.1). In calculating the surface integral of the magnetic induction and obtaining the rotor flux, one should take into account the magnetic induction B_r^R, created by the electrical currents in distributed rotor winding. In the expression for radial component of the rotor field

$$B_r^R(\theta) = \mu_0 \frac{J_{R0}R}{\delta} \sin(\theta - \theta_m).$$

J_{R0} represents the maximum line density of the rotor currents, while the angle θ_m represents the rotor position, that is, the rotor displacement with respect to the stator. At the same time, θ_m denotes the angular displacement between the stator and rotor windings (Fig. 9.1). The flux through the contour R1–R2 is

$$\Phi_{R1} = \int_{\theta_m}^{\pi+\theta_m} B_r^R(\theta)\,L \cdot R\,d\theta = \frac{\mu_0 LR^2}{\delta} J_{R0} \int_{\theta_m}^{\pi+\theta_m} \sin(\theta - \theta_m)d\theta \tag{9.10}$$

$$= \frac{\mu_0 LR^2}{\delta} J_{R0}[-\cos(\theta - \theta_m)]|_{\theta_m}^{\pi+\theta_m} = \frac{2\mu_0 LR^2}{\delta} J_{R0}.$$

Calculation of the flux in one turn may be done in a shorter way, avoiding the integration. Magnetic inductance B_r passes through the semicylindrical surface $S = \pi RL$ in radial direction. In cases where the magnetic inductance $B_r^S(\theta)$ does not change over the interval $\theta \in [0 .. \pi]$, the flux in one stator turn can be obtained by multiplying the inductance $B_r^S(\theta) = $ const. and the surface area πRL. Yet, in electrical machine with sinusoidally distributed conductors, magnetic inductance changes along the circumference. Both $B_r^S(\theta)$ and $B_r^R(\theta)$ change as sinusoidal functions of the angle θ. Notwithstanding variable magnetic inductance in the air gap, the surface integration can be avoided in all cases where the average value of $B_r(\theta)$ is known on one semicircle. The flux Φ_{S1} in the stator turn S1–S2 can be calculated as the product of the surface πRL of the semicylinder and the average value of the magnetic induction $B_r^S(\theta)$ over the interval $\theta \in [0 .. \pi]$. With $B_r^S(\theta) = B_{max} \sin(\theta)$, the average value is $\pi/2$ times lower than the maximum value; thus, $B_{av} = (2/\pi)\,B_{max} = 2\mu_0 RJ_{S0}/(\delta\pi)$. The result Φ_{S1} is obtained by multiplying the average value B_{av} of the magnetic induction and the surface area $S = \pi RL$, and it is in accordance with (9.9).

It should be noted that the contours S1–S2 and R1–R2 have been selected so as to have their conductors placed in the regions with the highest density of conductors. The stator flux *vector* is shown in Fig. 9.1, and it coincides with the normal on the contour. Other turns have their conductors displaced with respect to S1–S2, and their normals are inclined with respect to the stator flux *vector*. Namely, the lines of the stator magnetic fields pass through the inclined turns at an angle other than $\pi/2$. Therefore, the flux in other turns is smaller than the flux in the turn

S1–S2. As the angular displacement of the turn with respect to S1–S2 increases toward $\pi/2$, the flux decreases.

In the same way, the rotor flux *vector*, shown in Fig. 9.1, coincides with the normal to the contour R1–R2. Therefore, the flux in the turn R1–R2 has the maximum value of all rotor turns.

The flux in turns that are inclined with respect to S1–S2 (R1–R2) is smaller compared to the values given by (9.9) and (9.10). It is shown hereafter that this flux depends on the cosine of the angle between the flux *vector* and the normal on the relevant turn. As an example, the flux is calculated in one stator turn with conductor $\otimes$ in position $\theta = \theta_1$ and conductor $\odot$ in position $\theta = \pi + \theta_1$. The normal on the considered turn is shifted by θ_1 with respect to the normal on the turn S1–S2. The flux $\Phi_S(\theta_1)$ in the inclined turn is determined by calculating the surface integral of the magnetic induction (incited by the stator currents) over the semicylindrical surface reclining on the conductor $\otimes$ in position $\theta = \theta_1$ and reaching the conductor $\odot$ in position $\theta = \pi + \theta_1$:

$$
\Phi_S(\theta_1) = \int_{\theta_1}^{\pi+\theta_1} B_r^S(\theta)\,L \cdot R\,d\theta = \frac{\mu_0 L R^2}{\delta} J_{S0} \int_{\theta_1}^{\pi+\theta_1} \sin(\theta)\,d\theta
$$

$$
= \frac{\mu_0 L R^2}{\delta} J_{S0}(-\cos\theta)\big|_{\theta_1}^{\pi+\theta_1} = \frac{2\mu_0 L R^2}{\delta} J_{S0}\cos\theta_1 = \Phi_{S1}\cos\theta_1. \tag{9.11}
$$

Equation (9.11) shows that flux in the turn shifted by angle θ_1 is cosine function of the angle. When $\theta_1 > \pi/2$, the flux in this turn obtains negative value. With $\theta_1 = \pi$, the turn gets to positions S1 and S2, with directions $\otimes$ and $\odot$ exchanged. The flux in such turn reaches the same absolute value as the flux in the original turn S1–S2, but it has the opposite sign. In the same way, it can be shown that the flux created by the rotor currents in one rotor turn depends on the angle θ_2 between the normal on the considered turn (contour) and the vertical n_R on the turn R1–R2 (Fig. 9.1). The flux in the rotor turn $\Phi_R(\theta_2)$ is calculated from the surface integral of the rotor magnetic induction over the surface encircled by the conductor $\otimes$ in position $\theta = \theta_m + \theta_2$ and the conductor $\odot$ in position $\theta = \pi + \theta_m + \theta_2$ (9.12).

The results (9.11) and (9.12) show that the flux in a single stator or rotor turn depends on the cosine of the angle between the normal on the considered turn and the flux *vector* whose amplitude, course, and direction represent the field of magnetic induction.

The method of representing the field of magnetic induction by the flux *vector* has been discussed in Subsect. 4.4 and used in Figs. 9.1 and 9.2:

$$
\Phi_R(\theta_2) = \frac{\mu_0 L R^2}{\delta} J_{R0} \int_{\theta_2+\theta_m}^{\pi+\theta_2+\theta_m} \sin(\theta-\theta_m)\,d\theta
$$

$$
= \frac{\mu_0 L R^2}{\delta} J_{R0}(-\cos(\theta-\theta_m))\big|_{\theta_2+\theta_m}^{\pi+\theta_2+\theta_m}
$$

$$
= \frac{2\mu_0 L R^2}{\delta} J_{R0}\cos\theta_2 = \Phi_{R1}\cos\theta_2. \tag{9.12}
$$

Fig. 9.4 Flux in concentrated
winding

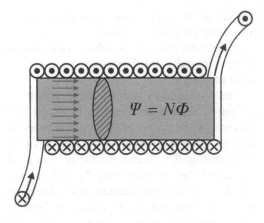

$$\Psi = N\Phi$$

Fig. 9.5 The surface
reclining on a concentrated
winding with three turns

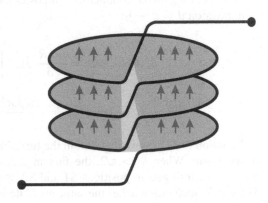

9.4.3   Winding Flux

Flux in a winding is the sum of fluxes in individual turns constituting the winding.
A sample winding consisting of N series connected turns wound around a straight
ferromagnetic bar is given in Fig. 9.4. Each turn of the sample winding has the same
flux Φ. Therefore, the total flux of the whole winding Ψ is determined as the product
of the number of turns N and the flux of one turn Φ; thus, $\Psi = N\Phi$. This is due to
the fact that the surfaces encircled by individual turns have equal areas while their
normals are oriented in the same direction. A winding where all the turns have the
same flux while their normals are collinear is called *concentrated* winding.

Like the flux in one turn, the flux in a winding can be determined as surface
integral of the magnetic induction over the surface reclining on the entire winding.
While the surface encircled by one turn is easily identified, the *surface of the
winding* is more difficult to identify. In Fig. 9.5, an attempt is made to illustrate
the surface of a concentrated winding. The three turns making this winding constitute a complex contour. The shadowed area shows the surface encircled by the

winding conductors. If distance between the turns is sufficiently small, it is justified to assume that the three turns have the same flux. Therefore, each of the lines of the considered magnetic field passes through the surface of the winding three times. With Φ designating the flux in one turn, the winding flux Ψ is equal to $N\Phi = 3\Phi$.

In a cylindrical machine with distributed windings, the stator and rotor turns are distributed along the circumference. The stator conductors are placed in slots on the inner surface of the stator magnetic circuit, while the rotor conductors are placed in slots on the rotor surface facing the air gap. Two diametrical conductors constitute one turn, that is, one contour. A winding consists of a number of series connected turns. The winding flux is the sum of the fluxes in individual turns. The flux in one turn depends upon its relative position with respect to the magnetic field. In cases where the field lines are perpendicular to the surface of the turn, the flux in the turn has maximum value. The flux becomes zero in cases where the turn surface runs parallel with the lines of the magnetic field.

It is proven that the flux in one turn is proportional to the cosine of the angle between the **vector** of magnetic induction and the normal on the turn surface. This normal is called magnetic axis of the turn. In Fig. 9.1, the normal is perpendicular to the straight line connecting the conductors $\otimes$ and $\odot$. Given the orientation of the magnetic field, the flux $\Phi(\theta)$ in each turn can be determined in terms of its angular position θ. Equations (9.11) and (9.12) provide the flux values for one stator and one rotor turn. They are expressed in terms of the angle between the normal of the relevant turn and the flux **vector** which represents the magnetic field.

Since the turns that constitute one distributed winding have their axes oriented in different directions, their relevant fluxes will assume different values. For this reason, the total winding flux cannot be obtained by multiplying the flux in one turn by the number of turns.

In general, the winding flux is determined by adding all the contributions of individual turns. In cases where the winding is concentrated, the winding flux **vector** has an amplitude of $\Psi = N\Phi$. In cases where the winding is distributed with the conductor line density of $N'(\theta)$, the winding flux is determined by integration. The number of conductors within a tiny segment of angular width $d\theta$ is

$$dN = N'(\theta)Rd\theta,$$

where R is diameter of the machine. Each of the conductors positioned on the interval $\theta \in [0 .. \pi]$ completes one turn with its diametrically positioned counterpart on the interval $\theta \in [\pi .. 2\pi]$. The flux in one turn is determined by the angle between the flux **vector**, representing the magnetic field and the axis of the turn. Eventually, the flux in one turn can be expressed in terms of the position θ of the turn. Contributions of all dN turns to the total flux of the winding is

$$d\Psi = N'(\theta)\Phi(\theta)Rd\theta,$$

while the total flux of the winding is

$$\Psi = \int\limits_{0}^{\pi} N'(\theta)\Phi(\theta)Rd\theta. \tag{9.13}$$

Equation (9.13) can be used for calculation of the winding fluxes of both stator and rotor windings.

An example of practical use of the (9.13) is calculation of the self-inductance of the stator winding. Self-inductance is coefficient that defines effect of winding currents on winding flux. In cases where the winding flux Ψ does not have any external originator and exists due to the winding current I only, the self-inductance can be calculated as $L_S = \Psi/I$. Before using (9.13), it is necessary to calculate the flux in one turn $\Phi(\theta)$. It is calculated as the surface integral of the magnetic induction B_r^S in the air gap, wherein B_r^S denotes the radial component of the magnetic inductance created by the stator currents. Dividing the flux in the stator winding by the stator current I gives the self-inductance of the stator winding.

Similar procedure can be used to determine the mutual inductance between the stator and rotor windings. The mutual inductance L_m defines the effect of the rotor currents on the flux in the stator winding. In cases where the rotor currents contribute to resultant magnetic induction in the air gap, they also change the resultant flux in the stator turns and, hence, the flux in the stator winding. The same coefficient defines the effects of the stator currents on the flux in the rotor winding. Calculation of L_m requires the previous procedure to be modified. When calculating the flux in one stator turn $\Phi(\theta)$, it is necessary to replace the magnetic inductance B_r^S, created by the stator currents, by the magnetic inductance B_r^R, created by the rotor currents. In this way, the value of $\Phi(\theta)$ corresponds to the flux that the rotor currents establish in one stator turn. At that time, calculation of the stator flux according to (9.13) results in the flux created in the stator winding by action of the rotor currents. Dividing this value by the rotor current gives coefficient of mutual inductance between stator and rotor windings.

To proceed, the flux in the stator winding of the electrical machine shown in Fig. 9.1 is calculated by using (9.13), assuming that the magnetic field in the air gap is excited by the stator currents. Therefore, the magnetic field in the air gap is calculated assuming that the rotor currents are equal to zero. Distributed winding of the stator can be considered as a set of $N_T = N_C/2$ contours, where N_C denotes the number of conductors in the stator winding while N_T is the number of turns. According to (8.2), the number of turns is equal to $N_T = 2R\, N'_{Smax}$, where N'_{Smax} is the maximum line density of the stator conductors, which exists at positions of conductors S1 and S2 in Fig. 9.1.

In order to calculate the winding flux, it is necessary to calculate the flux in one turn. For a turn with conductor $\otimes$ in position θ and with conductor $\odot$ in position $\pi + \theta$, the flux $\Phi_S(\theta)$ is determined from (9.11):

$$\Phi_S(\theta) = \Phi_{S1} \cos \theta$$

Since the line density of the stator conductors is

$$N'(\theta) = N_{S\,max} \cos\theta,$$

Equation (9.13) becomes

$$\Psi_S = \int_0^\pi N_{S\,max} \cos\theta \cdot \Phi_{S1} \cos\theta \cdot R \cdot d\theta$$

$$= N_{S\,max} \Phi_{S1} R \int_0^\pi \cos^2\theta \, d\theta$$

$$= \frac{\pi}{2} N_{S\,max} \Phi_{S1} R = \frac{\pi}{4} N_T \Phi_{S1}. \tag{9.14}$$

With $\pi/4 < 1$, it is concluded from (9.14) that the flux in a distributed winding is smaller than the flux in a concentrated winding having the same number of turns. A concentrated winding would be obtained by placing all the conductors in places S1 and S2 (Fig. 9.1). The flux in the concentrated winding is obtained by multiplying the number of turns N_T and the flux in one turn Φ_{S1}.

9.4.4 Winding Flux Vector

The winding flux can be represented by a *vector* denoting the course, direction, and amplitude of the flux. The winding flux *vector* can be obtained by adding the flux *vectors* representing the fluxes in individual turns. In cases where the flux *vectors* of individual turns have different orientations, it is necessary to determine their sum and find the course and direction for the flux *vector* of the winding.

The convention of representing the flux in one turn by *vector* is presented in Section 4.4. The course and direction of the flux *vector* in one turn are determined from the normal to the surface reclining on the relevant turn. In Sect. 5.5, magnetic axis of a winding has been defined on the basis of the course and direction of the lines representing the magnetic field created by electrical currents in the winding itself. The course of the winding axis is determined by positions of magnetic poles created in the magnetic circuit due to the winding currents. The convention of representing the magnetic field of a distributed winding by flux *vector* is given in Sect. 8.5. Once again, the course and direction of the flux *vector* are determined from spatial orientation of the magnetic field, that is, from positions of the magnetic poles. Practical example of calculating the course and direction of the flux *vector* is given for the machine presented in Fig. 9.1. It starts with determining the spatial distribution of the magnetic field created by the currents in distributed winding. Direction of the winding flux *vector* is determined on the basis of direction of the

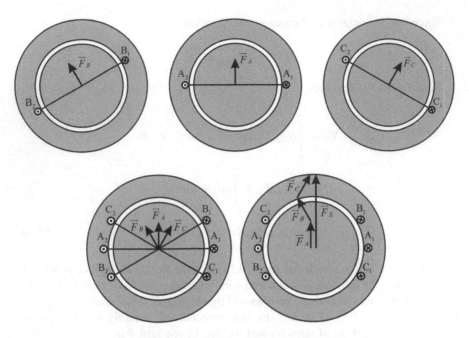

Fig. 9.6 Vector addition of magnetomotive forces in single turns and magnetic axis of individual turns

field lines, while the course is determined by the magnetic poles, wherein the poles are identified as diametrically positioned zones of the magnetic circuit where the magnetic induction reaches maximum values. By using this procedure, magnetic fields of the stator and rotor have been represented by flux *vectors* given in Fig. 9.2.

Derivation of the course of the flux *vector* and the magnetic axis of the winding can be performed otherwise, by *vector* addition of magnetomotive forces created by individual turns or by *vector* addition of their flux *vectors*. Figure 9.6 shows a stator winding comprising three turns, A1–A2, B1–B2, and C1–C2. The upper part of the figure shows individual *vectors* of magnetomotive forces for each turn. These magnetomotive forces are denoted by *vectors* F_A, F_B, and F_C. They are determined by the normals of corresponding turns. Magnetomotive force F_A would determine the course and direction of the resultant winding flux if the turns B and C did not exist. The resultant magnetomotive force F_S of the stator winding comprising all the three turns is shown in the lower part of the figure. The *vector* of the resultant magnetomotive force is obtained by *vector* addition of F_A, F_B, and F_C. Since each flux is determined by dividing the corresponding magnetomotive force by the magnetic resistance, the course and direction of the *vector* representing the flux in the winding are determined from the resultant magnetomotive force. As it is shown in the figure, the flux *vector* of the winding is collinear with the normal of the middle turn A1–A2.

9.5   Winding Axis and Flux Vector

Previous considerations provided details on determining the axis of a winding, the course of the flux *vector* in one turn, and the course of the flux *vector* in a winding comprising several turns. A brief survey of the conclusions is presented hereafter, aimed to be used in the subsequent considerations. The survey is illustrated by Fig. 9.7.

The figure presents a distributed winding with sinusoidal change of the conductor line density. The maximum line density is reached at positions where the conductors A1 and A2 are placed. The maximum value Φ_{S1} of the flux in one turn is reached in the turn A1–A2.

The course of the flux Φ_{S1} in the turn A1–A2 is determined by the normal n_{S1}. The normal n_{S1} is a unit *vector* perpendicular to the flat surface reclining on the contour A1–A2.

The course of the stator winding flux *vector* Ψ_S is determined by the unit *vector* n_S, representing the winding axis (magnetic axis of the winding). Therefore, in the case of a distributed winding with sinusoidal distribution of conductors, the flux *vector* and the winding axis have the same course and direction as the flux *vector* Φ_{S1} in the turn A1–A2, wherein the conductors A1 and A2 reside at positions where the conductor line density is maximum.

9.6   Vector Product of Stator and Rotor Flux Vectors

Figure 9.1 shows the lines representing magnetic field of the stator and magnetic field of the rotor in a cylindrical electrical machine with DC currents in both windings. Electromagnetic torque is generated by interaction of the two magnetic

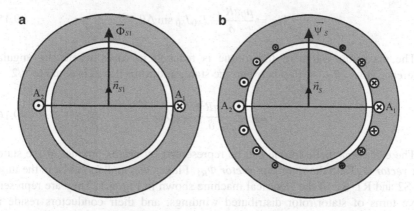

Fig. 9.7 Spatial orientation of flux vector of one turn (**a**), axis of the winding (**b**), and flux vector of the winding

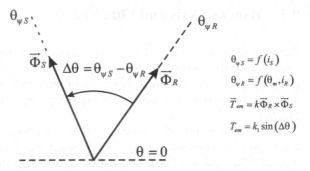

Fig. 9.8 Spatial orientation of the stator flux vector. Spatial orientation of the rotor flux vector. The electromagnetic torque as the vector product of the two flux vectors

$$\theta_{\psi S} = f(i_s)$$

$$\theta_{\psi R} = f(\theta_m, i_R)$$

$$\vec{T}_{em} = k\vec{\Phi}_R \times \vec{\Phi}_S$$

$$T_{em} = k_1 \sin(\Delta\theta)$$

fields, and it is expressed by (9.7), where the angle θ_m represents the position of the rotor relative to the stator. With DC currents in the windings, the angle θ_m defines as well the angle between the stator and rotor flux *vectors*.

In general, both stator and rotor may comprise several windings carrying DC or AC currents. Alternating currents may create rotating magnetic field which can be represented by a revolving flux *vector*. Such field revolves relative to the windings that give rise to the magnetomotive forces that originate the field. In cases where AC currents in the stator windings create rotating magnetic field, the orientation of the stator flux *vector* $\theta_{\psi S}$ depends on instantaneous values of electrical currents in the stator windings. In cases where AC currents in the rotor create rotating magnetic field, instantaneous values of the rotor currents determine the position of the rotor flux *vector* with respect to the rotor. Hence, the orientation of the rotor flux *vector* $\theta_{\psi R}$ with respect to the stator is determined by the rotor position θ_m and by instantaneous values of electrical currents in the rotor winding. Therefore, in general, the angle $\Delta\theta = \theta_{\psi S} - \theta_{\psi R}$ between the stator flux *vector* and the rotor flux *vector* is dependent on the rotor position θ_m and also on instantaneous values of electrical currents in the machine windings, as indicated in Fig. 9.8. On the basis of (9.6), expression for the electromagnetic torque assumes the form

$$T_{em} = \frac{\mu_0 \pi R^3 L}{\delta} J_{R0} J_{S0} \sin \Delta\theta. \tag{9.15}$$

The maximum value of the torque is reached in cases where the angular difference $\Delta\theta = \theta_{\psi S} - \theta_{\psi R}$ between the stator and rotor fluxes is equal to $\pi/2$.

$$T_{max} = \frac{\mu_0 \pi R^3 L}{\delta} J_{R0} J_{S0}. \tag{9.16}$$

The electromagnetic torque can be represented by *vector* product of the stator flux *vector* Φ_{S1} and the rotor flux *vector* Φ_{R1}. Fluxes Φ_{S1} and Φ_{R1} exist in the turns S1–S2 and R1–R2 of the electrical machine shown in Fig. 9.1. They are representative turns of stator/rotor distributed windings, and their conductors reside at positions with the maximum line density of stator/rotor conductors. The flux Φ_{S1} is also the resultant flux in the turn S1–S2 in cases when the electrical machine has only the stator currents, while the flux Φ_{R1} is also the resultant in the turn R1–R2 in cases where only the rotor currents exist in the machine. On the basis of (9.9) and (9.10), amplitudes of the two *vectors* are determined by expressions

$$\Phi_{S1} = \frac{2\mu_0 LR^2}{\delta} J_{S0}, \qquad \Phi_{R1} = \frac{2\mu_0 LR^2}{\delta} J_{R0}.$$

Since the torque is expressed by function $\sin \Delta\theta$, it can be calculated as the *vector* product of the stator and rotor flux *vectors*. Expression for the torque can be represented in the form

$$\frac{\mu_0 \pi R^3 L}{\delta} J_{R0} J_{S0} \sin \Delta\theta = \left(\frac{\pi\delta}{4\mu_0 LR}\right) \left(\frac{2\mu_0 LR^2}{\delta} J_{S0}\right) \left(\frac{2\mu_0 LR^2}{\delta} J_{R0}\right) \sin \Delta\theta$$

which comprises the amplitude of the *vector* product of the stator and rotor flux *vectors*, given by (9.17),

$$\vec{T}_{em} = \left(\frac{\pi\delta}{4\mu_0 LR}\right) \cdot \left[\vec{\Phi}_R \times \vec{\Phi}_S\right] = k\left[\vec{\Phi}_R \times \vec{\Phi}_S\right] \tag{9.17}$$

where the constant k is

$$k = \frac{\pi\delta}{4\mu_0 LR}.$$

Since the flux *vectors* Φ_{S1} and Φ_{R1} of the turns S1–S2 and R1–R2 have the same spatial orientation as the flux *vectors* of the stator and rotor windings, respectively, the electromagnetic torque can be expressed as the *vector* product $\Psi_S \times \Psi_R$ of the flux *vector* Ψ_S in the stator winding and the flux *vector* Ψ_R in the rotor winding. On the basis of (9.14), the amplitudes of *vectors* Ψ_S and Ψ_R are determined by expressions

$$\Psi_S = \frac{\pi}{4} N_{TS} \Phi_{S1}, \qquad \Psi_R = \frac{\pi}{4} N_{TR} \Phi_{R1},$$

where N_{TS} and N_{TR} denote the number of turns of the stator and rotor windings, respectively. Since the flux *vectors* of the representative turns are collinear with the flux *vectors* of respective windings, (9.17) takes the form

$$\begin{aligned}
\vec{T}_{em} &= \left(\frac{\pi\delta}{4\mu_0 LR}\right) \cdot \left[\vec{\Phi}_R \times \vec{\Phi}_S\right] \\
&= \left(\frac{\pi\delta}{4\mu_0 LR}\right) \cdot \left(\frac{4}{\pi N_{TS}}\right) \left(\frac{4}{\pi N_{TR}}\right) \left[\vec{\Psi}_R \times \vec{\Psi}_S\right] \\
&= \left(\frac{4\delta}{\mu_0 \pi LR N_{TS} N_{TR}}\right) \cdot \left[\vec{\Psi}_R \times \vec{\Psi}_S\right] \\
&= k_1 \cdot \left[\vec{\Psi}_R \times \vec{\Psi}_S\right],
\end{aligned} \tag{9.18}$$

where k_1 is constant equal to

$$k_1 = \frac{4\delta}{\mu_0 \pi LRN_{TS}N_{TR}}.$$

Equations (9.17) and (9.18) give the *vector* of electromagnetic torque. The course of the obtained *vector* is determined by axis z of the cylindrical coordinate system, that is, by the axis of rotor revolutions. Direction of the torque *vector* is in accordance with the reference direction of z-axis. The torque of a positive value acts toward increasing the rotor speed Ω_m and moves the rotor toward increasing the angle θ_m, which corresponds to the movement in counterclockwise direction. The torque amplitude is determined by equation

$$T_{em} = k\left|\vec{\Phi}_R \times \vec{\Phi}_S\right|.$$

Question (9.3): Expression for the torque (9.7) has the leading minus sign. Should the torque of a negative value be expected in the case represented in Fig. 9.1? Why (9.15) does not include negative sign?

Answer (9.3): The electromagnetic torque is determined by the function $\sin \Delta\theta$, where $\Delta\theta$ is the angle equal to $\theta_{\psi S} - \theta_{\psi R}$. The angles $\theta_{\psi S}$ and $\theta_{\psi R}$ determine the course of the stator and rotor flux *vectors*. In Fig. 9.1, courses of the two flux *vectors* are shown assuming that the windings carry a constant DC current. Then $\theta_{\psi S} = \pi/2$, while $\theta_{\psi R} = \pi/2 + \theta_m$, resulting in $\Delta\theta = -\theta_m$. In the considered case, the torque is proportional to function $\sin \Delta\theta = -\sin \theta_m$, which gives the minus sign in (9.7).

9.7   Conditions for Torque Generation

The electromagnetic torque can be calculated from the *vector* product of the flux *vector* Φ_{S1} and the flux *vector* Φ_{R1}, wherein the former is created by the stator currents in the stator turn S1–S2 while the latter is created by the rotor currents in the turn R1–R2 (Fig. 9.1). The expression for the electromagnetic torque is given by (9.17). Equation (9.18) reformulates the expression by introducing the *vector* product of the stator flux *vector* and the rotor flux *vector*.

The torque is proportional to the sine of the angle $\Delta\theta$ between the relevant flux *vectors*. With DC currents in both the stator and the rotor windings, the stator flux does not move with respect to the stator while the rotor flux does not move with respect to the rotor. In this case, the stator flux is leading with respect to the rotor flux by $\Delta\theta = -\theta_m$.

With the rotor revolving at a constant angular speed Ω_m, and with the initial rotor position $\theta_m(0) = 0$, the rotor position changes as $\theta_m(t) = \Omega_m t$. Therefore, the angle between the stator flux *vector* and the rotor flux *vector* is $\Delta\theta = -\theta_m = -\Omega_m t$.

Hence, electrical machine with both stator and rotor windings carrying DC current develops electromagnetic torque proportional to $\sin(\Omega_m t)$. The average value of this torque and the average value of the corresponding conversion power $P_{em} = T_{em}\,\Omega_m$ are both equal to zero. Therefore, an electrical machine with DC currents in the stator and in the rotor windings cannot provide an average power other than zero.

A nonzero average value of the torque is obtained in cases where the angle $\Delta\theta$ between the stator and rotor flux *vectors* is constant. For the given winding currents, the electromagnetic torque has the maximum value with $\Delta\theta = \pi/2$.

Condition $\Delta\theta = $ const. cannot be achieved with both the stator and rotor windings carrying DC currents. It will be proven later on that at least one of the windings on either stator or rotor must have AC currents and create the revolving magnetic field. There are three distinct cases when the constraint $\Delta\theta = $ const. is fulfilled. These three methods of accomplishing constant relative position of the two flux *vectors* have been shown in Fig. 7.8. These examples are reinstated hereafter and explained again in terms of the flux *vectors*.

It is of interest to notice that a nonzero average value of the electromagnetic torque is achieved only in cases where the angle $\Delta\theta = \theta_{\Psi S} - \theta_{\Psi R}$ is constant. Namely, the stator and rotor flux *vectors* must retain their relative position. Whether they revolve or stay firm, the angle between flux *vectors* representing the stator and rotor magnetic fields must not change. There are three cases where this condition is fulfilled:

(a) Stator field is still with respect to stator. Rotor field rotates with respect to rotor in the opposite direction of rotation of rotor; thus, rotor field does not move with respect to stator.

(b) Stator field rotates with respect to stator. Rotor field rotates with respect to rotor. Sum of rotor speed relative to stator and rotor field speed relative to rotor is equal to speed of rotation of stator field relative to stator.

(c) Stator field rotates with respect to stator at speed equal to rotor speed. Rotor field does not move with respect to rotor.

Generation of stator or rotor magnetic field which does not rotate with respect to originating windings can be done by one or more windings with DC currents. Generation of the field that revolves with respect to originating windings requires at least two windings of different spatial orientation and with AC currents of the appropriate frequency and initial phase. Conditions for creation of a rotating field have been discussed in the section devoted to rotating field.

Case (a) corresponds to direct current machines (DC machines), and it is shown in Fig. 7.8a. *Vector* of the stator flux does not move with respect to the stator because the stator conductors have constant DC currents. The torque generation requires the rotor flux to retain its relative position to the stator flux. This means that the rotor flux *vector* in Fig. 7.8a should not move either. For the rotor flux *vector* to remain still while the rotor windings revolve, electrical currents in rotor conductors should create magnetic field which rotates with respect to the rotor at the speed $-\Omega_m$. In this case, the rotor revolves in positive direction at the speed $+\Omega_m$, while the rotor flux revolves with respect to the rotor in the opposite direction. Therefore, the rotor flux remains still with respect to the stator. For the rotor windings to create revolving field, the rotor must have AC currents, the frequency of which is determined by the speed of rotation. Yet, DC machines are supplied from DC power sources. In order to convert DC supply currents into AC rotor currents, DC machines have a mechanical commutator, device with brushes attached to the stator and collector attached to the rotor. Collector has a number of isolated segments, connected to the rotor conductors. When the rotor revolves, the segments slide under the brushes, altering electrical connections and changing the way of injecting the supply current into the rotor winding. In such way, commutator directs DC current of the electrical power source into rotor conductors in such way that the rotor conductors have AC currents. The frequency of these currents is determined by the speed of rotation. The commutator will be explained in more detail in the chapter on *DC machines*. Thanks to the commutator, the flux *vector* Φ_{R1} remains still with respect to flux *vector* Φ_{S1}; thus, the angle $\Delta\theta$ remains constant.

Case (b) shown in Fig. 7.8b corresponds to asynchronous machines. Windings of the stator and rotor have AC currents of angular frequency ω_s and ω_k, respectively. The stator flux *vector* rotates with respect to the stator at the speed Ω_s, determined by the angular frequency ω_s. The rotor flux *vector* rotates with respect to the rotor at the speed of Ω_k, determined by the angular frequency ω_k. Difference $\omega_s - \omega_k$ in angular frequency determines the rotor speed Ω_m. The flux *vectors* of the stator and rotor rotate at the same speed (Ω_s). Consequently, their mutual position $\Delta\theta$ does not change.

Case (c) shown in Fig. 7.8c corresponds to synchronous machines. There are rotor windings with constant DC currents producing the rotor flux. Alternatively, the rotor does not have any windings. Instead, there are permanent magnets mounted on the rotor. In either case, the rotor flux *vector* Φ_{R1} rotates at the same speed as the rotor does. The stator of the machine has a system of windings with two or more phases carrying AC currents. Angular frequency ω_s of the stator currents creates the stator magnetic field which revolves at the speed Ω_s, determined by the angular frequency ω_s of the stator currents. In synchronous machines, the stator frequency ensures that the field rotates at the same speed as the rotor, that is, $\Omega_s = \Omega_m$. Therefore, the flux *vectors* Φ_{R1} and Φ_{S1} rotate at the same speed, and their mutual position $\Delta\theta$ does not change.

Question (9.4): Derive the expression for the torque acting on a contour with electrical current in a homogenous magnetic field, with the normal to the contour being inclined with an angle θ with respect to the *vector B*, as shown in Fig. 3.6. The contour is circular, with diameter D and with electrical current I. Assume that the contour revolves around the axis which is orthogonal to the direction of the field. The speed of rotation is known and constant. Determine the instantaneous and average value of the torque acting on the contour. Assuming that the magnetic field cannot be changed, but it is possible to have an arbitrary current in the contour, determine the current $i(t)$ which would result in a nonzero average value of the torque.

With the assumption that the induction $B(t)$ is variable while the electrical current $i(t) = I$ is constant, determine one solution for $B(\theta)$ which results in a nonzero average value of the torque.

Answer (9.4): The electromagnetic torque T_{em} acting on the contour is equal to $I{\cdot}B{\cdot}S{\cdot}\sin\theta$. When the contour rotates at angular speed Ω while both the current and magnetic inductance are constant, the torque varies according to function $\sin(\Omega\, t)$, and its average value is zero. In cases where the magnetic induction is constant while the current $i(t) = I\sin(\omega t)$ changes with angular frequency ω equal to the speed of rotation Ω, the torque is proportional to $(\sin(\omega t))^2$ and has a nonzero average value. If the contour has a constant current I, nonzero average of the torque can be obtained in cases when the magnetic induction changes as $B(t) = B_m{\cdot}\sin(\omega t)$, where the angular frequency ω corresponds to the speed of rotation Ω.

9.8 Torque–Size Relation

Expression for electromagnetic torque (9.16) shows that the torque is proportional to $R^3 L$, that is, to the axial length of the machine L and third power of its diameter D. Diameter and axial length are linear dimensions of the machine, and common notation l can be used for both. Therefore, the electromagnetic torque is proportional to fourth power of the linear dimensions l of the machine, $T \sim l^4$. Volume of the machine is proportional to the third power of linear dimensions, $V \sim l^3$. Therefore, the torque is proportional to $T \sim V^{4/3}$.

Electrical machines are made of iron and copper, materials of known specific masses.[1] Therefore, the mass m of an electrical machine is determined by the electromagnetic torque for which it has been designed. The mass m and torque T are related by $T \sim m^{4/3}$. As an example, a new machine with all the three dimensions doubled with respect to the original machine develops electromagnetic torque increase $2^4 = 16$ times.

Relation $T \sim m^{4/3}$ can be verified in another way. It has been proven that the torque can be expressed as **vector** product of two flux **vectors**. The flux amplitude depends on the surface ($S \sim l^2$) and magnetic induction ($B < B_{max}$), the latter being limited by magnetic saturation of the ferromagnetic material and not exceeding 1.5 . 1.7 T. The product of two fluxes depends on the fourth power of linear dimension l. Hence, the product of stator and rotor flux **vectors** depends on l^4. Hence, the electromagnetic torque available from electrical machine is proportional to l^4.

Power of electromechanical conversion in an electrical machine depends on the torque and speed of rotation Ω; therefore, $P \sim V^{4/3}\Omega$. Considering two machines with the same dimensions and different speeds, the one with the higher speed delivers more power. In cases requiring a constant power of electromechanical conversion P, while the speed of rotation of the electrical machine can be arbitrarily chosen, it is beneficial to select the machine with higher speed, resulting in a lower torque $T = P/\Omega$ and consequently smaller dimensions of the machine due to $T \sim l^4$. An example where the required load speed can be achieved with different machine

---

[1] $\gamma_{Fe} = \Delta m_{Fe}/\Delta V = 7{,}874$ kg/m^3, $\gamma_{Cu} = \Delta m_{Cu}/\Delta V = 8{,}020$ kg/m^3.

speeds is the case where the load and the machine are coupled by gears. While designing the system, the gear ratio can be selected so as to result in a higher speed of the machine. This will reduce the size and weight of the machine.

The torque expression (9.16) suggests that the torque is inversely proportional to the air gap width δ. With constant stator and rotor currents, a decrease in the air gap results in an increase in electromagnetic torque. The expression suggests that the torque can be increased with no apparent limits, provided that the air gap δ can get sufficiently small. This conclusion is incorrect as it overlooks the phenomenon of magnetic saturation. The expression for the torque has been derived as a result of an analysis where magnetic saturation in iron is neglected. The stator and rotor magnetic circuits are made of iron sheets of very high permeability μ_{Fe}, making the magnetic field in iron H_{Fe} negligible. The torque expression (9.16) is based on such an assumption. It holds in all the conditions with no magnetic saturation in iron parts of the magnetic circuit. With excessive values of B_{Fe} resulting in magnetic saturation, the value of H_{Fe} cannot be neglected, and this invalidates the (9.16). The preceding analysis finds the magnetic induction B_0 in the air gap inversely proportional to the air gap δ. Disregarding the slots, the magnetic induction in iron is roughly the same, $B_{Fe} \approx B_0$. Therefore, progressive decrease of the air gap leads to increased magnetic induction. As the magnetic induction B reaches $B_{max} = 1.5 \, .. \, 1.7$ T, the iron gets saturated, permeability μ_{Fe} drops, and the magnetic field H_{Fe} assumes considerable value that cannot be neglected. At this point, (9.15) and the consequential results, obtained by neglecting saturation, are not valid and cannot be used.

The air gap of electrical machines is designed to be as small as possible, in order to obtained the desired magnetic induction B_0 with smaller electrical currents and, consequently, smaller copper losses. However, there are limitations of mechanical nature which prevent the air gap of smaller machines from getting much below one millimeter. The air gap of large electrical machines is at least several millimeters. A lower limit of the air gap is required to prevent the revolving rotor from touching the stator. Undesired touching and scratching can happen due to finite tolerances in manufacturing the stator and rotor surfaces. Elastic deformation of the shaft in radial direction can result in rotor touching the stator. These phenomena prevent the use of electrical machines with very small air gaps.

Question (9.5): The expressions for the electromagnetic torque and power of electrical machine give values inversely proportional to the air gap δ. Based on these expressions, the power and torque can be increased with no apparent limits, keeping the electrical currents constant and reducing the air gap. There are reasons that invalidate such conclusion. Provide two reasons which indicate that such expectations are not realistic.

Answer (9.5): The expression for the electromagnetic torque suggests that reduction of the air gap δ results in higher torque and higher power of electromechanical conversion. Apparently, very high torque can be achieved with an adequate reduction of the air gap.

This conclusion overlooks the phenomenon of magnetic saturation. The torque expression comes from an analysis that starts with an assumption that the magnetic field H_{Fe} in iron is negligible. This assumption holds only in cases where the flux density B_{Fe} does not reach the saturation limit of $B_{max} = 1.5 .. 1.7$ T. Namely, given a magnetomotive force $Ni = 2H\delta$, the flux density $B = \mu_0 Ni/(2\delta)$ grows as the air gap decreases. As the flux density B reaches the saturation limit B_{max}, any further increase of B_{Fe} is determined by expression $\Delta B = \mu_0 \Delta H$. In other words, differential permeability $\Delta B/\Delta H$ of the saturated ferromagnetic material is close to μ_0. Considering the flux changes, magnetic saturation is equivalent to removing the iron parts of the stator and rotor magnetic circuits. Due to $\Delta B/\Delta H \approx \mu_0$, saturated iron behaves like air. Therefore, the magnetic saturation can be considered as a very large increase in the air gap. Therefore, the initial projections of the air gap reduction leading to large torque gains are not realistic. It is of interest to notice that most electrical machines are designed so as to get the most out of their magnetic circuits. For this to achieve, the flux density levels are close to saturation limits. Therefore, there is no margin to accommodate any further increase in B.

Another reason that prevents the torque increase is the fact that the air gap cannot be decreased below certain limits, roughly 1 mm, imposed by mechanical conditions.

9.9 Rotating Magnetic Field

According to previous considerations, conditions for generating a nonzero average electromagnetic torque include a constant relative position $\Delta\theta$ of the stator and rotor flux *vectors*. In DC machines, both flux *vectors* remain still with respect to the stator. In alternating current machines, whether asynchronous or synchronous, both flux *vectors* rotate at the same speed.

In order to meet the above condition and due to rotor revolution, at least one of the two fluxes (Φ_S or Φ_R) has to rotate with respect to the winding that originates the magnetomotive force resulting in the relevant flux. The magnetic field which rotates with respect to the originating windings is also called *rotating magnetic field*. In this section, it is shown that rotating magnetic field requires a system with at least two separate windings with appropriate spatial displacement of their magnetic axes. Alternating currents in the windings should have the same frequency and amplitude. Their initial phases are to be different and should correspond to the spatial displacement of the magnetic axes. In this case, the system of windings creates magnetomotive force and flux that revolve at the speed determined by the frequency of AC currents.

9.9.1 System of Two Orthogonal Windings

Electromagnetic torque is determined by the *vector* product of the stator and rotor flux *vectors*, and it depends on sine of the angle $\Delta\theta$ between the two *vectors*. A continuous conversion of energy with constant torque and constant power

Fig. 9.9 A system with
two orthogonal windings

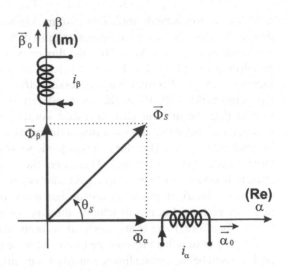

requires that the angle $\Delta\theta$ is constant. For this reason, it is necessary that the stator
and/or rotor windings create rotating magnetic field.

Figure 9.9 shows a stator with two windings, α and β. Each winding has N turns.
The winding conductors could be either concentrated or distributed. Their construction does not affect the subsequent analysis and conclusions. For brevity, it will be
considered that windings α and β are concentrated. Flux in one turn of a concentrated
winding is denoted by Φ. It is equal to the ratio of the magnetomotive force $F = Ni$
and the magnetic resistance R_μ. Due to high permeability of iron, magnetic field H_{Fe}
can be neglected. Considering concentrated winding with electrical current i, the
field strength in the air gap is obtained from relation $Ni = 2\delta H_0$, while magnetic
induction B in the air gap is equal to $B_0 = \mu_0 H_0$. Surface S_1 is encircled by one turn
of the considered concentrated winding. Assuming that the surface passes through
the air gap, it represents one half of a cylinder and it has the surface area $S_1 = \pi LR$.
Therefore, the flux in one turn is

$$\Phi = B_0 S_1 = \mu_0 H_0 \pi LR = \frac{\mu_0 \pi LR}{2\delta} Ni.$$

This expression can be verified by calculating the flux by dividing the
magnetomotive force and the magnetic resistance, $\Phi = F/R_\mu$. Magnetic resistance
R_μ is calculated considering that $H_{Fe} = 0$, and taking into account that each field
line passes twice through the air gap. Therefore,

$$R_\mu = \frac{1}{\mu_0} \frac{2\delta}{S_1} = \frac{1}{\mu_0} \frac{2\delta}{\pi LR},$$

where $\pi LR = S_1$ represents the surface area of the cross section of considered
magnetic circuit while μ_0 is the permeability in the air gap. Quantity 2δ represents

the length of the magnetic circuit where the field strength H assumes considerable values and the line integral of the field results in *magnetic voltage drop*. The air gap is passed twice, where the field lines enter and exit the rotor (or stator) magnetic circuit, that is, next to the north and south magnetic poles. Expression $\Phi = F/R_\mu$ becomes

$$\Phi = \frac{F}{R_\mu} = \frac{Ni}{\left(\frac{1}{\mu_0}\frac{2\delta}{\pi LR}\right)} = \frac{\mu_0 \pi LR}{2\delta} Ni. \tag{9.19}$$

Magnetic axes of the windings α and β reside on the abscissa and ordinate of the orthogonal coordinate system shown in Fig. 9.9. Axis of the winding α is horizontal, along the course defined by the unit *vector* $\boldsymbol{\alpha}_0$. By establishing electrical current i_α, magnetomotive force $F_\alpha = Ni_\alpha$ is produced along the course and direction of the unit *vector* $\boldsymbol{\alpha}_0$. The flux in one turn is obtained by dividing the magnetomotive force and the magnetic resistance, $\Phi_\alpha = Ni_\alpha/R_\mu$. It is assumed that the windings are concentrated; thus, the flux Ψ_α in the winding is equal to $N\Phi_\alpha = N^2 i_\alpha/R_\mu$. The axis of the winding β is orthogonal with respect to the α winding, and it extends along the course defined by the unit *vector* $\boldsymbol{\beta}_0$. The magnetomotive force F_β and the flux Φ_β of this winding are oriented in accordance with the ordinate axis β, and it is proportional to the winding current i_β. By using the unit *vectors* of the two axes, fluxes in the winding turns can be represented by expressions

$$\vec{\Phi}_\alpha = \frac{N}{R_\mu} i_\alpha \cdot \vec{\alpha}_0, \quad \vec{\Phi}_\beta = \frac{N}{R_\mu} i_\beta \cdot \vec{\beta}_0.$$

Electrical currents in windings α and β are alternating currents of the same amplitude I_m and the same angular frequency ω_s. The symbol ω_s denotes the angular frequency of electrical currents in the stator windings. The initial phases of the two currents are different. The current in winding α leads by $\pi/2$, the angle that corresponds to the spatial shift between α and β magnetic axes. Variation of currents in the windings is given by (9.20):

$$i_\alpha = I_m \cos(\omega_s t) = I_m \cos \theta_S,$$

$$i_\beta = I_m \cos\left(\omega_s t - \frac{\pi}{2}\right) = I_m \sin(\omega_s t) = I_m \sin \theta_S. \tag{9.20}$$

The resultant magnetomotive force F_S of the stator winding and the stator flux Φ_S are obtained by summing their α and β components. If the orthogonal windings α and β in Fig. 9.9 have electrical currents as given by (9.20), the magnetomotive force *vector* F_S is created, determined by expression (9.21), resulting in the flux *vector* given in (9.22). Since α and β components are proportional to functions $\cos\theta_s$ and $\sin\theta_s$, where the angle θ_s changes as $\omega_s t$, the latter equation describes the flux *vector* which rotates at the speed of ω_s and has an amplitude which is constant. Equation (29.22) gives the resultant flux corresponding to one turn. Since a concentrated winding is in consideration, the resultant *vector* of the flux of the windings is obtained by multiplying the flux in one turn by the number of turns, as defined in (9.23):

$$\vec{F}_S = R_\mu \vec{\Phi}_S = N I_m (\vec{\alpha}_0 \cos \theta_s + \vec{\beta}_0 \sin \theta_s)$$
$$= N I_m \left[\vec{\alpha}_0 \cos(\omega_s t) + \vec{\beta}_0 \sin(\omega_s t) \right], \tag{9.21}$$

$$\vec{\Phi}_S = \frac{N I_m}{R_\mu} (\vec{\alpha}_0 \cos \theta_s + \vec{\beta}_0 \sin \theta_s) \tag{9.22}$$

$$\vec{\Psi}_S = N \vec{\Phi}_S = \frac{N^2 I_m}{R_\mu} (\vec{\alpha}_0 \cos \theta_s + \vec{\beta}_0 \sin \theta_s). \tag{9.23}$$

The flux components Φ_α and Φ_β are the projections of the stator flux *vector* Φ_S (9.22) on the axes of the α–β coordinate system defined by the unit *vectors* α_0 and β_0. Projections Φ_α and Φ_β of the *vector* Φ_S on α- and β-axes represent, at the same time, the fluxes in one turn of the respective α and β windings. Axes of the windings shown in Fig. 9.9 are mutually orthogonal. Therefore, the currents in α winding do not cause variations of the flux in β winding. The same way, the currents in β winding do not affect the flux in α winding.

Since the revolving *vector* has α and β components of the flux, it can be concluded that creation of a rotating field requires the existence of at least two spatially displaced windings.

In (9.22), components Φ_α and Φ_β of the flux are accompanied by unit *vectors* α_0 and β_0. Written presentation can be simplified by substituting the plane α–β with the plane representing complex numbers, with α-axis being the real axis and β-axis being the imaginary axis. Formal translation of equations from α–β coordinate system into the complex plane is done by substituting the unit *vector* α_0 with 1 and substituting the unit *vector* β_0 with imaginary unit j. In this way, (9.22) changes into

$$\underline{\Phi}_S = \Phi_\alpha + j \Phi_\beta = \frac{N I_m}{R_\mu} (\cos \theta_s + j \sin \theta_s) = \frac{N I_m}{R_\mu} e^{j\theta_s}. \tag{9.24}$$

On the basis of the preceding analysis, it is concluded that a system of two orthogonal, mutually independent windings can create a rotating magnetic field. In cases when the windings carry sinusoidal currents of the same angular frequency ω_s, the same amplitude I_m, and with their initial phases shifted by $\pi/2$, the consequential magnetomotive force and flux in the machine revolve. Therefore, these quantities can be represented by rotating *vectors*. The *vectors* rotate at the speed Ω_S which is determined by the angular frequency ω_s. In the course of rotation, there is no change in amplitude of these *vectors*. For the system of two windings shown in Fig. 9.9, the speed Ω_S is equal to the angular frequency ω_s.

Question (9.6): Consider the stator winding shown in Fig. 9.9 and assume that the amplitudes of the two stator currents are equal. The difference of initial phases of currents i_α and i_β is denoted by φ.

- Determine and describe the stator flux *vector* in cases where $\varphi = 0$.
- Determine and describe the stator flux *vector* for $\varphi = \pi/2$.
- Show that in cases with $0 < \varphi < \pi/2$, the *vector* of the stator flux can be represented by the sum of two flux *vectors*, one of them rotating at the speed $\Omega_S = \omega_s$ and maintaining a constant amplitude while the other pulsating back and forth along the same course.

Answer (9.6): Equation (9.19) allows the flux in one turn to be calculated as the ratio of magnetomotive force $F = Ni$ and magnetic resistance R_μ,

$$\Phi = \frac{\mu_0 \pi L R}{2\delta} Ni = \frac{Ni}{\left(\frac{1}{\mu_0} \frac{2\delta}{\pi L R}\right)} = \frac{Ni}{R_\mu}.$$

When currents i_α and i_β are known, components of the stator flux are determined by expressions

$$\vec{\Phi}_\alpha = \frac{N}{R_\mu} i_\alpha \cdot \vec{\alpha}_0, \quad \vec{\Phi}_\beta = \frac{N}{R_\mu} i_\beta \cdot \vec{\beta}_0.$$

In cases with $\varphi = 0$, the instantaneous values of electrical currents $i_\alpha(t)$ and $i_\beta(t)$ are equal; thus, the resultant flux *vector* is

$$\vec{\Phi}_S = \frac{N}{R_\mu} I_m \cos(\omega_S t) \cdot \left(\vec{\alpha}_0 + \vec{\beta}_0\right).$$

Therefore, with $\varphi = 0$, the flux *vector* does not revolve, and it pulsates along the line inclined by $\pi/4$ with respect to the abscissa. The algebraic value of the *vector* oscillates at the angular frequency ω_s.

In cases where $\varphi = \pi/2$, the *vector* of the magnetomotive force is determined by (9.21), while the resultant flux *vector* of one turn is

$$\vec{\Phi}_S = \frac{N I_m}{R_\mu} \left[\vec{\alpha}_0 \cos(\omega_s t) + \vec{\beta}_0 \sin(\omega_s t)\right].$$

In general, electrical current in winding β can be written in the form

$$i_\beta = I_m \cos(\omega_S t - \phi) = I_m \cos(\omega_S t) \cos\phi + I_m \sin(\omega_S t) \sin\phi,$$

while the current in winding α can be written as the sum

$$i_\alpha = I_m \cos(\omega_S t) = I_m \cos(\omega_S t)(1 - \sin\phi) + I_m \cos(\omega_S t) \sin\phi.$$

The flux *vector* can be represented by the sum of two *vectors*:

$$\vec{\Phi}_S = \vec{\Phi}_{SO} + \vec{\Phi}_{SP},$$

where the elements of the sum are determined by

$$\vec{\Phi}_{SO} = \frac{NI_m}{R_\mu} \cdot \sin\phi \cdot \left[\vec{\alpha}_0 \cos(\omega_s t) + \vec{\beta}_0 \sin(\omega_s t) \right],$$

$$\vec{\Phi}_{SP} = \frac{NI_m}{R_\mu} \cdot \cos(\omega_s t) \cdot \left[\vec{\alpha}_0 (1 - \sin\phi) + \vec{\beta}_0 \cos\phi \right].$$

Vector Φ_{SO} represents a rotating field which revolves at the speed $\Omega_s = \omega_s$. The amplitude of this *vector* does not change in the course of rotation, and it is proportional to the sine of the angle φ. Therefore, with $\varphi = 0$, the rotating field of the machine does not exist. The *vector* Φ_{SP} has a course which does not change. This course is determined by the angle φ. With $\varphi = 0$, the course of the pulsating field Φ_{SP} is $\pi/4$ with respect to the abscissa. The flux Φ_{SP} does not rotate but pulsates instead along the indicated course.

9.9.2   System of Three Windings

In most cases, asynchronous and synchronous motors are fed from voltage sources providing a symmetrical three-phase system of voltages and currents. When operating as generators, the machines convert the mechanical work into electrical energy and produce a system of three-phase voltages available to electrical loads at the stator terminals. For this reason, the stator windings of asynchronous and synchronous machines usually have three phases. That is, there are three separate, spatially displaced windings on the stator. The three separate stator windings are called the phases and assigned letters a, b, and c. In three-phase machines, the axes of the phase windings are spatially displaced by $2\pi/3$. Figure 9.10 shows a machine with phase windings a, b, and c carrying sinusoidal currents of equal amplitudes I_m, equal angular frequency ω_s, and with difference in initial phases of $\pm 2\pi/3$. The phase shift of the phase currents corresponds to the spatial displacement between the magnetic axes of the phase windings.

The magnetomotive forces F_a, F_b, and F_c of the windings have amplitudes Ni_a, Ni_b, and Ni_c. Their orientation is determined by magnetic axes of respective windings, and their courses can be expressed in terms of unit *vectors* α_0 and β_0,

$$\vec{a}_0 = \vec{\alpha}_0, \qquad \vec{b}_0 = -\frac{1}{2}\vec{\alpha}_0 + \frac{\sqrt{3}}{2}\vec{\beta}_0, \qquad \vec{c}_0 = -\frac{1}{2}\vec{\alpha}_0 - \frac{\sqrt{3}}{2}\vec{\beta}_0. \tag{9.25}$$

Fig. 9.10 Positions of the vectors of magnetomotive forces in individual phases, position of their magnetic axes, and unit vectors of the orthogonal coordinate system

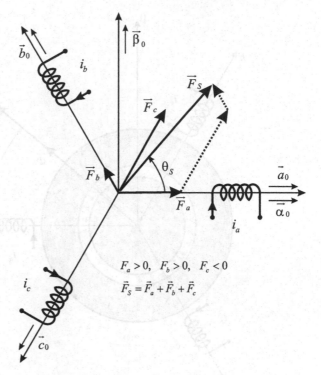

$$F_a > 0, \quad F_b > 0, \quad F_c < 0$$
$$\vec{F}_S = \vec{F}_a + \vec{F}_b + \vec{F}_c$$

Currents of the phase windings are determined by equations

$$i_a = I_m \cos \omega_S t,$$
$$i_b = I_m \cos(\omega_S t - 2\pi/3),$$
$$i_c = I_m \cos(\omega_S t - 4\pi/3),$$

(9.26)

in such way that **vectors** of the magnetomotive forces of individual phases become

$$\vec{F}_a = N i_a \vec{a}_0, \quad \vec{F}_b = N i_b \vec{b}_0, \quad \vec{F}_c = N i_c \vec{c}_0.$$

By using relation (9.25), the **vectors** representing magnetomotive forces in individual phases can be expressed in terms of unit **vectors** α_0 and β_0,

$$\vec{F}_a = N i_a \vec{\alpha}_0,$$
$$\vec{F}_b = N i_b \left(-\frac{1}{2}\vec{\alpha}_0 + \frac{\sqrt{3}}{2}\vec{\beta}_0\right),$$
$$\vec{F}_c = N i_c \left(-\frac{1}{2}\vec{\alpha}_0 - \frac{\sqrt{3}}{2}\vec{\beta}_0\right).$$

(9.27)

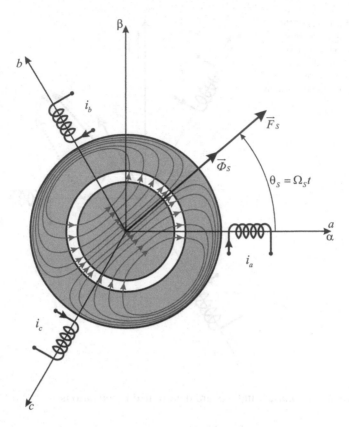

Fig. 9.11 Field lines and vectors of the rotating magnetic field

The resultant magnetomotive force of the stator windings is obtained by **vectors** summation of magnetomotive forces in individual phases, and it is given by equation

$$\vec{F} = \vec{F}_a + \vec{F}_b + \vec{F}_c = \frac{3}{2}NI_m\left[\vec{\alpha}_0 \cos \omega_s t + \vec{\beta}_0 \sin \omega_s t\right]. \qquad (9.28)$$

Summing the individual magnetomotive forces of the three phases, one obtains the rotating **vectors** of the resultant magnetomotive force with an amplitude of $3/2NI_m$. Projections of this **vectors** on axes α and β are proportional to functions $\cos(\omega_s t)$ and $\sin(\omega_s t)$, proving that the **vectors** revolves at the speed of $\Omega_s = \omega_s$ and that it has a constant amplitude. The ratio of the magnetomotive force F_S of the stator windings and the resistance R_μ of the magnetic circuit gives the flux **vectors** Φ_S of the stator which rotates at the same speed as the **vectors** F_S. Hence, the system of sinusoidal currents in three-phase stator winding results in a revolving magnetic field with the speed of rotation determined by the angular frequency of the phase currents, while the field magnitude depends on the maximum value I_m of the phase currents.

As already emphasized in the introduction, it is necessary to distinguish between the speed of rotation of the rotor, being mechanical quantity expressed in rad/s, and the angular frequency of electrical currents and voltages, appertaining to electrical circuits and being expressed in rad/s as well. Throughout this book, mechanical speed is denoted by Ω, while the angular frequency of voltages and currents is denoted by ω.

Figure 9.11 shows *vectors* of the resultant magnetomotive force and the resultant flux in electrical machine with three-phase system of stator windings.

Chapter 10
Electromotive Forces

Electromotive forces induced in windings of electrical machines are analyzed and discussed in this chapter. Analysis includes transformer electromotive forces and dynamic electromotive forces. The rms values, the waveforms, and harmonics are derived for concentrated and distributed windings. For real windings that have conductors distributed in a limited number of slots, the electromotive forces are calculated by introducing, explaining, and using chord factors and belt factors. Discussion includes design methods that suppress low-order harmonics in electromotive forces. This chapter concludes with the analysis of distributed windings with sinusoidal change of conductor density. Calculation of flux linkage and electromotive force in such windings shows that they achieve suppression of all harmonic distortions and operate as ideal spatial filters.

Variation of the flux in the machine windings results in induction of electromotive forces, proportional to the first time derivative of the flux. The voltage balance in each winding is given by equation $u = Ri + \mathrm{d}\Psi/\mathrm{d}t$, where u denotes the voltage across the winding terminals, i is the electrical current in the winding, R is the winding resistance, while the flux derivative represents the induced electromotive force. In the introductory courses of electrical engineering, the electromotive force is calculated as the first derivative of the flux with a leading negative sign, $e = -\mathrm{d}\Psi/\mathrm{d}t$. This convention indicates that the induced electromotive force and consequential change in electrical current oppose to the flux changes. Namely, in a short circuited winding ($u = 0$), the flux changes produce the electromotive force which, in turn, gives a rise to electrical current which opposes to the flux changes. Adopting another convention has its own advantages as well. By defining the electromotive force $e = +\mathrm{d}\Psi/\mathrm{d}t$, the current in the winding is determined as the ratio of the voltage difference ($u - e$) and the resistance R. The electromotive force defined as $e = +\mathrm{d}\Psi/\mathrm{d}t$ is *opposed* to the voltage. Therefore, it is also called *counter electromotive force*. In this book, the latter convention has been adopted with e denoting $+\mathrm{d}\Psi/\mathrm{d}t$ and with the voltage balance in each windings being $u = Ri + e$.

S.N. Vukosavic, *Electrical Machines*, Power Electronics and Power Systems,
DOI 10.1007/978-1-4614-0400-2_10, © Springer Science+Business Media New York 2013

10.1   Transformer and Dynamic Electromotive Forces

Electromotive forces are generated due to changes of the winding flux. The flux can change due to changes in electrical currents of the windings or due to motion of the rotor with respect to the stator. In cases where the flux changes take place due to motion, the consequential electromotive forces are called *dynamic electromotive forces*. In cases where the stator windings carry constant currents, dynamic electromotive force appears in the rotor winding which rotates with respect to stator. Constant stator currents create a stationary magnetic field which does not move with respect to the stator. One part of this flux encircles the rotor windings as well. The rotor flux caused by the stator current depends on the relative position between the stator and rotor. When the rotor moves, such rotor flux changes, and this leads to creation of a dynamic electromotive force.

Electromotive force can also arise in cases with no rotor movement. If electrical current in stator conductors is variable, the flux created by the stator winding is variable as well. In part, the rotor flux is a consequence of stator currents. The amount of the rotor flux caused by stator currents is determined by the mutual inductance between the stator and rotor windings. Even with the rotor that does not move, the rotor flux varies due to variable electrical currents in the stator winding. As a consequence, an electromotive force is induced in the rotor winding. It is called *transformer electromotive force*. In a power transformer, alternating currents in the primary winding produce a variable flux which also encircles the secondary winding. Variable flux leads to the *transformer* electromotive force in the secondary winding, providing the means for passing the electrical power from the primary to the secondary side.

Electromotive force $e = L di/dt$, which appears in a stand-alone winding due to variation of the electrical current i, is proportional to the coefficient L, the self-inductance of the winding. This electromotive force is called the *electromotive force of self-induction*.

10.2   Electromotive Force in One Turn

For the purpose of modeling electrical machines, it is necessary to calculate the electromotive force induced in concentrated and distributed windings. Figure 10.1 shows an electrical machine with permanent magnets in the rotor magnetic circuit. The magnets are shaped and arranged in the way to create sinusoidal distribution of the magnetic induction in the air gap,

$$B = B_m \cos(\theta - \theta_m). \tag{10.1}$$

Due to rotation of the rotor, the maximum induction $B_m = \mu_0 H_m$ is reached at position $\theta = \theta_m$, where θ_m represents relative position of the rotor with respect to

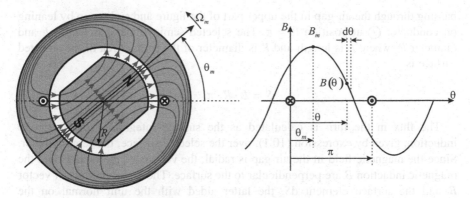

Fig. 10.1 Rotor field is created by action of permanent magnets built in the magnetic circuit of the rotor

the stator. In the case when rotor revolves at a constant speed Ω_m, position of the rotor changes as $\theta_m = \Omega_m t$, and magnetic induction at the observed position θ is equal to

$$B(\theta, t) = B_m \cos(\theta - \Omega_m t).$$

The expression represents a wave of sinusoidally distributed magnetic field which moves along with the rotor. It is of interest to determine the electromotive force induced in one turn of the stator. This turn is shown in Fig. 10.1, and it is made of two diametrical conductors. Conductor denoted by $\otimes$ is at position $\theta_\otimes = 0$, while conductor denoted by $\odot$ is at position $\theta_\odot = \pi$. The electromotive force in the stator turn can be determined in two ways:

1. By determining the first derivative of the flux encircling the turn
2. By calculating and summing the electromotive forces of individual conductors

10.2.1 Calculating the First Derivative of the Flux in One Turn

First derivative of the flux in one turn determines the counter electromotive force which is induced in the turn. The flux encircling the turn exists due to action of the permanent magnets mounted on the rotor, which create magnetic field in the air gap. Spatial distribution of the magnetic induction in the air gap, along the rotor circumference, is described by expression (10.1). The flux in the considered turn is equal to the surface integral of the magnetic induction over any surface leaning on conductors $\otimes$ and $\odot$. In order to facilitate the calculation, one should select the surface which passes through the air gap; the expression $B(\theta)$ is readily available. Such a surface is a semicylinder starting from conductor $\otimes$ at position $\theta = 0$, passing through the air gap in the upper part of the figure and ending up by leaning on conductor $\odot$ at position $\theta = \pi$. The selected semicylinder has length L and diameter R, where L is length and R is diameter of the rotor. Area of the selected surface is

$$S = L \cdot R \cdot \pi. \tag{10.2}$$

The flux in the turn is calculated as the surface integral of the magnetic induction, given by expression (10.1), over the selected surface, the semi cylinder. Since the magnetic field in the air gap is radial, the vectors of the field H and the magnetic induction B are perpendicular to the surface. The scalar product of vector B and the surface element dS, the latter aided with the unit normal on the semicylinder, becomes the product between the amplitudes of B and dS. Thus, the expression for the flux obtains the form

$$\Phi = \int_S \vec{B} \cdot d\vec{S} = \int_S B \cdot dS.$$

Elementary surface of the semicylinder is $dS = L \cdot R \cdot d\theta$, and expression for the flux in one turn of the stator due to action of permanent magnets assumes the form

$$\Phi = \int_0^\pi d\Phi = \int_0^\pi B \cdot dS = \int_0^\pi B \cdot LR d\theta$$
$$= (2LRB_m) \sin \theta_m = \Phi_m \sin \theta_m. \tag{10.3}$$

Flux in the turn is dependent on relative position between the stator and rotor. Maximum value of the flux in one turn is reached when the rotor comes to position $\theta = \pi/2$. In this position, the flux is equal to Φ_m, where

$$\Phi_m = 2 \cdot R \cdot L \cdot B_m. \tag{10.4}$$

In other positions, the flux in one turn has smaller values, $\Phi(\theta_m) = \Phi_m \sin \theta_m$. When the rotor revolves at a constant speed Ω_m, the rotor position changes as $\theta_m = \Omega_m t$, and the flux in one turn is

$$\Phi = \Phi_m \sin(\Omega_m t),$$

while the (counter) electromotive force in the turn is

$$e_1 = + \frac{d\Phi}{dt} = \Omega_m \Phi_m \cos(\Omega_m t). \tag{10.5}$$

10.2.2 Summing Electromotive Forces of Individual Conductors

It is possible to determine the electromotive force in one turn by summing the electromotive forces of individual conductors that make one turn. In the example presented in Fig. 10.1, the field of permanent magnets revolves with respect to the stator conductors. The peripheral speed of relative motion is $v = R\Omega_m$, where R is the rotor radius and Ω_m is the angular speed of the rotor. In the conductor denoted by $\otimes$, the electromotive force is

$$e_\otimes = L \cdot R \cdot \Omega_m \cdot B_m \cdot \cos(\Omega_m t).$$

At position $\theta = \pi$, where the diametrical conductor $\odot$ is placed, the magnetic induction has the same amplitude and the opposite direction. For this reason, the electromotive force $e_\odot$ induced in the conductor denoted by $\odot$ is of the opposite direction, $e_\otimes = -e_\odot$. The electromotive force in one turn is obtained by summing the electromotive forces in individual conductors. One turn is formed by connecting the two conductors in series. The terminals of the turn are made available at the front side of the cylinder. The other ends of the conductors $\otimes$ and $\odot$ are connected at the rear side of the cylinder. Therefore, when circulating along the contour made by the two conductors, the electromotive forces of conductors $e_\otimes$ and $e_\odot$ are summed according to $e = e_\otimes - e_\odot$. Finally, the electromotive force in one turn is

$$e_1 = 2 \cdot e_\otimes = 2 \cdot L \cdot R \cdot \Omega_m \cdot B_m \cdot \cos(\Omega_m t). \tag{10.6}$$

According to expression (10.4), the maximum value of the flux in one turn is equal to $\Phi_m = 2LRB_m$, and the previous expression can be written as

$$e_1 = \Omega_m \Phi_m \cos(\Omega_m t), \tag{10.7}$$

which is in accordance with (10.5).

In the considered example, the induced electromotive forces are harmonic functions of time; thus, it is possible to represent them by phasors. Summing the electromotive forces $e_\otimes$ and $e_\odot$ can be represented by phasors, as shown in Fig. 10.3.

10.2.3 Voltage Balance in One Turn

The voltage balance within one turn is given by equation

$$u = Ri + e_1 \approx \frac{d\Phi}{dt} = 2\Omega_m LRB_m \cos \theta_m$$
$$= \Omega_m \Phi_m \cos \theta_m = E_m \cos \theta_m, \tag{10.8}$$

where e_1 represents (counter) electromotive force. If the resistance R of the turn is sufficiently low, the voltage drop Ri can be neglected, and the electromotive force e_1 is equal to the voltage across the terminals of the turn. Since the electromotive force is a sinusoidal function of time, its rms (*root mean square*) value is given by (10.9), where f is the frequency in Hz, that is, the number of rotor revolutions per second. Therefore, the electromotive force is proportional to the maximum value of the flux and to the frequency,

$$E_{1rms}^{turn} = \frac{E_m}{\sqrt{2}} = \frac{(2\pi f)\Phi_m}{\sqrt{2}} = 4,44f\ \Phi_m. \tag{10.9}$$

10.2.4   Electromotive Force Waveform

Preceding analysis dealt with an electrical machine with permanent magnets on the rotor which create magnetic field with sinusoidal distribution $B(\theta)$ along the air gap circumference. With constant rotor speed, sinusoidal electromotive force is induced in stator turn shown in Fig. 10.1.

If the speed of rotation varies, the electromotive force induced in the turn may deviate from sinusoidal change. In cases when distribution of magnetic field is sinusoidal, but the speed of rotation changes in time, $\Omega_m(t)$, the electromotive force in the turn is

$$e_1 = +\frac{d}{dt}\left(\Phi_m \sin(\Omega_m t)\right)$$

$$= \Omega_m \Phi_m \cos(\Omega_m t) + t \cdot \Phi_m \cos(\Omega_m t) \cdot \frac{d\Omega_m}{dt}.$$

This expression represents a harmonic function of time if the speed of rotation is constant, that is, in cases where $d\Omega_m/dt = 0$.

In cases where the permanent magnets create a non-sinusoidal periodic distribution of the magnetic field in the air gap, the induced electromotive force assumes a non-sinusoidal function of time. Let $B(\theta - \theta_m)$ be a periodic function specifying the change of the magnetic induction along the air gap circumference. At position of the conductor $\otimes$, magnetic induction is equal to $B_\otimes = B(0 - \theta_m) = B(-\theta_m)$. If the rotor revolves at a constant speed, the electromotive force of the turn calculated according to expression (10.6) is equal to

$$e_1 = 2 \cdot e_\otimes = 2LR\Omega_m \cdot B_\otimes = 2LR\Omega_m \cdot B(-\Omega_m t). \tag{10.10}$$

The obtained expression shows that the electromotive force waveform is determined by the function $B(\theta - \theta_m)$, expressing the spatial distribution of the magnetic induction originating from the rotor permanent magnets. Therefore, the form of the function $B(\theta - \theta_m)$ determines the waveform $e_1(t)$ of the electromotive force in one turn. In cases with non-sinusoidal distribution $B(\theta - \theta_m)$ of the magnetic field in the air gap, the electromotive force induced in the turn (Fig. 10.1) will be non-sinusoidal as well.

Question (10.1): Assume that synchronous generator supplies electrical loads and comprises concentrated stator winding with all the conductors located in two slots on the inner surface of the stator. These slots are diametrically positioned grooves in the magnetic circuit facing the air gap. Permanent magnets of the rotor create the induction $B(\theta) = B_m \operatorname{sgn}[\cos(\theta - \theta_m)]$ in the air gap. Determine and sketch the form of the voltage supplied to the electrical load.

Answer (10.1): In accordance with (10.8), the voltage across terminals of the stator winding is equal to

$$u \approx 2NLR\Omega_m B(\Omega_m t) = (2NLR\Omega_m B_m) \operatorname{sgn}[\cos(\Omega_m t)].$$

10.2.5 Root Mean Square (rms) Value of Electromotive Forces

The AC voltages and currents in electrical engineering are characterized by their *rms* value (*root mean square*). Sinusoidal voltages and currents are mostly described in terms of their *rms* value instead of their peak values. *Root mean square (rms)* value of an AC voltage corresponds to DC voltage that would result in the same power when applied to resistive loads. Namely, when an AC voltage with the *rms* value of U is applied to the resistance R, the power dissipated in resistive load will be $P = U^2/R$. The same power is obtained when a DC voltage U is applied across the same resistance. Therefore, the *rms* value of AC voltages is also called *equivalent DC voltage*. For sinusoidal voltages, their *rms* value is obtained by dividing their peak value by the square root of two. The *rms* value can be defined as well for periodic non-sinusoidal voltages. For a voltage that changes periodically with a period T, the *rms* value is calculated according to

$$U_{rms} = \sqrt{\frac{1}{T} \int_0^T u^2 \mathrm{d}t}.$$

The *rms* value can be also defined for AC currents, using the previous expression. For sinusoidal currents, their *rms* value is $\sqrt{2}$ times lower than their peak value. Expressions (10.11) and (10.12) give the *rms* value of the electromotive force in one turn and *rms* value of the electromotive force in one conductor,

$$E_{1rms}^{turn} = \frac{E_m}{\sqrt{2}} = \frac{(2\pi f)\Phi_m}{\sqrt{2}} = 4,44 f \; \Phi_m \qquad (10.11)$$

$$E_{1rms}^{con} = 2,22f \ \Phi_m.$$ (10.12)

The angular frequency $\omega = 2\pi f$ of the electromotive force is determined by the speed of rotation Ω_m. The rms value of an induced electromotive force can be expressed in terms of the frequency f and the flux as $E_{rms} = 4.44 \ f\Phi_m$. For a concentrated winding consisting of N turns, the rms value of the electromotive force is given by equation

$$E_{rms}^{wind} = 4,44Nf \ \Phi_m.$$ (10.13)

10.3   Electromotive Force in a Winding

Electrical machines usually have a number of windings. Most induction and synchronous machines, also called AC machines, have their stator designed for the connection to a three-phase system of alternating voltages and currents. Therefore, most AC machines have three windings on the stator, also called phase windings. Some authors use the term *stator winding* to describe the winding system comprising three-phase windings.

Each winding has one or more turns. Individual turns are connected in series. The ends of this series connection are usually made available at machine terminals. In this section, the electromotive force induced in a winding is calculated for concentrated and distributed windings. This electromotive force determines the voltage across the machine terminals.

10.3.1   Concentrated Winding

Conductors making a winding can be concentrated in two diametrically positioned grooves, constituting a concentrated winding. Since all the turns of a concentrated winding reside in the same position, they all have the same flux and the same electromotive force. The rms value of the electromotive force in one turn, made of two diametrically positioned conductors, is given by expression (10.12). For a concentrated winding with N turns, the rms value of the induced electromotive force is given by expression (10.14).

10.3.2   Distributed Winding

Windings are usually made by placing conductors in a number of equally spaced slots along the machine circumference. Individual turns are spatially shifted and have different electromotive forces. For this reason, the electromotive force

Fig. 10.2 Distribution of
conductors of a winding
having fractional-pitch turns
and belt distribution in $m = 3$
slots

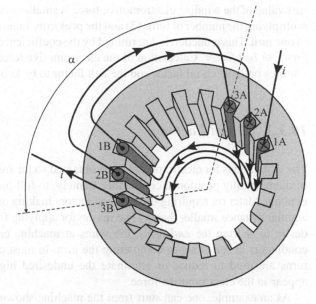

induced in a distributed winding is not equal to the product of the number of turns N
and the electromotive force induced in one turn. The process of calculating the
electromotive force of a distributed winding is explained in this section.

The turns making a winding are often made of conductor pairs that do not have
diametrical displacement. Namely, the two conductors making one turn may not
have angular displacement of π. Displacement of the two conductors is also called
pitch of the turn. With pitch lower than π, one obtains *fractional-pitch turn* or
fractional-pitch coil.

In distributed windings, conductors of a winding may be placed in several
adjacent slots. A group of conductor is called a *winding belt*. In Fig. 10.2, a sample
winding is depicted with three series-connected turns, 1A–1B, 2A–2B, and 3A–3B.
Conductors 1A and 1B belong to one turn, and they are placed in the slots at an
angular distance of α. With $\alpha = \pi$, the conductors are diametrically positioned,
making a *full-pitch coil*. In the case when $\alpha < \pi$, conductors 1A and 1B reside on a
chord. In this case, turn 1A–1B is *fractional-pitch turn*.

The electromotive force induced in a fractional-pitch turn is smaller than in the
case of a full-pitch turn. The reduction is determined by a coefficient called *chord
factor*.

Conductors 1A, 2A, and 3A are placed in three adjacent slots which make a
winding belt. In the same way, conductors 1B, 2B, and 3B are placed in other three
adjacent slots. The electromotive forces induced in the turns 1A–1B, 2A–2B, and
3A–3B are not equal due to spatial displacement of corresponding turns. Therefore,
electromotive forces in individual turns do not reach their maximum value at the
same instant. The spatial shift between magnetic axes of individual turns results in a
phase shift between corresponding electromotive forces. For this reason, the peak and

rms value of the winding electromotive force is smaller than the product obtained by multiplying the number of turns (3) and the peak/rms value of the electromotive force in one turn. This reduction is determined by the coefficient called the *belt distribution factor* or *belt factor*. Calculation of the electromotive force of a distributed winding requires both the chord factor and the belt factor to be known.

10.3.3   Chord Factor

The expressions for electromotive force obtained so far are applicable to turns made of diametrically positioned conductors, namely, to full-pitch turns. There is a need explained later on requiring the two conductors making one turn to be placed at an angular distance smaller than π. The reason for applying fractional pitch may be the desire to shorten the *end turns*, the wires at machine ends that connect the two conductors in series hence completing the turn. In most cases, the fractional-pitch turns are used to reduce or eliminate the undesired higher harmonics that may appear in the electromotive force.

As an example, one can start from the machine shown in Fig. 10.2 and assume that the spatial distribution of the magnetic inductance $B(\theta)$ in the air gap is not sinusoidal. For the purpose of discussion, the function $B(\theta)$ is assumed to be $B_{m1}\cos(\theta - \theta_m) + B_{m5}\cos5(\theta - \theta_m)$. Based upon that, the electromotive forces $e_\otimes$ and $e_\odot$ induced in the conductors constituting one turn comprise the component at the basic angular frequency $\omega = \Omega_m$, but also the fifth harmonic at frequency of $5\Omega_m$. In cases where conductors $\otimes$ and $\odot$ are positioned at angular distance π, the electromotive forces induced in them are of the same shape. The fundamental harmonic component of the electromotive force depends on B_{m1}. At the same time, higher harmonics of the magnetic induction determine the higher harmonics of the electromotive force. By connecting diametrically positioned conductors in series, the electromotive force at the fundamental frequency is doubled, but so is the unwanted electromotive force of the fifth harmonic.

If the conductors are placed so that the angular distance between them is $\alpha = 4\pi/5$, there is a phase shift between the electromotive forces $e_\otimes$ and $e_\odot$ which depends on the angular distance α. In the course of the rotor motion, the instant of passing of the north magnetic pole of the rotor against the stator conductor $\odot$ is delayed by $\Delta t = \alpha/\Omega_m$ with respect to the instant of passing of the same pole against the conductor $\otimes$. The electromotive force of conductor $\odot$ has angular frequency Ω_m, and it is phase shifted by α with respect to the electromotive force of conductor $\otimes$. In Fig. 10.1, the angular distance between the two conductors making the turn is $\alpha = \pi$. In Fig. 10.2, there are turns with $\alpha < \pi$. With $\alpha = \pi$, electromotive forces in conductors $\otimes$ and $\odot$ have the opposite sign. They are connected in series by the end turn which provides the current path between the ends of conductors at the same machine and. With such connection, the electromotive forces with opposite sign actually add up. For that reason, the electromotive force induced in a single turn in Fig. 10.1 is two times larger than electromotive forces of individual conductors.

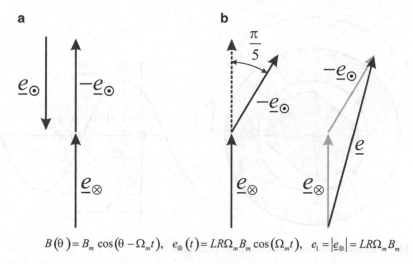

$$B(\theta)= B_m \cos(\theta - \Omega_m t), \quad e_\otimes(t)= LR\Omega_m B_m \cos(\Omega_m t), \quad e_1 =|\underline{e}_\otimes|= LR\Omega_m B_m$$

Fig. 10.3 Electromotive forces of conductors of a turn. (**a**) Full-pitched turn. (**b**) Fractional-pitch turn

Electromotive forces in individual conductors can be represented by phasors $\underline{e}_\otimes$ and $\underline{e}_\odot$. With peripheral rotor speed of $v = R\Omega_m$, and with $B(\theta) = B_m \cos(\theta - \theta_m)$, both phasors have the same amplitude $|\underline{e}_\otimes| = |\underline{e}_\odot| = e_1 = LvB_m$, where L denotes the axial length of the machine. The phasor of the electromotive force in the turn, shown in Fig. 10.3a, is obtained by summing the phasors $\underline{e}_\otimes$ and $\underline{e}_\odot$, $\underline{e} = \underline{e}_\otimes - \underline{e}_\odot$. With angular distance between conductors of α, $\underline{e}_\otimes = e_1$, $\underline{e}_\odot = e_1 e^{-j\alpha}$ and $(-\underline{e}_\odot) = e_1 e^{-j\alpha+j\pi}$. By placing the conductors at angular distance $4\pi/5$, phasors of electromotive forces $\underline{e}_\otimes$ and $\underline{e}_\odot$ are not be collinear, and this is illustrated in part (b) of Fig. 10.3. With respect to $e_\otimes$, electromotive force $-e_\odot$ is phase shifted by $\pi/5$. Therefore, the amplitude of the resulting electromotive force in one turn $e_\otimes - e_\odot$ is slightly smaller than the sum of amplitudes LvB_m of electromotive forces in individual conductors.

Phasor diagrams can be constructed for the fundamental harmonic but also for each of the higher harmonics. With $B(\theta) = B_{m1}\cos(\theta - \theta_m) + B_{m5} \cos 5(\theta - \theta_m)$, electromotive forces have the fundamental component of frequency Ω_m and amplitude determined by B_{m1} and the fifth harmonic of frequency $5\Omega_m$ and amplitude determined by B_{m5}. For the fifth harmonic of electromotive forces, the phase difference is multiplied by 5, as well as the frequency. Since $5 \times \pi/5 = \pi$, the fifth harmonic of the electromotive force $e_\otimes$ is in opposition with the fifth harmonic of the electromotive force $(-e_\odot)$. Thus, they will be mutually canceled. Despite the presence of the fifth harmonic in the spatial distribution of $B(\theta)$, the electromotive force induced in the considered turn will not contain the fifth harmonic. One of the consequences of shortening the turn pitch is reduction of the first, fundamental harmonic of the electromotive force. The electromotive force in one turn formed by the fractional-pitch conductors can be determined from the phasor diagrams of

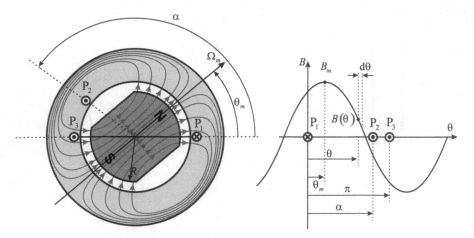

Fig. 10.4 Electromotive forces in a fractional-pitch turn

Fig. 10.3 but also by calculating electromotive force of turn P1–P2 in Fig. 10.4, which is the next step in the analysis.

Question (10.2): How do we place the conductors making one turn in order to eliminate the seventh harmonic of the induced electromotive force? It is assumed that the spatial distribution $B(\theta)$ comprises the seventh harmonic.

Answer (10.2): By placing conductors at the angular distance of $6\pi/7$ and by making their series connections in the way that leads to adding of electromotive forces at the fundamental frequency, phasors of electromotive forces at the fundamental frequency are shifted by $\pi/7$, while phasors of the seventh harmonic are shifted by π. Therefore, by summing the electromotive forces of the two conductors, the seventh harmonic is eliminated.

$$* * *$$

Electromotive force induced in the turn made of conductors P1 and P2, shown in Fig. 10.4, placed at angular distance $\alpha < \pi$ is smaller than the electromotive force induced in the turn P1–P3 which is obtained by connecting diametrically positioned conductors. Reduction of the electromotive force due to this *fractional-pitch* setting of conductors is determined by the coefficient k_T called *chord factor*. The coefficient k_T can be determined by investigating the electromotive force induced in the turn P1–P2, and its value is always $k_T \leq 1$.

Figure 10.4 shows cross section of an electrical machine with the stator turn P1–P2 having fractional-pitched conductors. It is assumed that the magnetic field in the air gap has sinusoidal distribution and that it rotates along with the rotor. Variation of the magnetic induction is depicted as well. The presented spatial distribution of the magnetic induction can be accomplished by insertion of permanent magnets into the rotor magnetic circuit.

The electromotive force of turn P1–P2 can be determined by calculating derivative of the flux encircled by the turn. The flux of the turn P1–P2 is determined by expression

$$\Phi = \int_0^\alpha B \cdot dS = \int_0^\alpha B \cdot LRd\theta = LRB_m \int_0^\alpha \cos(\theta - \theta_m) \cdot d\theta, \quad (10.14)$$

where L is axial length of the machine, R is radius of the rotor, while B_m is the maximum value of sinusoidally distributed magnetic induction which comes as a consequence of permanent magnets on the rotor. By calculating the integral, one obtains

$$\Phi = LRB_m \cdot \sin(\theta - \theta_m)|_0^\alpha = LRB_m[\sin(\alpha - \theta_m) - \sin(-\theta_m)]$$
$$= \left[2LRB_m \sin\frac{\alpha}{2}\right]\cos\left(\frac{\alpha}{2} - \theta_m\right) = \Phi_m^\alpha \cos\left(\frac{\alpha}{2} - \theta_m\right). \quad (10.15)$$

Φ_m^α denotes the maximum value of the flux in the turn P1–P2. This value is reached when the rotor is in position $\theta_m = \alpha/2$. Since the maximum value of the flux in the turn P1–P3 is $\Phi_m^\pi = 2LRB_m$, it is shown that the peak value for the fractional-pitch turn is reduced by factor of $\sin(\alpha/2)$. In other words, the peak value of the flux in the fractional-pitch turn (Φ_m^α) is smaller than the peak value of the flux of the full-pitch turn (Φ_m^π), and it is equal to $\Phi_m^\alpha = \Phi_m^\pi \sin(\alpha/2)$. The electromotive force induced in turn P1–P2 is given by expression

$$e_1(t) = \frac{d\Phi}{dt} = \Omega_m \Phi_m^\alpha \sin\left(\frac{\alpha}{2} - \theta_m\right) = E_m \sin\left(\frac{\alpha}{2} - \omega_m t\right). \quad (10.16)$$

The peak value of the electromotive force is equal to

$$E_m = 2\pi f \, \Phi_m^\pi \cdot \sin\frac{\alpha}{2}. \quad (10.17)$$

Coefficient $k_T = \sin(\alpha/2)$ determines reduction of the turn electromotive force due to fractional pitch. The rms value of the electromotive force of the turn is given by expression

$$E_{1rms}^{turn} = \frac{E_m}{\sqrt{2}} = 4,44f \, \Phi_m^\pi \cdot \sin\frac{\alpha}{2}. \quad (10.18)$$

The electromotive forces represented by means of corresponding phasors are depicted in Fig. 10.5. Using the phasors, the procedure of calculating the chord factor can be simplified. Hypotenuse AC of the right-angled triangle ABC represents

Fig. 10.5 Electromotive
force of a fractional-pitch
turn. The amplitude of the
electromotive force induced
in one conductor is denoted
by E_1. The amplitude of the
electromotive force induced
in one turn is determined by
the length of the phasor DC

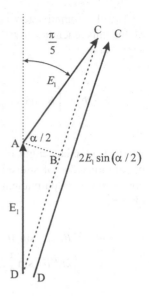

the rms value of the electromotive force E_1 induced in one of the conductors. Opposite to angle $\alpha/2$ is the side BC of the length $E_1 \cdot \sin(\alpha/2)$. The rms value of the electromotive force induced in the turn is represented by phasor DC. It has twice the length of the side BC, and it amounts $2E_1 \sin(\alpha/2) = 2E_1 k_T$. Therefore, due to fractional pitch, the electromotive force in one turn is smaller than $2E_1$. Chord factor $k_T = \sin(\alpha/2)$ is in accordance with the result (10.18).

Within the subsequent sections, winding design methods are discussed that result in reduction of harmonic distortions of the electromotive force and even elimination of certain harmonics of lower order. These methods include placing conductors of one turn along chord instead diameter and arranging winding turns in belts. To facilitate calculation of electromotive forces induced in such windings, analysis results in coefficients called *belt factor* and *chord factor*.

In Sect. 10.5, the electromotive force induced in ideal distributed winding is analyzed, proving that the induced electromotive force of this winding comprises only the fundamental harmonic, notwithstanding the non-sinusoidal distribution of the magnetic induction $B(\theta)$ and non-sinusoidal waveforms of electromotive forces induced in individual conductors. Basically, an appropriate series connection of all the winding conductors can be envisaged so as to cancel all the harmonics in induced electromotive force except for the fundamental. The subsequent analysis will consider the winding with sinusoidal distribution of conductors along the machine circumference. With an ideal sinusoidal distribution of conductors, all the distortions in are canceled, resulting in a sinusoidal electromotive force of the winding. The winding then acts as a *spatial filter*, removing distortions in the spatial distribution $B(\theta)$ and giving a sinusoidal electromotive force.

10.3.4 Belt Factor

Windings of practical machines are made by placing conductors in slots, axially cut grooves in the magnetic circuit. The stator and rotor have large numbers of slots. Figure 8.1 shows a cross section of the magnetic circuit of an electrical machine exposing the usual shapes of the slots and their number.

One turn consists of the two diametrical conductors or the two conductors placed at the ends of the chord. On the other hand, a three-phase stator winding comprises three phases, three separate windings having their magnetic axes displaced by $2\pi/3$. Therefore, the minimum number of slots contained by the stator magnetic circuit of a three-phase machine is $2 \times 3 = 6$. This number of slots is used in cases where each of the phase windings is concentrated, that is, where all conductors in a winding are placed in two diametrical slots. The number of slots is usually higher than 6. One of the reasons for using higher number of slots is the problem associated with placing one half of all the conductors of one winding into only one slot. Such slot would have an extremely large cross section that is not practical for the machine construction. Thus, the conductors are usually distributed in a number of neighboring slots. In this way, a *winding belt* is formed. A belt comprises two, three, or more adjacent slots. Each slot of the belt can accommodate one or more winding conductors. In order to simplify further considerations, one may assume that each slot of the belt contains only one conductor. In cases where each slot has M conductors, all the subsequent conclusions hold, except for the electromotive forces amplitude which has to be multiplied by M.

An example of a winding belt having conductors placed in three adjacent slots is presented in Fig. 10.6 The figure shows the cross section of such winding and phasor diagram showing relation between the electromotive forces induced in the individual windings. The turns 1A–1B, 2A–2B, and 3A–3B have full pitch, namely, these turns are made of diametrically positioned conductors. Phasors E_1, E_2, and E_3 represent the electromotive forces of turns 1A–1B, 2A–2B, and 3A–3B, respectively. Angle γ in Fig. 10.6 denotes the distance between the two adjacent slots. The spatial shift between the turns results in the phase shift γ between electromotive forces induced in corresponding turns.

Placing of conductors in winding belts allows elimination or reduction of higher harmonics in the induced electromotive force. Due to the spatial displacement between conductors, the electromotive forces in individual conductors are phase shifted, as explained in the preceding section discussing the chord factor. In cases where a winding belt ranges over m neighboring slots, the electromotive force of the winding can be determined according to equation

$$\vec{E}^{phase} = \vec{E}_1 + \vec{E}_2 + \dots + \vec{E}_m, \tag{10.19}$$

where letters $E_1 \div E_m$ denote the electromotive forces induced in individual turns. Due to the phase shift, the sum of individual electromotive forces gives a resultant

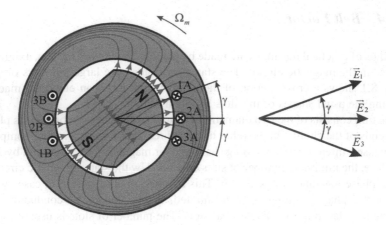

Fig. 10.6 Three series-connected turns have their conductors placed in belts. Each of belts has three adjacent slots (*left*). Phasor diagram showing the electromotive forces induced in the turns 1, 2, and 3 (*right*)

electromotive force with a rms value which is smaller than the sum of the rms values of individual electromotive forces,

$$\left|\vec{E}_1 + \vec{E}_2 + \ldots + \vec{E}_m\right| < \left|\vec{E}_1\right| + \left|\vec{E}_2\right| + \ldots + \left|\vec{E}_m\right|.$$

Adjacent slots are seen from the center of the rotor at an angle of γ, which determines at the same time the phase shift of the electromotive forces in individual turns, said turns comprising conductors placed in the adjacent slots. By using (10.19) and phasor diagram given in Fig. 10.7, it can be shown that distribution of conductors within the winding belts leads to elimination of certain higher harmonics in the induced electromotive force.

10.3.5 Harmonics Suppression of Winding Belt

According to (10.10), variation of the conductor electromotive force is determined by spatial distribution of the magnetic induction in the air gap. If the spatial distribution of magnetic induction contains a higher spatial harmonic of the order n, the time change of the electromotive force induced in a single conductor contains a higher harmonic of the order n. In a full-pitch turn, conductors are placed in diametri cally positioned slots, and the electromotive force is twice the electro-motive force of a single conductor. Therefore, the harmonic of the order n is present as well by the electromotive force of the turn.

The electromotive force of the winding shown in Fig. 10.6 is equal to the sum of electromotive forces in spatially shifted turns 1A–1B, 2A–2B, and 3A–3B. It is possible to cancel higher harmonics by the proper selection of the angle γ. Diagram

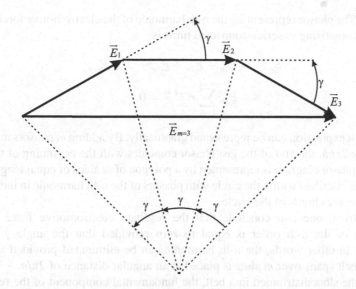

Fig. 10.7 Phasor diagram of electromotive forces in individual turns for the winding belt comprising $m = 3$ adjacent slots

in Fig. 10.6b shows phasors E_1, E_2, and E_3 which represent the fundamental harmonic of the electromotive forces in turns 1A–1B, 2A–2B, and 3A–3B. The phasors are shifted by angle γ which represents the spatial shift of the two adjacent turns. Higher harmonics of the electromotive forces can also be represented by phasors. Harmonics of the order n are of the same amplitude in all considered turns, but their initial phases differ due to their space shift. The angular frequency is increased n times for the n-th harmonic. In adjacent conductors, the spatial shift γ produces the phase delay for the n-th harmonic of $n\gamma$. In the phase diagram representing n-th harmonics of the electromotive forces of the turns, phasors E_{1n}, E_{2n}, и E_{3n} are shifted by angle $n\gamma$.

If angle γ in Fig. 10.6a is equal to $2\pi/(3{\cdot}n)$, phasors on the n-th harmonic E_{1n}, E_{2n}, и E_{3n} are shifted by $2\pi/3$. Since amplitudes of these phasors are all equal, their phase diagram is represented by an equilateral triangle where the beginning of the first and end of the last phasor coincide. That is, the sum of the phasors of the n-th harmonic in all the three turns is equal to zero. Therefore, in cases where $\gamma = 2\pi/(3{\cdot}n)$, harmonics of the electromotive force of the order n are eliminated.

In general, two winding belts making a winding may cover m consecutive slots each. The winding then consists of m turns connected in series. If the angle γ equals $2\pi/(m{\cdot}n)$, the phase shift of the fundamental component of the electromotive force in the adjacent turns is equal to $2\pi/(m{\cdot}n)$, while the phase shift of the n-th harmonic is $2\pi/m$. The electromotive force is induced in the winding with n-th harmonic obtained by adding m phasors, each representing the n-th harmonic in a single turn. These m phasors are of the same amplitude, with their initial phases shifted by $2\pi/m$. The phasor representing the n-th harmonic of the electromotive force of the winding comprising m series-connected turns is

$$\underline{E}_n^{wind} = \underline{E}_{1n} + \underline{E}_{2n} + \underline{E}_{3n} + \dots + \underline{E}_{mn}$$

$$= \underline{E}_{1n} \sum_{k=0}^{m-1} e^{-j\cdot k \cdot \frac{2\pi}{m}} = 0.$$

The last expression can be represented graphically. By adding m phasors mutually shifted by $2\pi/m$, the end of the last phasor coincides with the beginning of the first one. The phasor diagram is represented by a polygon of m sides of equal length. The polygon is inscribed within the circle with phasors of the n-th harmonic in individual turns being the chords of the circle.

Therefrom, one can conclude that the resultant electromotive force of the harmonic of the n-th order is equal to zero provided that the angle γ equals $2\pi/(m\cdot n)$. In other words, the n-th harmonic can be eliminated provided that the winding belt spans over m slots is placed at an angular distance of $2\pi/n$.

With the slots distributed in a belt, the fundamental component of the resultant electromotive force is slightly smaller than what would be the electromotive force in a concentrated winding. By using the diagram in Fig. 10.7, it can be noticed that the amplitude of the sum of phasors $E_1 + E_2 + E_3$ is smaller than the sum of individual amplitudes $|E_1| + |E_2| + |E_3|$. This difference appears since the phasors being added are not collinear. Factor k_P is equal to the ratio

$$k_P = \frac{|\vec{E}_1 + \vec{E}_2 + \vec{E}_3|}{|\vec{E}_1| + |\vec{E}_2| + |\vec{E}_3|}$$

and it is called *belt factor*.

In cases where the width of the winding belt is $m = 3$, the fundamental component of the resultant electromotive force is obtained by adding three phasors, as shown in Fig. 10.7, resulting in equation

$$\vec{E}_{m=3}^{wind} = \vec{E}_1 + \vec{E}_2 + \vec{E}_3. \tag{10.20}$$

Phasors E_1, E_2, and E_3 are lying on a circle of radius R. For each of the isosceles triangles of Fig. 10.7, it is known that the base of the triangle is equal to twice the product of the triangle side and the sine of one half of the opposite angle. Therefore, the expression $R \sin(\gamma/2) = E_1/2$ links the radius of the circle, the electromotive force E_1, and the angle γ. At the same time, the amplitude of the resultant phasor is

$$E_{m=3} = 2R \cdot \sin\frac{3\gamma}{2} = 2\left(\frac{E_1}{2 \cdot \sin\frac{\gamma}{2}}\right) \sin\frac{3\gamma}{2}$$

$$= E_1 \frac{\sin\frac{3\gamma}{2}}{\sin\frac{\gamma}{2}} = \left(\frac{\sin\frac{3\gamma}{2}}{3 \cdot \sin\frac{\gamma}{2}}\right) \cdot 3E_1 = k_P \cdot 3E_1,$$

where k_P stands for the belt factor of the winding with the belt width of $m = 3$ slots. For belt width $m \neq 3$, the amplitude of the resultant phasor representing the fundamental harmonic component is $E_m = 2R \sin(m\gamma/2)$, where radius R and electromotive force E_1 in one turn are related by $R \sin(\gamma/2) = E_1/2$. With E_1 representing the rms value of the fundamental component of the electromotive force induced in a single turn, the rms value of the fundamental component in the winding comprising m series-connected turns is

$$E_m = E_1 \frac{\sin \frac{m\gamma}{2}}{\sin \frac{\gamma}{2}} = \left(\frac{\sin \frac{m\gamma}{2}}{m \cdot \sin \frac{\gamma}{2}} \right) \cdot mE_1 = k_P \cdot mE_1.$$

The belt factor k_P is determined by the expression

$$k_P = \frac{|\vec{E}_{m=3}|}{m|\vec{E}_1|} = \frac{\sin \frac{m\gamma}{2}}{m \cdot \sin \frac{\gamma}{2}}, \tag{10.21}$$

and the rms value of the winding electromotive force can be determined from

$$E_{rms} = E^{phase} = 4,44 k_P f N \, \Phi_m^\pi, \tag{10.22}$$

where $\Phi_m^\pi = 2LRB_m$ is the maximum value of the flux in one full-pitched turn, N is the number of series-connected turns, while $f = \Omega_m/(2\pi)$ is the frequency.

Question (10.3): Due to non-sinusoidal distribution of magnetic induction $B(\theta)$, the electromotive forces induced in conductors comprise higher harmonics. A winding of 3 turns consists of 6 conductors. The conductors are placed at angular distance of $\gamma = 24°$. The resultant electromotive force measured at the winding terminals does not have some of the higher harmonics that are present in electromotive forces of single conductor. What is the order of these harmonics?

Answer (10.3): The resultant electromotive force of the harmonic of the order n is equal to zero in cases where $\gamma = 2\pi/(m \cdot n)$ or $\gamma = q \cdot 2\pi/(m \cdot n)$, where q is an integer. Since $m = 3$ and $\gamma = 2\pi/15$, all harmonics of the order $n = 5q$ are eliminated. These are all the harmonics with the order n being an integer multiple of five.

10.4   Electromotive Force of Compound Winding

Windings are usually made by series connecting the fractional-pitch turns, namely, the turns made of conductors that are not diametrically placed, but reside at the ends of a chord. In addition, conductors of several turns are distributed in adjacent slots that make up one winding belt. For this reason, calculation of the fundamental (first) harmonic component in the electromotive force of the winding should include the belt and pitch factors. These factors are given by expressions

$$k_T = \sin\frac{\alpha}{2}, \tag{10.23}$$

$$k_P = \frac{\sin\frac{m\gamma}{2}}{m \cdot \sin\frac{\gamma}{2}},$$

while the rms value of the electromotive force of a compound winding can be determined from

$$E_{rms} = 4,44 \cdot k_P k_T N \cdot f \cdot (2L \cdot R \cdot B_m), \tag{10.24}$$

where $N = N_C/2$ is the number of turns, B_m is the maximum value of sinusoidally distributed magnetic induction in the air gap, while L and R stand for the machine length and the rotor radius. Quantity f is the frequency of the induced electromotive forces. In the example given in Fig. 10.6, the frequency f is equal to the number of rotor revolutions per second.

The preceding expression can be written in the form

$$E_{rms} = 4,44 k_P k_T N f \Phi_m^\pi,$$

where $\Phi_m{}^\pi = 2LRB_m$ represents the maximum value of the flux which would have existed in a single full-pitched turn. In distributed windings with fractional-pitch turns, the quantity $\Phi_m{}^\pi$ is hypothetic, and expression (10.24) is more suitable.

10.5   Harmonics

The waveform of dynamic electromotive force is determined by distribution of the magnetic field in the air gap. In cases where the magnetic field H and induction B vary sinusoidally along the machine circumference, the electromotive force is a sinusoidal function of time, and it does not contain distortions and higher harmonics. The fundamental or basic frequency component is also called *first* harmonic. The term *higher harmonics* refers to any other harmonic of the order $n > 1$. The presence of higher harmonics distorts the waveform and makes it non-sinusoidal. The field in the air gap appears as a consequence of the magnetomotive forces created by electrical currents in conductors but also due to the presence of permanent magnets on the rotor. In both cases, one of the goals which is set up in the course of machine design is to achieve near-sinusoidal distribution of the magnetic field in the air gap, so as to obtain sinusoidal electromotive forces. This goal cannot be accomplished in full for a number of reasons. For one, conductors making the windings do not have sinusoidal distribution as they have to be placed in slots. Magnetic circuit of electrical machines usually has several tens of slots. Thus, there is a relatively small number of slots available for placing conductors. Electrical current in such conductors produces magnetomotive force and magnetic field in the air gap. Deviation from harmonic distribution of conductors leads to appearance of higher spatial harmonics of the magnetic induction $B(\theta)$ in the air gap. The presence of higher harmonics makes the spatial distribution of $B(\theta)$ a non-sinusoidal function of the angle θ. Similarly, permanent magnets built in the rotor cannot produce an ideal, sinusoidal distribution of the magnetic field, but they create magnetic field comprising higher spatial harmonics.

With the magnetic field $B(\theta)$ in the air gap and with the rotor rotating at the speed of Ω_m, electromotive forces are induced in stator conductors, proportional to the magnetic inductance and the speed. In the conductor at position $\theta = 0$, the induced electromotive force is $E = R\Omega_m LB(0 - \theta_m)$, and it can be written as $E = kB(0 - \Omega_m t)$. In cases when the spatial distribution $B(\theta)$ contains the fifth harmonic, magnetic induction at position $\theta = 0$ is

$$B_5 \, \cos 5(\theta - \theta_m) = B_5 \, \cos 5(0 - \theta_m) = B_5 \, \cos(5\Omega_m t);$$

thus, the fifth harmonic of the electromotive force induced in the conductor placed at $\theta = 0$ is

$$E_5 = R\Omega_m LB_5 \, \cos(5\Omega_m t).$$

Therefore, higher spatial harmonics of the function $B(\theta)$ result in higher harmonics of the induced electromotive force.

Higher harmonics of the electromotive force create electromagnetic disturbances and contribute to pulsations in electromagnetic torque. They increase the maximum and rms values of the electrical current with respect to the case with sinusoidal electromotive forces. As a consequence, power of losses in electrical machine increases. For this reason, the windings of electrical machines are designed and built with the aim of minimizing the influence of higher harmonics of the magnetic field to the induced electromotive forces. Most often, it is not possible to obtain an ideal, sinusoidal distribution of the magnetic field. For this reason, higher harmonics are reduced by the proper design of the windings.

A winding consists of a number of conductors connected in series. In each conductor, induced electromotive force depends on the rotor speed and the magnetic inductance in the air gap. Whether sinusoidal or not, these electromotive forces are periodic, AC waveforms with their frequency determined by the rotor speed. The initial phase of electromotive force induced in a conductor depends upon angular position of the slot where the conductor is placed. The electromotive force of the winding is the sum of phase-shifted electromotive forces of individual conductors. The conductors may be connected in the way that higher harmonics of the electromotive force are in counter phase; thus, they will mutually cancel in the process of summing. The phase shift is further dependent upon the order n of harmonic, and this makes the process of harmonic elimination more involved. Namely, in cases where a winding is made in the way that one higher harmonic is canceled, it is possible that in the same process, the other higher harmonic is summed up and augmented.

Practical methods of designing the windings specify the way of placing individual conductors in various slots. In most cases, all the series-connected conductors along with their end turns are made of one single, uninterrupted wire. In such cases, the winding design provides a scheme or a map which indicates position and sequence of slots where the wire is to be inserted. The winding design relies on the fact that the most harm comes from the low-order harmonics. Therefore, most winding design schemes are focused on suppressing low-order harmonics.

The waveform $B(\theta)$ of the spatial distribution of magnetic induction is usually symmetrical with respect to the maximum B_m, and it does not contain even harmonics. Therefore, the induced electromotive forces could contain only odd harmonics. Moreover, the odd harmonics of the order $3n$ are not relevant either. This claim is briefly explained below.

For star-connected three-phase windings, the sum of electrical currents is equal to zero ($i_a + i_b + i_c = 0$). The fundamental components of electrical currents are phase shifted by $2\pi/3$, which drives their sum to zero. Considering harmonics of the order $3n$ (*triplian*), their mutual phase shift is $3n \times 2\pi/3 = n \times 2\pi$; hence, they have the same phase. Therefore, higher triplian harmonics in electrical currents cannot exist, as their sum would not be zero. Without the path for electrical currents, the higher triplian harmonics in electromotive forces are not relevant as they do not produce electrical currents. In a star connection, the electromotive force of the phase winding may comprise a triplian harmonic, but it cannot produce any electrical current. The phase of triplian harmonics of phase electromotive forces is equal due to $3n \times 2\pi/3 = n \times 2\pi$. Therefore, the line voltage, being the difference between the two phase voltages, will be free from triplian harmonics.

For the above reasons, practical approaches to winding design are primarily focused on suppressing the fifth and seventh harmonic. Where possible, the next harmonics to be targeted are the eleventh and thirteenth.

10.5.1 Electromotive Force in Distributed Winding

Practical windings of electrical machines are formed by series connecting of the conductors placed in slots. Usually, there are several tens of slots, meaning that the conductors could be placed at one of several tens of discrete positions. Practical windings are designed to have spatial distribution of their conductors as close to sinusoidal as possible. Winding design techniques include the use of fractional pitch of conductors making one turn (fractional-pitch turns) in order to eliminate or reduce some of the higher harmonics. In addition, conductors of the winding are distributed in winding belts, comprising a number of adjacent slots, and this approach contributes as well to elimination or reduction of higher harmonics in the winding electromotive force. For this reason, further considerations are made in order to analyze the resultant electromotive force in windings with near-to-sinusoidal distribution of their conductors along the machine circumference.

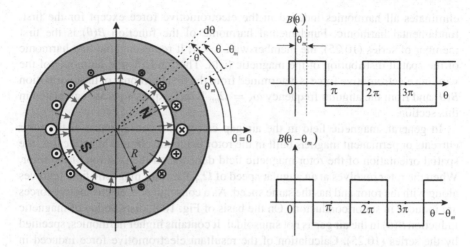

Fig. 10.8 Cross section of an electrical machine comprising one stator winding with sinusoidal distribution of conductors and permanent magnets in the rotor with non-sinusoidal spatial distribution of the magnetic inductance

It will be shown hereafter that an ideal winding with sinusoidal distribution of conductors acts as a *spatial filter* which eliminates completely all the higher harmonics of the resulting electromotive force. In cases where the magnetic induction $B(\theta)$ in the air gap has a non-sinusoidal distribution along the machine circumference, electromotive forces in individual harmonics are non-sinusoidal as well, and they comprise higher-order harmonics. Notwithstanding distortions of B (θ), the winding with sinusoidal distribution of conductors has a sinusoidal-induced electromotive force, free from higher harmonics and distortions. This statement will be proved at the end of this chapter. Hence, sinusoidal distribution of conductors is an ideal worth striving for. Yet, in practice, it cannot be accomplished since each machine has a relatively small number of slots. Therefore, electromotive forces in practical machines deviate from an ideal sinusoidal form due to the fact that higher harmonics are not completely eliminated.

In the following considerations, the induced electromotive force is calculated for windings where distribution of conductors along the machine circumference is assumed to be perfectly sinusoidal. Analysis is focused on electrical machine shown in Fig. 10.8.

It has permanent magnets on the rotor. The magnets produce the magnetic inductance in the air gap with spatial distribution $B(\theta)$. The function $B(\theta)$ is non-sinusoidal, and it has higher harmonics. The machine under scope has a stator winding with conductors distributed along the inner surface of the stator magnetic circuit. It is assumed that each of these conductors can be placed at an arbitrary location and that the distribution of the conductors along the circumference is sinusoidal. It is going to be proved that in this case, the winding has an induced electromotive force that is sinusoidal, even though the field has a non-sinusoidal distribution in the air gap. The winding plays the role of a spatial filter which eliminates all harmonics induced in the electromotive force except for the first, fundamental harmonic. Fundamental harmonic of the function $B(\theta)$ is the first member of series (10.25), the member with $i = 1$. It represents the first harmonic of the spatial distribution of the magnetic field. The fundamental harmonic of the winding electromotive force is determined from the first harmonic of the distribution $B(\theta)$ and from the angular frequency $\omega_S = \Omega_m$. Exact expressions are calculated in this section.

In general, magnetic field in the air gap can be the consequence of the rotor currents or permanent magnets built in the rotor magnetic circuit. In both cases, the spatial orientation of the rotor magnetic field depends on the position of the rotor. When the rotor revolves at an angular speed of Ω_m, the rotor magnetic field revolves along with the rotor and has the same speed. As a consequence, electromotive forces are induced in stator conductors. On the basis of Fig. 10.8, distribution of magnetic induction $B(\theta)$ in the air gap is not sinusoidal. It contains higher harmonics, specified by the series (10.25). Calculation of the resultant electromotive force induced in the winding will be carried out with the aim of proving that it contains only the fundamental harmonic.

Magnetic induction created by permanent magnets built in the rotor magnetic circuit can be described by function $B(\theta - \theta_m) = B_m \mathrm{sgn}[\cos(\theta - \theta_m)]$. Over the interval $-\pi/2 < (\theta - \theta_m) < +\pi/2$, the magnetic induction is equal to $+B_m$, while for $+\pi/2 < (\theta - \theta_m) < +3\pi/2$, the induction is $-B_m$. By expanding this function to Fourier series, one obtains

$$B(\theta - \theta_m) = \sum_{i=1}^{\infty} \frac{4}{\pi} \frac{B_m}{2i - 1} (-1)^{i+1} \cos[(2i - 1)(\theta - \theta_m)].   \quad (10.25)$$

The function contains all odd harmonics, while even harmonics are equal to zero. The amplitude of specific harmonics decrease with their order, $A \sim 1/(2n - 1)$. The absence of even harmonics could have been predicted from the fact that the function $B(\theta - \theta_m)$ is symmetrical, $B(x) = B(-x)$. The amplitude of the first harmonic is equal to $4B_m/\pi$.

10.5.1.1   Flux in One Turn

Calculation of the electromotive force induced in the stator winding requires the flux in the winding to be determined first. Then, the electromotive force can be found from the first derivative of the winding flux. Since winding consists of a number of series-connected turns, each one in a different position, the total flux is obtained by adding (integrating) fluxes in individual turns. First of all, it is necessary to determine the flux $\Phi(\theta)$ in one turn. It is assumed that the turn is made of conductors A and B. It is also assumed that the conductor A, denoted by $\otimes$, resides at position θ, while the conductor B of the same turn, denoted by $\odot$, resides at

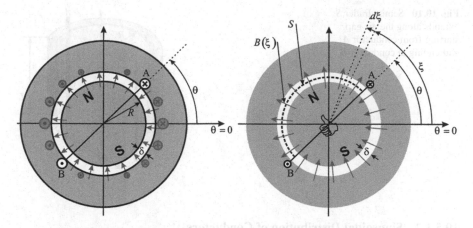

Fig. 10.9 Calculation of the flux in turn A–B (*left*). Selection of the surface S (*right*)

position $\theta + \pi$. The considered turn has full pitch. Conductors A and B are shown in Fig. 10.9.

Flux $\Phi(\theta)$ is equal to the surface integral of the vector of magnetic induction **B** over the surface S which is leaning on the conductors A and B of the considered turn (Fig. 10.10). The integral is calculated over the surface which lies in the air gap starting from conductor A up to conductor B. The integration surface passing through the air gap is selected because there is an analytical expression for the change of magnetic induction B in terms of the angle θ. The reference direction is determined by the right-hand rule. The flux is determined by integrating the quantity $B(\xi)dS = B(\xi)LRd\xi$ between the limits θ and $\theta + \pi$, as indicated by expression

$$
\Phi(\theta) = \int_{\theta}^{\theta+\pi} B(\xi) L \cdot R \, d\xi
$$

$$
= \int_{\theta}^{\theta+\pi} \left\{ \sum_{i=1}^{+\infty} \frac{4}{\pi} \frac{B_m}{2i-1} (-1)^{i+1} \cos[(2i-1)\cdot(\xi-\theta_m)] \right\} L \cdot R \, d\xi. \quad (10.26)
$$

The integration is carried out over the surface S passing through the air gap, the elements of which are $dS = LRd\xi$, where L is the axial length of the machine and $R = D/2$ is radius of the rotor cylinder. Expression $B(\xi)$ represents radial component of the magnetic induction at an arbitrary position ξ within the interval [θ .. $\theta + \pi$). The result of the integration is the flux determined by equation

$$
\Phi(\theta) = LRB_m \frac{8}{\pi} \sum_{i=1}^{\infty} \frac{(-1)^i}{(2i-1)^2} \sin[(2i-1)(\theta-\theta_m)]. \quad (10.27)
$$

Fig. 10.10 Semicylinder S
extends along the air gap
starting from conductor A
and ending at conductor B

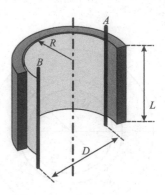

10.5.1.2 Sinusoidal Distribution of Conductors

Resultant flux ψ_S in the winding is calculated by summing all the fluxes in individual turns. The distributed winding shown in Fig. 10.9 has the line density of conductors along the machine circumference expressed by $N'(\theta)$,

$$N'(\theta) = N_m \cos \theta.$$

The highest density of conductors is at positions $\theta = 0$ and $\theta = \pi$. Within a very small element of the machine circumference $Rd\theta$, there are dN conductors,

$$dN = N'(\theta) \cdot R \cdot d\theta.$$

All the conductors with reference direction $\otimes$ are placed within the interval $[-\pi/2 .. \pi/2)$. Each conductor having reference direction $\otimes$ is connected in series with diametrically placed conductor of direction $\odot$. These two conductors are connected at machine ends, and they make one turn. The flux $\Phi(\theta)$ within one turn is determined by expression (10.27). Considered turn has the conductor $\otimes$ in position θ and the conductor $\odot$ positioned at $\theta + \pi$.

Resultant flux Ψ_S of the winding is determined by integrating

$$d\Psi_S = N_m \Phi(\theta) R cos(\theta) \cdot d\theta$$

over the interval $[-\pi/2 .. \pi/2)$,

$$\Psi_S = \int\limits_{-\pi/2}^{\pi/2} N_m \Phi(\theta) R cos(\theta) \cdot d\theta. \tag{10.28}$$

By introducing $\Phi(\theta)$ from expression (10.27), one obtains that the flux of the winding is

$$\Psi_S = N_m LR^2 B_m \frac{8}{\pi} \sum_{i=1}^{\infty} \frac{(-1)^i}{(2i-1)^2} \int_{-\pi/2}^{+\pi/2} \sin[(2i-1)(\theta - \theta_m)]\cos\theta \cdot d\theta \qquad (10.29)$$

If the integrand of the integral (10.29) is written in the form

$$f(\theta, i) = \sin[(2i-1)(\theta - \theta_m)]\cos(\theta)$$
$$= 1/2[\sin(2i\theta - (2i-1)\theta_m) + \sin((2i-2)\theta - (2i-1)\theta_m)],$$

while the value of $(N_m LR^2 B_m \cdot 8/\pi)$ is denoted by k, the result (10.29) can be represented as the sum in which each element is an integral of the function $f(\theta, i)$ within limits $-\pi/2$ and $\pi/2$,

$$\Psi_S = k \sum_{i=1}^{\infty} \frac{(-1)^i}{(2i-1)^2} \int_{-\pi/2}^{+\pi/2} f(\theta, i) \cdot d\theta.$$

For $i \geq 2$, each of the elements of the function $f(\theta, i)$ is a sine function with an integer number of its periods on the interval $[-\pi/2 .. \pi/2)$. Therefore, the integral of $f(\theta, i)$ over the interval has a non zero value only for $i = 1$. Any other of the sum but the first is equal to zero. For the first member of the sum, for $i = 1$, the integrand becomes

$$f(\theta, 1) = [\sin(2\theta - \theta_m) + \sin(-\theta_m)]/2.$$

Introducing this expression in (10.29), for $i = 1$, one obtains

$$\Psi_S^1 = N_m LR^2 B_m \frac{8}{\pi} \frac{(-1)^1}{(2-1)^2} \int_{-\pi/2}^{+\pi/2} \sin(\theta - \theta_m)\cos(\theta)\,d\theta,$$

which gives the resultant flux in the winding. It has been demonstrated that only the first (fundamental) harmonic of the non-sinusoidal distribution $B(\theta)$ of the magnetic induction produces the flux in the winding which has sinusoidal distribution of its conductors. Since

$$\int_{-\pi/2}^{+\pi/2} \sin(\theta - \theta_m)\cos(\theta) \cdot d\theta = \frac{1}{2} \int_{-\pi/2}^{+\pi/2} [\sin(2\theta - \theta_m) + \sin(-\theta_m)] \cdot d\theta$$

$$= -\frac{1}{4}\cos(2\theta - \theta_m)\Big|_{-\pi/2}^{+\pi/2} + \frac{1}{2}\sin(-\theta_m) \cdot \theta \Big|_{-\pi/2}^{+\pi/2} = -\frac{\pi}{2}\sin\theta_m,$$

the value of the flux Ψ_S^1 is equal to

$$\Psi_S^1 = N_m L R^2 B_m \frac{8(-1)}{\pi} \left(-\frac{\pi}{2}\sin\theta_m\right) = 4N_m L R^2 B_m \sin\theta_m. \qquad (10.30)$$

The remaining elements of the sum in expression (10.29) have $i > 1$, and they are all equal to zero. For $i = 2$, the integrand is

$$f(\theta, 2) = [\sin(4\theta - 3\theta_m) + \sin(2\theta - 3\theta_m)]/2.$$

The integral of this function on the interval $[-\pi/2 .. \pi/2)$ is equal to zero. The same hold for any $i > 1$. Therefore, it can be concluded that the induced electromotive force does not have any higher harmonics.

10.5.1.3 Flux of the Winding with Arbitrary Distribution of Conductors

The previous calculation has been carried out in order to demonstrate that a winding with sinusoidal distribution of conductors acts as a spatial filter and removes all the higher harmonics from the electromotive force waveform. It is also of interest to derive the expression for the winding flux in a more general case, where distribution of conductors is described by an arbitrary function $N'(\theta)$.

The first step in calculating the winding flux is getting the flux in the turn placed at position θ. This flux is calculated according to expression (10.26),

$$\Phi(\theta) = \int\limits_{\theta}^{\theta+\pi} B(\xi)L \cdot R\,d\xi,$$

where function $B(\xi)$ determines distribution of the magnetic induction in the air gap.

The total flux is obtained by summing the fluxes of all individual turns. In the case of a distributed winding, this summing is performed by integration. In the case where line density of conductors is $N'(\theta) = N_m\cos\theta$, the flux of the winding is calculated according to expression

$$\Psi_S = \int\limits_{-\pi/2}^{\pi/2} N_m \Phi(\theta) R\cos\theta \cdot d\theta.$$

From the obtained results, the expression for flux of the winding takes the form

$$\Psi_S = \int\limits_{-\pi/2}^{\pi/2} N_m \left\{ \int\limits_{\theta}^{\pi+\theta} LRB(\xi)d\xi \right\} R\cos\theta \cdot d\theta. \qquad (10.31)$$

In general, line density of conductors constituting the winding may have an arbitrary distribution of conductors, described by the function $N'(\theta)$. The total flux of the winding is then calculated by using expression

$$\Psi_S = \int_{-\pi/2}^{\pi/2} N'(\theta) \left\{ \int_{\theta}^{\pi+\theta} LRB(\xi)d\xi \right\} R \cdot d\theta. \qquad (10.32)$$

Expression (10.32) can be applied for calculation of the flux in a distributed winding for an arbitrary field distribution $B(\xi)$ and an arbitrary line density of conductors $N'(\theta)$.

10.5.2 Individual Harmonics

The calculation carried out in this section is focused on deriving the electromotive force induced in the stator winding which has sinusoidal distribution of its conductors along the machine circumference. The electromotive force is determined for the case where permanent magnets of the rotor generate the air gap field with non-sinusoidal distribution $B(\theta)$, comprising higher harmonics. It is started with the expression for magnetic induction in the air gap (10.25), which contains the first, fundamental harmonic but also all odd harmonics. Since this non-sinusoidal distribution of the magnetic induction is symmetrical, the function $B(\theta - \theta_m)$ does not comprise even harmonics. In the considered case, the line density of stator conductors is $N'(\theta) = N_m\cos\theta$, while the variation of the magnetic induction in the air gap is given in (10.25). This expression represents a development of the function $B(\theta)$ into a series comprising only odd members. The first element of the series has $i = 1$, and it represents the fundamental harmonic of the spatial distribution of magnetic induction. For $i > 1$, elements of the series represent higher harmonics of the spatial distribution of the field. On the basis of expression (10.25), spatial harmonic of function $B(\theta)$ of the order $(2i - 1)$ is equal to

$$\frac{4}{\pi} \frac{B_m}{2i - 1} (-1)^{i+1}.$$

By introduction of the latter into expression (10.31), one obtains the quantity Ψ_S^{2i-1} which represents the contribution of the harmonic $(2i - 1)$ to the total winding flux. The value of Ψ_S^{2i-1} is related to $(2i - 1)^{\text{th}}$ harmonic of the spatial distribution of magnetic induction.

$$\Psi_S^{2i-1} = \frac{4N_m LR^2 B_m (-1)^{i+1}}{\pi(2i-1)}$$

$$\times \int_{-\pi/2}^{\pi/2} \left\{ \int_{\theta}^{\pi+\theta} \cos[(2i-1)(\xi - \theta_m)]d\xi \right\} \cos\theta \cdot d\theta. \qquad (10.33)$$

By integration of the function $\cos[(2i-1)(\xi - \theta_m)]$ in terms of ξ one obtains

$$\int \cos[(2i-1)(\xi - \theta_m)]d\xi = \frac{1}{2i-1} \sin[(2i-1)(\xi - \theta_m)].$$

The obtained result can be used in expression (10.33) in order to calculate definite integral of the integrand in terms of ξ, within limits from θ up to $\theta + \pi$. The calculation results in

$$\left. \frac{1}{2i-1} \sin[(2i-1)(\xi - \theta_m)] \right|_{\theta}^{\theta+\pi} = -\frac{2}{2i-1} \sin[(2i-1)(\theta - \theta_m)].$$

By introducing developed results into previous expression, it becomes

$$\Psi_S^{2i-1} = \frac{8N_m LR^2 B_m (-1)^i}{\pi(2i-1)^2} \int_{-\pi/2}^{\pi/2} \sin[(2i-1)(\theta - \theta_m)] \cos\theta \cdot d\theta. \qquad (10.34)$$

Since

$$\sin[(2i-1)(\theta - \theta_m)] \cos\theta = \frac{1}{2} \sin[2i\theta - (2i-1)\theta_m]$$

$$+ \frac{1}{2} \sin[(2i-2)\theta - (2i-1)\theta_m],$$

the obtained result can be separated into two definite integrals, I_A and I_B

$$\Psi_S^{2i-1} = \frac{4N_m LR^2 B_m (-1)^i}{\pi(2i-1)^2} \int_{-\pi/2}^{\pi/2} \sin((2i-2)\theta - (2i-1)\theta_m) \cdot d\theta$$

$$+ \frac{4N_m LR^2 B_m (-1)^i}{\pi(2i-1)^2} \int_{-\pi/2}^{\pi/2} \sin(2i\theta - (2i-1)\theta_m) \cdot d\theta = I_A + I_B. \qquad (10.35)$$

Since index i varies from 1 to $+\infty$, the integral I_B is equal to zero since the interval $[-\pi/2 \, .. \, \pi/2)$ comprises an integer multiple of periods of the function $\sin(2i\theta)$.

The same conclusion applies for the integral I_A if $i > 1$. Therefore, a nonzero value of flux Ψ_S^{2i-1} exists only for the first (fundamental) harmonic, namely, for $i = 1$. With any $i > 1$, the integrand function is a sine wave with its period comprised an integer number of times within the integration domain $[-\pi/2 .. \pi/2)$.

On the basis of the obtained results, it is concluded that a sinusoidal distribution of the winding conductors along the machine circumference prevents all the higher harmonics of $B(\theta)$ from affecting the winding flux. For that reasons, the induced electromotive force of the winding remains unaffected by higher harmonics in $B(\theta)$ waveform.

Hence, sinusoidal distribution of conductors results in a winding which eliminates all the higher harmonics of the induced electromotive force, retaining only the fundamental harmonic.

10.5.3 Peak and rms of Winding Electromotive Force

On the basis of the obtained results, the following passages provide the expressions for the instantaneous, peak and rms values of the electromotive force induced in the considered stator winding. Expressions (10.37), (10.38), and (10.39) apply for windings with distribution of conductors $N'(\theta) = N_m \cos\theta$ and for distribution of the magnetic induction in the air gap shown in Fig. 10.8 and described by (10.25).

10.5.3.1 Suppression of Higher Harmonics

Results obtained so far indicate that sinusoidal distribution of conductors eliminates higher harmonics of the electromotive force induced in a winding. In a winding with an ideal, sinusoidal distribution of conductors, a sinusoidal electromotive force is induced notwithstanding the higher harmonics in the spatial distribution of magnetic induction $B(\theta)$. Any harmonic of the order $(2i - 1)$ in the spatial distribution of $B(\theta)$ gives its contribution of Ψ_S^{2i-1} to the total flux of the winding, and this contribution is given in expression (10.35). For any $i > 1$, contribution Ψ_S^{2i-1} is equal to zero. Therefore, the winding performs the role of the spatial filter which eliminates the effects of higher harmonics of the spatial distribution of $B(\theta)$, and it passes only the fundamental harmonic.

10.5.3.2 Winding Flux

Only the fundamental harmonic of distribution $B(\theta)$ contributes to the winding flux. The winding flux is determined by the following equation:

$$\Psi_S = \Psi_S^1 = 4B_m N_m L R^2 \sin\theta_m. \tag{10.36}$$

10.5.3.3   Electromotive Force

If the rotor revolves at a constant speed Ω_m, position of the rotor varies according to the law $\theta_m = \Omega_m t$. The flux in the stator winding is then a sinusoidal function of time with angular frequency ω_S determined by the angular speed Ω_m of the rotor. With $\theta_m = \Omega_m t$, variation of the flux is given by equation

$$\Psi_S(t) = 4B_m N_m LR^2 \cdot \sin\Omega_m t = \Psi_{S\max} \sin\Omega_m t.$$

A sinusoidal (counter) electromotive force is induced in the winding, and it is equal to the first derivative of the flux,

$$e_S(t) = \frac{d}{dt}\Psi_S = 4\Omega_m B_m N_m LR^2 \cos\Omega_m t = \Psi_{S\max}\Omega_m \cos\Omega_m t. \tag{10.37}$$

The maximum value of this sinusoidal electromotive force of the stator winding is

$$e_{max} = \Psi_{S\max}\Omega_m = 4\Omega_m B_m N_m LR^2, \tag{10.38}$$

while its rms value is equal to

$$e_{rms} = \frac{1}{\sqrt{2}}\Psi_{S\max}\Omega_m = 2\sqrt{2}\Omega_m B_m N_m LR^2.$$

The maximum and rms values of the electromotive force are expressed in terms of N_m, the maximum density of the stator conductors along the machine circumference. Instead, they can be expressed as functions of the number of turns N_N. For the winding with sinusoidally distributed conductors, the expression (8.2) relates the number of turns N_N to the maximum line density of its conductors N_m,

$$N_T = 2R \cdot N_m;$$

thus, the maximum value of the electromotive force can be calculated from

$$e_{max} = 2\Omega_m B_m N_T LR,$$

while the rms value of the electromotive force in the winding can be calculated from

$$e_{rms} = \sqrt{2}\Omega_m B_m N_T LR. \tag{10.39}$$

Performed analysis shows that the electromotive force induced in a winding with sinusoidal distribution of conductors does not contain higher harmonics, which proves that such a winding performs the role of a spatial filter.

Question (10.4): Consider the electrical machine which is the subject of the previous analysis, shown in Fig. 10.8. Assume that the number of conductors does not change but that they are concentrated at positions $\theta = 0$ and $\theta = \pi$. Determine the shape and amplitude of the electromotive force induced in the stator winding.

Answer (10.4): Electromotive force $e_{1C} = LvB(\theta)$ is induced, in each conductor, where L is the machine axial length, $B(\theta)$ is magnetic induction at angular position θ of the conductor placement, while $v = R\Omega_m$ is the peripheral rotor speed. The speed v reflects the *relative* movement of conductors with respect to the field. The spatial distribution of magnetic induction $B(\theta)$ is shown in Fig. 10.8. Therefore, a complex periodic electromotive force is induced with rectangular shape and with the period and frequency determined by the rotor speed. The conductors of the concentrated winding are placed at positions $\theta = 0$ and $\theta = \pi$. Since $B(0) = -B(\pi)$, the electromotive forces induced in diametrical conductors are of the opposite signs. The way of connecting a pair of conductors into one turn leads to subtracting of the respective electromotive forces. Subtracting the two values of the same amplitude and of the opposite sign results in electromotive force in one turn e_{1T} which is twice larger than e_{1C}. Hence, the electromotive force induced in one turn is equal to $e_{1T} = 2\ e_{1C} = 2\ LR\Omega_m\ B(0)$. For a concentrated winding with N turns, the electromotive force of the winding is equal to $e_w = 2\ NLR\Omega_m\ B(0)$. It has the shape of a train of rectangular pulses with an amplitude of

$$e_w^{\max} = 2\Omega_m B_{\max} N_k LR.$$

Using the relations expressed in (10.25), the maximum value of the first harmonic of this train of rectangular pulses is $4/\pi$ times higher than the amplitude of the pulses. Therefore, the rms value of the first harmonic of the electromotive force induced in the concentrated winding is equal to

$$
\begin{aligned}
e_w^{rms} &= \frac{4}{\pi} \cdot \frac{1}{\sqrt{2}} 2\Omega_m B_{max} N_T LR \\
&= \frac{4\sqrt{2}}{\pi} \Omega_m B_{max} N_T LR \approx 1,8\ \Omega_m B_{max} N_T LR.
\end{aligned}
$$

It is of interest to compare this result with the rms value of the electromotive force obtained in the winding with the same number of turns but with sinusoidal distribution of the conductors. On the basis of (10.39), this value is

$$e_{rms} = \sqrt{2}\Omega_m B_{max} N_T LR \approx 1,41\ \Omega_m B_{max} N_T LR.$$

It can be concluded that the rms value of the induced electromotive force in the winding with distributed conductors is $4/\pi$ times smaller compared to the electromotive force in the winding with concentrated conductors. The former amounts approximately 78.5% of the latter.

Question (10.5): Consider electrical machine where permanent magnets on the rotor create magnetic induction with sinusoidal distribution in the air gap, $B(\theta) = B_m \cos(\theta - \theta_m)$, where θ_m is displacement of the rotor with respect to the stator. The stator winding has N_T turns, hence $2N_T$ conductors. The winding can be realized in two ways. The first way is forming a concentrated winding having conductors located at positions $\theta = 0$ and $\theta = \pi$. The other way of making the winding is to have sinusoidal distribution of conductors, with the conductor density $N'(\theta) = N_{sm} \cos(\theta)$ and with the total number of conductors being $2N_T$. Determine the maximum value of the stator flux $\Psi_S(\theta_m)$ in both cases. Geometry of the machine is the same as the one considered in Question 10.4.

Answer (10.5): Relation between the peak conductor density N_{sm} and total number of turns N_T is

$$N_T = \int_{-\pi/2}^{+\pi/2} |N_S'(\theta)| R \, d\theta = 2R N_{Sm},$$

where R denotes the radius of the rotor. The maximum flux achievable in one turn is equal to the product of the average value of magnetic induction wave $B(\theta)$, $B_{av} = 2B_m/\pi$, and the surface area $S = \pi RL$ of the semicircular surface encircled by the turn, wherein the turn is made of two diametrical conductors. The maximum value of the flux is obtained as $\Phi_m = 2B_m LR$.

In cases where the stator winding is concentrated, the stator flux reaches the maximum value of $\Psi_1 = 2B_m LRN_T$. This value is achieved with rotor in position $\theta_m = \pi/2$, when the vector of the rotor flux gets collinear with the magnetic axis of the stator winding.

If the turns of the stator winding are distributed, the stator flux is denoted by $\Psi_2(\theta_m)$. Conductor density is denoted by $N'(\theta) = N_{sm} \cos(\theta)$, and the stator flux Ψ_2 with rotor in position $\theta_m = \pi/2$ is calculated from

$$\Psi_2 = \int_{-\pi/2}^{+\pi/2} \Phi(\theta) \, dN = \int_{-\pi/2}^{+\pi/2} \left\{ \int_{\theta}^{\theta+\pi} RLB_m \cos(\xi - \pi/2) \, d\xi \right\} RN'_S(\theta) d\theta$$

$$= \int_{-\pi/2}^{+\pi/2} \left\{ \int_{\theta}^{\theta+\pi} RLB_m \sin(\xi) \, d\xi \right\} RN_{S\max} \cos(\theta) d\theta$$

$$= \frac{RN_{S\max}}{2} \int_{-\pi/2}^{+\pi/2} \left\{ \int_{\theta}^{\theta+\pi} \Phi_m \sin(\xi) \, d\xi \right\} \cos(\theta) d\theta$$

$$= \frac{RN_{S\max}\Phi_m}{2} \int_{-\pi/2}^{+\pi/2} \{2\cos(\theta)\} \cos(\theta) d\theta$$

$$= R\Phi_m \frac{N_T}{2R} \frac{\pi}{2} = \frac{\pi}{2} B_m LRN_T = \frac{\pi}{4} \Psi_1.$$

Based on the above calculation, conclusion is drawn that, all the remaining conditions being equal, the machine with sinusoidally distributed stator winding has the peak stator flux which is $\pi/4$ times lower than the peak stator flux in the machine with concentrated stator winding.

Chapter 11
Introduction to DC Machines

Prior to commissioning the first electrical power stations, electrical energy was mostly obtained from batteries, chemical sources of electrical current. The batteries provide DC voltages and currents at their output terminals. It is for this reason that the first experiments and applications of electrical machines have been made with DC current electrical machines. Electrical engineers have studied the principles of operation of these machines and analyzed their characteristics, and they found the way of designing and manufacturing DC machines.

At first, the processes of production, transmission, and application of electrical energy were based on DC voltages and currents. All the tasks of electromechanical conversion were employing DC machines. Electric power stations were built in close vicinity of industrial facilities, cities, and other major consumers of electrical energy. The energy obtained from water or steam turbines used to be converted to electrical energy by means of electrical machines providing DC voltages and currents, also called DC generators. At their output terminals, most DC generators produced DC voltages of several hundred volts. By using a pair of conductors, electrical power was transmitted over short distances of 1–2 km and delivered to consumers. Early consumers of electrical energy have been designed to operate with DC voltages and currents. Electrical lighting bulbs have been made to convert electrical energy into light, while DC motors have been producing controlled mechanical work put to use in production processes.

Designing DC generators and motors for voltages in excess of 1,000 V involves technical difficulties that will be explained later on. As a consequence, DC generators and motors were manufactured and used for relatively low DC voltages U. Therefore, transmission and distribution of electrical power P involved very high currents due to $I = P/U$. Transmission of power of 1 MW required electrical currents in excess of 1,000 A. Electrical conductors in transmission lines were designed with very large cross sections in order to reduce the line resistance R. Such transmission was accompanied with considerable losses (RI^2) and large voltage drops (RI). Higher transmission voltage U leads to lower line current $I = P/U$ and, hence, lower losses and lower voltage drop. Yet, at that time, the maximum DC

S.N. Vukosavic, *Electrical Machines*, Power Electronics and Power Systems,
DOI 10.1007/978-1-4614-0400-2_11, © Springer Science+Business Media New York 2013

voltages available at generator terminals were rather limited. At the same time, there were no DC/DC power converters capable of transforming a low voltage, obtained from DC generators, into a high DC voltage, suitable for power transmission.

Contemporary systems for production, transmission, and distribution of electrical energy make use of alternating currents (AC) with frequencies of 50 or 60 Hz. By using power transformers, a relatively low voltage produced by AC generators is increased to several hundreds of thousands of volts. Most AC transmission lines are three-phase, with line-to-line voltages of 110, 220, 400, or 700 kV. Electrical currents in transmission lines are therefore reduced, along with the losses and voltage drops. With $P = 100$ MW and $U_l = 400$ kV, the line current is lower than 150 A. In the vicinity of consumers, high voltage at transmission lines is transformed by means of power transformers and scaled down to the level suitable for consumers (220 V). A power transformer changes voltage level according to transformation ratio $m = N_1/N_2$, defined by the number of turns of primary and secondary windings.

DC voltage cannot be transformed by power transformers. Recent developments in power electronics over the past couple of decades resulted in high power, high voltage static power converters required for DC power transmission. These devices were not available at the wake of electrical power systems. Therefore, DC voltage across generator terminals in the early power stations used to be fed to transmission lines without any conversion. The same voltage was made available to electrical loads, connected by distribution lines. Low DC voltages at transmission lines resulted in large currents, large voltage drops, and heavy losses. In order to keep the losses and voltage drops relatively low, transmission of electrical power over DC lines was feasible only at short distances.

Nowadays, electrical power generation, transmission, distribution, and consumption are based on AC voltages and currents. Processes of electromechanical conversion involve AC generators and motors. Therefore, the use of DC generators and motors is declining. DC electrical machines are being replaced by AC machines. Nevertheless, it is of interest to study DC machines as the first electrical machines that were widely used. Moreover, their relatively simple model makes them suitable for introducing basic principles, notions, and characteristics of electrical machines. While studying DC machines within the following three chapters, the reader gets acquainted with mechanical characteristics, steady-state operating area, transient characteristics, steady-state equivalent circuits, dynamic models, analysis of losses, power supply and controls, and other tasks, problems, and phenomena involved with application of electrical machines.

This chapter starts with description and principles of operation of DC motors and generators. Some basic information concerning the design of DC machines is presented as well. Analysis includes the operation of mechanical commutator, key component of DC machines, which converts DC currents into AC currents. This device receives DC current from electrical source and conducts them into rotor conductors. Due to mechanical commutation, rotor currents depend on rotor position. At constant speed, rotor conductors have AC current with their angular frequency determined by the speed. Some basic ways of forming the rotor winding and connecting the rotor conductors with the commutator are shown as well. This chapter ends with analytical expressions for electromagnetic torque and electromotive force.

Mathematical model of the machine is developed within the next chapters, describing the machine behavior during transients. On the basis of the steady-state analysis, the equivalent circuit is introduced and explained. It allows calculation of the current, flux, and torque of DC machine in the steady state, where the supply voltage and the rotor speed are constant and known. Within these chapters, mechanical characteristic is derived, expressing the steady-state relation of the speed and the torque. Losses in windings and magnetic circuits are analyzed along with processes of heating and the ways of heat removal (cooling). The maximum permissible steady-state current, power, and torque are introduced and explained, as well as the rated values of relevant. The field-weakening operation is introduced and explained, as well as relevant relations and characteristics. Transient and steady-state operating areas are determined from the torque-speed pairs attainable during transients and in the steady state.

11.1 Construction and Principle of Operation

DC machines consist of the stator magnetic circuit, rotor magnetic circuit, and rotor winding. The stator may have stator winding, called *excitation winding*, or permanent magnets. The stator flux is created either by permanent magnets in stator magnetic circuit or by DC currents in the stator winding. Currents in rotor conductors create the rotor magnetomotive force. In the preceding chapter, it has been shown that the vector product of two fluxes

$$\vec{T}_{em} = k\left[\vec{\Phi}_R \times \vec{\Phi}_S\right]$$

determines the electromagnetic torque of an electrical machine. Thus, the torque of a DC machine is determined by the vector product of the stator and rotor fluxes. The torque vector is collinear with the axis of the machine.

11.2 Construction of the Stator

The stator flux is called *excitation flux*, and it is obtained from direct electrical currents in the stator winding. The excitation flux can also be created by permanent magnets built in the stator magnetic circuit. The case when the excitation flux is obtained by the stator excitation winding is called *electromagnetic excitation*. Stator winding carries a direct current (DC) which creates stator magnetomotive force and stator flux. Since the stator carries a DC current, the stator flux does not move.

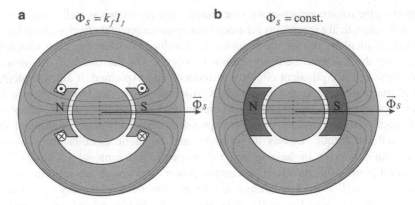

Fig. 11.1 Position of the stator flux vector in a DC machine comprising stator winding with DC current (**a**) and in DC machine with permanent magnets (**b**)

Instead of stator winding, DC machine can have permanent magnets built in the magnetic circuit of the stator. Permanent magnets have significant remanent induction B_r even in cases with no external field H. With permanent magnets, stator flux is obtained without any need to build a stator winding. In Fig. 11.1, lines of the stator field start from the north magnetic pole and propagate toward the south magnetic pole, passing along their way through the rotor.

11.3   Separately Excited Machines

Many DC machines have excitation that does not change with rotor currents. These machines are called *separately excited machines*. They include DC machines with permanent magnets, where the rotor currents do not affect the excitation flux caused by the magnets. They also include DC machines with the stator excitation winding fed from a separate electrical source, decoupled from the rotor supply.

In other types of DC machines, the stator excitation gets affected by the rotor currents. If the stator winding (i.e., the excitation winding) is connected in series with the rotor winding, the excitation current equals the rotor current. This type of electrical machine is called *series excited DC machine* or *series DC motor*. The excitation can depend on the rotor current in other ways. The excitation (stator) winding can be connected in parallel with the rotor winding. As the rotor voltage changes with the rotor currents, the excitation voltage and current would change with the rotor current as well. There are also DC machines where both series and parallel excitations are present.

Separately excited DC machines are the main subject of the study within this chapter.

11.4   Current in Rotor Conductors

Rotor of DC machine has axially set conductors carrying electrical currents. Through interaction with magnetic induction of the excitation field, the rotor conductors are exposed to electromagnetic force. A couple of forces create mechanical torque that acts on the rotor and incites motion. Due to specific construction of DC machines and the presence of mechanical commutator, which directs the electrical current into rotor conductors, direction of the current in conductors below the north magnetic pole (N) does not change. It remains the same despite the fact that the rotor revolves. Direction of the current in the conductors below the south pole (S) is opposite to the one in the conductors which are below the north pole.

Conductors 1 and 2 are built in the rotor slots, and they revolve at the same speed as the rotor does. With rotor making one half turn (Fig. 11.2b), conductors 1 and 2 exchange their places. In order to have a positive torque, it is necessary to change directions of the currents in conductors, as shown in the Figure. In conductor 1, the direction was $\otimes$ while it was in the zone of the north pole of the stator. When this conductor comes to the zone of the south pole, it has to carry a current of direction $\odot$ so that the torque remains positive. Similar conclusion may be drawn for conductor 2.

Conductors 1 and 2 constitute one contour (turn) of the rotor. When the rotor revolves at a speed Ω_m, it makes one revolution each $T = 2\pi/\Omega_m$. For the purpose of creating a torque which would not change sign but remain positive instead, current

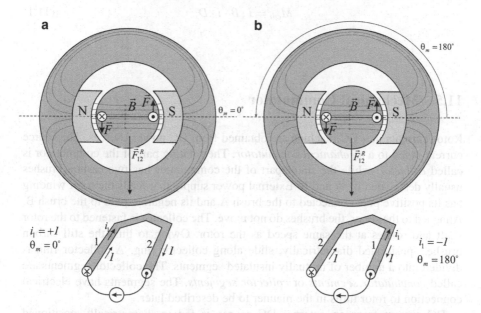

Fig. 11.2 Position of rotor conductors and directions of electrical currents. (**a**) Rotor at position $\theta_m = 0$. Rotor conductor 1 in the zone of the north pole of the stator and conductor 2 below the south pole of the stator. (**b**) Rotor shifted to position $\theta_m = \pi$. Conductors 1 and 2 have exchanged their places

through the considered contour should change sign synchronously with the rotor movement. The current should be positive during time interval $T/2 = \pi/\Omega_m$ and then negative during the next interval of $T/2$. Therefore, current in the rotor has to be periodic with the period T determined by the rotor speed.

The ways of directing the rotor currents into conductors so as to obtain periodic electrical currents will be put aside at this time. By considering Fig. 11.2 and assuming that, notwithstanding the rotor motion, electrical current in rotor conductor below the north pole retains direction $\otimes$, while the conductor below the south pole always has direction of the current $\odot$, the rotor magnetomotive force and flux can be represented by vectors directed downward. On the other hand, vector of the stator field is horizontal, directed from left to right. It is concluded that the angle between the two fluxes is constant and equal to $\pi/2$, irrespective of the speed and direction of the rotor motion. This fulfills the optimal condition for creating constant torque.

Considered pair of rotor conductors is in the region under the stator poles. Coupled forces producing positive torque act upon conductors 1 and 2 (Fig. 11.2a). This torque supports counterclockwise movement. The electromagnetic torque is determined by (11.1) where l is length of one conductor, B is magnetic induction in the zones of the stator magnetic poles, and D is rotor diameter.

$$\vec{F} = i(\vec{l} \times \vec{B}),$$

$$M_{em} = l \cdot B \cdot i \cdot D. \tag{11.1}$$

11.5 Mechanical Commutator

Rotor currents in a DC machine are obtained from DC power sources. The source current is fed to a *mechanical commutator*. The rotating part of the commutator is called *collector*, while the stator part of the commutator has two carbon brushes usually designated by A and B. External power supply that feeds the rotor winding has its positive pole connected to the brush A and its negative pole to the brush B. Attached to the stator, the brushes do not move. The collector is fastened to the rotor shaft and rotates at the same speed as the rotor. Owing to this, the still carbon brushes, positioned diametrically, slide along collector ring. A collector ring is divided into a number of mutually insulated segments. The collector segments are called *commutator segments* or *collector segments*. The segments have electrical connection to rotor turns in the manner to be described later.

DC current from an external DC source is fed to diametrically positioned brushes A and B. Immobile brushes lean on the collector segments, directing in this way current into the rotor conductors. When the rotor moves, the brushes slide from one pair of the collector segments to the next pair. As a consequence, a change occurs in distribution of electrical current in rotor conductors. The end result is that, notwithstanding the rotor motion, the rotor currents create the magnetomotive force vector $F^R{}_{12}$ which does not move with respect to the stator, as shown in Fig. 11.2. The rotor magnetomotive force gives a rise to the rotor flux, which does not move either.

Mechanical commutator converts DC currents, obtained from the power supply of the rotor winding, to periodic currents carried by rotor conductors. Frequency of these currents is determined by the speed of rotation. The role of the mechanical commutator is similar to that of static power converters called inverters, which employ power transistor switches and convert DC voltages and currents into AC voltages and currents. The change of voltages and currents in conductors of the rotor equipped with mechanical commutator is similar to the change of voltages and currents in a system comprising DC supplied inverter which feeds AC currents and supplies asynchronous or synchronous machines.

In modern applications of electrical machines, DC machines with mechanical commutator are replaced by static power converters (transistor inverters) feeding asynchronous or synchronous machines, also called AC machines.

The method of making of rotor winding as well as the method of connecting this winding to collector may be relatively involved. A detailed study of different methods of realization of rotor windings of DC machines is beyond the scope of this book. For the purpose of understanding the operation of mechanical commutator, further discussion presents an analysis of some relatively simple examples, intended for illustration of basic functions of mechanical collector.

11.6 Rotor Winding

Figure 11.3 shows a rotor having only one turn. It is made of conductors 1 and 2 connected in series. The conductors are in electrical connection with collector which has segments S1 and S2. Such collector can be made from a metal cylinder by cutting it in two mutually insulated halves. The front end of conductor 1 is connected to segment S1, while the front end of conductor 2 is connected to segment S2. At the rear end of the rotor, the ends of conductors 1 and 2 are brought together by the end turn. Brushes A and B are connected to a current source supplying electrical current $i(t)$. In the given position, brushes A and B lean on segments S1 and S2, respectively. Therefore, there is electrical contact between the brush and the segment that gets in touch with it. At position presented in the figure, current in conductor 1 has direction $\otimes$, while current in conductor 2 has direction $\odot$. When the rotor moves by π, conductors exchange their places. At the same time, the segments S1 and S2 change their places as well, as they are fastened to the rotor and revolve along with the rotor. Now conductor 1 gets under the south pole (left), but direction of the electrical current in this conductor is changed. Therefore, the electrical current in rotor conductor under the north pole retains direction $\otimes$, while the current in rotor conductor under the south pole retains direction $\odot$. Mechanical commutator insures that the

Fig. 11.3 Mechanical collector. *A*, *B*, brushes; *S1*, *S2*, collector segments

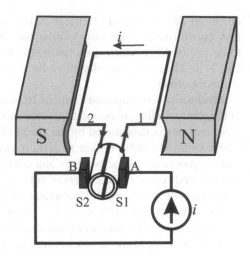

current distribution under the stator poles remains the same for any position of the rotor, notwithstanding its relative motion.

Current in each of conductors changes direction synchronously with the rotor motion. The commutator with its brushes and two-segment collector converts DC current of the source into periodic current in the rotor conductors. One revolution of the rotor corresponds to one period of alternating currents in rotor conductors. As a consequence, the observer at the stator side (i.e., the observer which does not move with respect to the stator) does not see any motion of the rotor magnetomotive force. Namely, the rotor conductors below the north magnetic pole, designated by N in Fig. 11.4, retain the current direction, while the conductors below the south pole S retain direction ⊙. This distribution remains unaltered in spite of the fact that the rotor and rotor conductors move.

In practice, rotor winding has a number of turns evenly distributed along the rotor perimeter. The conductors are connected to respective segments of the collector. A collector ring may have a number of mutually insulated segments which are galvanically connected to two or more conductors. Current i_a is fed to the rotor by means of a couple of carbon brushes which are in touch with the collector and which pass the electrical current to the segments. The commutator directs the current to the rotor conductors in such way that distribution of the rotor currents corresponds to the one shown in Fig. 11.4. Said distribution does not change notwithstanding the rotor motion. It should be noted that the rotor turns along with the rotor conductors. Due to the action of mechanical commutator, the rotor has alternating currents and they create the rotor current sheet which does not revolve but remains still with respect to the stator. In Fig. 11.4, direction of the rotor magnetomotive force and the rotor flux is vertical, irrespective of the rotor position. Therefore, the rotor flux remains still with respect to the stator, and the angle between the stator and rotor flux vectors is $\pi/2$. DC machines usually have a relatively large number of the rotor conductors and corresponding number of collector segments. Appearance of the rotor of a typical DC machine is shown in Fig. 11.5.

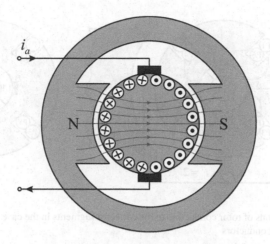

Fig. 11.4 Position of the rotor current sheet with respect to magnetic poles of the stator

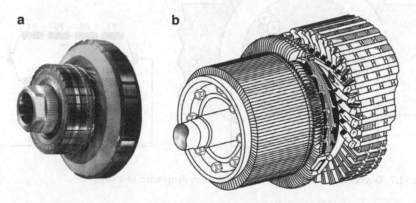

Fig. 11.5 Appearance of the rotor of a DC machine. (**a**) Appearance of the collector. (**b**) Appearance of the magnetic and current circuits of a DC machine observed from collector side

Figure 11.6 shows the method of connecting rotor conductors to the collector segments for DC machine with 4 rotor slot, 4 collector segments, and 8 conductors. Given example is seldom seen in practice. The number of rotor slots is usually much higher. Yet, the case in Fig. 11.6 is selected as an introduction to making the rotor winding. Conductors P1–P8 are placed in slots. Each slot houses two conductors. The rotor is observed from the side where the mechanical collector is mounted on the shaft. The ends of conductors P1–P8 are connected as well on the rotor side opposite to mechanical collector, the rare side of the machine. This side is not visible. Therefore, relevant wire connections at the rare side are shown in the right-hand side of Figure 11.6 by dotted lines. Connected by rare side connections, conductors make four turns, P1–P2, P3–P4, P5–P6, and P7–P8. At the front side of the rotor, where the mechanical collector is mounted, wire connections are represented in the left-hand side of Fig. 11.6 by solid lines.

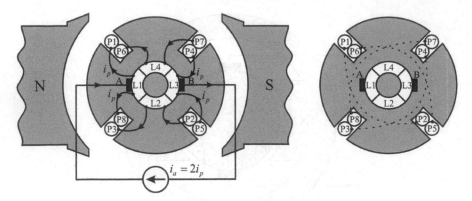

Fig. 11.6 Connections of rotor conductors to the collector segments in the case when 4 rotor slots contain a total of 8 conductors

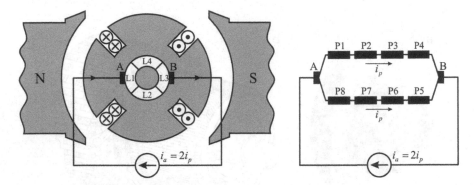

Fig. 11.7 Direction of currents in 8 rotor conductors distributed in 4 slots

Segment L1 of the collector is connected to conductors P8 and P1, segment L2 to conductors P2 and P3, segment L3 to conductors P4 and P5, and segment L4 to conductors P6 and P7. Connections of the segments with relevant conductors are at the front side, represented by solid lines, while the connections of conductor ends at the rear side of the rotor are drawn by dotted lines. In the considered rotor position, brushes A and B are in touch with segments L1 and L3; thus, the current of the source i_a is split in two parallel paths, as shown in Fig. 11.7. Each of the rotor conductors carries electrical current of $i_a/2$.

In all the conductors below the north magnetic pole of the stator, direction of electrical current is $\otimes$, while in conductors below the south magnetic pole, direction is $\odot$. If the rotor is turned by $\pi/2$, brush A comes in touch with segment L4, while brush B touches segment L2. Connections between the conductors and segments shown in Fig. 11.6 can be used to determine direction of currents in rotor conductors after the rotor moves. Due to rotation, conductors will change their position. At the

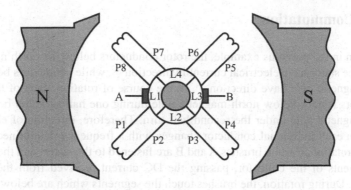

Fig. 11.8 Wiring diagram of the rotor current circuit

same time, direction will change in some of them due to mechanical commutator. Finally, direction of currents in conductors below the north magnetic pole remains ⊗ even after the rotor has moved, while in conductors below the south pole direction remains ⊙. The way of making the rotor winding and connecting the segments ensures that direction ⊗ is preserved in all conductors under the north magnetic pole, notwithstanding the rotor motion. In every single conductor, direction changes and becomes ⊙ as the conductor passes from the zone under the north pole into the zone below the south magnetic pole. In the course of rotation, the rotor conductors have alternating currents with frequency determined by the speed of rotation.

Wiring diagram of the rotor current circuit can be presented in the manner shown in Fig. 11.8. The figure shows the rotor with rather simple construction, having 4 slots, 8 conductors, and mechanical commutator with 4 segments. Commutator allows creation of the rotor current sheet which does not rotate with respect to the stator, producing in this way vectors of the rotor magnetomotive force and flux which do not move with respect to the stator. The power supply to the rotor winding is shown in Figs. 11.6 and 11.7 as a constant current source connected to brushes A and B.

In the course of the rotor motion, the brushes direct the current to collector segments and subsequently to the rotor conductors. As the rotor conductor moves below the north pole and passes under the south pole, direction of electrical current changes. For that reason, each rotor conductor has alternating current with a frequency determined by the speed of rotation. Observed from the stator side, distribution of the rotor currents remains unaltered. Therefore, the rotor currents create a current sheet which does not move with respect to the stator. Rotor currents are shown by signs ⊗ and ⊙ in Fig. 11.7, and they create the rotor flux which can be represented by the vector of vertical direction, standing at an angle of $\pi/2$ with respect to the stator field. According to the right-hand rule, the rotor flux is directed downward.

11.7   Commutation

As shown in the previous example, the rotor conductors below the north magnetic pole of the stator have electrical current in direction ⊗, while conductors below the south magnetic pole have direction ⊙. In the course of rotation, each of the rotor conductors resides below north magnetic pole during one half turn and below the south magnetic pole under the second half turn. Therefore, direction of electrical current in each individual conductor changes with a frequency determined by the speed of rotation. Carbon brushes A and B are fastened to the stator, and they touch the segments of the collector, passing the DC current received from the power supply i_a. During rotation, the brushes touch the segments which are below them at each particular instant. Hence, the segments slide under the brushes. Brush transition from one segment to the next is followed by change in electrical current in rotor conductors attached to relevant segments. Directing DC current i_a by collector action results in alternating currents in the rotor conductors. Transition of the brush from one segment to the other and consequential change in electrical current in rotor conductors is called *commutation*. In the course of transition, one brush touches two segments at the same time, bringing them into short circuit and short circuiting the rotor turns connected to relevant segments. The case when brush A simultaneously touches segments L1 and L2 is shown in Fig. 11.9.

Advancing from position given in Fig. 11.8 in clockwise direction, toward position given in Fig. 11.9, the rotor moves by $\pi/4$. Since the brush A in Fig. 11.9 makes a short circuit between segments L1 and L2, while the brush B makes a short circuit between segments L3 and L4, turns P1–P2 and P5–P6 are short-circuited during commutation. If the rotor makes further move by $\pi/4$ in the same direction, it arrives at position shown in Fig. 11.10. With respect to Fig. 11.8, the rotor position is changed by $\pi/2$ in clockwise direction, and now the brush A has contact with segment L2. Wiring diagram of Fig. 11.10 should be compared with the wiring diagrams in Figs. 11.6 and 11.7.

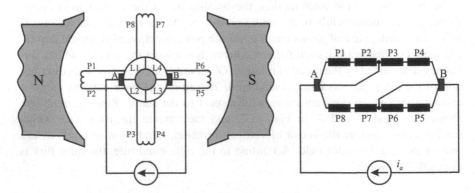

Fig. 11.9 Short circuit of rotor turns during commutation

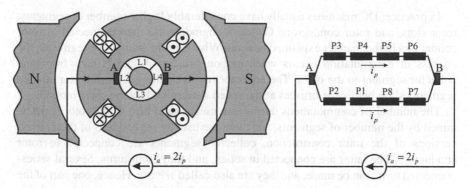

Fig. 11.10 Rotor position and electrical connections after the rotor has moved by $\pi/4 + \pi/4$ with respect to position shown in Fig. 11.8

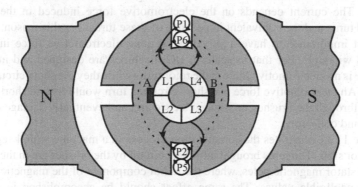

Fig. 11.11 Position of short-circuited rotor turns during commutation. The turns P1–P2 and P5–P6 are brought into short circuit by the brushes A and B, respectively

For the considered rotor, commutation repeats four times per each mechanical turn. It is necessary to analyze the problems associated with periodic short circuits of the rotor turns in the process of commutation. Starting from the scheme of placing the rotor conductors into slots, shown in Fig. 11.6, it is concluded that, during commutation shown in Fig. 11.9, the turns P1–P2 and P5–P6 get short-circuited. At the same time, conductors P1, P2, P5, and P6 pass through the zone between the stator magnetic poles, halfway between the north and south pole, where the radial component of magnetic induction is small and changes sign. The place where the short-circuited turns are found during commutation is shown in Fig. 11.11, where the remaining turns are omitted. Contribution of stator excitation to magnetic induction in the air gap is the highest in the middle of stator magnetic poles. At places where conductors P1, P2, P5, and P6 are found in Fig. 11.11, the radial component of magnetic induction is close to zero. Therefore, the electromotive forces induced in these conductors are close to zero. As a consequence, short circuiting these turns does not produce any significant short circuit currents.

In practice, DC machines usually have considerably higher number of segments, rotor slots, and rotor conductors. Collector segments are then connected to rotor conductors in the manner explained later on. Whatever the number of segments, the process of commutation occurs when carbon brushes A and B pass from one collector segment to the other. The collector revolves along with the rotor, and its segments slide below the brushes at the speed determined by the rotor motion.

The number of commutations during one mechanical turn of the rotor is determined by the number of segments, and there are usually several tens of them. In all versions of the rotor construction, collector segments are connected to rotor conductors. The latter are connected in series, and they form turns. Several series-connected turns can be made, and they are also called *section*. Hence, one part of the rotor winding is connected between each pair of neighboring segments, and this part can be a single turn or a multiple turn section. Whenever the brush touches two adjacent segments, considered part of the rotor winding gets short-circuited. The short circuit current between the adjacent segments is established through the brushes. The current depends on the electromotive force induced in the short-circuited turns and the equivalent impedance of these turns. For this reason, it is of uttermost importance to have a very low or none electromotive force in short-circuited windings. For that to achieve, DC machines are designed and made so that there is no electromotive force in the rotor turns while they get short-circuited by brushes. An electromotive force in a short-circuited turn would lead to short circuit currents through the brushes, sparking, electric arc, and eventually damage of both brushes and collector.

Figure 11.11 illustrates the commutation process in a machine with 4 segments and 4 rotor slots. The turns brought into short circuit by the brushes are in the region between stator magnetic poles, where the radial component of the magnetic induction has negligible values. The same effect should be accomplished in all DC machines. In machines having a large number of rotor segments, brushes may be wider and extend over two or more segments. In this case, several segments are brought into short circuit by one brush. All conductors belonging to short-circuited turns in the course of commutation have to be away from the stator magnetic poles, in the region between the poles, where the induction is negligible. The area between the magnetic poles is called *neutral zone*. This will be dealt with in the subsequent sections.

11.8   Operation of Commutator

Mechanical commutator of DC machines performs the function of converting DC currents, supplied from DC power source via brushes, into periodic currents which exist in rotor conductors. Change of the current in rotor conductors is shown in Fig. 11.12b. Direction of this current changes synchronously with the rotor motion. Between the commutation intervals, denoted by shaded areas in the figure, the current i_c is equal either to $+i_a/2$ or $-i_a/2$. The commutation intervals are relatively

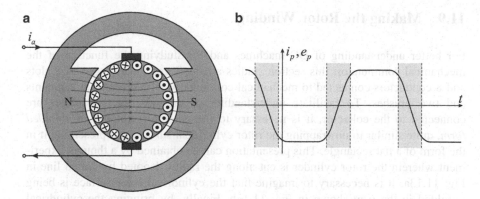

Fig. 11.12 Commutator as a DC/AC converter. (**a**) Distribution of currents in rotor conductors. (**b**) Variation of electromotive force and current in a rotor conductor. *Shaded* intervals correspond to commutation

short with respect to the period. Thus, the shape of rotor currents is close to a train of rectangular pulses having amplitude $i_a/2$. Rotor currents are not sinusoidal. Current i_p in Fig. 11.12b is a symmetrical, periodic current with average value equal to zero.[1]

Therefore, the commutator shown in Fig. 11.12a is a mechanical converter which converts DC currents to alternating currents (AC). Frequency of currents carried by rotor conductors is determined by the rotor speed.

In cases where DC machine is used as generator, rotor is put to motion by means of driving torque obtained from a water turbine or steam turbine. Rotor conductors revolve in magnetic field created by stator excitation. Relevant magnetic induction is proportional to the stator flux. Electromotive forces induced in rotor conductors are proportional to magnetic induction and peripheral speed. Under the north pole, magnetic induction has opposite direction with respect to that under the south pole. For this reason, electromotive forces induced in conductors below the two poles have different signs. While the rotor turns, each rotor conductor passes below stator poles with a period determined by the rotor speed. Therefore, electromotive force induced in a single conductor changes periodically. Its average value is equal to zero while its frequency depends on the rotor speed. Sample electromotive force induced in one conductor due to rotor motion is shown in Fig. 11.12b. It is shown in following sections that AC electromotive forces in rotor conductors result in a DC electromotive voltage measured between brushes A and B. This AC/DC conversion of voltages takes place due to the action of mechanical commutator which adds the AC electromotive forces in individual rotor conductors in such way that a DC electromotive force appears between the brushes. Hence, the commutator acts as a rectifier.

---

[1] In a broader sense, it is possible to call it *alternating current*. Strictly speaking, only sinusoidal functions of time are understood as alternating currents, called sinusoidal or harmonic currents.

11.9 Making the Rotor Winding

For better understanding of DC machines and for studying the function of the mechanical commutator, this section studies a sample rotor winding with 8 slots and 8 conductors connected to mechanical commutator with 4 collector segments and two brushes. To facilitate understanding of the way the conductors are connected to the collector, it is necessary to present rotor winding in *unfolded form*, quite similar to unwrapping the rotor cylindrical surface and presenting it in the form of a flat rectangle. This presentation can be obtained by a thought experiment wherein the rotor cylinder is cut along the radius denoted by dotted line in Fig. 11.13a. It is necessary to imagine that the cylindrical rotor surface is being unfolded in the way shown in Fig. 11.13b. Finally, by bringing the cylindrical surface to a plane, *unfolded form* is obtained, given in Fig. 11.13c. Rotor conductors are shown on this unfolded drawing with magnetic poles of the stator shown on the top. It should be noted that conductors P1–P4 are shown under the north magnetic pole, as they are in Fig. 11.13a.

In Fig. 11.15, the segments are denoted by L1–L4, while the conductors are denoted by P1–P8. Conductors P1–P4 carry electrical currents in direction ⊗, and they reside under the north stator pole. Conductors P5–P8 carry electrical currents in direction ⊙, and they reside under the south stator pole. While denoting direction of electrical currents, it is assumed that the reader is at the front side of the rotor, looking at the mechanical collector, as shown in Fig. 11.14. Sign ⊗ designates electrical current directed from front part (collector) to the rear part of the rotor, while sign ⊙ designates electrical current directed from the rear part of the rotor toward the front side and toward the reader.

The four segments shown in the figure can be obtained by splitting a metal ring into four equal arcs and by putting electric insulation layers in between. The segments can also be presented in unfolded form, using the same approach (Fig. 11.15). It should be noted that the width of the unfolded drawing corresponds to the rotor circumference, namely, the unfolded drawing is 2π wide. One can also consider that the horizontal axis in Fig. 11.15 corresponds to angular change from 0 (left) to 2π (right). Unfolded presentation in Fig. 11.15 shows the conductors, brushes, and collector segments. On the top is the position of the magnetic poles. The poles (N, S)

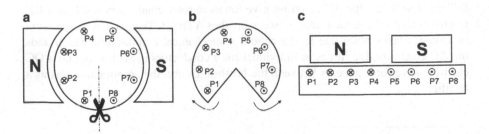

Fig. 11.13 Unfolded presentation of the rotor

Fig. 11.14 Rotor of a DC
machine observed from the
front side

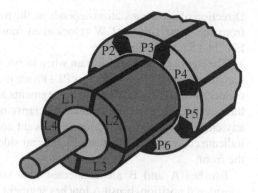

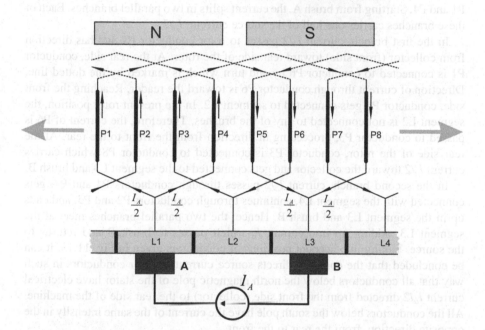

Fig. 11.15 Unfolded presentation of rotor conductors and collector segments. The brushes *A* and
B touch the segments *L1* and *L3*

and the brushes (A, B) do not move. In Fig. 11.15, the rotor motion can be envisaged
as a parallel transition of conductors P1–P8 and segments L1–L4 toward left or right.

At considered position, conductors P1–P4 are below the north magnetic pole.
The brushes A and B touch the segments L1 and L3, respectively. When the rotor
moves, there is relative movement of the collector segments and rotor conductors
with respect to the stator. Magnetic poles of the stator and brushes are fastened
to the stator, and they do not move. The effects of the rotor motion in Fig. 11.15
are manifested as translation of the rotor conductors and collector segments.

Direction of such translation depends on the rotor speed. The rotor motion observed from the front side as CW (clockwise) moves the conductors and segments in Fig. 11.15 toward right.

The figure presents the instant when brush A touches segment L1, while brush B touches segment L3. Conductors P1÷P8 are drawn by thick lines while their internal connections and connections to the segments are drawn by thin lines. The change in the line thickness is used to enhance clarity of the drawing, and it does not imply any change in the cross section of relevant conductors and wires. The dotted lines indicate connections effectuated at the rear side of the rotor and hence invisible from the front.

Brushes A and B are connected to a source of constant current I_a. In the considered position, brush A touches segment L1 which is connected to conductors P1 and P4. Starting from brush A, the current splits in two parallel branches. Each of these branches carries one half of the source current, $I_a/2$.

In the first branch, current $I_a/2$ passes to rotor conductor P1 and has direction from collector (front side) toward rear side of the rotor. At the rear side, conductor P1 is connected to conductor P6 by end turn which is marked by the dotted line. Direction of current through conductor P6 is toward the reader. Reaching the front side, conductor P6 gets connected to segment L2. In the present rotor position, the segment L2 is not connected to any of the brushes. Therefore, the current of P6 is passed to conductor P3, proceeding in direction from the front to the rear. At the rear side of the rotor, conductor P3 is connected to conductor P8 which carries current $I_a/2$ toward the collector and gets connected to the segment L3 and brush B.

In the second branch, current $I_a/2$ passes through conductors P4 and P7, gets connected with the segment L4, continues through conductors P2 and P5, and ends up in the segment L3 and brush B. Hence, the two parallel branches meet at the segment L3, adding up into current I_a which passes to brush B and returns to the source. Taking into account positions of conductors given in Fig. 11.13, it can be concluded that the collector directs source current to rotor conductors in such way that all conductors below the north magnetic pole of the stator have electrical current $I_a/2$ directed from the front side (collector) to the rear side of the machine. All the conductors below the south pole have the current of the same intensity in the opposite direction, from the rear to the front.

As the rotor turns, conductors P1–P8 and segments L1–L4 on unfolded drawing in Fig. 11.5 move toward left or right, while the brushes and the stator poles remain still. When the rotor moves by $\pi/4$ toward left, the brush A gets in touch with segment L2, while the brush B gets in touch with segment L4. At this new position of the rotor, distribution of currents in rotor conductors is given in Fig. 11.16. Conductors P3, P4, P5, and P6 are below the north pole of the stator. Direction of current in these conductors is from the reader toward the rear side of the rotor. Conductors P7, P8, P1, and P2 are under the south magnetic pole, and they carry currents in the opposite direction. By comparing this with the previous case (Fig. 11.15), it can be concluded that rotation of the rotor leads to variation of electrical currents in individual rotor conductors, but it does not change distribution of rotor currents observed from the stator side. In other words, irrespective of the

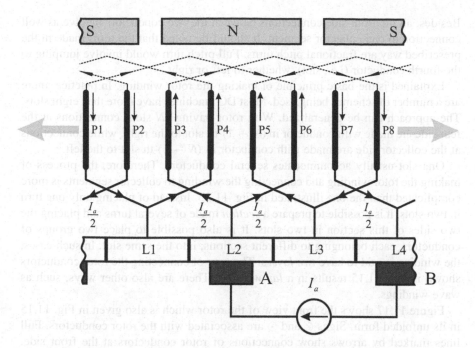

Fig. 11.16 Directions of currents in rotor conductors at position where brush *A* touches the segment *L2*

rotor motion, currents in conductors below the north magnetic pole retain direction ⊗, while currents in conductors below the south magnetic poles retain direction ⊙. In this way, rotation of the rotor does not change the course and direction of the rotor magnetomotive force and the rotor flux which remain unmoving with respect to the stator. Hence, the mechanical commutator insures that the rotor flux vector remains still with respect to the stator flux vector.

Selecting one of rotor conductors and tracking the change in its electrical current as the rotor completes one mechanical turn, one comes to conclusion that the conductor has alternating current and that one mechanical turn corresponds to one period of the current. Hence, mechanical commutator can be envisaged as a device which converts DC source current I_a into an alternating current.

The method of making the rotor winding is shown in Fig. 11.15. It starts by connecting the conductor P1 with the segment L1 at the front side of the machine and proceeds with putting an end turn at the rear side which connects P1 to P6. It is of interest to notice that the connection P1–P6 at the rear of the machine is realized by connecting the end of P1 to the *fifth conductor to the right*. Further on, the front side of the conductor P6 is connected to the segment L2, and this connection involves the front end of the conductor P3, the *third to the left*. The making of the winding proceeds in the same manner until all the conductors are connected. All the rear side connections are made by jumping to the *fifth conductor to the right*. All the front side connections are made by jumping to the *third conductor to the left*.

Besides, all the front side connections between the two conductors involve as well connection to one collector segment. It should be noted that the turns made in the prescribed way are fractional pitch turns. Full pitch turn would involve jumping to the fourth conductor (i.e., slot), whether to left or right.

Explained is the basic principle of making the rotor winding. In practice, there are a number of schemes being used. Most DC machines have more than eight slots. The approach can be generalized. With rotor having $2N$ slots, connections at the rear side are made with conductor in $(N + 1)$-th slot to the right, while connections at the collector side are made with conductor in $(N - 1)$-th slot to the left.

One slot usually accommodates several conductors. Therefore, the process of making the rotor winding and connecting the winding to collector segments is more complicated than the one illustrated in Fig. 11.15. Instead of placing only one turn in two slots, it is possible to prepare a *section* made of several turns and placing the two sides of this section in two slots. It is also possible to place two groups of conductors, each belonging to different sections, into the same slot. In such cases, the winding is said to have *two layers*. The way of connecting the rotor conductors shown in Fig. 11.15 results in a *lap winding*. There are also other ways, such as wave windings.[2]

Figure 11.17 shows the front view of the rotor which is also given in Fig. 11.15 in its unfolded form. Signs $\otimes$ and $\odot$ are associated with the rotor conductors. Full lines marked by arrows show connections of rotor conductors at the front side. Dotted lines show the connections between rotor conductors at the rear side of the machine. Designation P1P6 next to the dotted line marks that this is connection between conductors P1 and P6 made at the rear side of the rotor. A comparison of this presentation with the one in Fig. 11.15 illustrates the merits of the unfolded scheme.

11.10   Problems with Commutation

By considering the example of a rotor winding having 8 conductors and 4 collector segments analyzed in the preceding section, it is concluded that at each instant, there are two parallel branches between brushes A and B. In each of them, there are 4 conductors connected in series. Current in the rotor conductors is equal to one half of the current taken from the source which is connected to the collector brushes and which feeds the rotor winding.

When brush A passes from segment L1 to segment L2, brush B passes from segment L3 to segment L4. Passage of brushes from one segment to another leads to changes of direction of electrical currents in individual rotor conductors. In the course of commutation, shown in Fig. 11.18, brush A makes a short circuit between

---

[2] More details on windings of electrical machines can be found in publication Pyrhonen J, Jokinen T, Hrabovcova V (2008) Design of rotating electrical machines. Wiley, ISBN: 978-0-470-69516-6

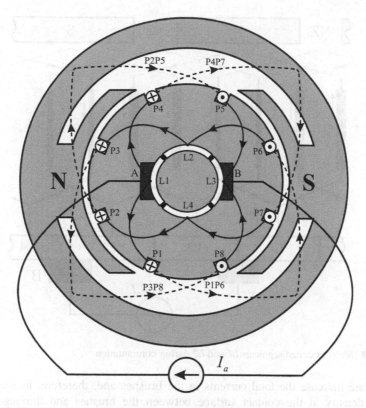

Fig. 11.17 Front side view of the winding whose unfolded scheme is given in Fig. 11.15

segments L1 and L2 while brush B makes a short circuit between segments L3 and L4. Owing to the short circuit between the adjacent collector segments, one rotor turn made of conductors P1 and P6 is short-circuited by the brush A, while the turn P2–P5 is short-circuited by the brush B.

At position presented in Fig. 11.15, direction of electrical current in conductors P1 and P2 is ⊗, while in conductors P5 and P6 direction of current is ⊙. When the rotor moves by π/2 and arrives at position presented in Fig. 11.16, direction of electrical current in conductors P1 and P2 is changed to ⊙, while direction of current in conductors P5 and P6 is changed to ⊗. Therefore, during commutation, direction of current is changed in those parts of the rotor winding which are short-circuited by the brushes. In Fig. 11.18, conductors of short-circuited turns are drawn by thicker lines.

In the case where, at the same time, the electromotive force in turn P1–P6 assumes significant value, a short circuit current will be established through the brush A, limited only by the impedance of the turn. In the same way, electromotive force in turn P2–P5 results in short circuit current through the brush B. The short circuit current is determined by the ratio of the electromotive force and the equivalent impedance of the short-circuited turns. Short circuit currents in the turns that

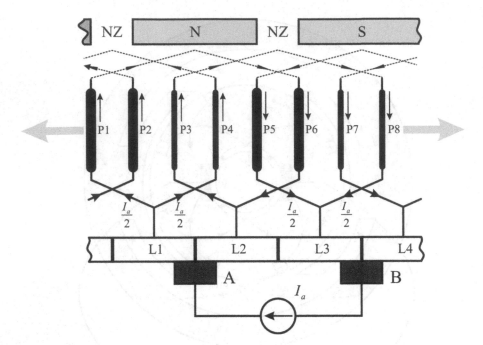

Fig. 11.18 Short-circuited segments *L1* and *L2* during commutation

commutate increase the total currents in the brushes and, therefore, increase the current density at the contact surface between the brushes and the segments. Excessive overheating of the brushes can result in an electric arc between the brushes and the segments, as well as between the adjacent segments. Sparking and arc can result in accelerated ware of the brushes and eventual damage to the collector. If the commutation is inadequate, there is a permanent electric arc between the brushes and collector segments. In such case, the collector and brushes overheat. Sparking and arc produce considerable quantity of ionized particles in close vicinity of brushes. Particles of ionized gas created under the brush A adhere to the collector surface. Due to rotation, they get carried away toward the brush B. In cases where the commutation is severely impaired, the electric arc may extend between brushes A and B. In this state, called *circular arcing*, the brushes and the rotor power supply I_a are in short circuit. At the same time, the rotor winding gets short-circuited. Prolonged operation in this mode leads to permanent damage to the winding and to mechanical collector and presents a fire risk.

Inadequate commutation leads to an increase of losses, damages collector and brushes, and may result in circular arcing and permanent damage to the machine. For this reason, it is significant that the electromotive forces in short-circuited turns are kept close to zero during their commutation (contours P1–P6 and P2–P5 in Fig. 11.18). In the position shown in this figure, relevant conductors are found between the stator magnetic poles in the neutral zones denoted by NZ. In the cross section of the machine, shown in Fig. 11.17, the neutral zones are in the upper part

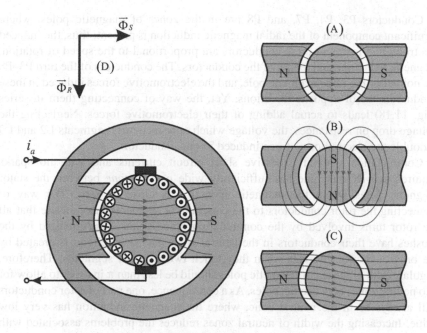

Fig. 11.19 Armature reaction and the resultant flux

of the rotor cylinder, at an angle of $\pi/2$ with respect to the brushes. The distance between the rotor and stator magnetic circuit is very large in the neutral zone. The radial component of the magnetic induction faces a very large magnetic resistance. At the same time, magnetomotive force of the stator excitation does not contribute to radial fields in the neutral zone. For this reason, it can be considered that radial component of magnetic induction in neutral zones is very small. With negligible magnetic induction, electromotive forces induced in rotor conductors passing through neutral zones are of little consequence as well. Therefore, the assumption is justified that electromotive forces in rotor turns involved in commutation process can be neglected.

Cross section of the machine is shown in Fig. 11.19, along with the lines of the stator magnetic field. The lines come out of the north magnetic pole, pass through the rotor magnetic circuit, and enter the south magnetic pole of the stator. It is justified to assume that there are some lines of the field passing by the conductors which are located in the neutral zone. They extend in tangential direction which is collinear with the vector of the peripheral speed. Electrical field induced due to motion depends on the vector product of the speed and magnetic induction. As these two vectors are collinear, the vector product is equal to zero, as well as induced electrical field. In the absence of induced electrical field, electromotive force in relevant conductors is equal to zero. Hence, tangential component of the magnetic field in the neutral zone does not induce any electromotive force in rotor turns that are short-circuited by brushes in the course of commutation.

Conductors P3, P4, P7, and P8 are in the zones of magnetic poles, where significant component of the radial magnetic induction is present; thus, the induced electromotive forces in these conductors are proportional to the speed of rotation, magnetic induction, and length of the conductors. The conductors of the turn P3–P8 are not below the same magnetic pole, and the electromotive forces induced in these conductors are of opposite directions. Yet, the way of connecting them in series (Fig. 11.18) leads to actual adding of their electromotive forces. Neglecting the voltage drop on resistances, the voltage which appears across segments L2 and L3 is double the electromotive force induced in one conductor.

Commutation without excessive short circuit currents and with no sparks requires that DC machine has sufficiently wide *neutral zone* between the stator magnetic poles where the magnetic induction is close to zero. The way of connecting the rotor conductors to the commutator segments must ensure that all the rotor turns involved by the commutation process and short-circuited by the brushes have their conductors in the neutral zone. The short circuit is created by the brushes during intervals when they touch two adjacent segments. Therefore, angular width of the stator magnetic poles should be less than π in order to allow for two neutral zones between the poles. As a consequence, one part of rotor conductors will always be in the neutral zone, where the magnetic induction has very low value. Increasing the width of neutral zones reduces the problems associated with commutation. At the same time, it reduces the number of rotor conductors which are encircled by the stator magnetic field and which contribute to the electrome-chanical conversion and torque generation.

Question (11.1): In Fig. 11.18, short-circuited turns P2–P5 and P1–P6 contain the conductors placed at the edges of neutral zones, in the vicinity of magnetic poles. Discuss the risk that, due to vicinity of magnetic poles, electromotive forces are induced in short-circuited windings.

Answer (11.1): Figure 11.18 is drawn in the way that conductors P1, P2, P5, and P6 are at the edges of neutral zones, in the vicinity of magnetic poles. It is justified to expect that, at this position, the radial component of magnetic induction is higher than in the middle of the neutral zone. Conductors P2 and P5 make one turn which is short-circuited by the brush B (Fig. 11.18). They are laid at the edges of the north magnetic pole, symmetrically with respect to the pole. Vicinity of the magnetic pole contri-butes to an increased magnetic induction. Because of the symmetry, in positions where conductors P2 and P5 are placed, radial component of the magnetic induction has the same value. For this reason, the electromotive forces induced in conductors P2 and P5 are of equal amplitude and direction. Conductors P2 and P5 are series connected and make short-circuited turn P2–P5. Connections of conductors P2 and P5 are such that their electromotive forces subtract and cancel. Therefore, the electromotive force of the turn P2–P5 in position shown in Fig. 11.18 is equal to zero. This is due to the fact that conductors P2 and P5 are symmetrical with respect to the north magnetic pole. In all cases where short-circuited conductors, such as P2 and P5, come at the very edge of the neutral zone, in close vicinity of the stator magnetic poles, it is possible to have considerable electromotive forces induced in such conductors. Yet, their symmetrical placement with respect to the pole ensures cancelation of their electromotive forces. The electromotive force in short-circuited turn P2–P5 is equal to zero. The same conclusion can be drawn for conductors P1 and P6.

11.11 Rotor Magnetic Field

Maintaining low intensities of the magnetic induction in neutral zones is hindered by the presence of the rotor magnetomotive force. Figure 11.19 shows a simplified presentation of the stator field, which propagates horizontally (a), and the rotor field, created by the current sheet, which propagates in vertical direction (b). It can be concluded that the rotor currents create the rotor flux which has its north and south poles in neutral zones, in the area comprising the rotor conductors involved in commutation and short-circuited by the brushes. The resultant magnetic field of the machine is the sum of the stator and the rotor field. It can be presented in the way shown in Fig. 11.19c. The stator and rotor fluxes can be represented by the two mutually orthogonal vectors designated by Φ_S and Φ_R.

Rotor of DC machine is also called *armature*, owing to the appearance of the rotor conductors which are shown in Fig. 11.5b. In the relevant literature, the term *induct* is also used to designate rotor of DC machines. In rotor conductors, electromotive forces are *induced,* proportional to the angular speed of rotation and to the stator flux, hence the term *induct*. Electrical current I_a fed to the brushes is often called *armature current* while the voltage U_a between the brushes A and B is called *armature voltage*. The magnetomotive force and flux created by the rotor currents are called *reaction of induct* or *armature reaction*. The term *reaction* is used due to the fact that the rotor electromotive force comes as a consequence of the stator flux. At the same time, the rotor currents get affected by this electromotive force. Since the rotor currents create the rotor flux, such flux is considered to be a *reaction* to the excitation coming from the stator. In a way, the stator flux *induces* electromotive forces in rotor conductors and affects the rotor current. For this reason, the stator is also called *inductor*.

The rotor flux (i.e., armature flux) is relatively small. The lines of the rotor field come out of the rotor magnetic circuit and enter a very large air gap in neutral zones. Therefore, the rotor flux passes through regions of very low permeability (μ_o) and very high magnetic resistance. For this reason, the value of magnetic induction in neutral zone is relatively small. Nevertheless, even a relatively small field in neutral zone may have undesirable influence on commutation. The presence of magnetic induction in neutral zone results in induced electromotive forces in rotor conductors passing through the neutral zone. These conductors are involved in the process of commutation. The turns made of such conductors are connected to adjacent collector segments, and they get short-circuited by the brushes. Short-circuited loop created in the prescribed way involves the rotor conductors, collector segments, and brushes. Induced electromotive forces create short circuit currents which may cause an electric arc at the contact between brushes and segments. For this reason, DC machines make use of additional elements intended to reduce magnetic induction in the neutral zones. DC machines may have *compensation winding* and *auxiliary poles* which are designed and made to suppress the armature reaction. They reduce magnetic induction in the neutral zone and ensure that commutation takes place with no sparks and no arcing.

11.12   Current Circuits and Magnetic Circuits

Structural elements of any electrical machine can be divided into magnetic and current circuits, the latter also called windings. In general, it is possible to identify four main items:

- Stator magnetic circuit
- Rotor magnetic circuit
- Stator current circuits
- Rotor current circuits

Figure 11.20 shows cross section of a DC machine presenting basic elements of current circuits and magnetic circuits of DC machines. The figure does not show the commutator which is described in the preceding sections. The rotor magnetic circuit (A) contains an opening in the center, intended for the shaft, and it has axial slots along the perimeter. Parts of the stator magnetic circuit are the main poles (B), yoke (C), and auxiliary poles (D). Rotor current circuit (F) includes conductors which are laid in slots on the rotor. They are connected in the way described in the preceding sections. Stator current circuits comprise the excitation winding (G), compensation winding (E), and auxiliary poles winding (H). A more detailed description and functions of these elements will be presented further on.

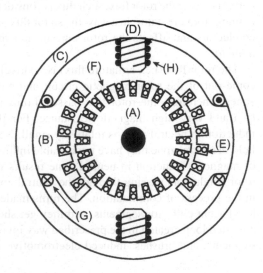

Fig. 11.20 Construction
of a DC machine

11.13 Magnetic Circuits

The stator magnetic circuit contains the main poles, auxiliary poles, and yoke. The main poles direct the stator flux, also called *excitation flux*. Starting from the north magnetic pole of the stator, the flux passes through the air gap, goes through the rotor magnetic circuit, makes another passage through the air gap, enters the south magnetic pole of the stator, and then, via yoke, returns to the north pole. Within the stator magnetic poles, the flux does not change the course and direction; thus, the magnetic induction in the stator iron is constant. For that reason, there are no losses in iron. The stator magnetic circuit does not have to be *laminated*, that is, it does not have to be made by stacking iron sheets. Instead, it can be made of solid iron. The auxiliary stator poles are used to reduce the magnetic induction in the neutral zone, which will be explained later.

The rotor magnetic circuit is of cylindrical shape. The rotor flux does not move with respect to the stator. It remains still with respect to the stator and the stator flux. The rotor revolves in magnetic field created by the stator and rotor windings. As a consequence, direction of the magnetic field relative to the rotor magnetic circuit varies as the rotor turns. Namely, observer that revolves with the rotor experiences revolving magnetic field. As the field pulsates with respect to the rotor magnetic circuit, there are eddy currents and iron losses in the rotor. The frequency of the field pulsations depends on the rotor speed. Variable magnetic field produces both hysteresis and eddy current losses in rotor iron. In order to reduce these losses, the rotor is built by stacking iron sheets (lamination). The shape of these sheets is given in Fig. 11.20. Along the rotor perimeter, there are slots where the rotor conductors are placed. At the center of the rotor sheets, there is a round opening intended for fastening the shaft. Cylindrical magnetic circuit of the rotor is formed by stacking a large number of iron sheets of thickness less than 1 mm.

11.14 Current Circuits

Current circuits of the stator include the excitation winding, compensation winding, and winding of the auxiliary poles. The excitation winding has N_f turns around the main poles. Excitation current creates the magnetomotive force of excitation $F_f = N_f I_f$. All the quantities and parameters related to the excitation winding have the subscript f for *field*. Namely, the excitation winding provides the magnetic field of DC machine. Dividing the magnetomotive force by magnetic resistance of the magnetic circuit, one obtains the stator flux Φ_f, also called excitation flux. This flux is equal to the surface integral of magnetic induction B over cross section of the main poles. At the same time, it is equal to the surface integral of the magnetic induction over the surface extending in the air gap below the main poles. In Figs. 11.19 and 11.20, the excitation flux is directed from left to right and passes twice through the air gap of width δ. By neglecting H_{Fe} and the magnetic voltage drop in iron, the magnetic induction in the air gap below the main poles can be estimated as $B = \mu_o N_f I_f/(2\delta)$. Therefore, one can change the excitation flux Φ_f by varying the excitation current. In addition to excitation winding, stator has compensation winding and winding of auxiliary poles.

Compensation winding consists of conductors laid in slots made on the inner side of the main poles of the stator. This winding is connected in such way that its conductors carry electrical current $I_a/2$, the same current that is carried by rotor conductors. Direction of the current in conductors pertaining to compensation winding is opposite to direction of rotor currents. Since conductors of the rotor winding and those of the compensation windings are in close vicinity, their magnetomotive forces mutually cancel due to opposite directions of their electrical currents. Therefore, current in the compensation winding *compensates* and cancels the magnetomotive force of the rotor. This is done in order to reduce the magnetic induction in the neutral zone and to avoid short circuit currents in rotor turns involved in the process of commutation. The compensation winding cancels the magnetomotive force of all the rotor conductors located under the main stator poles. This does not include all the rotor conductors, as some of them are found in neutral zones, away from the main poles. The magnetomotive force of these conductors is not fully compensated by action of the compensation winding. Therefore, the auxiliary poles are also built and placed against neutral zones with the purpose to restrain the magnetic induction in these zones. Auxiliary poles have corresponding winding.

Winding of auxiliary poles has turns around the auxiliary poles magnetic core, denoted by (D) in Fig. 11.20. The air gap below the auxiliary poles is considerably wider compared to the air gap below main magnetic poles. This is done to increase the magnetic resistance encountered by the armature reaction. Current in the winding of the auxiliary poles is made proportional to the rotor current. Direction of this current and the number of turns are adjusted to achieve compensation of the magnetomotive force of the rotor which has not been compensated by action of the compensation winding. By joint action of the compensation winding and auxiliary poles, it is possible to control and suppress the magnetic induction in neutral zones, that is, below auxiliary poles. Essentially, the magnetic induction in the neutral zones should be reduced to a value close to zero. Detailed analysis of the process of commutation is not included in this book. Further study shows that the stress and ware of the brushes and collector segments is reduced in cases with *linear commutation*, where the current in short-circuited turns involved in commutation process varies from $+I_a/2$ to $-I_a/2$ in linear fashion. Prerequisite for linear commutation is establishing magnetic induction in neutral zones which has a very small value that varies in proportion to armature current I_a. Detailed analysis of the process of commutation and method of designing the compensation winding and the winding of auxiliary poles are beyond the scope of this book.

Rotor winding is formed by connecting rotor conductors which are placed in corresponding slots to the collector segments in the manner prescribed earlier. With N_R rotor conductors connected to segments, there are two parallel branches between brushes A and B at each instant, each branch having $N_R/2$ conductors. As stated before, the rotor winding is also called *armature winding*, while current I_a, fed to the brushes from an external source, is also called *armature current*. The terms *inductor* (for the stator) and *induct* (for the rotor) are also in use, since the electromotive force in the rotor is *induced* due to the stator flux. Thus, excitation current I_f is also called inductor current, while armature current I_a is called *induct current*. The magnetomotive force of the rotor and the rotor flux can be represented by the vectors of vertical direction (Fig. 11.19d) and are called the *magnetomotive force of induct, flux of induct* but also *reaction of induct* or *armature reaction*. Term *armature reaction* can be explained by taking example of a DC machine operating as a generator. If the brushes A and B of the generator are connected to a resistive load, the armature current and current in the rotor conductors are obtained by dividing the rotor electromotive force by the equivalent resistance of the electrical circuit. Hence, the armature current is proportional to the induced electromotive force. The electromotive force is proportional to the rotor speed and to the inductor (stator) flux. As a consequence, the rotor currents, magnetomotive force, and flux are all proportional to the stator excitation. Therefore, the rotor current and flux are apparent reaction to the stator excitation, which brings up the term *armature reaction*.

Question (11.2): Determine the excitation flux Φ_f of a DC machine. The excitation current is I_f, the number of turns in the excitation winding is N_f, the axial length of the machine is L, the rotor radius is R, the main north pole of the machine is seen from the center of the rotor at an angle α, while the air gap under the main poles is δ.

Answer (11.2.): The excitation current creates magnetomotive force $F_f = N_f I_f$. The excitation flux passes through the yoke and main poles, through rotor magnetic circuit, and it passes twice through the air gap under the main poles. Since magnetic field H_{Fe} in iron is negligible due to a very high permeability of iron, it is justified to assume that significant values of magnetic field H exist only in the air gap. Therefore, radial component of magnetic field in the air gap is $H = N_f I_f/(2\delta)$. Radial component of the magnetic induction in the air gap is $B = \mu_o N_f I_f/(2\delta)$. Under circumstances, magnetic induction under the main poles has a constant value. Therefore, the excitation flux is obtained multiplying the magnetic induction by the surface of the main poles. The surface of each main magnetic pole is $S = \alpha R \cdot L$. Eventually, $\Phi_f = \alpha R \cdot L \cdot \mu_o N_f I_f/(2\delta)$.

Question (11.3): For the machine described in the previous question, it is known that the rotor conductors carry electrical current $I_a/2$. There are 10 rotor conductors under the north pole of the stator and 10 conductors under the south pole of the stator. Calculate the electromagnetic torque of the machine.

Answer (11.3): The vector of magnetic induction is of radial direction; it is orthogonal to the conductor. The electromagnetic force acting on the straight conductor depends on the vector product of the magnetic induction B and the conductor length l. Thus, the force acts in tangential direction. Relevant vectors are perpendicular, and the force $F = LBI_a/2$ acts on each of conductors. Contribution of each conductor to the total electromagnetic torque is $T_1 = RF = R\,LBI_a/2$.

The electromagnetic force acts only upon the conductors below the main poles. Namely, there is no force on conductors in the neutral zone where magnetic induction has negligible values. Therefore, the electromagnetic torque is $T = 20$ $T_1 = 20 \cdot R \cdot L \cdot B \cdot I_a / 2$.

Question (11.4): The machine described in the preceding questions rotates at a constant speed of Ω_m. Assume that the brushes are disconnected from the power supply and that the voltage between the brushes is measured by a voltmeter. If the rotor winding is made to have two parallel branches between the brushes, determine the voltmeter reading.

Answer (11.4): The electromotive force is $E_1 = L \cdot v \cdot B$, where $v = R \cdot \Omega_m$ is the peripheral rotor speed. It is induced in conductors which are under the main stator poles. In neutral zones, magnetic induction is negligible, and the electromotive force induced in rotor conductors passing through neutral zones is very small and should not be taken into account. The rotor conductors are connected in series, and their electromotive forces add up. In Fig. 11.15, it is shown that all the conductors are split into two parallel branches. Series connection of conductors P1, P6, P3, and P8 has its ends connected to the brushes. At the same time, series connection of conductors P4, P7, P2, and P5 is connected between the brushes as well. Both branches with series-connected conductors are made in such way that the electromotive forces of individual conductors are added. The question concerns the machine having 10 conductors under each of the stator magnetic poles. Therefore, the total number of conductors having electromotive force E_1 is equal to 20. Since the conductors are split in two parallel branches, the electromotive force E_a is equal 10 E_1, namely, $10 \, L \cdot R \cdot \Omega_m \cdot B$.

11.15 Direct and Quadrature Axis

Stator flux is also called excitation flux, or flux of the inductor. Magnetic axis that corresponds to the excitation flux is called *direct axis*. Within previous figures, the direct axis is set horizontally. As a rule, direct axis is determined by the position of the main stator poles. In Figs. 11.19 and 11.20, the armature reaction is directed along vertical axis. The rotor current sheet has electrical currents in the left-hand side of the cross section, directed away from the reader. In the right-hand side of the cross section, the currents are of the opposite direction, toward the reader. For this reason, the rotor magnetomotive force and flux, also called the armature reaction, can be represented by vectors in vertical direction. The axis of the armature reaction is called *quadrature axis*.

The stator auxiliary poles act along the quadrature axis, compensating the effects of the armature reaction. The same is the role of the compensation winding, whose conductors create a magnetomotive force along the lateral axis in the opposite direction of the reaction of induct. The purpose of the auxiliary poles and compensation winding is reducing the magnetic induction in the neutral zone, along the

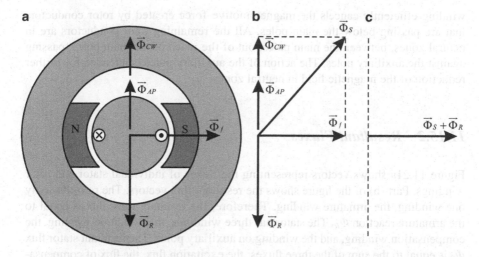

Fig. 11.21 Vector representation of the stator and rotor fluxes. (**a**) Position of the flux vectors of individual windings. (**b**) Resultant fluxes of the stator and rotor. (**c**) Resultant flux of the machine

quadrature axis. The compensation winding cancels the magnetomotive force of the rotor conductors passing under the main poles, while the auxiliary poles are built in neutral zones where they affect the magnetic induction and electromotive force induced in the turns involved in the process of commutation. By joint action of the compensation winding and auxiliary poles, the neutral zone has very low values of the resultant magnetic induction. The flux vectors representing the stator and rotor windings are shown in Fig. 11.21. Quadrature axis flux of the rotor is compensated by the quadrature axis flux of the stator.

11.15.1 Vector Representation

The rotor flux of a DC machine is also called armature reaction and it is represented by the flux vector Φ_R in Fig. 11.21. Along the quadrature axis there are vectors Φ_{AP} and Φ_{CW}, which represent fluxes of auxiliary poles and compensation winding. Direct axis of the machine is set horizontally, while quadrature axis is vertical. Directions of fluxes Φ_{AP} and Φ_{CW} are opposite to direction of the rotor flux Φ_R. This is due to the need to reduce the magnetic induction in neutral zones, accomplishing in this way an efficient commutation. In other words, it is necessary to reduce the resultant flux along the quadrature axis. Ideally, the compensation winding and auxiliary poles completely eliminate the armature reaction Φ_R, making the resultant flux along the quadrature axis equal to zero. As already stated, the compensation winding has conductors laid in the immediate vicinity of the rotor conductors, and they carry electrical currents of the same intensities but of opposite directions. The conductors are separated by a relatively small air gap; thus, the compensation winding efficiently cancels the magnetomotive force created by rotor conductors that are passing below the main poles. All the remaining rotor conductors are in neutral zones, between the main poles, out of the reach of the main poles, passing against the auxiliary poles. The action of the auxiliary poles is intended for further reduction of the magnetic field in neutral zones.

11.15.2 Resultant Fluxes

Figure 11.21a shows vectors representing the fluxes of individual stator and rotor windings. Part (b) of the figure shows the resultant flux vectors. The rotor has only one winding, the armature winding. Therefore, the resultant rotor flux is equal to the armature reaction Φ_R. The stator has three windings, the excitation winding, the compensation winding, and the winding on auxiliary poles. The resultant stator flux Φ_S is equal to the sum of the three fluxes: the excitation flux, the flux of compensation winding, and the flux of auxiliary poles.

11.15.3 Resultant Flux of the Machine

Resultant flux of the machine is equal to the vector sum of the fluxes in all the windings of the machine. Therefore, the resultant flux vector of the machine is equal to the sum of the resultant stator flux Φ_S and the rotor flux Φ_R. In cases where fluxes Φ_{AP} and Φ_{CW} compensate the armature reaction in full, the resultant flux along the quadrature axis is equal to zero, while the resultant flux along the direct axis is equal the excitation flux Φ_f, as shown in Fig. 11.21c.

Question (11.5): Figure 11.21 shows the case when the compensation winding and auxiliary poles eliminate the rotor flux Φ_R in full. By adding these two fluxes, one obtains the stator flux along the quadrature axis $\Phi_{AP} + \Phi_{CW}$ which has the same amplitude as the rotor flux Φ_R, but it has the opposite direction. For this reason, the total flux of the machine along the quadrature axis is equal to zero. From previous chapters, it is known that the electromagnetic torque of the machine is determined by the vector product of the stator and rotor fluxes. Does the fact that equivalent quadrature flux of the machine equals zero leads to conclusion that the electromagnetic torque of the machine is equal to zero as well?

Answer (11.5): It is necessary to note that the electromagnetic torque depends on the vector product of vectors Φ_R and Φ_S. Vector Φ_R represents the flux created by all the rotor windings, while vector Φ_S represents the flux created by all the stator windings. In the case of a DC machine, the rotor has only one winding, the armature winding, and the flux Φ_R is equal to the armature reaction flux. The stator has three windings. Therefore, the vector Φ_S represents the sum of the fluxes in these three stator windings, namely, the excitation winding, the compensation winding, and the winding of auxiliary poles. Electrical currents in conductors of the compensation winding and the winding of auxiliary poles create a quadrature component of the stator flux. In the considered case, the compensation winding and the winding of auxiliary poles create the flux along the quadrature axis which is of the same amplitude as the rotor flux but of the opposite direction. The sum of the fluxes created by the compensation winding and the auxiliary poles is $-\Phi_R$. Therefore, the resultant flux of the machine in quadrature axis is equal to zero. It is necessary to notice that the resultant flux of the machine comes as the sum of the fluxes of all the machine windings, residing on both stator and rotor. Although the resultant flux along the quadrature axis is zero, there still exists the rotor flux along the quadrature axis. For that reason, there still exists the possibility for the machine to generate the electromagnetic torque. The torque is determined by the vector product of the stator and rotor flux vectors. It depends on flux vectors Φ_R and Φ_S, shown in Fig. 11.21b. The angle between these two flux vectors is not equal to $\pi/2$, but the sine of this angle has a non zero value. The torque is determined by the product of amplitudes Φ_R and Φ_S and the sine of the angle between them. Therefore, the torque assumes a nonzero value in the case under consideration. The torque can be calculated as the product of the direct component of the stator flux Φ_f and quadrature component of the rotor flux Φ_R, both different than zero in the considered case. Therefore, despite the fact that the resultant quadrature flux of the machine is equal to zero, the electromagnetic torque of the machine is different from zero.

11.16 Electromotive Force and Electromagnetic Torque

Further study of DC machines requires the expressions for calculating the electromotive force and electromagnetic torque from the machine flux, current, and speed. For purposes of modeling, deriving the steady-state equivalent schemes, and constructing mechanical characteristics, it is necessary to derive the torque expression and the electromotive force expression for DC machines. The electromotive force E_a in armature winding is also called the rotor electromotive force and denoted by E_a, and it can be measured between the brushes A and B in conditions where the armature current I_a is equal to zero (no load condition). The electromagnetic torque and electromotive force should be expressed in terms of the armature current, excitation flux, angular speed of the rotor, and the machine parameters.

11.16.1 Electromotive Force in Armature Winding

In each of the rotor conductors passing under the main poles of the stator, there is an induced electromotive force $E_1 = R \cdot \Omega_m \cdot B \cdot L$, where Ω_m is angular speed of the rotor, R is the rotor radius, L is length of the rotor cylinder, while B is the radial

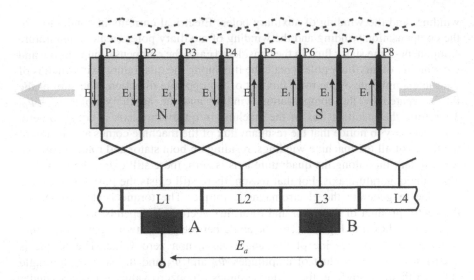

Fig. 11.22 Calculation of electromotive force E_a

component of the magnetic induction under the main poles. In conductors passing through the neutral zone, there are no electromotive forces because magnetic induction in neutral zones is negligible.

Conductors passing under the north magnetic pole have an electromotive force of the opposite sign with respect to conductors passing under the south magnetic pole. Therefore, as the rotor revolves, each conductor slides under opposite magnetic poles and has an electromotive force that changes its sign periodically, in synchronism with rotor mechanical turns. The frequency of the sign changes is determined by the rotor speed. An example of the change in electromotive force induced in a single conductor is given in Fig. 11.12b. The change of this electromotive force resembles a train of pulses. The pulse amplitude is E_1 while the sign changes in synchronism with the rotor motion. The sign changes as the considered conductor leaves the region under the north magnetic pole of the stator and enters the region under the south pole. It is shown hereafter that connections of rotor conductors and collector segments results in adding individual electromotive forces and provides a DC voltage between the brushes A and B. In the prescribed way, the mechanical commutator converts the AC electromotive forces of individual conductors into DC voltage available between the brushes.

Rotor conductors are divided in two parallel branches, and each of the branches is connected between the brushes. As the rotor revolves, individual conductors slide under the stator magnetic poles. At the same time, they pass from one of the branches to another. Namely, conductors do not appertain to any of the parallel branches for more than one half of the rotor mechanical turn. During the next half turn, the same conductor belongs to the other parallel branch. This occurs due to mechanical commutator and the process of commutation. Figure 11.23a shows wiring diagram of rotor conductors in rotor position given in Fig. 11.22, where

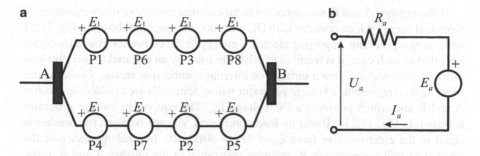

Fig. 11.23 (a) Addition of electromotive forces of individual conductors. (b) Representation of armature winding by a voltage generator with internal resistance

the brush A touches the segment L1. The two parallel branches are connected between the brushes A and B. These branches are P1-P6-P3-P8 and P4-P7-P2-P5. The electromotive force induced in conductors belonging to each of these branches is added to produce E_a, the electromotive force of the armature (rotor) winding. This electromotive force is equal to voltage U_a, measured between the brushes in no load condition, when the armature current is $I_a = 0$. In the position shown in Figs. 11.22 and 11.23, the electromotive forces in conductors P1, P2, P3, and P4 are of the same direction. These conductors pass under the north magnetic pole. Conductors P5, P6, P7, and P8 are passing under the south magnetic pole. In these conductors, the electromotive force has direction which is opposite with respect to the one induced in P1, P2, P3, and P4. When considering one of parallel branches, such as P1-P6-P3-P8, and taking into account the way of making their series connection in Figs. 11.22 and 11.23, it is concluded that electromotive forces of the four conductors are added, resulting in $E_a = 4E_1$.

It is of interest to show that E_a is a DC quantity and that the rotor motion does not lead to variation of E_a sign. When the rotor turns by $\pi/4$ with respect to position in Fig. 11.22, the brushes A and B get in touch with segments L2 and L4, as shown in Fig. 11.16. In this new position, conductors P3, P4, P5, and P6 pass below the north magnetic pole and therefore have their induced electromotive forces of the same direction. Conductors P7, P8, P1, and P2 pass below the south magnetic pole. Their electromotive forces have the opposite direction with respect to conductors passing under the north pole. At the same time, distribution of conductors between the two parallel branches changes. According to Fig. 11.16, conductors P3, P8, P5, and P2 make one parallel branch, while conductors P6, P1, P4, and P7 make the other parallel branch. The figure shows that the conductors are connected in the way that their electromotive forces are actually added; thus, the electromotive force E_a remains equal to $4E_1$, with brush A being at the higher potential than brush B. In other words, a DC voltage exists between the brushes A and B, notwithstanding the rotor motion. This shows that the mechanical commutator has the role of a rectifier which converts the alternating electromotive forces into DC electromotive force E_a. This electromotive force is called *rotor* or *armature* electromotive force.

If the brushes A and B are connected to an external resistor or other consumer of electrical energy which operates with DC currents, the machine shown in Fig. 11.23 will run as a generator, supplying electrical energy to the consumer. It will be shown later that in such case, it is required to drive the rotor by an external torque obtained from a hydro turbine, steam turbine, or internal combustion motor. The armature winding then represents a voltage generator whose terminals are available at brushes A and B and which produces a DC voltage U_a. The equivalent voltage generator is shown in Fig. 11.23b. Under no load conditions, voltage between the brushes is equal to the electromotive force $E_a = 4E_1 = 4R\Omega_mBL$. Internal resistance of the equivalent voltage generator R_a includes resistance of the brushes A and B, resistance of rotor conductors connected in two parallel branches, and a relatively small resistance of collector segments. It can be determined by measuring resistance between the brushes in conditions where the rotor speed is equal to zero ($\Omega_m = 0$). The electromotive force is then equal to zero, and the equivalent voltage source is reduced to internal resistance R_a. The value of R_a can be calculated for DC machine with eight rotor conductors, divided in two parallel branches with four conductor each, as shown in Fig. 11.23. If resistance of one rotor conductor is R_1 while the equivalent resistance of the brushes and collector is ΔR, then the internal resistance of the equivalent source is $R_a = 2R_1 + \Delta R$. Previous considerations show that the armature winding of DC machine can be represented by a voltage generator having no load electromotive force E_a and internal resistance R_a.

11.16.2 Torque Generation

It is required to determine relation between the electromagnetic torque, excitation flux, and *armature* current. The excitation flux from the main poles passes to the rotor magnetic circuit via air gap, where the magnetic induction is of radial direction. The surface separating internal side of the main pole from the air gap is of the form of a bent rectangle which represents a sector of the cylinder. The surface of this sector is $S = WL$, where L is length of the machine while $W = \alpha R$ is the width of the main pole, measured along its internal side which faces the air gap. The width W is one section of the circle having the radius R. To an observer positioned at the rotor center, the surface S is seen at the angle α. Magnetic induction B in the air gap below the main poles is equal to the ratio of flux Φ_f and surface S. The current in rotor conductors is $I_a/2$, where I_a is the *armature* current, fed to the brushes from an external source. Figure 11.24a gives an unfolded scheme of the rotor winding and shows forces acting upon its conductors. Part (b) of this figure shows directions of electrical currents in rotor conductors, seen from the collector side. At the given position of the rotor, conductors P1, P2, P3, and P4 are below the north pole and carry currents of direction $\otimes$. Lines of the field of magnetic induction come out of the main pole, denoted by N; they pass through the air gap, come across the rotor conductors, and enter the rotor magnetic circuit. In the considered zone below the north pole, the vector product of radial component

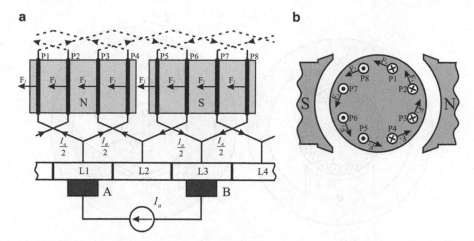

Fig. 11.24 (a) Forces acting upon conductors represented in an unfolded scheme. (b) Forces acting upon conductors. Armature winding is supplied from a current generator

of magnetic induction and coaxially directed current ($\otimes$) gives tangential force, denoted by F_1. Below the south pole, direction of the current in conductors changes. At the same time, direction of the magnetic induction changes as well. It comes out of the rotor magnetic circuit, passes through the air gap, and enters the stator magnetic circuit (S). Change in directions of both the current and magnetic induction results in tangential force $F1$ which retains its original direction and acts in counterclockwise direction.

Individual forces contribute to electromagnetic torque. The arm of each force is equal to R. The torque contribution of each individual conductor is $T_1 = RF_1 = RL(I_a/2)B$. Since the total number of conductors is 8, the electromagnetic torque is equal to $T_{em} = 8RL(I_a/2)B = 2DLI_a\Phi_f/S$, where D, L, I_a, Φ_f, and S denote the rotor diameter, axial length of the machine, armature current, excitation flux, and cross-section area of the main poles, respectively.

11.16.3 Torque and Electromotive Force Expressions

In this section, induced electromotive force of the armature winding E_a is expressed in terms of the rotor speed and the excitation flux. Further on, the electromagnetic torque T_{em} is expressed in terms of the excitation flux and the armature current. To begin with, it is necessary to establish relation between the magnetic induction in the air gap, excitation current, and the excitation flux.

Figure 11.25 provides the form and dimensions of the main poles. The electromotive force and forces upon conductors depend on position where the conductors are located. Rotor conductors are positioned at the external surface of the rotor magnetic circuit; thus, it is necessary to determine magnetic induction in

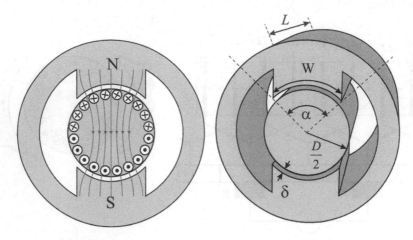

Fig. 11.25 Dimensions of the main magnetic poles

the air gap. The width δ of the air gap is much smaller than the radius R of the rotor. Boundary conditions for magnetic field can be applied to the surface separating the air gap from the ferromagnetic material. According to boundary conditions, magnetic induction on the air gap side of the surface is oriented in radial direction. As already shown, the cross-section S of the main poles is WL. Variable $W = \alpha D/2$ is the width of the main poles, while α is the angle covered by the main pole to an observer positioned at the rotor center. The excitation flux passes through the stator and rotor magnetic circuits where the magnetic field H_{Fe} is very small. The flux passes twice through the air gap; thus,

$$F_f = N_f I_f = 2 \cdot \delta \cdot H_f, \tag{11.2}$$

where $F_f = N_f I_f$ is the magnetomotive force of the excitation winding, while H_f is the intensity of magnetic field below the main poles. Magnetic induction B_f created by the excitation winding in the air gap, under the main poles is equal to

$$B_f = \mu_0 H_f = \mu_0 \frac{N_f I_f}{2 \cdot \delta}. \tag{11.3}$$

Magnetic induction in the air gap has the same value throughout the main pole cross section. Therefore, the excitation flux is equal to the product of the magnetic induction and surface area

$$\Phi_f = S B_f = L W \mu_0 H_f = \mu_0 \frac{L W N_f I_f}{2 \cdot \delta}, \tag{11.4}$$

where $W = \alpha D/2$ is the width of the main pole. Magnetic resistance along the path of the excitation flux is equal to the ratio of magnetomotive force F_f and flux Φ_f. Magnetic resistance R_μ is

$$R_\mu = \frac{F_f}{\Phi_f} = \frac{2 \cdot \delta}{\mu_0 \cdot L \cdot W}. \tag{11.5}$$

Total flux of the excitation winding is $\Psi_f = N_f \Phi_f$. Inductance L_f of the excitation winding is the ratio of the total flux and the excitation current, and it is determined according to expression

$$L_f = \frac{\psi_f}{I_f} = \mu_0 \frac{LWN_f^2}{2 \cdot \delta} = \frac{N_f^2}{R_\mu}. \tag{11.6}$$

Coefficient of proportionality between the excitation flux Φ_f, which represents the flux in one turn of the excitation winding, and excitation current I_f is equal to

$$L_f' = \frac{\Phi_f}{I_f} = \frac{\psi_f}{N_f I_f} = \mu_0 \frac{LWN_f}{2 \cdot \delta} = \frac{L_f}{N_f}. \tag{11.7}$$

The rotor comprises a total of N_R conductors, but some of them are not under the main poles, and they pass through the neutral zone between the main poles. The remaining rotor conductors are in the zone of the main poles, within the reach of the magnetic induction B_f. In the neutral zone, the conductors are not subject to any force and no electromotive force is induced in them. The rotor conductors are evenly distributed along the machine perimeter. Below the north pole of the width W, there are $N_R W/(\pi D)$ rotor conductors. The same number of conductors is found below the south magnetic pole.

11.16.4 Calculation of Electromotive Force E_a

In the rotor conductors influenced by the field of the main poles, the induced electromotive force E_1 is

$$E_1 = B_f \cdot L \cdot v = B_f \cdot L \cdot R \cdot \Omega_m, \tag{11.8}$$

where v is rotor peripheral speed, Ω_m is angular speed of rotor rotation, and R is radius of the rotor. Rotor conductors are connected in series in the way that the induced electromotive forces are added. Total number of conductors containing induced electromotive force is equal to the sum of conductors below the north and south poles, $2N_R W/(\pi D)$. Since all rotor conductors are connected in two parallel branches, while the branch terminals are brought to brushes A and B, electromotive force E_a is equal to the sum of electromotive forces in one of the branches. Induced electromotive forces exist in conductors under the main poles. Therefore, E_a is calculated as the product of the electromotive force E_1 in one conductor, and the number of conductors positioned below one of magnetic poles,

$$E_a = \frac{1}{2}\left(\frac{2N_RW}{\pi D}\right)E_1 = \frac{N_RW}{2\pi R}B_fLR\Omega_m = \frac{N_R}{2\pi}\left(LWB_f\right)\Omega_m. \tag{11.9}$$

By using relation (11.4) between the magnetic induction and excitation flux, previous expression takes the form

$$E_a = \frac{N_R}{2\pi}\Phi_f\Omega_m = k_e\Phi_f\Omega_m, \tag{11.10}$$

where coefficient $k_e = N_R/(2\pi)$ is determined by the number of rotor conductors and is called *coefficient of electromotive force*.

11.16.5   Calculation of Torque

Each conductor which passes through the zone of the main poles is subject to the force

$$F_1 = L\frac{I_A}{2}B_f = \frac{\Phi_f}{W\cdot L}\cdot\frac{I_A}{2}\cdot L = \frac{\Phi_f}{W}\cdot\frac{I_A}{2}. \tag{11.11}$$

The arm of the considered force is $R = D/2$, and its contribution to the total torque is equal to $T_1 = F_1\,D/2$. Since the force F_1 acts only on conductors positioned below the main poles, there are $2N_RW/(\pi D)$ conductors contributing to the torque. The sum of their contributions is

$$T_{em} = \left(\frac{2N_RW}{\pi D}\right)\cdot T_1 = \frac{2N_RW}{\pi D}\cdot\frac{D}{2}\cdot\frac{\Phi_f}{W}\cdot\frac{I_A}{2}$$
$$= \frac{N_R}{2\pi}\cdot\Phi_f\cdot I_A = k_m\cdot\Phi_f\cdot I_A,$$

where $k_m = N_R/(2\pi)$ is the *coefficient of electromagnetic torque*.

Equation (11.10) shows that the electromotive force of the armature winding is determined by the product of the excitation flux and the rotor angular speed, while (11.12) shows that the electromagnetic torque of the machine is determined by the product of the excitation flux and the armature current.

Chapter 12
Modeling and Supplying DC Machines

In this chapter, mathematical model is developed for DC machines with excitation windings and DC machines with permanent magnet excitation. The block diagram of the model is used to provide a brief introduction to the torque control. Steady-state equivalent circuits are derived and explained for armature and excitation windings. These circuits are used to introduce and analyze mechanical characteristic of separately excited DC machine and determine the steady-state speed. The chapter provides basic elements for the control of the rotor speed. Steady-state operation of DC generators is explained along with basic output characteristics. Typical applications of DC machines are classified on the basis of the speed and torque changes within the four quadrants of T_{em}–Ω_m plane. On that ground, the basic requirements are specified for the power supply of the armature windings. The operation of switching power converter with H-bridge is briefly explained, along with the basic notions on pulse-width modulation (PWM). The impact of pulsed power supply on the machine operation is considered by studying the ripple of the armature current. The chapter closes with an overview of most common power converter topologies used in supplying DC machines.

Analysis of electrical and mechanical characteristics of a DC machine is based on mathematical model. The model contains differential equations and algebraic relations describing transient processes in the machine. In DC electrical machines, the excitation flux is established along *direct* axis, while the rotor flux (*armature reaction*) appears along quadrature axis. The two axes are orthogonal, and the mutual inductance between the excitation winding and the armature windings is

Fig. 12.1 Connections of
a DC machine to the power
source and to mechanical load

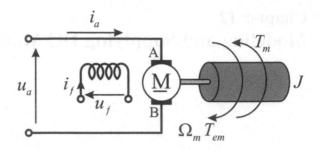

equal to zero.[1] Namely, changes in excitation current do not have an immediate impact on the rotor flux. At the same time, changes of the rotor current do not affect the excitation flux. The absence of interaction between direct and quadrature axes makes the transient phenomena of these axes decoupled. Transients in armature winding do not affect[2] the excitation winding. Hence, differential equation describing the changes of the excitation flux and current does not have factors proportional to the rotor flux and the armature current. For this reason, mathematical model of DC machine is relatively simple and clear. The flux control loop is decoupled from the torque control loop, and their design and application are quite straightforward.

The subject of modeling is a DC machine connected to the source u_a, feeding the armature winding, and to the source u_f, feeding the excitation winding. Connections of DC machine to electrical sources and mechanical load are shown in Fig. 12.1. Notations u_a and u_f are used in the figure since the winding voltages can be variable and change in time. In steady state, when the supply voltages are constant, these quantities are denoted by U_a and U_f. The shaft of the machine rotates at the angular speed of Ω_m. Revolving masses of inertia J are accelerated or decelerated by electromagnetic torque T_{em} and load torque T_m. Reference directions of the two torques are opposite, and they are shown in the figure.

The analysis of transient phenomena in a DC machine presented here results in differential equations that make up the mathematical model. According to conclusions of the preceding section, the mathematical model includes:

---

[1] *Note*: Mutual inductance of orthogonal windings is equal to zero if magnetic circuit is linear, that is, in cases where magnetic saturation does not occur. Otherwise, flux in one of the two orthogonal axes changes the operating point (B,H) on nonlinear magnetizing curve of ferromagnetic material, which affects magnetic resistance and flux in the other axis. Namely, the lines of the excitation flux and the lines of the rotor flux pass through the same magnetic circuit. Orthogonal fluxes share the same ferromagnetic material on both stator and rotor. Variation of one of these fluxes changes degree of saturation of ferromagnetic material (iron), acting indirectly upon the other flux. Nonlinearity of magnetic circuit leads to coupling of the orthogonal axes in all cases where their flux linkages share the same magnetic circuit.

[2] Due to nonlinear $B(H)$ characteristic of iron, magnetic circuit may saturate. In cases where the saturation level is altered by the armature current, there is a change in magnetic resistance on the path of the excitation flux. Through this secondary effect, called *cross saturation*, the armature current may affect the excitation flux.

- Differential equations of voltage balance in the windings
- Differential equation describing changes of angular speed (Newton equation)
- Algebraic relations between fluxes and currents (inductance matrix)
- Expression for electromagnetic torque

Unless otherwise stated, the process of modeling electrical machines throughout this book includes the four approximations discussed in the preceding sections:

- Parasitic capacitances are neglected (as well as the energy of electrical field).
- Spatial distribution of the energy of magnetic field is neglected. It is assumed that the energy is concentrated in discrete elements such as inductances. Thus, the equivalent circuits are represented as lumped parameter networks.
- Losses in magnetic circuit are neglected (i.e., losses in iron).
- Nonlinearity of magnetic circuit is neglected, that is, there is no magnetic saturation.

12.1 Voltage Balance Equation for Excitation Winding

The magnetomotive force along quadrature axis is created by the rotor conductors, and it has no influence on the excitation flux. Therefore, instantaneous value of the flux ψ_f in the excitation winding is $\psi_f = N_f \Phi_f = L_f i_f$. In further considerations, instantaneous values of currents, flux linkages, and voltages are dealt with. Therefore, notation i_f is used, denoting the variables that change in time, such as the excitation current $i_f(t)$. For brevity, further expressions are written by using representation such as i_f, without an explicit specification such as $i_f(t)$, showing that the considered variable is a time-varying function.

Coefficient of self-induction of the excitation winding is given by (11.6). The excitation winding has a finite electrical resistance R_f, and the voltage balance in this winding is expressed by the equation

$$u_f = R_f i_f + \frac{\mathrm{d}\psi_f}{\mathrm{d}t} = R_f i_f + L_f \frac{\mathrm{d}i_f}{\mathrm{d}t}. \tag{12.1}$$

Excitation flux Φ_f passes through the main poles. At the same time, Φ_f is the flux in a single turn of the excitation winding. This flux is proportional to the excitation current. On the basis of (11.7), the excitation flux is

$$\Phi_f = L'_f i_f = \left(\frac{L_f}{N_f}\right) i_f. \tag{12.2}$$

Therefore, the excitation winding can be represented by an R-L circuit, as shown in the left-hand side of Fig. 12.2. In the case when a DC voltage U_f is fed to the terminals of the excitation winding, the excitation current increases exponentially toward the final value $i_f(\infty) = I_f = U_f/R_f$, which is reached in the steady state. It is

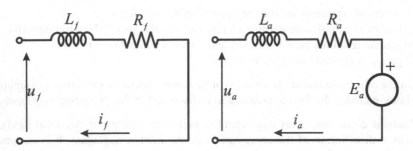

Fig. 12.2 Voltage balance in the excitation winding (*left*) and in the armature winding (*right*)

of interest to determine the change in the excitation current in the case where the initial value is $i_f(0)$ and the excitation voltage is $u_f(t) = U_f$ for $t > 0$. During transient, instantaneous value of the excitation current is

$$i_f = i_f(0)e^{-\frac{t}{\tau}} + i_f(\infty)\left(1 - e^{-\frac{t}{\tau}}\right) = \frac{U_f}{R_f}\left(1 - e^{-\frac{t}{\tau}}\right), \tag{12.3}$$

where $\tau = L_f/R_f$ is the electrical time constant of the excitation winding. In steady state, relation between the excitation flux and voltage across the excitation winding is

$$\Phi_f = L_f' I_f = L_f' \frac{U_f}{R_f}. \tag{12.4}$$

Question (12.1): Determine time constant of the excitation winding of a DC machine with the main poles cross section $S = 0.01\ \mathrm{m^2}$, with the air gap $\delta = 1$ mm, $N_f = 4{,}000$, and with resistance of the excitation winding of $R_f = 400\ \Omega$.

Answer (12.1): Magnetic resistance for the excitation flux is equal to

$$R_\mu = \frac{2 \cdot \delta}{\mu_0 \cdot S} = 159155\ \mathrm{H^{-1}}.$$

Inductance of the excitation winding is equal to

$$L_f = \frac{N_f^2}{R_\mu} = 100,5\ \mathrm{H}.$$

Time constant of the excitation winding is equal to

$$\tau = \frac{L_f}{R_f} = 0,251\ \mathrm{s}.$$

12.2   Voltage Balance Equation in Armature Winding

Rotor winding has two parallel branches connected between the brushes A and B. The equivalent internal resistance of armature winding $R_a = N_R R_1/4 + \Delta R$ can be measured between the brushes at standstill. The number of rotor conductors is N_R, while R_1 is resistance of a single conductor. The equivalent resistance of the commutator, including the brushes and collector, is denoted by ΔR. In addition to resistance, rotor (armature) winding has self-inductance L_a. Namely, the presence of electrical currents in rotor conductors creates magnetomotive force called armature reaction. Consequently, the rotor flux is created, inversely proportional to the magnetic resistance R_μ. The magnetic resistance along the path of the rotor flux is relatively high, since the rotor flux passes through the neutral zone between the main poles, where the lines of magnetic field face a very large air gap. Inductance of the armature winding L_a is proportional to the square of the number of turns and inversely proportional to the magnetic resistance of the magnetic circuit containing the rotor flux. This magnetic resistance is significantly higher than the one encountered by the excitation flux. This is due to the fact that the excitation flux passes through a very small air gap δ, while the armature reaction faces a very large air gap under the auxiliary poles. Smaller DC machines are made with no auxiliary poles at all, and they have even larger magnetic resistance in quadrature axis. In most DC machines, the number of turns in excitation winding is much larger than the number of turns in armature winding. As a consequence, armature inductance L_a is two or three orders of magnitude smaller compared to the inductance L_f of the excitation winding. In every coil, electrical current changes at the rate $di/dt \sim u/L$, proportional to the applied voltage and inversely proportional to the coil inductance. Therefore, the armature current in DC machines changes at a rate which is two or three orders of magnitude higher than the rate of change of the excitation current.

Question (12.2): Determine the equivalent internal resistance R_a of armature winding having a total of $N_R = 40$ conductors. Resistance of each conductor is $0.1\ \Omega$, while the equivalent resistance of the mechanical commutator with two brushes is equal to $\Delta R = 0.2\ \Omega$. Determine the self-inductance of the rotor winding L_a. The equivalent cross section of the magnetic circuit comprising the rotor flux is $S = 0.1\ \mathrm{m}^2$. Distance between the rotor and stator in the neutral zone is $d = 20\ \mathrm{mm}$. Determine time constant of the armature winding circuit.

Answer (12.2): Resistance of the armature winding is equal to $R_a = 0.1\ \Omega \cdot 40/ 4 + 0.2\ \Omega = 1.2\ \Omega$.

Magnetic resistance along the rotor flux path is

$$R_\mu = \frac{2 \cdot d}{\mu_0 \cdot S} = 318\,310\,\mathrm{H}^{-1}.$$

The armature winding has 40 conductors and 20 turns. Inductance of the armature winding is equal to

$$L_a = \frac{20^2}{R_\mu} = 1.256 \, \text{mH}.$$

Time constant of the excitation winding is equal to

$$\tau_a = \frac{L_a}{R_a} = 1.047 \, \text{ms}.$$

***

In addition to the voltage drop due to resistance R_a and inductance L_a, the rotor circuit has induced electromotive force E_a, proportional to the rotor speed Ω_m and to the excitation flux Φ_f. The voltage balance equation for the armature winding takes the following form:

$$u_a = R_a i_a + L_a \frac{di_a}{dt} + E_a = R_a i_a + L_a \frac{di_a}{dt} + k_e \Phi_f \Omega_m. \tag{12.5}$$

Equivalent circuit of the armature winding is shown in the right-hand side of Fig. 12.2. There are no changes of the electrical current in steady-state conditions, and the first derivative of the current is equal to zero. In steady-state conditions, (12.5) assumes the form

$$U_a = R_a I_a + E_a = R_a I_a + k_e \Phi_f \Omega_m. \tag{12.6}$$

Model of the machine includes expressions for the induced electromotive force and electromagnetic torque derived earlier:

$$E_a = k_e \Phi_f \Omega_m, \quad T_{em} = k_m \Phi_f i_a. \tag{12.7}$$

$$k_m = \frac{N_R}{2\pi}, \quad k_e = \frac{N_R}{2\pi}.$$

12.3   Changes in Rotor Speed

In addition to modeling transients in the windings, which represent the electrical subsystem, it is necessary to model the mechanical subsystem of the machine and to derive differential equation describing changes in the rotor angular speed. The rotor is coupled to a work machine or a driving machine by means of its shaft. Equivalent inertia of all rotating parts is denoted by J. It comprises inertia of the rotor, shaft, work machine, coupling elements, transmission elements, and of all the parts moving along with the rotor at speed Ω_m. The rotor speed is affected by:

- Electromagnetic torque T_{em}
- Friction torque $k_F\Omega_m$
- Load torque T_m
- Inertial torque $Jd\Omega_m/dt$

According to notation presented in Fig. 12.1, reference direction for electromagnetic torque is positive, meaning that positive value of this torque acts in direction of increasing algebraic value of the rotor speed. Load torque T_m represents mechanical load of the work machine which resists to motion and affects the rotor speed. Reference direction of this torque is negative, meaning that positive value of this torque acts in the direction of reducing the algebraic value of the speed. Friction torque resists the motion in either direction; thus, it acts in the direction of reducing the speed absolute value. Inertial torque $Jd\Omega_m/dt$ represents the torque required to change the speed and provide the acceleration $d\Omega_m/dt$. Equation (12.8) expresses the balance of all the torque components mentioned above. As a matter of fact, (12.8) is Newton's second law of motion applied to rotation. Since the friction torque can be two orders of magnitude lower than T_{em} and T_m, it is often neglected:

$$J\frac{d\Omega_m}{dt} = T_{em} - T_m - k_F\Omega_m. \tag{12.8}$$

12.4 Mathematical Model

Equations derived so far represent the mathematical model of DC machine. The model can be used for analysis of transient processes and steady states, and it is also called *dynamic model*. A concise review of these equations is presented here.

Voltage balance in the excitation winding:

$$u_f = R_f i_f + \frac{d\psi_f}{dt} = R_f i_f + L_f \frac{di_f}{dt}.$$

Voltage balance in the armature winding:

$$u_a = R_a i_a + L_a\frac{di_a}{dt} + E_a.$$

Expressions for the electromotive force and torque:

$$E_a = k_e\Phi_f\Omega_m,$$
$$T_{em} = k_e\Phi_f i_a.$$

Relation of excitation current to excitation flux:

$$\Phi_f = L'_f i_f.$$

Newton equation:

$$J \frac{d\Omega_m}{dt} = T_{em} - T_m - k_F \Omega_m.$$

12.5 DC Machine with Permanent Magnets

Figure 11.1 shows cross sections of DC machine with excitation winding on the stator and DC machine with permanent magnets on the stator. Excitation flux can be obtained by using a DC current I_f in excitation winding, which creates the magnetomotive force and flux along direct axis of the machine. The machine can also be made without an excitation winding. Instead, permanent magnets can be inserted instead of main poles to provide the excitation flux. The advantage of having permanent magnets is the absence of the excitation winding. There is no need to have a separate power supply for the excitation winding. At the same time, the overall efficiency of the machine is increased due to absence of copper losses in the excitation winding. A disadvantage of DC machines with permanent magnet excitation is that the flux cannot be changed. The flux is defined by $B(H)$ characteristics of the magnets and by the magnetic resistance of the magnetic circuit. Specifically, the flux is closely related to the remanent magnetic induction of permanent magnets. Hence, the permanent magnet excitation is not suitable for applications requiring variable flux. In machines with excitation winding, excitation flux can be varied by changing the excitation voltage and current.

Mathematical model of DC machines with permanent magnets is obtained by removing one differential equation from the model derived in the preceding section. The flux Φ_f is constant and determined by characteristics of the magnet. Since a DC machine with permanent magnets does not have an excitation winding, the differential equation describing the voltage balance in this winding is omitted.

12.6 Block Diagram of the Model

The mathematical model can be presented in the form of a diagram, shown in Fig. 12.3. Individual blocks in this diagram contain transfer functions obtained by applying Laplace transform to differential equations of the model. As an example, voltage balance differential equation of the excitation winding has time domain form of

$$u_f = R_f i_f + L_f \frac{di_f}{dt}.$$

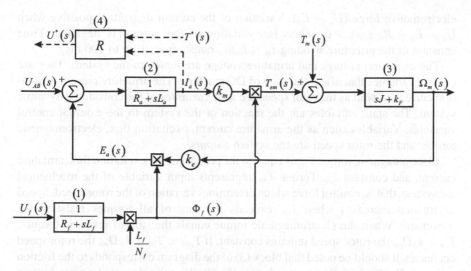

Fig. 12.3 Model of a DC machine presented as a block diagram

Application of Laplace transform to this differential equation results in an algebraic equation with complex images of the excitation current and excitation voltage. In equation

$$U_f(s) = R_f I_f(s) + sL_f I_f(s) - i_f(0),$$

s denotes *Laplace operator*, that is, *differentiation operator*, while $U_f(s)$ and $I_f(s)$ are complex images of the originals $u_f(t)$ and $i_f(t)$. Assuming that initial value of the excitation current is $i_f(0) = 0$, the equation takes the form

$$U_f(s) = (R_f + sL_f)I_f(s). \tag{12.9}$$

Excitation winding is a subsystem with excitation voltage at the input. The voltage is used as the control variable which determines the change of excitation current. The excitation current comes as a consequence or reaction to the excitation voltage. Therefore, the voltage is a control input, while the current is the output of the considered system. On the basis of the previous equation, the transfer function of block (1) in Fig. 12.3 is $I_f(s)/U_f(s) = 1/(R_f + sL_f)$. Block (2) represents the transfer function of the armature winding, while block (3) is Newton differential equation.

The torque and flux of DC machine depend on the excitation and armature voltages, and this is shown on the left-hand side of the diagram. The excitation current and flux are dependent on the excitation voltage, and they vary according to transfer function $I_f(s)/U_f(s) = 1/(R_f + sL_f)$. Electrical time constant $\tau_f = L_f/R_f$ of the excitation winding ranges between 200 ms and 10 s. Variation of the armature current depends on the voltage difference between the external voltage and induced electromotive force $(U_a - E_a)$. Variation of the current $di_a(t)/dt$ is positive when $U_a - E_a - R_a \, i_a(t) > 0$; otherwise, variation of the current is negative. Time constant of the armature winding $\tau_a = L_a/R_a$ ranges from 1 up to 100 ms.

The excitation voltage and armature voltage are *inputs* to the system. They are *control* variables that affect the state of DC machine. The armature current, excitation current, as well as the rotor speed are *state variables* of the considered dynamic system. The state variables are the reaction of the system to the external control variables. Variables such as the armature current, excitation flux, electromagnetic torque, and the rotor speed are the system *outputs*.

Electromagnetic torque T_{em} is equal to the product of the excitation flux, armature current, and constant k_m. Torque T_{em} represents input variable of the mechanical subsystem, that is, control force which determines variation of the rotor speed. Speed of rotation increases when T_{em} exceeds the sum of all torques resisting the movement. When the electromagnetic torque equals the sum of resisting torques, $T_m + k_F \Omega_m$, the rotor speed remains constant. If $T_{em} < T_m + k_F \Omega_m$, the rotor speed decreases. It should be noted that block (3) of the diagram corresponds to the friction torque $k_F \Omega_m$. The friction torque is usually smaller than the rated torque by two orders of magnitude. Therefore, friction is often neglected, and the transfer function is represented by $1/(J \cdot s)$.

12.7   Torque Control

Block (4) in Fig. 12.3 is denoted by R, and it does not belong to the mathematical model of DC machine. Connections of this block are made by dotted lines. This block illustrates the possibility of controlling the torque of the machine, which is discussed here. DC motors are often used in motion control applications, where they provide the means for controlling the speed and position of tools and workpieces in automated production lines. They are also used for running elevators, conveyors, and similar devices. In motion control tasks, electrical motors are used to provide a variable torque T_{em} which should be equal to the torque reference T^*, calculated within the motion controller which is not shown in the figure. The torque reference T^* is determined so as to overcome the motion resistances and ensure desired speed and/or position. Its change depends on desired speed changes and on forces and torques resisting the motion. The torque T_{em} should be as close to the reference T^* as possible. *Torque control* implies a set of actions and measures conceived to maintain the electromagnetic torque T_{em} at the desired reference value T^*. In cases where the reference changes, controlled variable T_{em} should track these changes. The torque T_{em} is proportional to the product of the armature current and the flux. At the first glimpse, the torque control can be done either by changing the flux or by changing the armature current. Yet, only the later approach is used in practice. This is due to the fact that the flux changes are rather slow. Moreover, flux control is not available with DC machines having permanent magnet excitation. With DC machines having an excitation winding, the time constant of the armature winding τ_a is considerably smaller than the time constant of the excitation winding. While only slow variations of the flux are possible, the current i_a can be changed quickly. Thus, regulation of the torque implies regulation of the armature current. The speed of the torque response is defined by the speed of response of the armature current. Starting from the voltage balance equation of the armature winding, variation of the current is determined by equation

$$\frac{\mathrm{d}i_a}{\mathrm{d}t} = \frac{1}{L_a}\left(u_a - R_a i_a - E_a\right). \tag{12.10}$$

Therefore, variations of the armature current can be accomplished by varying the armature voltage u_a. For this reason, DC motors are supplied from static power converters, power electronic devices that provide variable armature voltage. For the purpose of current control, it is necessary to measure the armature current and compare it to the reference in order to establish the error $\Delta i_a = i^* - i_a$. If the armature current is below the reference ($\Delta i_a > 0$), armature voltage should be increased in order to obtain $\mathrm{d}i_a/\mathrm{d}t > 0$. From (12.10), it can be concluded that increasing armature voltage leads to increasing armature current; thus, error Δi_a is reduced. In a like manner, if the current is too high ($\Delta i_a < 0$), the voltage should be reduced. The algorithm that calculates the control variable u^* from the error Δi_a is called *control algorithm*. Device or block diagram which implements such algorithm is called *regulator* or *controller*. Regulator can often be described by transfer function. The error Δi_a is an input to the regulator, while the control variable u^* is the output. Control algorithm affects the speed and character of the system dynamic response. Block (4) in Fig. 12.3 indicates the method of connecting such regulator. The regulator output u^* represents the desired armature voltage. This voltage reference is fed to the static power converter which supplies the armature winding. A more detailed analysis of the regulation problem is beyond the scope of this book. Hence, design of the regulator structure and setting of its parameters are left out of discussion.

Design of regulators and controllers requires some basic knowledge on transient processes in electrical machines and their mathematical models. These models and processes are studied and exercised in this book.

12.8   Steady-State Equivalent Circuit

It is of interest to analyze the steady-state operation of DC machines. In steady state, there are no changes in the rotor speed nor in electrical currents in the windings. During transients, instantaneous value of electrical current is denoted by $i(t)$, while in steady state it is denoted by I. Steady state in excitation winding is defined by equation $U_f = R_f I_f$. This relation is represented by the equivalent circuit

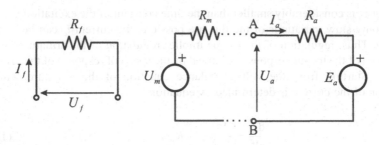

Fig. 12.4 Steady-state equivalent circuits for excitation and armature winding

given in the left-hand part of Fig. 12.4. The voltage balance equation of the armature winding is given by expression

$$U_a = R_a \cdot I_a + k_e \cdot \Phi_f \cdot \Omega_m.$$

Therefore, the steady-state value of the armature current is $I_a = (U_a - E_a)/R_a$, where U_a is the voltage fed to the brushes. This equation can be represented by the steady-state equivalent circuit given in the right-hand part of Fig. 12.4. The circuit can be used for determination of the current, torque, and power of a DC machine. Given the rotor speed and the excitation flux, one can calculate the electromotive force, find the difference $U_a - E_a$, determine the armature current I_a, and find the torque and power. In cases where the voltage and current are known, the equivalent circuit can be used to calculate the electromotive force and determine the rotor speed according to expression $\Omega_m = E_a/(k_e\Phi_f)$.

The voltage U_a between the brushes A and B in Fig. 12.4 is equal to $U_a = R_a I_a + E_a$. When DC machine is used as a motor, it is supplied from an external source of DC voltage. This source is shown in Fig. 12.4. It has internal resistance R_m and no load voltage U_m. With $R_m \approx 0$, the armature voltage U_a is approximately equal to U_m. Further on, whenever the armature voltage is supplied from an external source U_m, it is assumed that $U_a = U_m$.

Question (12.3): For a DC machine, it is known that $k_e\Phi_f = k_m\Phi_f = N_R\Phi_f/(2\pi) = 1$ Wb. Rotor shaft is coupled to a work machine which resists the motion and provides the load torque T_m. Machine runs in steady state, where the electromagnetic torque T_{em} is equal to the load torque T_m. The rotor speed is constant and equal to $\Omega_m = 100$ rad/s. The armature winding is fed from a voltage source $U_a = 110$ V. Equivalent resistance of the armature winding is $R_a = 1$ Ω. (1) Determine the electromagnetic torque, power delivered by the source, and power of electromechanical conversion. (2) Assuming that the rotor shaft is decoupled from the work machine, and a new steady state is reached, determine the rotor speed. (3) Assume that the rotor shaft is coupled to the work machine which maintains the rotor speed at $\Omega_m = 100$ rad/s, notwithstanding changes in electromagnetic torque T_{em}. If the source voltage is reduced to 90 V, determine the electromagnetic torque and power of electromechanical conversion in new steady-state conditions.

Answer (12.3):

(1) $I_a = 10$ A, $T_{em} = 10$ Nm, $P_{source} = 1,100$ W, $P_{em} = 1,000$ W.

(2) $T_{em} = 0 \rightarrow I_a = 0 \rightarrow U_a = E_a \rightarrow \omega_m = 110$ rad/s.

(3) $I_a = -10$ A, $T_{em} = -10$ Nm, $P_{em} = -1,000$ W; machine is operating in the generator mode.

12.9 Mechanical Characteristic

Mechanical characteristic of a DC machine is a curve in T_{em}–Ω_m plane which relates the torque and speed in steady-state operation, where the load torque T_m, the armature voltage U_a, and the excitation voltage U_f remain constant and do not change. It can be expressed either as $T_{em}(\Omega_m)$ or as $\Omega_m(T_{em})$.

The following considerations assume that a DC machine runs in the steady state. It has constant excitation flux and constant voltage U_a applied to the armature winding. In Fig. 12.5, it is shown that an external voltage source U_m is connected to the brushes, providing the required armature voltage. The voltage balance equation is $U_m = U_a = R_a I_a + k_e \Phi_f \Omega_m$. In conditions where the armature current is equal to zero, the electromotive force is equal to the supply voltage. Therefore, with $I_a = 0$, the rotor angular speed is $\Omega_0 = U_m/(k_e \Phi_f)$. In this condition, the electromagnetic torque T_{em} is equal to zero as well. Therefore, the speed Ω_0 is called *no load speed*. With constant supply voltage U_m and with $E_a = k_e \Phi_f \Omega_m = U_m - R_a I_a$, any increase in armature current reduces the electromotive force. With constant flux, the electromotive force is proportional to the rotor speed. Therefore, an increase in armature current decreases the rotor speed. On the other hand, the electromagnetic torque T_{em} is proportional to the armature current. In steady-state conditions, $T_{em} = T_m = k_m \Phi_f I_a$. Hence, any increase in the load torque decreases the rotor speed.

The following considerations assume that DC machine runs in the steady state. Therefore, the rotor speed is constant; hence, $J \cdot d\Omega_m/dt = 0$. With friction torque being neglected, the load torque T_m is equal to the electromagnetic torque T_{em}. Therefore, the Newton equation reduces to $T_{em} = T_m + k_F \Omega_m \approx T_m$. Hence, in the steady state and with no friction, the electromagnetic torque matches the load torque. Therefore, the armature current $I_a = T_{em}/(k_m \Phi_f)$ is proportional to the

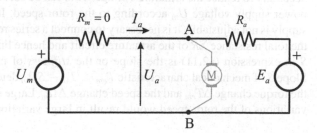

Fig. 12.5 Supplying armature winding from a constant voltage source

load torque T_m. For that reason, an increase in the load torque T_m results in an increase of the armature current, hence decreasing the rotor speed.

Previous considerations are drawn for steady-state operation of DC machine with constant supply voltages and constant load torque. They show that the rotor speed decreases with the load torque and, hence, with the electromagnetic torque. Steady-state relation of the speed and torque is called *mechanical characteristic*, and it can be represented by a curve in T_{em}–Ω_m or Ω_m–T_{em} plane. While the steady-state equivalent circuit relates the voltages and currents at electrical terminals of a DC machine, the mechanical characteristic relates the rotor speed and torque at the rotor shaft, wherein the shaft represents the mechanical access to the machine. The mechanical characteristic depends on the supply voltages and DC machine parameters. Variations in armature voltage U_a affect the no load speed Ω_0 and alter the mechanical characteristic $T_{em}(\Omega_m)$. Variation in excitation voltage changes the excitation current and flux, changing in such way the mechanical characteristic.

Mechanical characteristic can be determined by starting from voltage balance equation for the armature winding, given in (12.6). Calculation of function $T_{em}(\Omega_m)$ starts with $T_{em} = k_m \Phi_f I_a$. It is required to express the armature current in terms of the rotor speed. From voltage balance equation $U_a = R_a I_a + k_e \Phi_f \Omega_m$, depicting the voltage balance in armature winding in the steady state, the armature current is found to be

$$I_a = \frac{U_m - E_a}{R_a} = \frac{U_m - k_e \, \Phi_f \Omega_m}{R_a},$$

and the electromagnetic torque is calculated according to expression

$$
\begin{aligned}
T_{em} &= k_m \, \Phi_f \frac{U_m - k_e \, \Phi_f \Omega_m}{R_a} = k_m \, \Phi_f \frac{U_m}{R_a} - \frac{k_m k_e \, \Phi_f^2}{R_a} \Omega_m \\
&= T_0 - S \cdot \Omega_m.
\end{aligned}
\tag{12.11}
$$

Torque $T_0 = k_m \Phi_f U_m / R_a$ in (12.11) is *start-up torque*, the electromagnetic torque developed by DC machine when the rotor is at standstill. If the rotor speed is zero, the induced electromotive force is zero as well, and the armature winding has *start-up* current $I_0 = U_m / R_a$. It will be shown later that the start-up current has very large values which could damage the machine or power supply feeding the machine. In order to restrain the armature current to acceptable values, at low speeds, it is necessary to reduce the armature voltage. This can be accomplished by adjusting the power supply voltage U_m according to the rotor speed. In cases where the power supply is not adjustable, it is necessary to connect a series resistor in order to increase the total resistance ΣR of the armature circuit and hence limit the current. Parameter S in expression (12.11) is the slope or the *stiffness* of mechanical characteristic. Slope S of mechanical characteristic $T_{em} = T_0 - S\Omega_m$ determines the ratio between the torque change ΔT_{em} and the speed change $\Delta \Omega_m$. Large stiffness means that small variations of the rotor speed would result in large variations of the torque.

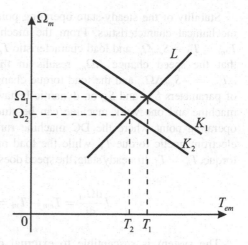

Fig. 12.6 Steady state at the intersection of the machine mechanical characteristics and the load mechanical characteristics

Similarly to electrical machines which have their own mechanical characteristic $T_{em}(\Omega_m)$, work machines connected to the rotor shaft have mechanical characteristics $T_m(\Omega_m)$ of their own, determining the change of the load torque with the rotor speed. Characteristic $T_m(\Omega_m)$ is called *load characteristic*. Since the steady state is reached with $T_m = T_{em}$, the steady-state operating point in T–Ω_m plane is at the intersection of the two mechanical characteristics, the one of the electrical machine and the one of the load. In Fig. 12.6, the load characteristic is represented by a straight line L. There are two different mechanical characteristics of a DC machine shown in the figure, K_1 and K_2. The characteristics K_1 and K_2 are obtained for different supply voltages of the armature winding. If DC machine has characteristic K_1, steady state is reached at point (T_1, Ω_1). With characteristic K_2, steady state is reached at point (T_2, Ω_2). Therefore, it is possible to change mechanical characteristic of the machine and change the steady-state speed and torque by varying the armature supply voltage $U_m = U_a$.

12.9.1 Stable Equilibrium

The equilibrium reached at the intersection of the two mechanical characteristics can be stable or unstable. When the operating point is displaced from the stable equilibrium by action of external disturbances, it returns to the same point after certain transient phenomena. The unstable equilibrium is retained only in the absence of disturbances. When the operating point is displaced from the unstable equilibrium, it does not return to the same point. An example of an unstable equilibrium is a ball positioned precisely at the peak of the hill. Left alone, it remains at the peak. Any disturbance would move the ball OFF the peak and make it roll all the way down the slope.

Stability of the steady-state operating point depends on the stiffness of the two mechanical characteristics. From the mechanical characteristic of DC machine $T_{em} = T_0 - S_{em}\Omega_m$ and load characteristic $T_m = T_{0m} - S_m\Omega_m$, it can be concluded that the speed change $\Delta\Omega_m$ results in the electromagnetic torque change of $\Delta T_{em} = -S_{em}\Delta\Omega_m$ and the load torque change of $\Delta T_m = -S_m\Delta\Omega_m$. The influence of parameters S_{em} and S_m on dynamic behavior of the system comprising one DC machine and one work machine can be studied by starting from the steady-state operating point where the DC machine runs at angular speed Ω_1 and develops electromagnetic torque T_1, while the load machine resists to motion by the same torque, $T_m = T_1$. In steady state, the speed does not change, and Newton equation reads

$$J\frac{d\Omega_m}{dt} = T_{em} - T_m = T_1 - T_1 = 0.$$

The system is susceptible to external disturbances that may produce small changes of the torque and speed. If a small variation of the rotor speed $\Delta\Omega_m$ occurs for any reason, the rotor speed becomes $\Omega_1 + \Delta\Omega_m$, while the electromagnetic torque changes to $T_1 - S_{em}\Delta\Omega_m$. At the same time, the load torque becomes $T_1 - S_m\Delta\Omega_m$. Since Ω_1 is a constant, Newton equation becomes

$$J\frac{d(\Omega_1 + \Delta\Omega_m)}{dt} = J\frac{d\Delta\Omega_m}{dt} = T_{em} - T_m$$
$$= T_1 - S_{em}\Delta\Omega_m - T_1 + S_m\Delta\Omega_m$$
$$= (S_m - S_{em})\Delta\Omega_m. \tag{12.12}$$

With $S_m - S_{em} > 0$, a positive value of $\Delta\Omega_m$ gives a positive value of the first derivative $d(\Delta\Omega_m)/dt$. Therefore, disturbance $\Delta\Omega_m$ will progressively increase. A negative disturbance $\Delta\Omega_m$ gives a negative value of the first derivative $d(\Delta\Omega_m)/dt$. In this case, disturbance will progressively advance toward negative values of ever larger magnitude. Hence, the steady-state operating point with $S_m - S_{em} > 0$ is unstable. Namely, any disturbance, whatever the size and however small, puts the system into instability.

With $S_m - S_{em} < 0$, a positive value of $\Delta\Omega_m$ gives a negative value of the first derivative $d(\Delta\Omega_m)/dt$. Therefore, disturbance $\Delta\Omega_m$ will decrease and gradually converge toward zero, bringing the system to the original steady-state operating point. This dynamic behavior is called *stable* since the system returns to the initial steady state after being disturbed and moved from the equilibrium. On the other hand, systems that progressively move away from the initial state and do not return are called *unstable*.

Question (12.4): Starting from (12.12) and assuming that stiffness of the characteristic and inertial torque are known, and that $\Delta\Omega_m(0) = A$, determine the change $\Delta\Omega_m(t)$.

Answer (12.4): Solution of differential equation $dy/dx = ay$ is $y(x) = y(0) \cdot e^{ax}$. In (12.12), $y = \Delta\Omega_m$, $y(0) = A$, $x = t$, and $a = (S_m - S_{em})/J$. Therefore, the change of $\Delta\Omega_m$ is determined by expression

$$\Delta\Omega_m(t) = \Delta\Omega_m(0) \cdot e^{\frac{S_m - S_{em}}{J}t}.$$

12.10 Properties of Mechanical Characteristic

Mechanical characteristic of a DC machine is shown in Fig. 12.7, including the intersection with the abscissa and ordinate. The intersection with the ordinate represents no load speed Ω_0. This speed is achieved when the electromagnetic torque is equal to zero. With zero torque, the armature current is equal to zero as well. In the absence of the voltage drop $R_a I_a$, the electromotive force $k_e \Phi_f \Omega_m$ is equal to the armature voltage. On the basis of (12.11), no load speed is

$$\Omega_0 = \frac{T_0}{S} = \frac{U_m}{k_e \Phi_f}.$$ (12.13)

The intersection with the abscissa represents the initial torque T_0 which is developed when the rotor is at standstill. The initial torque is equal to

$$T_0 = k_m \Phi_f \frac{U_m}{R_a}.$$ (12.14)

The slope of the mechanical characteristic determines the ratio of ΔT_{em} and $\Delta\Omega_m$, as shown in Fig. 12.7. Mechanical characteristic is often represented by the function $T_{em}(\Omega_m) = T_0 - S\Omega_m$. In other words, the stiffness S is considered positive if the

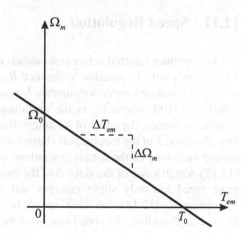

Fig. 12.7 No load speed
and nitial torque

torque drops as the speed increases. Therefore, the slope is determined according to expression

$$S = -\frac{\Delta T_{em}}{\Delta \Omega_m}. \tag{12.15}$$

Slope of the mechanical characteristic of a DC machine with the armature power supply as shown in Fig. 12.5 is equal to

$$S = \frac{k_m k_e \, \Phi_f^2}{R_a}. \tag{12.16}$$

Mechanical characteristic can be also represented by function $\Omega_m = f(T_{em})$. Equation (12.11) can be presented in the form

$$\Omega_m = \Omega_0 - \frac{1}{S} \, T_{em}. \tag{12.17}$$

Power supply of the armature winding has a finite internal resistance, as well as the conductors connecting the supply to the brushes. In addition, the armature circuit may have a resistor inserted in series with the purpose of reducing the initial current and initial torque. At the same time, series resistor may be used to alter the mechanical characteristic and change the rotor speed. In practice, total resistance in the armature circuit is higher than the equivalent resistance R_a of the armature winding and mechanical collector. For this reason, expressions (12.11), (12.14), and (12.16) should use ΣR instead of R_a. Notation ΣR represents the sum of all resistances in the armature circuit, namely, the sum of internal resistance of the power source, resistance of wiring and connections, inserted series resistances, and the equivalent resistance of the armature winding, collector, and brushes.

12.11 Speed Regulation

In cases without inserted series resistances, total resistance of the armature circuit ΣR is very small. In practice, resistance R_a of DC machines in conjunction with usually encountered armature currents I_a results in a voltage drop $R_a I_a$ of only $U_a/1{,}000 \, .. \, U_a/100$, where U_a is the armature voltage in most common operating conditions. Hence, the value of R_a ranges from $(U_a/I_a)/1{,}000$ to $(U_a/I_a)/100$. Therefore, the slope S of the mechanical characteristic is relatively high. This means that, during variations of the torque, variations of the speed will be very small. From (12.17), a high value of the slope S of the mechanical characteristic ensures that the rotor speed has only slight changes and remains close to the no load speed. According to (12.13), no load speed is determined by the armature voltage $U_a = U_m$. Therefore, the speed can be changed by varying the armature voltage.

In cases where the rotor speed is to be varied while the power source voltage U_m is constant, the armature voltage and the speed can be changed by inserting a variable series resistance R_{ext} in the armature circuit. In this way, the armature voltage is reduced to $U_a = U_m - R_{ext}I_a$, and this reduces the rotor speed. Insertion of a variable series resistance is a simple but inefficient way of controlling the rotor speed. The power losses due to Joule effect in series resistance are proportional to the square of the armature current. More efficient way of controlling the speed is the use of power source that provides variable voltage U_m. Continuous and lossless change of the armature voltage is feasible with static power converters that employ semi-conductor power switches.

In conditions where $R_{ext} = 0$, variation of the supply voltage U_m changes the no load speed and maintains the slope of the mechanical characteristic. Equation (12.16) proves that changes in U_m do not affect the slope S of the mechanical characteristic. On the other hand, no load speed Ω_0 is proportional to the supply voltage (12.13). By changing the supply voltage U_m, a family of mechanical characteristics is obtained, all of them having the same slope. Supply voltage U_m determines the intersection of each of these characteristics with the Ω_m axis of the T_{em}–Ω_m plane, as shown in Fig. 12.8. As a matter of fact, changes in the armature supply voltage result in translation of the mechanical characteristic in direction of Ω_m axis. Translation of the mechanical characteristic can be used to change the intersection with the load characteristic and, hence, change the running speed for the given load. In other words, the rotor speed can be changed by altering the armature supply voltage. In Fig. 12.8, mechanical characteristics K_1, K_2, K_3, and K_4 are given, each one obtained with different armature voltage. Characteristic K_4 is obtained for the case when the armature supply voltage is equal to zero. This characteristic passes through the origin.

Diagram in Fig. 12.8 is divided in four quadrants. In quadrant I, the rotor speed and electromagnetic torque both have positive values. Their product represents the power of electromechanical conversion, and it has positive value in the first quadrant, where the electrical machine operates as a motor. In the second and fourth quadrants, direction of the electromagnetic torque is opposite to direction of the rotor speed. In these quadrants, torque and speed have opposite signs, and their product assumes a negative value. In these quadrants, the power of electromechanical conversion is negative, and the machine operates as a generator. In generator mode, electrical machine creates electromagnetic torque which resists the motion, namely, it *brakes* and acts toward decreasing the rotor speed. To keep the rotor running, generator requires water turbines, steam turbines, or other similar devices that provide the driving torque that runs the rotor and maintains the rotor speed. In the third quadrant, the machine operates in the motor mode, quite like in the first quadrant. The difference is that both torque and speed in the third quadrant are negative.

The need for DC machines to operate in one or more quadrants depends upon the mechanical load or work machine used in actual application. A DC machine can be used to run a fan in a blower application. Direction of the air flow does not change. For that reason, DC motor runs in the same direction, without a need to change direction of the rotor speed. The air resistance produces the load torque which is

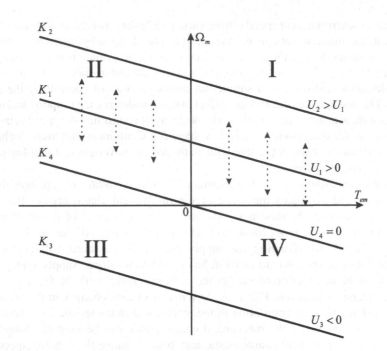

Fig. 12.8 The impact of armature voltage on mechanical characteristic

opposite to the rotor speed. Hence, direction of the torque $T_m = T_{em}$ does not change either. Hence, a DC machine used in typical blower application operates in the first quadrant.

A number of DC machines are used to control the motion of various parts, tools, objects, or vehicles. The motion is usually made in both directions. The speed has positive sign while moving in one direction and negative sign while moving in the other direction. Moreover, each motion cycle starts with an acceleration phase, where the speed increases, and ends with a braking phase, when the speed is decreased and brought back to zero (Fig. 12.10). In acceleration, the torque has the same direction as the speed, while in braking phase, the torque changes direction and acts against the speed. Hence, motion from one position to another involves the torque of both directions and the speed in only one direction. Coming back to the original position (Fig. 12.10) involves the speed of the opposite direction. Therefore, a forth-and-back motion requires the speed and torque changes with all the four possible combinations of signs, $(T_{em} > 0, \Omega_m > 0)$, $(T_{em} < 0, \Omega_m > 0)$, $(T_{em} > 0, \Omega_m < 0)$, and $(T_{em} < 0, \Omega_m < 0)$. In other words, it is required to accomplish the *four-quadrant operation*.

Question (12.5): A work machine resists the motion by developing the torque $T_m = 0{,}001\ \Omega^2$. For a DC motor with independent excitation and with constant excitation flux, the following parameters are known: $R_a = 0.1\ \Omega$, $k_m \Phi_f = 1$ Wb, and $U_a = 100$ V. Determine speed of rotation in steady state.

Answer (12.5): Steady-state values of the electromagnetic torque T_{em} and the load torque T_m are equal. The steady-state speed corresponds to the intersection of the mechanical characteristic of the motor and the load characteristic. On the basis of (12.11), the electromagnetic torque is $T_{em} = T_0 - S\Omega_m$. For the given parameters, $T_0 = 1\cdot100/0.1$ Nm $= 1,000$ Nm,   while $S = 1\cdot1/0.1$ Nm·s/rad.   Equation $T_{em}(\Omega_m) = T_m(\Omega_m)$ results in a quadratic equation in terms of Ω_m. Positive solution of this quadratic equation is 99.02 rad/s, which is the speed of the considered system at steady state.

12.12   DC Generator

DC machines can operate in the generator mode. If the rotor is put to motion by means of a steam or hydroturbine, the machine receives mechanical work which is converted to electrical energy. Mechanical power supplied to the shaft is the product of the rotor speed and the turbine torque T_T which keeps the rotor in motion and maintains the speed. With rotor in motion, the electromotive force $E_a = k_e\Phi_f\Omega_m$ is induced in the armature winding, and it is available between the brushes. The voltage U_a can be used to supply DC electrical loads such as the light bulbs, heaters, and similar. The armature voltage of DC generator is often denoted by U_G. With a resistive load connected between the brushes, the load current is established in direction which is opposite to the adopted reference direction of the armature current I_a. Respecting the reference direction of the armature current, electrical current in generator mode has negative sign. For this reason, analysis of DC generators is often made by assuming a new reference direction of the current, opposite to the one used in motoring mode. Equivalent circuit in Fig. 12.9 includes electrical current $I_G = -I_a$ which circulates from brush B to brush A. Brush A represents positive pole of the voltage supplied to electrical consumers. Starting from equation

$$U_G = U_a = R_aI_a + k_e\Phi_f\Omega_m,$$

and introducing substitution $I_G = -I_a$, one obtains

$$U_G = k_e\Phi_f\Omega_m - R_aI_G = E_a - R_aI_G, \tag{12.18}$$

which determines variation of the generator voltage as function of consumer current I_G. Starting from (12.18), the current–voltage characteristic is obtained, given in Fig. 12.9. No load voltage is equal to E_a. Slope $\Delta U/\Delta I$ determines the voltage drop experienced by electrical consumers. The voltage drop ΔU is proportional to the consumer current. The slope $\Delta U/\Delta I$ is equal to the armature resistance R_a. In cases when the electrical load is connected over long lines with considerable resistance, the slope $\Delta U/\Delta I$ is equal to the sum of the armature resistance and the line resistances.

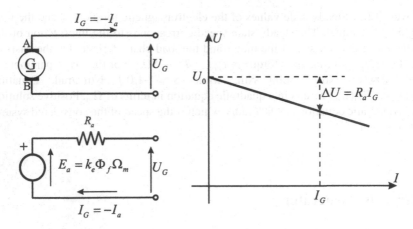

Fig. 12.9 Voltage–current characteristic of a DC generator

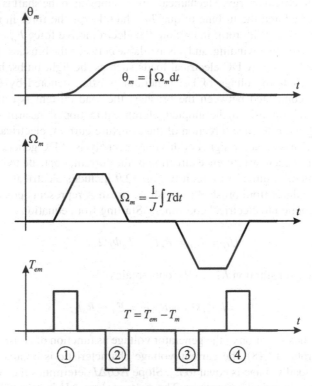

Fig. 12.10 Variations of the position, speed, and torque within one cycle

When supplying electrical consumers and loads operating with DC currents, it is of interest to keep the supply voltage constant. Due to the voltage drop ΔU caused by a finite resistance of the armature winding, the voltage across consumers depends on the load current. In order to keep the voltage U_G at the desired value notwithstanding the changes in electrical current, it is required to increase the electromotive force and maintain the voltage $E_a - R_a I_G$. In practical applications, it is necessary to measure the voltage U_G, to compare the measurement with the desired value, and then to adjust the excitation voltage and current in order to obtain the excitation flux which results in desired electromotive force. The voltage across the load $U_G = E_a - R_a I_G$ remains constant if the changes in the voltage drop $R_a I_G$ are matched and compensated for by contemporary changes of the electromotive force E_a. In the prescribed way, it is possible to achieve the voltage regulation of a DC generator.

When a DC generator supplies resistive load which absorbs the current $I_G > 0$, the armature current is then $I_a = -I_G < 0$. Since the armature current I_a is negative, the electromagnetic torque T_{em} is negative as well. With $T_{em} < 0$, the electromagnetic torque resists the motion and acts against the rotor speed. In other words, DC machine provides a *braking* torque. Therefore, the power of electromechanical conversion $P_{em} = T_{em}\Omega_m$ is negative. This means that the mechanical work is being converted to electrical energy. The generator receives mechanical work from a driving turbine by means of the shaft. The product of the driving torque T_T of the turbine and the rotor speed represents mechanical power which is delivered to the machine. The torque T_T acts in the direction opposite to the previously adopted reference direction for the load torque T_m. With reference directions shown in Fig. 12.1, the generator mode implies $T_{em} < 0$ and $T_m < 0$.

Question (12.6): A hydroturbine drives DC generator at angular speed of Ω_m = 100 rad/s. The parameters $R_a = 1\ \Omega$ and $k_m\Phi_f = 1$ Wb are known, while the resistance of the load connected between the brushes is $R_L = 4\ \Omega$. Determine the voltage across the load, the torque $T_T = -T_{em}$ delivered to the rotor by the turbine, the turbine power $P_T = T_T\omega_m$, and the power $P_G = U_a I_G$ delivered to the load. Why is $P_T > P_G$?

Answer (12.6): Electromotive force of the generator is $E_a = 100$ V. Generator current is $I_G = -I_a = 100$ V$/(1\ \Omega + 4\ \Omega) = 20$ A. Voltage across the consumer is 80 V. The electromagnetic torque is $-1 \cdot 20$ Nm $= -20$ Nm. The turbine torque is $T_T = -T_{em} = 20$ Nm. The turbine power is $P_T = 2,000$ W. The power delivered to the consumer is $P_G = 1,600$ W. The difference $R_a I_a{}^2 = 400$ W is converted to heat in the rotor windings.

12.13 Topologies of DC Machine Power Supplies

Whether used as electrical motors or generators, DC machines are often connected to static power converters. Variable speed applications require continuous voltage change of the power supply connected to the armature winding. On the other hand, DC voltages obtained from DC networks or batteries are mostly constant. Cases where a variable voltage DC load such as DC machine has to be connected to a constant voltage source are frequently encountered. In such cases, it is necessary to use a DC/DC static power converter which conditions the armature voltage according to needs. Moreover, DC machines are often supplied from AC mains. In these cases, it is necessary to use static power converter which converts constant AC voltages in adjustable DC voltage. Most common topologies of static power converters used in conjunction with DC machines are discussed in the following section.

12.13.1 Armature Power Supply Requirements

A DC motor takes electrical energy from power source, performs electromechanical conversion, and delivers mechanical work to the output shaft. Electrical motors are usually cylindrical rotating machines which deliver the driving torque to work machines by means of the rotor shaft and subsequent mechanical couplings. Mechanical power delivered to work machine is determined by the product of the driving torque and the speed of rotation.

It is of interest to specify the required characteristics of the power source supplying the armature winding. A DC motor is mainly used for controlling the motion of tools, workpieces, semifinished articles, finished articles, packaging machines, manipulators, vehicles, and other objects. A typical motion cycle includes start from initial position, motion toward the targeted position, reaching the target and resting at the target position, and then turning back to the initial position. Representative motion cycle is depicted in Fig. 12.10 by typical changes in the position θ_m, speed Ω_m, and torque T_{em} in the course of moving from the start position to the target and coming back. In order to get a closer specification for the armature power supply, it is of interest to observe the torque and speed changes during this motion.

Characteristic phases of the motion cycle are denoted by numbers 1 to 4. In phase 1, the torque has positive value, and it accelerates the motor, increasing the speed and initiating the motion toward target position. It is observed in Fig. 12.10 that the desired speed is reached soon and then the torque reduces while the speed remains constant. In constant speed interval between the phases 1 and 2, the torque is very low. With constant speed and with no need to provide the acceleration torque $Jd\Omega_m/dt$, the torque reduces to a very small friction, and it is considered as equal to zero. In phase 2, position θ_m gets close to the target position. For this reason, it is necessary to brake and reduce the speed. Negative torque is developed in order to reduce the speed to zero and eventually stop at the target position. In the course of coming back to the initial position, the speed and torque required in phases 3 and 4 are of the opposite direction compared to the speed and torque required in phases 1 and 2. It can be concluded that the motion cycle given in Fig. 12.10 comprises the following four combinations of the speed and torque directions:

- $T_{em} > 0$, $\Omega_m > 0$ (phase 1)
- $T_{em} < 0$, $\Omega_m > 0$ (phase 2)

- $T_{em} < 0$, $\Omega_m < 0$ (phase 3)
- $T_{em} > 0$, $\Omega_m < 0$ (phase 4)

In the first and third phases, the electrical machine operates in motor mode, while in the second and fourth phases, it operates in generator mode. Hence, throughout the motion cycle depicted in Fig. 12.10, the operating point $(T_{em}–\Omega_m)$ has to pass through all the four quadrants of the torque-speed plane. This can be used to specify the armature power supply and define the required voltages and currents. The torque $T_{em} = k_m\Phi_f I_a$ is determined by the armature current. Direction of the electromagnetic torque is determined by the sign of the armature current I_a. At the same time, the voltage drop $R_a I_a$ is often neglected, and the armature voltage U_a is assumed to be close to the induced electromotive force $E_a = k_e\Phi_f\Omega_m$. Therefore, the sign of the voltage U_a is determined by direction of the rotor speed. With $U_a \approx k_e\Phi_f\Omega_m$ and $I_a = T_{em}/(k_m\Phi_f)$, the change of the operating point in $T_{em}–\Omega_m$ plane can be used to envisage the required voltages and currents in $(I_a - U_a)$ plane. In this way, it is possible to specify the characteristics of the power source intended for supplying the armature winding. From the previous conclusions and from relations $T_{em} = k_m\Phi_f I_a$ and $U_a \approx E_a = k_e\Phi_f\Omega_m$, it can be concluded that, in the course of motion cycle depicted in Fig. 12.10, the voltage and current of the armature winding change signs in the following way:

- $I_a > 0$, $U_a > 0$ (phase 1)
- $I_a < 0$, $U_a > 0$ (phase 2)
- $I_a < 0$, $U_a < 0$ (phase 3)
- $I_a > 0$, $U_a < 0$ (phase 4)

Therefore, the power source for supplying the armature winding should provide voltages and currents of both directions and in all the four combinations. These requirements are crucial for the topology of the static power converter intended for supplying the DC machine.

There are applications of DC motors where the speed and torque do not change the sign. In these cases, the power source supplying the armature winding is more simple. In earlier mentioned example of a fan driver, the machine operates in the first quadrant, and armature winding can be supplied by a static power converter with strictly positive voltages and currents, $I_a > 0$ and $U_a > 0$.

12.13.2   Four Quadrants in T–Ω and U–I Diagrams

If a DC machine is used to effectuate the motion shown in Fig. 12.10, in different phases of this motion, it passes through all the four quadrants of the $T–\Omega$ plane. If direction of the excitation flux does not change, direction of the torque is determined by direction of the armature current, while direction of the electromotive force is determined by direction of the angular rotor speed. Applying the

Fig. 12.11 Four quadrants
of the T–Ω and U–I diagrams

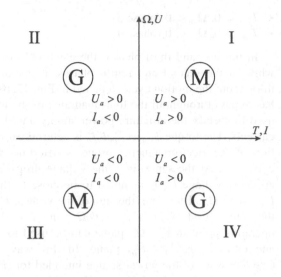

torque-current relation ($T_{em} = k_m \Phi_f I_a$) and the voltage-speed relation ($U_a \approx E_a = k_e \Phi_f \Omega_m$), the quadrants of the T–Ω plane and the quadrants of the U–I plane can be shown by the common Fig. 12.11.

12.13.2.1 I Quadrant

The machine develops a positive torque and rotates in reference direction; thus, $T_{em} > 0$ and $\Omega_m > 0$. The armature voltage and current are positive due to $U_a \approx E_a$ and $E_a \sim \Omega_m$ and due to $I_a \sim T_{em} > 0$. Power taken from the source is positive, $P_i = U_a I_a > 0$. Power of electromechanical conversion is also positive, $P_{em} = T_{em} \Omega_m = E_a I_a > 0$. The machine operates in motor mode.

12.13.2.2 II Quadrant

The machine develops a negative torque, while the rotor speed is positive considering the reference direction; thus, $T_{em} < 0$ and $\Omega_m > 0$. The armature voltage is positive, but the current is negative. Power taken from the source is negative because the voltage and current do not have the same sign. Power of the electromechanical conversion is also negative because the direction of the torque and speed does not have the same sign. The machine operates in the generator mode; therefore, it resists the motion and brakes.

Fig. 12.12 Topology of the converter intended for supplying the armature winding

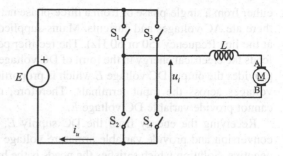

12.13.2.3 III Quadrant

The machine develops a negative torque and rotates opposite to the reference direction; thus, $T_{em} < 0$ and $\Omega_m < 0$. The armature voltage and current are negative because $U_a \approx E_a \sim \Omega_m < 0$, while $I_a \sim T_{em} < 0$. Power taken from the source is positive, $P_i = U_a I_a > 0$. Power of the electromechanical conversion is also positive, $P_{em} = T_{em}\Omega_m = E_a I_a > 0$. The machine operates in the motor mode.

12.13.2.4 IV Quadrant

The machine develops a positive torque and rotates opposite to the reference direction; thus, $T_{em} > 0$ and $\Omega_m < 0$. The armature current is positive, but the voltage is negative. Power taken from the source is negative because the voltage and current do not have the same signs. Power of the electromechanical conversion is also negative because the torque acts in direction opposite to the speed. The machine operates in the generator mode. It resists the motion and brakes.

12.13.3 The Four-Quadrant Power Converter

Topology of the static power converter which is used for supplying armature winding of a DC machine supporting the motion cycle shown in Fig. 12.10 is presented in this section. Electrical circuit of the power converter is shown in Fig. 12.12. The basic requirements are described in previous section. The converter should supply the armature winding by variable voltages and variable currents in all four possible combinations of their polarities.

In the left-hand part of the figure, E denotes a DC supply with constant voltage which feeds the static power converter. The voltage E is obtained either from a battery or a *rectifier*. The rectifier is a static power converter comprising diodes or other semiconductor power switches, and it converts electrical energy of AC voltages and currents to electrical energy of DC voltages and currents. It is supplied either from a single-phase or from a three-phase network. At the input of a rectifier, there are AC voltages and currents. Mains-supplied rectifier has the AC quantities at the line frequency (50 or 60 Hz). The rectifier performs AC/DC conversion and feeds the electrical energy in the form of DC voltages and currents. A diode rectifier provides the output DC voltage E which is proportional to the rms value of the AC voltages across the input terminals. Therefore, mains-supplied diode rectifiers cannot provide variable DC voltage E.

Receiving the energy from the DC supply E, it is necessary to perform the conversion and provide variable armature voltage of both polarities, positive and negative. Solution which satisfies the needs is the bridge comprising four switches, S_1, S_2, S_3, and S_4. Within these preliminary considerations, the switches are considered to be ideal. This means that they do not carry any electrical current when turned OFF (open). At the same time, the voltage drop across the switch which is turned ON (closed) is considered negligible and equal to zero. This means that any switch in the state of conduction (closed) does not have any *conduction losses*. Moreover, it is also assumed that the processes of closing and opening the switch do not involve any losses. Hence, there are no *commutation losses*. The transients of changing the switch state are called *commutation*.

12.13.3.1 Power Switches

Real mechanical switches as well as semiconductor power switches carry a small amount of *leakage current* even when switched OFF. Besides, in their state of conduction (ON, closed), they have a small voltage drop across the switch. Hence, real switches do have a certain amount of conduction losses. Each process of turning ON or OFF a semiconductor power switch and each process of closing or opening a mechanical switch involve energy losses. The energy loss incurred in each commutation is multiplied by the number of commutations per second to obtain the commutation losses.

Mechanical switches have contacts which close (get in touch) or open (get detached) in order to operate the switch. Turn-OFF commutation losses in mechanical switches arise due to an intermittent electrical arc which appears during separation of contacts. Even though the contacts are being detached, the current continues for a short while through an electric arc which breaks up in the space between contacts. The contacts are disengaged quickly, and the arc is very brief. Yet, it contributes to energy losses. Turn-ON commutation losses of mechanical switches arise due to electrical current being established prior to proper closing of the contacts, which contributes to the commutation losses.

Contemporary static power converters feeding the armature winding do not use mechanical switches. Instead, semiconductor power switches are used, such as BJT (bipolar junction transistors), MOSFET (metal oxide field effect transistors), and IGBT (insulated gate bipolar transistors). In semiconductor power switches, commutation losses arise due to phenomena of a different nature. Due to transient processes within semiconductor power switches, the change from the OFF state, characterized by $u = E$ and $i \approx 0$, into the ON state, characterized by $u \approx 0$ and $i = I_a$, a brief *commutation* interval Δt_c exists where considerable voltage and considerable current exist at the same time. Commutation time Δt_c is different for BJT, MOSFET, and IGBT transistors, and it ranges from 100 ns up to 1 μs. The time integral of the *ui* product during the commutation interval represents the energy loss incurred during one commutation event. In both cases of mechanical and semiconductor switches, power of commutation losses is dependent on the energy loss of single commutation and on the number of commutations per second.

By closing switches S_1 and S_4, the voltage $+E$ is established between brushes A and B. By closing switches S_2 and S_3, the voltage $-E$ is established between brushes A and B. Therefore, the switching bridge shown in Fig. 12.12 provides voltages of both polarities. Current through the closed switches is equal to the armature current. As the armature current has both directions, the switches should be capable of conducting the current in both directions. According to previous considerations, the voltage and current direction depend on the quadrant where the operating point of the machine resides.

12.13.3.2 Switching States

Each of the four switches is closed (ON) or opened (OFF). Assuming that all the switches can be controlled independently, the number of switching states for the four switches is $2^4 = 16$. When the switches are connected to the switching bridge, shown in Fig. 12.12, the number of available switching states is reduced.

Considering switches S_1 and S_2, the switching state $S_1 = S_2 = $ ON would bring the power source E into short circuit. At the same time, the switching state $S_1 = S_2 = $ OFF would leave no path for the armature current and cannot be used either. Hence, the branch (arm) S_1–S_2 has only two available switching states, and these are ($S_1 = $ ON, $S_2 = $ OFF) and ($S_1 = $ OFF, $S_2 = $ ON). The same holds for branch S_3–S_4. With two branches (arms) and with two possible switching states in each branch, the number of distinct switching states for the entire switching bridge is four.

The same conclusion regarding the number of possible switching states can be obtained by reasoning whether the number of switches being turned ON at any given instant should be 0, 1, 2, 3, or 4. Considering the switching bridge in Fig. 12.12, it has to be noted that the number of switches is turned ON 2 at each instant. First of all, it can be neither 4 nor 3. If the number of turned-ON switches is 3, then either the branch S_1–S_2 or the branch S_3–S_4 would bring the source E into short circuit. The source E is either a battery or a diode rectifier, and it has a very small internal resistance. Therefore, turning ON of one entire branch would lead to very high current through the source and through the switches, leading very quickly to their permanent damage. On the other hand, the number of turned-ON switches cannot be less than 2. This is due to the fact that the switching bridge must provide the path for the armature current at any instant, due to the fact that the armature current cannot be interrupted. The armature current gets from the branch S_1–S_2 to the brush A and then from the brush B to the branch S_3–S_4. For the purpose of

Table 12.1 Switching states

S_1	S_2	S_3	S_4	u_i
0	1	0	1	0
0	1	1	0	$-E$
1	0	0	1	$+E$
1	0	1	0	0

Fig. 12.13 Notation for semiconductor power switches. IGBT transistor switch (**a**), MOSFET transistor switch (**b**), and BJT (bipolar) transistor switch (**c**)

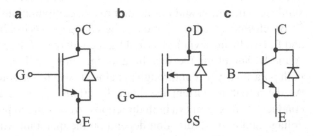

providing the path for the armature current, it is necessary that one of the switches in branch S_1–S_2 and one of the switches in branch S_3–S_4 gets turned ON. The last statement confirms the hypothesis that the number of switches turned ON at each instant is 1 in each branch (arm) and 2 in the switching bridge as a whole.

Available switching states for the bridge in Fig. 12.12 are given in Table 12.1. For each of the four switches, notation 0 represents the OFF state, while notation 1 represents the ON state. The column on the right-hand side shows voltage u_i obtained at the output of the switching bridge in conditions where the given switching state is applied.

The switching bridge shown in Fig. 12.12 makes use of semiconductor power switches. They are mostly power transistors applying BJT, MOSFET, or IGBT technology. The type of transistor to be used in static converter depends upon the operating voltage, operating current, commutation frequency, cooling conditions, required reliability, price, and also upon other factors. Each of transistor technologies has its advantages, disadvantages, and characteristic application area. The most frequently used notation for the mentioned transistors is given in Fig. 12.13.

12.13.3.3 MOSFET, BJT, and IGBT Transistors

The outline of most salient features of contemporary semiconductor power switches is included so that the reader may have an overview of practical voltage drops, commutation characteristics, and switch control requirements. Further study of power electronics is out of the scope of this book.

The state of power transistors is controlled by the third, control electrode. In BJT transistor, control electrode is called base. Positive base current brings the BJT power transistor in ON state, while ceasing the base current and exposing the base to negative voltage turns the transistor OFF. Switching of IGBT and MOSFET transistors is accomplished by varying the voltage of the control electrode called gates. The gate voltage of +15 V is larger than the threshold $V_T \in$ [+4 V .. +6 V], and it brings the transistor into conduction state (ON). Turning OFF is achieved by applying -15 V to the gate. Supplying the base current to large BJT transistors may involve considerable amount of power, while the gate control of IGBT and MOSFET transistors is virtually lossless.

All the three families of power transistors have very small currents while in OFF state. Their ON behavior is different. The voltage drop across power transistor in the state of conduction (ON) is rather small. Roughly, it varies between 100 mV and 3 V. Bipolar junction transistor (BJT) is turned ON by feeding the base which is sufficiently high current to bring the transistor to the state of *saturation*, when the voltage $V_{CE} = V_{BE} - V_{BC}$ across collector and emitter terminals is very small. Small BJT transistors in the state of saturation may have V_{CE} as low as 200 mV. Power BJT transistors have their internal voltage drops and may have the values of V_{CE} anywhere between 500 mV and 1 V. On the other hand, large-current BJT transistors have relatively low current gain $\beta = I_C/I_B$ and require very large base current. For that reason, most semiconductor power switches in BJT technology have two transistors connected in *Darlington* configuration where the voltage drop in ON state is $V_{CE} = 2V_{BE} - V_{BC}$, ranging between 1.5 and 3 V.

The MOSFET and IGBT transistors are turned ON by applying +15 V to the gate. MOSFET transistors in ON state behave as a resistor and have voltage drop of $R_{ON}I_{DS}$, where R_{ON} is the "ON" of the MOSFET channel. Transistors made for operating voltages below 100 V may have R_{ON} of only 1 mΩ, resulting in very low voltage drops. Therefore, these transistors are preferred choice for all applications with low operating voltages. Due to specific properties of power MOSFET switches, their resistance R_{ON} increases with the maximum sustainable voltages. Due to $R_{ON} \sim U^{2.5}$, transistor made to sustain twice the voltage would have 5.6 time larger resistance in ON state. For that reason, high-voltage MOSFET transistors are rarely used due to their large voltage drop. IGBT power transistors are developed as a hybrid of BJT and MOSFET technologies, combining positive characteristics of the both. Therefore, they are widely used and made available for voltages up to several kilovolts and currents above 1 kA.

12.13.3.4   Freewheeling Diodes

Electrical current in armature winding changes direction to provide both motoring and braking torques. Therefore, each of the switches has to be ready to conduct electrical currents in both directions. Power switches in Fig. 12.13 are mostly made with power transistors. Placing one power transistor in place of the switches S_1, S_2, S_3, and S_4 is not sufficient since power transistors operate with only one direction of current. When turned ON, bipolar transistor conducts the current that enters collector and goes to emitter. An attempt to establish emitter current of opposite direction is of little use. Power transistors are suited for bidirectional currents. Any inverse current may result in significant losses and eventually damage semiconductor device. Therefore, the use of transistors with inverse current is not of interest in static power converters. Transistors such as BJT, IGBT, and MOSFET are used to conduct electrical current only in one direction. For this reason, each of the switches $S_1 .. S_4$ has one power transistor and one semiconductor power diode. Element denoted by (C) in Fig. 12.13 is a parallel connection of one bipolar transistor conducting in CE direction and one power diode conducting in EC direction. All the four switches shown in Fig. 12.12 are constructed in the prescribed way. Therefore, each of $S_1 .. S_4$ switches should be considered as a parallel connection of one power transistor and one power diode.

12.13.3.5   Available Output Voltages

According to Table 12.1, a positive voltage across the armature winding is obtained by turning ON the switches S_1 and S_4. In this switching state, positive armature current circulates through power transistors within switches S_1 and S_4. Otherwise, with negative armature current, the current is established through power diodes within switches S_1 and S_4, connected in parallel with power transistors. Negative voltage across the armature winding is obtained by turning ON the switches S_2 and S_3. The same way as the previous, this switching state can be used for armature currents in both directions. There are also the switching states $S_1 = S_3 =$ ON and $S_2 = S_4 =$ ON which provide the armature voltage $u_i = 0$.

The switching structure in Fig. 12.12 with four available switching states allows feeding the armature winding by voltages and currents of both polarities. Therefore, it is compatible with the need to operate DC machine in all the four quadrants in T_{em}–Ω_m plane. Prescribed method cannot provide continuous change in armature voltage. Namely, there are only four available switching states, and they provide the output voltages of $+E$, $-E$, or 0. Hence, the armature winding can be supplied by the voltage that assumes one of the three discrete values. Instantaneous value of the armature voltage cannot have a continuous change. On the other hand, the armature voltage can be supplied by the train of pulses. The change (modulation) of the pulse-width changes the average value of the armature voltage.

Question (12.7):   What are the switching states that provide armature voltage equal to zero?

Answer (12.7):   By turning ON switches S_2 and S_4, armature voltage is made equal to zero. Armature winding is short circuited also when switches S_1 and S_3 are switched ON.

12.13.4   Pulse-Width Modulation

According to analysis summarized in Table 12.1, the output voltage u_i may have one of the three available values, $+E$, $-E$, or 0. It has been shown that application of DC electrical motors requires continuous variation of the supply voltage.

The switching structure in Fig. 12.12 cannot produce continuous change of the output voltage. On the other hand, it is possible to devise a sequence of switching states that would repeat in relatively short periods called *switching periods*. Each switching state may have adjustable duration and provide the armature voltages $+E$, $-E$, or 0. Sequential repetition of discrete voltages $+E$, $-E$, and/or 0 would result in an average voltage $+E > U_{av} > -E$. The voltage u_i would assume the form of a train of pulses of variable width. The width of the pulses would affect the average voltage U_{av} within each switching period. It is obvious, though, that variation of pulse width cannot result in continuous change of the instantaneous voltage. It is possible to change only the average value of the armature voltage within each switching period. Variation of pulse width is called *pulse-width modulation*.

12.13.4.1 Armature Voltage Requirements

In the areas of industrial robots, electrical vehicles, and in majority of applications involving motion control, there is a need for continuous variation of the speed of electrical motors. The steady-state armature voltage is equal to $U_a = R_a I_a + k_e \Phi_f \Omega_m$. The voltage drop $R_a I_a$ is usually much lower than the electromotive force. Windings of electrical machines are made to have small resistance, so as to reduce losses due to Joule effect and increase the efficiency. Therefore, it is justified to assume that $U_a \approx k_e \Phi_f \Omega_m$. With constant flux, the armature voltage is proportional to the rotor speed. As the speed changes continuously, the voltage must have continuous changes as well. The available voltage sources are usually batteries or diode rectifiers with constant voltage E. There is a possibility to use a series resistance ΔR in the armature circuit and to reduce the voltage by $\Delta R I_a$. The *rheostat* approach to the voltage regulation allows the voltage $U_a = E - \Delta R I_a$ to be changes varying resistor ΔR. This regulation is not convenient as it has poor energy efficiency. It is accompanied by the losses due to Joule effect in the resistor. In cases where $U_a = \frac{1}{2}E$ is required, one half of the input power is converted into heat in series resistor ΔR, while the other half is transferred to DC machine.

12.13.4.2 Pulse-Width Modulation

Electrical machines are supplied by variable voltage from switching power supplies. An example of power supply based on switching bridge is shown in Fig. 12.12. It does not contain series resistors or similar elements which would bring in power losses. Neglecting rather small conduction and commutation losses incurred in power switches, the switching bridge in Fig. 12.12 is virtually lossless. According to Table 12.1, the switching state ($S_1 = S_4 = $ ON and $S_2 = S_3 = $ OFF) provides the armature voltage $U_a = +E$, where E is DC voltage fed to the input terminals of the switching bridge. The source E is often called *primary source*. If the switching

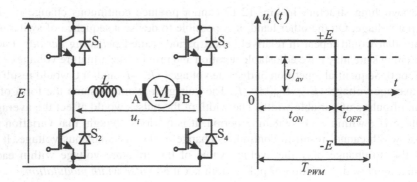

Fig. 12.14 Pulse-width modulation

state is changed and the other *diagonal* is activated where ($S_1 = S_4 =$ OFF and $S_2 = S_3 =$ ON), the armature voltage becomes $U_a = -E$. None of the two[3] considered states allows continuous voltage variations. However, by fast, periodic change of switching states, the output voltage resembles a train of pulses. The width of such voltage pulses can be altered by changing the dwell time of corresponding switching states. The average voltage of the pulse train depends on the amplitude and width of individual pulses. By a continuous variation of the pulse width, it is possible to accomplish a continuous variation of the average voltage.

12.13.4.3 Average Voltage

If the armature voltage obtained at the output of the switching bridge changes periodically, intervals with $S_1 = S_4 =$ ON are replaced by intervals when $S_2 = S_3 =$ ON. Within one switching period T, the switching bridge assumes the first switching state and then changes to the second switching state. The switching period is usually close to 100 μs. During one period, the switching state with diagonal S_1–S_4 turned ON is retained over the interval t_{ON}, where $0 < t_{ON} < T$. During the remaining part of the period, diagonal S_2–S_3 is turned ON. The form of the output voltage obtained across the armature winding is shown in Fig. 12.14. The average voltage within the period T is proportional to the pulse width t_{ON}:

$$U_{av} = \frac{1}{T} \int_0^T u_i(t) \cdot dt = \frac{2t_{ON} - T}{T} E. \qquad (12.19)$$

---

[3] There are two more switching states that provide $U_a = 0$. One of them is ($S_1 = S_3 =$ ON and $S_2 = S_4 =$ OFF), while the other is ($S_2 = S_4 =$ ON and $S_1 = S_3 =$ OFF). They are not considered in further discussion, so as to keep the introduction to pulse-width modulation principles as simple as possible. It has to be noticed, though, that there exist practical reasons to use these *zero-voltage* states in practical implementation.

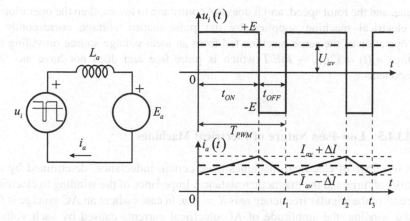

Fig. 12.15 Change of the armature current during one switching period

By continuous variation of the pulse width over the range $0 < t_{ON} < T$, the average value of the output voltage varies from $-E$ to $+E$. Since $U_a \approx k_e \Phi_f \Omega_m$, variation of the pulse width t_{ON} can be used to change the rotor speed within the range $-E/(k_e \Phi_f)$ up to $+E/(k_e \Phi_f)$. Variation of the pulse width is called *pulse-width modulation* (PWM). Switching bridge in Fig. 12.14 plays the role of a power amplifier whose operation is controlled by the variable t_{ON}. It provides the output voltage u_i with an average value determined by the pulse width t_{ON}. Whenever there is a need to make a continuous change of the armature voltage, this can be accomplished by changing the pulse width t_{ON} in a continuous manner. The switching bridge allows variation of the voltage with almost no losses.

12.13.4.4 AC Components of the Output Voltage

In addition to the average value, the armature voltage depicted in Fig. 12.15 also has an AC component. The voltage shape is periodic, and it contains a number of harmonic components. The basic frequency component, that is, the one with the lowest frequency, has the period T and frequency $f = 1/T$. The period T is the time interval comprising one positive voltage pulse and one negative voltage pulse. Repetition of such periods makes the pulse train providing the output voltage. Frequency f can be close to 10 kHz.

DC machines require the armature voltage that can change continuously. The instantaneous value of the armature voltage does not satisfy this requirement, as it takes one of the two discrete values, either $+E$ or $-E$. The voltage fed to the brushes is pulse-shaped voltage which, in addition to the average value, comprises parasitic AC components. It is necessary to envisage the consequences of such AC components of the voltage and analyze whether the switching bridge is a suitable power supply for electrical machines. If the AC component of the supply voltage does not have any significant effect on the armature current, electromagnetic torque, and the rotor speed, and it does not contribute to losses, then the operation of an electrical machine supplied by the pulse-shaped voltage corresponds to the operation of the same machine fed from an ideal voltage source providing the voltage $u_a(t) = (2t_{ON} - T)E/T$ which is pulse-free and does not have any AC components.

12.13.4.5   Low-Pass Nature of Electrical Machines

The windings of electrical machines have certain inductance, determined by the number of turns and the magnetic resistance. Impedance of the winding to electrical currents of the angular frequency ω is $X = L\omega$. In cases where an AC voltage is fed to the winding, the amplitude of AC electrical currents caused by such voltage decreases at elevated angular frequencies ω. Therefore, at very large frequencies, the impact of AC voltages on armature current is negligible.

Differential equation $u_a(t) = R_a i_a(t) + L_a di_a(t)/dt + E_a$ describes transient processes in the armature winding. Laplace transform can be applied to obtain the equation $U_a(s) = R_a I_a(s) + sL_a I_a(s) + E_a(s)$ with complex images of voltages and currents. The armature current is obtained as $I_a(s) = (U_a(s) - E_a(s))/(R_a + sL_a)$. The function $W(s) = 1/(R_a + sL_a)$ represents *transfer function* of the armature winding, and it describes the response of the armature current to changes in the armature voltage. The transfer function is obtained by dividing the complex images of relevant currents and voltages. Since the armature voltage is the cause while the armature current change is the consequence, the voltage is considered an input and the current an output. The function $W(s)$ is used to establish the response of armature current to the excitation by armature voltage, wherein the latter comprises certain harmonic components. The ratio of currents $I(j\omega)$ and voltages $U(j\omega)$ having certain angular frequency $\omega = 2\pi/T$ is obtained by replacing $s = j\omega$ in $W(s)$:

$$\frac{I(j\omega)}{U(j\omega)} = W(j\omega) = \frac{1}{R_a + j\omega L_a} \approx \frac{1}{j\omega L_a}. \tag{12.20}$$

At frequencies of the order of several kHz, it is justified to assume that $R_a \ll \omega L_a$; thus, the ratio of AC components of currents and voltages is determined as $1/(L_a\omega)$. Therefore, the voltage components at higher frequencies produce corresponding components of the armature current with rather small amplitude. At frequencies of the order of 10 kHz, reactance $L_a\omega$ is so high that the response of the armature current is negligible. Therefore, the pulsating nature of the supply voltage has no significant effect on the armature current. Therefore, for all practical uses, the presence of AC components in the armature voltage can be neglected. Therefore, the analysis of operation of DC machines fed from PWM-controlled switching bridge supplies can be simplified by modeling the switching bridge in a way that omits the high-frequency AC components. If the switching frequency $f = 1/T$ is sufficiently high, the switching supply shown in Fig. 12.12 can be

Fig. 12.16 Change of armature voltage, armature current, and source current for a DC machine supplied from PWM-controlled switching bridge

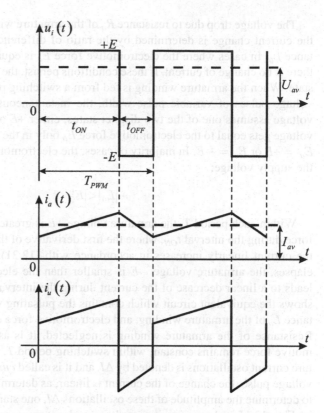

represented by an ideal voltage source providing adjustable output voltage, free from AC components. By varying the pulse width t_{ON}, the voltage of such source is changed according to the law $U_{av} = (2t_{ON} - T)E/T$.

12.13.5 Current Ripple

In the preceding subsection, it is shown that, at sufficiently high frequencies $f = 1/T$, it is justifiable to neglect the AC component of the train of voltage pulses fed to the armature winding. In this section, the changes of the armature current and voltage are analyzed by taking into account the high-frequency aspects. Moreover, electrical current of the primary source supplying the switching bridge is analyzed as well. In Fig. 12.12, the current of the source E is denoted by i_u. These quantities are shown in Fig. 12.16.

Variation of current in armature winding is determined by differential equation

$$\frac{di_a}{dt} = \frac{1}{L_a}(u_a - R_a i_a - E_a) \approx \frac{1}{L_a}(u_a - E_a). \tag{12.21}$$

The voltage drop due to resistance R_a of the armature winding is neglected; thus, the current change is determined by the ratio of difference $(u_a - E_a)$ and inductance L_a. In cases where the electromotive force E_a is equal to the supply voltage, there is no change of current. If these conditions persist, the system enters the steady state. When the armature winding is fed from a switching power supply feeding the voltage pulses of variable pulse width, the instantaneous value of the armature voltage assumes one of the two distinct states, either $+E$ or $-E$. The power supply voltage gets equal to the electromotive force E_a only in the exceptional cases where $E_a = +E$ or $E_a = -E$. In majority of cases, the electromotive force is smaller than the supply voltage:

$$|E_a| < |E|.$$

Within each period T, the armature voltage $+E$ is greater than the electromotive force during the interval t_{ON}, where the first derivative of the current is positive and the current linearly increases in accordance with (12.21). When the interval t_{ON} elapses, the armature voltage $-E$ is smaller than the electromotive force, which leads to a linear decrease of the current during the interval $T - t_{ON}$. Figure 12.15 shows the equivalent circuit which contains the pulsating voltage source u_i, inductance L_a of the armature winding, and electromotive force of the armature winding. Resistance of the armature winding is neglected. It is assumed that the electromotive force remains constant within switching period T. The amplitude of armature current oscillations is denoted by ΔI, and it is called *current ripple*. During each voltage pulse, the change of the current is linear, as determined by (12.21). In order to determine the amplitude of these oscillations ΔI, one starts from instant t_1, shown in Fig. 12.15, when a positive voltage pulse commences. At this instant, the armature current is $i_a(t_1) = I_{av} - \Delta I$. Over the interval $[t_1 .. t_2]$, the output voltage of the switching supply is equal to $u_i = +E$, and the current i_a increases with the slope $(E - E_a)/L_a$. The current change is linear, and it reaches the value of $i_a(t_2)$ at instant $t_2 = t_1 + t_{ON}$, which marks the end of the positive voltage pulse:

$$i_a(t_2) = i_a(t_1) + t_{ON}\frac{E - E_a}{L_a}. \tag{12.22}$$

Over the interval $[t_2 .. t_3]$ in Fig. 12.15, the power supply feeds negative voltage pulse. The output voltage of the switching supply is $u_i = -E$, and the current i_a decreases linearly with the slope of $(-E -E_a)/L_a$. During this interval, the current decreases linearly. At instant $t_3 = t_2 + t_{OFF} = t_2 + T - t_{ON} = t_1 + T$, which marks the end of the negative voltage pulse, the armature current reaches

$$i_a(t_3) = i_a(t_2) + (T - t_{ON})\frac{-E - E_a}{L_a}$$

$$= i_a(t_1) + t_{ON}\frac{E - E_a}{L_a} + (T - t_{ON})\frac{-E - E_a}{L_a}$$

$$= i_a(t_1) + E\frac{2t_{ON} - T}{L_a} - E_a\frac{T}{L_a}. \tag{12.23}$$

Prolonged operation in the prescribed way implies a constant width t_{ON} and a constant average value I_{av} of the armature current. Under circumstances, this operation can be characterized as the steady state, notwithstanding the fact that the current exhibits periodic oscillations ΔI. In such state, the instantaneous values of the armature current at the end of each negative pulse must be equal. Therefore, at steady state, $i_a(t_1) = i_a(t_3)$. On the basis of (12.23), the steady state is reached when

$$E\frac{2t_{ON} - T}{L_a} - E_a\frac{T}{L_a} = 0,$$

when the electromotive force is equal to

$$E_a = E\frac{2t_{ON} - T}{T}. \tag{12.24}$$

On the basis of (12.19), the previous expression represents the average value of the output voltage provided by the switching power supply. With the assumption of $R_a \approx 0$, the steady state in armature circuit is reached when the average value of the supply voltage is equal to the electromotive force E_a. This steady state is represented in Fig. 12.15. The armature current oscillates around an average value which is maintained constant. The amplitude of oscillations ΔI depends on the switching frequency, supply voltage, and armature inductance.

The ratio of the positive pulse width and period t_{ON}/T is called *modulation index*, and it is denoted by m. By replacing m in expressions (12.19) and (12.24), the following steady-state relation is obtained:

$$U_{av} = E_a = E(2m - 1). \tag{12.25}$$

The amplitude of oscillations of the armature current around its average value can be calculated from the modulation index, winding inductance, supply voltage, and switching frequency. According to Fig. 12.15, the current is changed by $i_a(t_2) - i_a(t_1) = 2\Delta I$ over interval $[t_1 .. t_2]$. From (12.22) it follows that

$$2 \cdot \Delta I = i_a(t_2) - i_a(t_1) = t_{ON}\frac{E - E_a}{L_a} = \frac{TE}{L_a}m(2 - 2m).$$

which leads to

$$\Delta I = \frac{TE}{L_a}(m - m^2). \tag{12.26}$$

The analysis shows that steady state in the armature circuit is established when the electromotive force and modulation index satisfy condition $(2m - 1)E = E_a$. Then, the average value of current does not change between the successive switching periods. The instantaneous value of the current oscillates around its average value with the period T and amplitude ΔI given in expression (12.26). The operating frequency of the switching bridge $f = 1/T = \omega/(2\pi)$ determines the frequency of oscillations of the armature current. The frequency f and angular frequency $\omega = 2\pi f$ are called *commutation frequency* and also *switching frequency*. When commutation frequency is sufficiently high, the current ripple is very small. Then, it is justified to neglect the AC component of the pulse-width-modulated train of pulses and consider that machine is fed from an ideal power source feeding a variable voltage $U_{av} = (2\ m - 1)E$, determined by the modulation index and free from AC components.

A close estimate of the ripple can be obtained by considering only the voltage and current components at the switching frequency $f = 1/T$. The ratio between the AC current of frequency $\omega = 2\pi/T$ and the voltage of the same frequency is determined by the inductance of the armature winding. Namely, $|I(j\omega)/U(j\omega)| \approx 1/(L_a\omega)$. Considering the train of voltage pulses where the values $+E$ and $-E$ repeat with period T and with duration of the positive pulse $t_{ON} = T/2$, the amplitude of the harmonic component of the frequency $f = 1/T$ is $V_1 = (4/\pi)E$. Corresponding harmonic component of the armature current has the amplitude of $I_1 = V_1/(L_a\omega) = (4/\pi)E/(2\pi L_a f) = (2/\pi^2)\ E/(L_a f) \approx 0.2026\ E/(L_a f)$. This approach neglects harmonic components at higher frequencies and overlooks the fact that the considered voltage does not change as a sinusoidal function. Instead, it is a train of pulses which comprises harmonic component at the frequency $f = 1/T$, but it also has harmonic components at frequencies that are odd multiples of f.

A more accurate estimate of the current ripple can be obtained by using expression (12.26). Ripple ΔI has the maximum value with modulation index of $m = 0.5$. In this case, the ripple is

$$\Delta I = \frac{TE_i}{4L_a}. \tag{12.27}$$

When using the above results, one should take into account that the preceding analysis assumes that the switching bridge in Fig. 12.12 uses only two switching states: the state with diagonal S_1-S_4 turned ON and the state with diagonal S_2-S_3 turned ON. In Table 12.1, there are two more available states, $S_1 = S_3 = ON$ and $S_2 = S_4 = ON$, which both produce the output voltage of zero. Control of the switching bridge can be organized by using additional two states and inserting the time intervals when the voltage is equal to zero. In this case, the relevant expressions for t_{ON} time change as well as the definition of the modulation index. In cases where DC machine requires a positive armature voltage, the train of voltage pulses is made by sequencing $+E$ and 0, providing an average value between these two values. Whenever a negative armature voltage is needed, the train of voltage pulses is made by sequencing $-E$ and 0. More detailed analysis of the pulse-width modulation technique is not studied in this book.

Input current taken from the source by the switching bridge is shown in Fig. 12.16, and it is denoted by i_u. This current depends on the instantaneous value of armature current i_a and on the switching state. If diagonal S_1-S_4 is ON, positive terminal of the source E is connected to brush A through the switch S_1, while negative terminal of the source is connected to brush B through the switch S_4. Therefore, in this switching state, $i_u = i_a$. If diagonal S_2-S_3 is ON, $i_u = -i_a$. As a consequence, the input current of the switching bridge has the shape of a train of pulses with an amplitude determined by the armature current and with the sign determined by the switching state of the bridge.

Question (12.8): Determine the change and the amplitude of oscillations of the armature current in the case when DC motor is supplied from the voltage shown in Fig. 12.26, with $E_a = 0$ and $R_a = 0$ and with known L_a, T, and E.

Answer (12.8): Since the average voltage has to match the electromotive force, duration of the positive voltage pulse is $t_{ON} = T/2$. During the first half period, $L_a di_a/dt = + E$. Therefore, the current change is linear. The same applies for the second half period. The amplitude of oscillations of the armature current around its average value is ΔI. Within the first half period, the current increases from $-\Delta I$ to $+\Delta I$. The change is linear, $di_a/dt = 2\Delta I/(T/2) = E/L_a$, and therefore, $\Delta I = ET/(4L_a)$.

Question (12.9): Control of the switching bridge of Fig. 12.12 is carried out by keeping switch S_4 permanently closed and switch S_3 permanently open. At the beginning of period T, the switch S_1 is turned ON. After the on time $t_{ON} = mT$ elapses, the switch is turned OFF. In the remaining part of the period $T - t_{ON}$, the switch S_2 is turned ON. The switching bridge is supplied from a constant voltage source E. Determine the average value of the output voltage, and find the expression for the current ripple.

Answer (12.9): The output voltage u_i is a periodic train of pulses that repeat with period T. In the first part of each period, during interval t_{ON}, the instantaneous value of the output voltage is $+E$. During the remaining part of the period $T - t_{ON}$, switches S_2 and S_4 are turned ON and the output voltage is equal to zero. The average value of the voltage is $U_{av} = [(t_{ON}) \cdot E + (T - t_{ON}) \cdot 0]/T = (t_{ON}/T)E = mE$. The current ripple is determined by repeating the calculation included in the previous analysis, starting with (12.23) and ending with (12.26). Compared to the previous analysis, where the instantaneous values of the output voltage were $+E$ and $-E$, in this example, they are $+E$ and 0. With this in mind, the armature current ripple is

$$\Delta I = \frac{TE}{2L_a}\left(m - m^2\right).$$

12.13.6   Topologies of Power Converters

DC machines are to be supplied by continuously variable DC voltage. The voltage should be suited for the desired operating mode of the machine. The switching bridge in Fig. 12.12 illustrates the principle of operation of static power converters which allow lossless conversion of DC voltages and currents. Practical static converter may have other parts that are not shown in Fig. 12.12. It has electronic circuits that control the state of the power switches, microprocessor-based controller, auxiliary power supplies for electronic circuits, diode rectifier which converts the AC mains into DC voltage E, protection devices, communication devices, and other auxiliary parts. Primary source of electrical energy for supplying electrical machines is usually low-voltage distribution network with line-frequency AC voltages. The line frequency of AC distribution networks is 50 Hz. Three-phase connections have line-to-line voltages of 400 V rms. Low-power converters can be supplied from single-phase connection with phase voltage of 220 V rms. The voltages of AC distribution network do not correspond to the needs of DC machines. Therefore, it is necessary to use static power converters. Their task is to convert the electrical energy of AC voltages and currents to electrical energy of DC voltages and currents to be fed to the armature winding and the excitation winding. It is necessary to provide continuous change of DC voltages fed to the windings. In some cases, the primary source of electrical energy is a battery, and it gives a constant DC voltage. This is the case in autonomous vehicles and autonomous devices and systems that do not have connection to AC mains. Batteries provide constant DC voltages that cannot be adjusted to meet the needs of DC machines. In such cases, it is necessary to use static power converter that converts constant DC voltage of the battery to variable DC voltage to be fed to the armature winding. The latter is continuously changed according to the rotor speed.

Generally speaking, DC machines can receive electrical energy from primary sources which include batteries, single-phase AC supplies, and three-phase AC supplies but also other voltage and current systems and forms. Primary source voltages are rarely compatible with the machine needs and therefore should be adjusted. For this reason, it is necessary to use static power converter between the primary source connections and the terminals of electrical machine. The role of static power converter is to convert the voltages and currents of the primary source to the form and amplitude required by the actual operating mode of electrical machine.

Figure 12.17 shows a simplified scheme of a switching power converter with transistors, intended for feeding and controlling DC machines. This converter is often met in practice. Part (D) shows a switching bridge comprising four power transistors. The bridge is entirely the same as the one shown in Fig. 12.12 and analyzed previously. Each IGBT transistor has a diode in parallel, called *freewheeling diode*, aimed to conduct the switch currents in direction from emitter to collector. Diode rectifier, shown in part (A) of Fig. 12.17, converts three-phase system of AC voltages, provided from the mains, into DC voltage E. Part (B) of Fig. 12.17 contains a series inductance and parallel capacitor used for filtering the rectifier output voltage. This part of static power converter is called *intermediate DC circuit* or *DC link*. The voltage of the intermediate DC circuit is constant, and it represents the input voltage to the switching bridge, previously denoted by E.

Part (C) contains an additional transistor switch which, as required, may be turned ON and thus connects a resistor in parallel to the capacitor. By turning ON this fifth transistor, DC link voltage appears across the resistor. The resistor current acts toward reducing the DC link voltage and converting a certain amount of energy

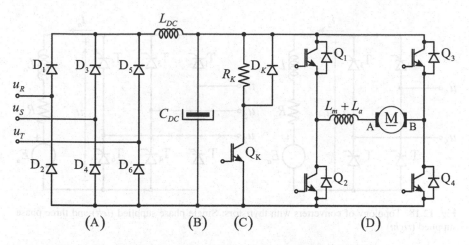

Fig. 12.17 Topology of switching power converter with transistors

into heat. This may be required when DC machine operates in the second or fourth quadrant, namely, when the torque and speed have opposite directions and the machine operates as a generator. In generator mode, the armature winding does not consume the electrical energy. Instead, it acts as a source and passes the energy back toward the switching bridge. In other words, the power flow changes and, in Fig. 12.17, it goes from the right to the left.

In even quadrants, DC machine breaks and converts mechanical work into electrical energy. The energy obtained in this way is called *braking energy*, since it is obtained by deducting mechanical work and/or kinetic energy from the mechanical subsystem. Direction of the current in the circuit is reversed, and the switching bridge does not take the energy from the intermediate DC circuit anymore. Instead, it delivers power and supplies the current to the elements of the intermediate DC circuit. Due to the sign change of the average value of the current i_u, this current is directed from the switching bridge to the DC link capacitor. Therefore, the voltage across DC link capacitor increases. The obtained energy cannot be returned to the AC mains. For this to achieve, direction of the rectifier current should be changed. Semiconductor diodes of the three-phase rectifier (A in Fig. 12.17) conduct the current from anode to cathode and cannot have the currents in the opposite direction. Hence, the braking energy cannot return to the mains and remains in DC link circuit. The excess energy is accumulated in the capacitor, increasing its energy to $W_C = \frac{1}{2}CE^2$. Excessive increase of E may damage circuit elements. Therefore, the process of accumulation of the braking energy has to be stopped. By turning the fifth transistor ON (C), a high-power resistor is connected in parallel to the intermediate DC link circuit, and the excess of the braking energy is dissipated in the resistor. The elements used in the process are called *braking device* or *dynamic braking device*. In Fig. 12.17, dynamic braking device is denoted by (C).

Static power converters with power transistors have advantageous characteristics compared to other solutions. Therefore, they are widely used. Their use is limited

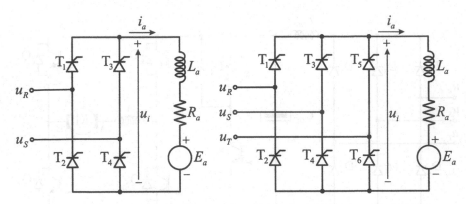

Fig. 12.18 Topology of converters with thyristors: Single phase supplied (*left*) and three phase supplied (*right*)

only by the voltage and current rating of the available power transistors. At present, commercially available transistors cover the voltages up to 6 kV and currents up to 2–3 kA. Practical switching power converters with power transistors reach the power levels in excess of 1 MW, covering virtually all practical applications of DC machines.

A couple of decades ago, at the end of the twentieth century, power transistor technology was sufficient for building switching power converters up to of several tens of kilowatts. At that time, there were no power transistors with sufficient voltage and current ratings to cover larger powers. In order to build large power static power converters, thyristors were used, four-layer semiconductor devices invented and put to practical use years before power transistors. Now and then, the voltage and current ratings of available thyristors exceed greatly the ratings of available power transistors. Yet, designing static power converters with thyristors is not an easy task. While power transistors can be turned ON or OFF at will, thyristors behave differently. They can be turned ON by gate pulses, but they cannot be turned OFF[4] unless the anode-to-cathode current does not return to zero. For those reasons, electrical schematics of thyristor-based static power converters do not resemble the ones with power transistors.

Characteristic topologies of thyristor converters for supplying large DC machines are shown in Fig. 12.18. Although the thyristor topologies are primarily of historical significance, one can encounter previously installed systems based on thyristor converters in industrial and other applications requiring controlled DC machines of large power.

---

[4] Thyristors have three electrodes. Their anode and cathode conduct the switch current, while the third electrode, the gate, serves as the control electrode. A small positive pulse of gate current turns ON the thyristor, provided that $u_{AK} > 0$. Conventional thyristor cannot be turned OFF by operating the gate. There are *gate turn-OFF thyristors* made in such way that a very large spike of negative gate current may result in turn OFF. Yet, their use and the associated auxiliary circuits are rather involved. Therefore, their use is rather limited.

Chapter 13
Characteristics of DC Machines

Working with DC machines requires the knowledge on their electrical and mechanical properties, parameters, and limitations. This chapter introduces and explains the concept of rated quantities and discusses the maximum permissible currents in continuous, steady-state service of DC machines. It also defines the safe operating area of DC machines in T_{em}–Ω_m plane, both in steady-state operation and during transients. For the sake of readers that meet electrical machines for the first time, the concepts of rated[1] current, rated voltage, mechanical characteristics, natural characteristics, rated speed, rated torque, and rated power are introduced and explained in this chapter. The need to use machines at higher speeds and with reduced flux is discussed and explained, introducing at the same time the constant flux operating region and the field-weakening operating region. The problems of removing the heat caused by the conversion losses are analyzed along with the performance restrictions imposed by temperature limits. Besides, an insight is given into possible short-term overload operation of DC machines. Principal conversion losses in DC machines are analyzed, discussed, and included in power balance. This chapter closes by discussing permissible operating areas in torque-speed plane. The steady-state safe operating area in T_{em}–Ω_m plane is also called exploitation characteristics. It is introduced and explained along with the transient safe operating area, also called the transient characteristic. Discussion and examples within this chapter are focused on separately excited DC machines.

---

[1] In electrical engineering, the concept of rated voltages, currents, and other similar quantities is widely used. Considered quantities are usually the ones that contribute to thermal, mechanical, dielectric, or other stress that may have potential of damaging electrical machine, transformer, or other electrical device or to increase its ware and reduce the expected lifetime. The rated value is most usually set by the manufacturer as a maximum value to be used with the considered device. Continued operation with voltages and currents that exceed the rated values causes permanent damage to machine windings, magnetic circuits, or other vital parts. For some quantities such as electrical current, the rated value can be surpassed during very short intervals of time without causing damages. The rated values are usually set somewhat below the level that damages the device. This is done to allow a certain safety margin.

S.N. Vukosavic, *Electrical Machines*, Power Electronics and Power Systems,
DOI 10.1007/978-1-4614-0400-2_13, © Springer Science+Business Media New York 2013

13.1   Rated Voltage

The winding rated voltage is the highest voltage that can be permanently applied to the winding terminals without causing breakdown or accelerated aging of electrical insulation. In some cases, the rated voltage can be briefly exceeded without causing any harm.

Electrical insulation separates the winding conductor from the walls of the slot where the conductor is placed. In addition, the electrical insulation separates this conductor from other conductors. A loss of insulation leads to a short circuit between individual conductors, between the winding terminals, as well as between the winding and the magnetic circuit or the machine housing. Each of the accidents mentioned results in permanent damage and interrupts the operation of the machine. In most cases, it has to be sent to the workshop or repair service.

The electrical field which exists in insulation is proportional to the supply voltage. The insulation is characterized by the rated voltage U_n as well as by the breakdown voltage U_{max}.

The breakdown voltage causes instantaneous damage of insulation. With breakdown voltage applied, the electrical field strength in critical parts of the insulation material exceeds the dielectric strength of the material and destroys insulation. Cumulative ionization of otherwise nonconductive dielectric provides a virtual short circuit between two conductors, or between the winding terminals, or a short circuit between the winding and earthed metal parts of the machine.

The rated voltage is lower than the breakdown voltage. Supplying the winding with a voltage which is higher than the rated but lower than the breakdown voltage does not necessarily lead to breakdown. Increased electrical field strength established with voltages above the rated lead to accelerated degradation and aging of the insulation material. This phenomenon reduces the expected lifetime of the insulation. Insulation aging is related to electrical, chemical, and thermal processes within dielectric materials. In most cases where the voltage is maintained within the limits of the rated voltage, expected lifetime of insulation materials and systems reaches 20 years. Continued operation with voltages exceeding the rated by 8% may halve the insulation lifetime.

13.2   Mechanical Characteristic

Mechanical characteristic is function $T(\Omega)$ or $\Omega(T)$ which gives relation between the angular speed of rotation and electromagnetic torque in steady-state operating conditions, with no variations of the speed, current, or flux of electrical machine. For separately excited DC machine, where the excitation winding and armature windings have separate supply, mechanical characteristic is given by expression

$$T_{em} = k_m\, \Phi_f \frac{U_a}{R_a} - \frac{k_m k_e\, \Phi_f^2}{R_a} \Omega_m$$

13.3 Natural Characteristic

Natural characteristic is mechanical characteristic obtained in the case when all the voltages fed to the machine are equal to the *rated voltages*.

13.4 Rated Current

Rated current I_n is the maximum permissible armature current in continuous operation. Namely, it is the largest current that can be maintained permanently, which at the same time does not cause any overheating, damages, faults, or accelerated aging. In AC machines, the rated current implies the rms value of the winding current.

Electrical currents in windings of electrical machine produce losses and develop heat, increasing the temperature of conductors and insulation. Current in the windings creates Joule losses which are proportional to the square of the current. The temperature of the machine is increased in proportion to generated heat. Increased *temperature difference* between the surface of an electrical machine and the environment gives a rise to heat transfer from the machine to the environment. The heat can be passed by conduction, convection, and radiation. Heat conduction takes place through the machine parts that are in touch with cold external solids such as the machine basis or flange. Heat convection relies on natural or forced streaming of air along the machine sides. Heat radiation is electromagnetic process caused by thermal motion of charged particles on the surface of electrical machine, and it depends on absolute temperature. The heat radiated from a warm machine to cold environment is larger than the heat absorbed by the machine due to radiation caused by the environment.

When the surface temperature of electrical machine exceeds the temperature of the environment, the heat is transferred to the environment, and the machine is cooled. The power P_T of the heat transfer defines how many joules of heat are transferred in each second. The processes of the heat transfer are nonlinear. Yet, for the range of temperatures encountered in operation of electrical machines, the power P_T can be considered proportional to the temperature difference, $P_T = \Delta\theta / R_T$, where R_T is *thermal resistance* of an electrical machine with respect to the environment, proportional to the surface of the machine and characteristics of this surface. Power P_T is expressed in W, temperature $\Delta\theta$ in °C, while thermal resistance R_T is expressed in °C/W. The equilibrium is established when the machine temperature reaches the value that results in heat transfer P_T which is equal to the total losses within the machine. Total losses of electrical machine are denoted by P_γ. With $P_T = P_\gamma$, machine temperature remains constant. The higher the losses in a machine, the higher the temperature reached in the steady state. Considering a DC machine, an increase of armature current I_a increases Joule effect losses. They are proportional to I^2, and they increase temperature of DC machine. Excessive temperature increase may damage some critical machine parts, such as the insulation of the windings.

Electrical insulation is made of paper, fiberglass, lacquer, or other materials. It separates copper conductors of machine windings from other conductive parts of machine, as well as from other conductors, thus preventing short circuits of the windings or their individual turns. The insulation can be damaged if temperature exceeds the maximum permissible limit for given insulating material. The insulation of thermal class A is damaged if the temperature exceeds 105 °C. For insulation of class F, the temperature limit is 155 °C. Therefore, DC machine with class F insulation would have higher permissible temperatures and higher rated current. When DC machine operates in steady state with the rated armature current, the losses are expected to heat the machine up to the temperature limit. With currents in excess to the rated, DC machine in continuous service would overheat. At the same time, temperature would exceed the limits and damage insulation or some other vital part of the machine. Other than insulation, there are other machine parts that are sensitive to elevated temperatures. Permanent magnets, ferromagnetic materials, and even the elements of steel construction of the machine could deteriorate and fail due to excessive temperatures. Ferromagnetic materials lose their magnetic properties when heated up to Curie temperature. Permanent magnets could be permanently damaged (demagnetized) by overheating. The elements of steel construction such as the shaft and bearings could be damaged due to thermal dilatation, changes of steel properties, and failure in lubrication of the bearing at high temperatures.

The rated current I_n is the highest armature current I_a in continuous service that does not cause any damage or fault and does not shorten the expected lifetime of the machine.

13.5 Thermal Model and Intermittent Operation

Conversion losses in an electrical machine lead to an increase of temperature of the magnetic circuit and windings. Machine is warmer than the environment, and it transfers heat to the environment. If the heat generated by the losses within the machine is equal to the heat transferred to the environment, the system is in steady-state conditions, and the temperature does not change. The temperature remains constant in cases where the heat generation remains in equilibrium with the heat emission. In other words, the *heating* has to be equal to *cooling* in order to achieve a constant temperature. The maximum permissible temperatures of vital parts of the machine are determined by endurance of the electrical insulation, magnetic circuit, windings, bearings, and housing. Variation of temperature in a machine is determined by thermal resistance R_T and thermal capacity C_T, the two machine parameters discussed further on. The former determines the heat emission from the machine into the environment, while the latter determines the heat accumulated within the machine. If thermal capacity is sufficiently large, the machine can endure

Fig. 13.1 Simplified thermal
model of an electrical
machine

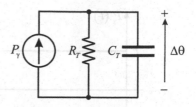

a short-term overload. In overload conditions, the current exceeds the rated, losses
in the machine are increased, and the heat is generated in excess to the heat removed
by cooling. Excess heat is accumulated in thermal capacity of the machine, and the
temperature rises. With sufficiently large thermal capacity, the temperature does
not reach the limits for significant interval of time. During that time, the armature
current exceeds the rated value, but it does not cause any damage. Hence, depending on thermal parameters of the machine and the current amplitude, an overload
condition is permissible for certain interval of time. The impact of thermal resistance, thermal capacity, and power losses on temperature change is shown in
Fig. 13.1, which represents a simplified thermal model of an electrical machine.

Electrical machines are made of copper, iron, aluminum, and insulating materials.
Each material used in manufacturing electrical machines has its own specific heat.
Specific heat represents the energy that raises by 1 °C the temperature of the unit
mass. By multiplying specific heat and mass of the part, one obtains thermal capacity
of considered part, expressed in terms of J/°C. Based on the assumption that all parts
of the machine are at the same temperature, simplified thermal model is obtained and
shown in Fig. 13.1. Parameter C_T of the thermal model represents the sum of thermal
capacities of all machine parts. Total thermal capacity C_T determines the heat which
causes the machine temperature to increase of one degree, under condition with no
heat being released into environment. The part of the loss power liable to the
temperature rise is

$$C_T \frac{d(\Delta\theta)}{dt}.$$

The remaining part of the loss power is transferred to the environment. Whenever the machine temperature exceeds the environment temperature by $\Delta\theta$, the
power of heat emission to the environment, also called cooling power, assumes the
value of

$$\frac{\Delta\theta}{R_T}.$$

Thermal resistance R_T is expressed in °C/W. It determines temperature rise $\Delta\theta$
required to obtain the cooling power $\Delta\theta/R_T$. The heat is transferred by convection,
conduction, and radiation. Thermal resistance depends on the surface area exposed
toward the environment, on the airspeed, on properties of the surface which radiates

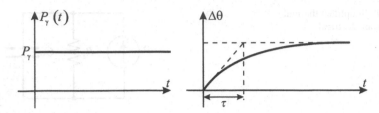

Fig. 13.2 Temperature change with constant power of losses

the heat, and also on other circumstances and parameters. The change of the machine temperature is determined by differential equation

$$P_\gamma = \frac{\Delta\theta}{R_T} + C_T \frac{d(\Delta\theta)}{dt}, \qquad (13.1)$$

where the first factor describes the cooling power, while the second factor represents the rate of the heat accumulation in thermal capacity of the machine. The equation is derived under assumption that all the machine parts have the same temperature. It should be noted that the equation resembles the one describing the voltage change in parallel RC circuit with resistor R_T and capacitor C_T, supplied from current source of the current P_γ. Hence, thermal model of the machine of Fig. 13.1 is dual with the electrical RC circuit supplied from a current source. The voltage corresponds to the temperature increase $\Delta\theta$, while the current corresponds to the power of losses.

In cases where $\Delta\theta(0) = 0$, and where the power of losses is represented by $P_\gamma(t) = P_1 h(t)$, the temperature of the machine changes according to expression

$$\Delta\theta = R_T P_1 \left(1 - e^{-\frac{t}{\tau}}\right), \qquad (13.2)$$

where $\tau = R_T C_T$ is *thermal time constant*. The thermal time constant of small electrical machines reaches several tens of seconds. Large machines may have their thermal time constants of several tens of minutes. Hence, the thermal processes within the machine are relatively slow. In Fig. 13.2, temperature change is presented for the case when the machine starts from the standstill, with the initial temperature equal to the ambient temperature. The operation proceeds with constant losses, and the temperature rises exponentially, according to the law given in (13.2). The temperature increase $\Delta\theta$ approaches to the steady-state value $\Delta\theta = P_\gamma R_T$ after 3τ .. 5τ , where R_T is the thermal resistance and τ is the thermal time constant.

If power of machine losses changes due to variations of the current, torque, or the rotor speed, the machine temperature follows variations of power losses with certain delay, determined by the thermal time constant. Figure 13.3 shows temperature changes of an electrical machine having periodic changes of the armature current. The intervals with considerable armature current are followed by the intervals when the current is equal to zero. Former intervals are associated with losses, while the

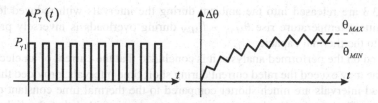

Fig. 13.3 Temperature change with intermittent load

latter interval passes with no Joule effect losses. Presented operating mode is called *intermittent mode*. At steady state, temperature oscillates between θ_{MAX} and θ_{MIN}. In hypothetical prolonged no load conditions, the machine would reach the ambient temperature θ_a, that is, the temperature of the environment. The value of θ_{MIN} is higher than the ambient temperature θ_a. Assuming that the interval with the armature current is prolonged indefinitely, the temperature will reach the value $\theta_\infty = \theta_a + \Delta\theta = \theta_a + P_{\gamma 1} \cdot R_T$, where $P_{\gamma 1} = R_a I_1{}^2$. The value of θ_{MAX} is lower than θ_∞. The power of losses $P_{\gamma 1}$ is shown in Fig. 13.3, and it corresponds to the operation with armature current I_1. In intermittent mode, this current can exceed the rated current and do no harm.

With $I_a > I_n$, power losses of $P\gamma 1$ in continuous service cause the overheating. Eventually, the machine temperature would reach θ_∞ and damage some vital parts of the machine. Nevertheless, the thermal capacity of the machine permits overload condition $I_a > I_n$ with higher loss power $P_{\gamma 1}$, but only for a short interval of time, determined by the thermal time constant $\tau = R_T C_T$. Hence, any electrical machine can withstand certain overload of limited duration. If such short time overloads are followed by intervals with no losses, the heat accumulated during overload pulses is released into ambient during prolonged intervals of time. With an adequate cooling, short time overloads can be repeated in the manner shown in Fig. 13.3. The overload intervals can be also followed by the intervals with loads that are sufficiently small to ensure a sufficient decrease in the machine temperature (cooling) before the next overload pulse.

When the load torque and the machine current exhibit periodic changes, it is possible to identify load cycles that comprise one overload interval followed by an interval with reduced losses. If the period of the load cycling is shorter than the thermal time constant τ, then the temperature oscillations $\theta_{MAX} - \theta_{MIN}$ are relatively small. With load cycle periods significantly shorter than τ, the difference $\theta_{MAX} - \theta_{MIN}$ becomes negligible. In such cases, the temperature increase $\Delta\theta$ depends on average losses within each load cycle. Notwithstanding periodic overloads, machine does not get damaged if the average power of losses does not exceed the rated loss power $P_{\gamma n}$, permissible in continuous operation. Neglecting all the losses except the ones in the armature winding, the rated power of losses can be estimated as $P_{\gamma n} = R_a I_n{}^2$. The losses during the overload interval of the load cycle may be considerably larger than the rated losses and yet maintain safe operation of electrical machine. The heat impulses generated during the overload intervals in

Fig. 13.3 are released into the ambient during the intervals with reduced losses. Intermittent temperature rise $\theta_{MAX} - \theta_{MIN}$ during overloads is inversely proportional to the thermal capacity C_T.

Based on the performed analysis, it is concluded that the current of an electrical machine may exceed the rated current during short time intervals, provided that the overload intervals are much shorter compared to the thermal time constant of the machine. The overload pulses can come as a train, provided that the light load intervals in between secure sufficient cooling and sufficient drop of the machine temperature prior to arrival of the next pulse. With load that comes as a train of pulses, machine operation is safe if the repetition period is shorter than the thermal time constant, and provided that the average power of losses does not exceed the losses incurred with nominal current.

Question (13.1): While operating with the rated load, the steady-state temperature of an electrical machine is only slightly increased above the ambient temperature. Is it designed properly?

Answer (13.1): No. Small steady-state operating temperatures show that the power of losses within the machine is rather small. This means that there is plenty of room for increasing the current density and increasing the magnetic induction. The flux increase accompanied by the increase in the armature current raises the output torque and power. Hence, the machine under consideration can produce significantly higher torque and power compared to the power declared as the rated. Consequently, too much copper and too much iron are used to make the machine. Other than being more expensive, the machine is also larger and heavier. The same load requirements can be met by electrical machine of much smaller size and weight.

Question (13.2): Electrical machine has not been used for a long time. After turning on, the armature current reaches the value of $2I_n$, twice the rated current. Prevailing losses are the winding losses. These are proportional to the square of the current. The iron losses can be neglected. The thermal time constant is 60s. Determine the longest permissible time the machine can be operated with $2I_n$ with no damages.

Answer (13.2): Permanent operation with twice the rated current produces the power of losses which is four times larger than the rated power of losses. Therefore, the machine would heat up rather quickly. Theoretical value of the steady-state temperature in this mode is four times higher than the permissible temperature. The limit temperature is reached after t_1, where $1 - \exp(-t_1/\tau_T) = \frac{1}{4}$. It follows that $t_1 = 17.26$ s.

Question (13.3): Electrical machine is loaded in such a way that every 10s a pulse of armature current appears having the amplitude of $2I_n$. This pulse is followed by an interval with no current. Prevailing losses in the machine are the winding losses, which are proportional to the square of the armature current, while the iron losses can be neglected. It is known that the thermal time constant is $\tau_T \gg 10$s. Determine the longest permissible width of the armature current pulse which does not cause damages to the machine.

Answer (13.3): Thermal time constant is significantly higher than the pulse period ($\tau_T \gg 10$ s). Therefore, temperature variations are slow, and the temperature changes within the pulse period are insignificant. For that reason, it is possible to consider the average power of losses. With any periodic current $i(t)$, the average power of Joule losses is calculated as

$$P_{\gamma(av)} = \frac{1}{T} \int_0^T R_a i_a^2 dt = R_a \left(\frac{1}{T} \int_0^T i_a^2 dt \right) = R_a I_{a(rms)},$$

where $I_{a(rms)}$ is the rms (*root mean square*) current. The rms current is also called *thermal equivalent* of periodic current $i(t)$. A resistor with constant $I_{a(rms)}$ will heat up very much the same as it does with periodic current $i(t)$. If the rms value does not exceed the rated value I_n, the machine does not overheat. A pulse of amplitude $2I_n$ that repeats every 10 s has the rms value of I_n provided that the pulse width is 2.5 s.

13.6 Rated Flux

Excitation flux Φ_f determines the current which has to be established in the armature winding so as to develop the desired torque, $I_a = T_{em}/(k_m \Phi_f)$. With higher values of flux, one and the same torque is obtained with lower armature current, and this reduces the winding losses. Therefore, it is of interest to increase the flux in order to reduce the required armature current. The excitation flux is limited by magnetic saturation of the ferromagnetic material. If the region of saturation is reached, any further increase of the excitation current results in a very small change of the flux. While operating in saturation region, the only effect of increasing the excitation current is the increase in power of excitation winding losses $R_f I_f^2$. Magnetic saturation becomes pronounced at the knee of the magnetizing curve $\Phi_f(I_f)$. The knee of the curve is the point where the initial, linear slope ends and the curve bends toward the abscissa, resulting in very small $\Delta\Phi/\Delta I_f$ ratio. The flux at the knee point is denoted by Φ_{fmax}. Since any further increase of the excitation current is of little practical effect, the value Φ_{fmax} represents the maximum flux. This flux can be used in order to reduce the armature current $I_a = T_{em}/(k_m \Phi_f)$ required to achieve desired torque T_{em}. Analysis performed in subsequent sections shows that, in certain operating conditions, the value of Φ_{fmax} cannot be used. An example is the operation at very high speeds, where the electromotive force $k_e \Phi_f \Omega_m$ must not exceed the rated voltage. In other cases, the flux Φ_{fmax} can be advantageously used, providing reduction in armature current. The knee point flux Φ_{fmax} is also called *rated flux*, and it is denoted by Φ_n.

13.7 Rated Speed

Rotor speed of electrical machine influences electromotive force induced in the windings. Rated speed is defined as the rotor speed where electrical machine with rated flux has electromotive forces equal to the rated voltage. Hence, when the machine with rated flux accelerates from zero to the rated speed, the electromotive forces increase from zero up to the rated voltage.

Electromotive forces should not exceed the rated voltage in order to avoid excessive voltage that could damage the insulation. Therefore, definition of the rated speed implies that the machine cannot exceed the rated speed yet maintaining the rated flux. In order to contain the electromotive forces between the rated limits, the flux has to be reduced as the speed goes beyond the rated speed.

Considering a DC machine with rated excitation flux and with a negligible resistance R_a, the electromotive force $E_a = k_e \Phi_n \Omega_m$ is induced, equal to the armature voltage U_a. At the rated speed, the electromotive force is equal to the rated voltage. Therefore, rated speed of DC machine is determined by relation

$$\Omega_n = \frac{U_n}{k_e \Phi_n} \qquad (13.3)$$

Since the armature voltage should not exceed the rated value, the operation at speeds larger than the rated is possible only with a reduced flux. Otherwise, the electromotive force $k_e \Phi_n \Omega_m$ would exceed the nominal voltage $U_n = k_e \Phi_n \Omega_n$.

For small electrical machines, the voltage drop due to winding resistances cannot be neglected. Since $U_a = R_a I_a + E_a$, the difference between the armature voltage and the electromotive force cannot be neglected. In such cases, definition of rated speed is made more precise, and it includes the voltage drop across the armature resistance. The rated speed can be defined as the one that results in rated voltage across the armature winding of electrical machine which operates with the rated flux and the rated current. In such conditions, the electromotive force is equal to $E_a = k_e \Phi_n \Omega_n = U_n - R_a I_n$, while the rated speed is defined by expression

$$\Omega_n = \frac{U_n - R_a I_n}{k_e \Phi_n}. \qquad (13.4)$$

13.8 Field Weakening

Operation of DC machines may require changes in the excitation flux. One example is the operation at speeds above the rated speed. In order to keep the electromotive force within the limits of the rated voltage, the flux should be reduced so as to maintain the relation $k_e \Phi_n \Omega_m < U_n$. Flux reduction at high speeds is called *field weakening*.

In addition, there are cases when the flux reduction is beneficial even at speeds below the rated. When electrical machine runs with relatively low electromagnetic torque, current is also low, as well as the power of losses in the winding. If at the same time the flux is kept at the rated value, the iron losses in the magnetic circuit become the principal conversion losses. In order to reduce the losses, it is necessary to reduce the flux. Lowering the flux increases the armature current required to develop desired torque, due to $I_a = T_{em}/(k_m \Phi_f)$. Operation with low T_{em} allows for relatively large flux reduction without any significant increase in armature current. For this reason, flux reduction at light load condition reduces the conversion losses.

13.8.1 High-Speed Operation

The electromotive force induced in the winding determines the voltage across the winding terminals. Therefore, the electromotive force must not exceed the rated voltage. The rated flux, voltage and speed are related by $U_n \approx k_e \Phi_n \Omega_n$. Therefore, electrical machine running with the rated flux Φ_n has the electromotive force that reaches the rated voltage as the speed approaches the rated speed. Any increase of the speed above the rated would result in excessive voltages. Therefore, the excitation flux has to be decreased as the rotor speed goes beyond Ω_n. Hence, electrical machines can maintain the operating speed $\Omega_m > \Omega_n$, provided that the flux is reduced so as to prevent the electromotive force from exceeding the rated voltage. For that to achieve, the flux should be varied according to the rotor speed. This change can be described by the function $\Phi(\Omega_m)$. Below the rated speed, $\Phi(\Omega_m) = \Phi_n$.

At speeds $\Omega_m > \Omega_n$, the electromotive force $E_a = k_e \cdot \Phi(\Omega_m) \cdot \Omega_m$ is induced in the machine. Neglecting the armature resistance and assuming that $E_a = U_n$, the flux to be used beyond the rated speed is

$$\Phi(\Omega_m)|_{\Phi > \Phi_n} = \frac{U_n}{k_e \Omega_m}. \tag{13.5}$$

From expression (13.3), which defines the rated speed, one obtains

$$\Phi(\Omega_m) = \Phi_n \frac{\Omega_n}{\Omega_m}, \tag{13.6}$$

which defines desired change of the flux at speeds above the rated speed. The flux is inversely proportional to the speed and varies according to $1/\Omega_m$. Variation of the flux which is necessary at speeds beyond the rated speed is defined by (13.6). The operating region where the speed is higher than the rated speed is called *field-weakening region*. If machine operates at speeds below the rated speed, it is possible to have the rated excitation flux Φ_n. For this reason, the operating region where $\Omega_m < \Omega_n$ is called *constant flux region*.

Fig. 13.4 Permissible
current, torque, and power
in continuous service in
constant flux mode (*I*) and
field-weakening mode (*II*)

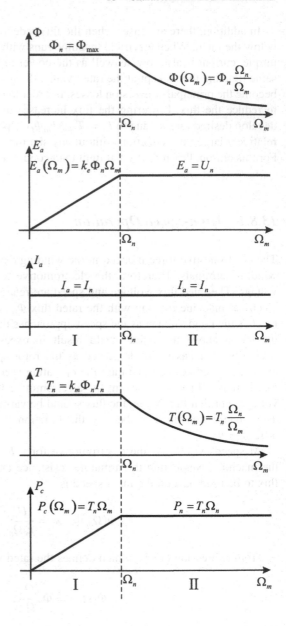

13.8.2   Torque and Power in Field Weakening

The five diagrams in Fig. 13.4 illustrate the change of flux, electromotive force,
power, torque, and current that can be maintained in field-weakening operation of
DC machines. The rotor speed is on the abscissa of all diagrams.

It should be noted that diagrams in Fig. 13.4. represent the change in *available* values or *limit* values, that is, the values that can be used at steady state without causing any damage to the machine. Hence, any value below these limits can be used as well. Due to thermal capacity of the machine, the instantaneous values of the current, torque, and power may exceed the limit values for a brief interval of time. At the same time, it is of interest to notice that presented diagram represents the absolute values. The torque, current, and power may also take negative values. Therefore, conclusions derived here are applicable to both motors and generators. Diagrams of Fig. 13.4 and the corresponding conclusions are applicable to any machine studied in this book, with exception of symbols which is slightly different for AC machine.

13.8.3 Flux Change

Abscissa of all the diagrams in Fig. 13.4 represents the rotor speed. At speeds below the rated speed, the excitation flux is maintained at the maximum value, which is also the rated value. As the speed exceeds the rated speed, the flux decreases according to hyperbola $\Phi(\Omega_m) = \Phi_n \Omega_n / \Omega_m$. Machine can also use lower flux values, but they cannot exceed the values shown by the curve $\Phi(\Omega_m)$.

13.8.4 Electromotive Force Change

The electromotive force varies according to the law $E_a = k_e \cdot \Phi(\Omega_m) \cdot \Omega_m$. Below the rated speed, the flux is equal to the rated flux, and the electromotive force increases in proportion to the speed of rotation. In the region of the field weakening where $\Omega_m > \Omega_n$, the flux is inversely proportional to the rotor speed. If the speed increases, the electromotive force remains constant and equal to the rated voltage. If the flux is below the value determined by (13.5), the electromotive force will be smaller than the rated voltage. In the region of field weakening, the flux $\Phi(\Omega_m) = \Phi_n \Omega_n / \Omega_m$ has to be applied. If, at the same time, machine operates with very small torque and current, it is beneficial to reduce the flux even below the limit $\Phi_n \Omega_n / \Omega_m$, so as to reduce the iron losses.

13.8.5 Current Change

At steady state, the armature current must not exceed its rated value, $I_a^2 \leq I_n^2$. The current can assume any value below the rated, $|I_a| < I_n$. Exceeding the rated value in continuous operation results in overheating and may damage magnetic and/or current circuits of the machine. For this reason, diagrams in Fig. 13.4 indicate that the armature current applicable over long time intervals is limited by $|I_a| < I_n$ at all speeds.

13.8.6   Torque Change

Electromagnetic torque is determined by the product of the flux and current. During long-term operation at speeds below the rated speed, the flux and current can hold their rated values. Therefore, in these conditions, available electromagnetic torque is $T_n = k_m \Omega_n I_n$, also called *rated torque*. At speeds above the rated speed, the flux decreases according to expression $\Phi_n \Omega_n / \Omega_m$. Therefore, the available torque in the region of flux weakening is equal to $T(\Omega_m) = T_n \Omega_n / \Omega_m$. The product of torque which decreases with the speed and the speed results in a constant power. For this reason, the change of the available torque in the field-weakening mode is called *hyperbola of constant power*.

13.8.7   Power Change

Electrical power converted to mechanical power is equal to the product of the torque and the rotor speed. In constant flux region, at speeds below the rated, the available torque is constant, thus the available power increases proportionally with the speed. This region is also called *the region of constant torque*. In the region of field weakening, the power is equal to the rated power $P_n = T_n \Omega_n$, since $P_c = T_{em} \Omega_m = \Omega_m (T_n \Omega_n / \Omega_m) = T_n \Omega_n = P_n$. Therefore, the available power in field-weakening region is constant and equal to the rated power P_n. Another name for the field-weakening region is *the region of constant power*. Power P_n is called *rated* power.

13.8.8   The Need for Field-Weakening Operation

Applications of electric motors often require high values of electromagnetic torque at small speeds and small torques at high speeds. A DC motor used for propulsion of electrical vehicles can serve as an example. While setting in motion a heavily loaded vehicle which has to manage a very steep slope, the motor has to deliver a very large torque. In such case, it is important to develop the torque required to get over the hill, while it is acceptable to operate the vehicle and the motor at a low speed. The same vehicle may have to move unloaded over prolonged, flat path, where the motion resistances are low, and where the motor delivers relatively small torque. At the same time, it could be required to complete such motion quickly, and this calls for high vehicle speeds and high rotor speeds.

The first example requires high-torque, low-speed operation, while the second example calls for low-torque, high-speed operation. These requirements correspond to a hyperbola $T(\Omega)$ in T–Ω plane. The curve $T(\Omega_m) = T_n \Omega_n / \Omega_m$ is called hyperbola of constant power. DC machines with permanent magnets cannot operate in the field-weakening mode. There are no practical ways to reduce the flux of the magnets. Therefore, these machines cannot go beyond the rated speed. Hence, the constant power mode is inaccessible for this kind of DC machines. Whenever the need exists for such machines to provide the constant power at high speed, the problem cannot be solved on electrical side of the system. Instead, it is necessary to use mechanical coupling with variable transmission ratio. In cases when the electrical motor cannot increase the speed, as it has reached the rated rotor speed, transmission ratio of mechanical coupling can be changed so as to obtain higher load speeds with the same motor speed. Variable transmission ratio is used in road vehicles as well. Automobiles with internal combustion engines (ICE) are usually equipped with variable transmission gears. Torque-speed characteristics of internal combustion engines do not include constant power range hyperbola in $T(\Omega)$ plane. In order to provide for both the low-speed, high-torque operation and the high-speed, low-torque operation, it is necessary to change the transmission ratio of the gears that pass the ICE torque to the wheels. The use of electrical machines capable of providing constant power operation in field-weakening region removes the need for additional gears.

13.9 Transient Characteristic

The transient characteristic is the area in $T(\Omega)$ plane which comprises all T–Ω points attainable in short time intervals. That is, it is a collection of all the operating regimes the machine can support for a short while. Peak values of the torque which can be developed at a given speed depend on the excitation flux $\Phi_f(\Omega_m)$ and on the peak value of the armature current. In DC machines, instantaneous value of the current is limited by characteristics of mechanical commutator and on the maximum current of the semiconductor power switches used to build the switching power converter that supplies the motor. An example of transient characteristic is shown in Fig. 13.5.

13.10 Steady-State Operating Area

The steady-state operating area includes all the T-Ω points in $T(\Omega)$ plane where the machine can provide continuous service for a very long time. In the field-weakening region, the area is limited by the hyperbola of constant power, $T_{em}(\Omega_m) = T_n \Omega_n / \Omega_m$, while in the constant flux region the limit is $T_{em}(\Omega_m) = T_n$. An example of steady-state operating area is shown in Fig. 13.5. Since operation of electrical machines includes all four quadrants, the steady-state operating area exists in all four quadrants as well.

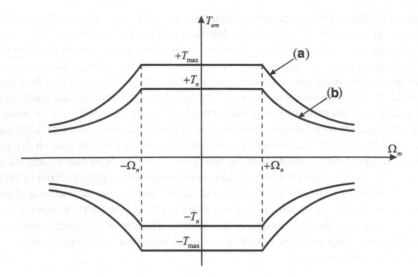

Fig. 13.5 (a) Transient characteristic. (b) Steady-state operating area

13.11 Power Losses and Power Balance

For the purpose of getting a better insight into the process of electromechanical energy conversion in DC electrical machines, it is required to study the power flow and the power of losses in the machine. Power balance equation includes the factors such as the iron losses in magnetic circuits, Joule losses in windings (also called *copper losses*), and mechanical losses due to rotation, also called losses in mechanical subsystem.

Losses in the compensation winding and the auxiliary poles windings are neglected, as these parts are not represented in each DC machine. While deriving the power balance, it is considered that the machine operates at steady state, with no variation of the armature current, excitation flux, torque, or the rotor speed. The power balance is shown in Fig. 13.6. The individual power components and losses are explained hereafter.

13.11.1 Power of Supply

The electrical sources feed the excitation and armature windings and supply the electrical power to the machine. The electrical power is $P_f + P_a = U_f I_f + U_a I_a = R_f I_f^2 + (R_a I_a^2 + E_a I_a)$.

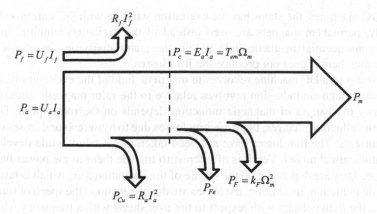

Fig. 13.6 Power balance

13.11.2 Losses in Excitation Winding

Due to Joule effect, the power of losses in the excitation winding is converted to heat. It is equal to $R_f I_f^2$, and it is also referred to as *copper losses* in excitation winding.

13.11.3 Losses Armature Winding

Due to Joule effect, the power of losses in the armature winding is converted to heat. It is equal to $R_a I_a^2$, and it is also called *copper losses* in armature winding.

13.11.4 Power of Electromechanical Conversion

Power $P_c = E_a I_a = (k_e \Phi_f \Omega_m) I_a = (k_m \Phi_f I_a) \Omega_m = T_{em} \Omega_m$ is power of the electromechanical conversion. Electrical power $E_a I_a$ is converted to mechanical power $T_{em} \Omega_m$, and both of them are equal to P_c.

13.11.5 Iron Losses (P_{Fe})

The iron losses depend on magnetic induction B and the frequency of its changes. Hence, the iron losses take place in those parts of the magnetic circuit where the magnetic field pulsates or revolves. On the other hand, in magnetic circuits where the magnetic field does not change neither its strength nor its orientation, the iron losses are equal to zero.

In DC machines, the stator has the excitation winding with DC currents. Alternatively, permanent magnets are used instead of the excitation winding. In both cases, consequential magnetic field within the stator magnetic circuit does not change and, hence, does not produce the iron losses.

The rotor of a DC machine revolves in magnetic field of the excitation winding. Therefore, magnetic induction revolves relative to the rotor magnetic circuit. The frequency of changes of magnetic induction depends on the rotor speed. Due to magnetic induction changes, there are iron losses due to hysteresis and losses due to eddy currents. The iron losses have not been taken into account while developing the mathematical model. Yet, it is of interest to include them in the power balance equation. They are dependent on the square of the excitation flux, which determines magnetic induction. In addition, the losses are dependent upon the speed of rotation, because the field pulsates with respect to the iron sheets with a frequency which is equal to the angular speed of the rotor. The iron losses within the rotor are denoted by P_{Fe}. They exist even in cases when the armature winding is disconnected and does not have any current.

The energy that accounts for these losses comes through the shaft, from the mechanical subsystem. When DC machine operates as a motor, braking torque $T_{Fe} = P_{Fe}/\Omega_m$ is subtracted from the electromagnetic torque T_{em}, reducing the torque passed to the work machine. Creation of the torque T_{Fe} can be explained in terms of joint action between the eddy currents in rotor iron and the excitation field that passes through the rotor. The torque T_{Fe} resists the motion in both directions of rotation.

13.11.6 Mechanical Losses (P_F)

In the process of rotation, one part of the energy is spent on overcoming the friction in bearings and the air resistance.[2] The rotor ends are supported by ball bearings which carry the rotor weight and provide support. The bearings are made in such way that the rotor can revolve freely. The friction between bearings and the rotor is very low, and it has a minor contribution to the machine losses. In the course of rotation, the rotor surface slides with respect to the air at certain peripheral speed, creating in such way the air resistance. Electrical machines could have their own cooling, provided by fixing a fan to one end of the rotor shaft. In the course of rotation, the fan creates an axial component of airstream which helps in removing the heat and provides better cooling of the machine. In this case, the air resistance is significantly higher. Besides the air resistance of the rotor, fan-cooled machines also have the losses due to the braking torque of the fan.

---

[2] Resistance of the air can be modeled by expression $T_{air} = k_{air}\Omega_m^2$. If the air resistance prevails among internal motion resistances, corresponding power is proportional to the third degree of the rotor speed, that is, $P_{air} = k_{air}\Omega_m^3$.

13.11.7 Losses Due to Rotation ($P_{Fe} + P_F$)

The sum of all the losses caused by rotation is called *rotation losses*, notwithstanding whether specific loss is of electrical or mechanical nature. Considering DC machines, losses due to rotation are equal to $P_{Fe} + P_F$. Friction in the ball bearings, the air resistance, and the braking torque T_{Fe} due to the iron losses[3] belong to the internal motion resistances, namely, to phenomena that resist the rotation and originate from the electrical machine alone. The sum of internal motion resistances is often modeled by an approximate expression $T_F = k_F \Omega_m$.

The iron losses are due to hysteresis and eddy currents. The angular frequency ω_m of the pulsation of magnetic induction in the rotor magnetic circuit is determined by the rotor speed Ω_m. For DC machines considered since,[4] $\omega_m = \Omega_m$. Since the power of iron losses in the rotor is $P_{Fe} = k_V \omega_m^2 + k_H \omega_m$, the corresponding braking torque is $T_{Fe} = k_V \omega_m + k_H$. The losses due to hysteresis are usually much lower than the eddy current losses, thus $T_{Fe} \approx k_V \omega_m$. Braking torque T_{Fe} due to eddy currents in rotor magnetic circuit of DC machines corresponds entirely to the model $T_F = k_F \Omega_m$, but this is not the case with the air resistance torque which depends on $(\Omega_m)^2$. If torque T_{Fe} prevails, it is then justified to consider that the sum of motion resistances gets proportional to the speed. In this case, corresponding power of losses due to rotation is modeled by expression $P_F = k_F \Omega_m^2$.

13.11.8 Mechanical Power

Electrical machine delivers to work machine mechanical power P_m. Mechanical power is obtained by subtracting the losses due to rotation from the power of electromechanical conversion. Mechanical power is equal to $P_m = T_m \Omega_m = P_c - P_{Fe} - P_F = T_{em} \Omega_m - T_{Fe} \Omega_m - k_F \Omega_m^2$. This power is delivered to the work machine via shaft. The work machine resists the motion by the torque of the same magnitude ($T_m = P_m/\Omega_m$), acting in the opposite direction.

Question (13.4): Consider a DC machine having rated flux and the armature current $I_a = 0$. The machine rotates at a constant, rated speed Ω_n. The torque required for maintaining the rotation is provided by a driving machine coupled via shaft. Power of conversion $P_c = E_a I_a$ is equal to zero. Are there any losses in the

---

[3] Losses in the rotor magnetic circuit and their place in the power balance are different in DC, asynchronous, and synchronous machines.

[4] Namely, DC machines analyzed in this chapter have two magnetic poles of the stator (and two magnetic poles of the rotor). Hence, they have one pair of magnetic poles. Electrical machines with multiple pole pairs are described in the subsequent chapters. DC machines with $p > 1$ pairs of magnetic poles and with the rotor speed of Ω_m have the angular frequency of the magnetic induction pulsations of $\omega_m = p\Omega_m$.

rotor? Is there any torque acting on the rotor? Describe behavior of the rotor in the case that the coupling with the driving machine is broken.

Answer (13.4): The rotor revolves in the magnetic field of the stator excitation. Relative motion of the field with respect to the rotor magnetic circuit gives a rise to losses due to hysteresis and eddy current within the rotor iron sheets. Eddy currents that exist in the rotor interact with the magnetic field created by excitation winding. Therefore, minute forces are generated, resulting in torque T_{Fe} that acts against the motion. The torque which resists the motion is equal to $T_{Fe} = P_{Fe}/\Omega_n$, where P_{Fe} are the iron losses in the rotor. In order to keep the rotor spinning, this torque has to be fed from the driving machine. If the shaft is not coupled to the driving machine, the rotor would, due to the braking torque, gradually slow down. The losses in iron of the rotor would heat the rotor on account of its kinetic energy $\frac{1}{2} J\Omega_n^2$.

13.12   Rated and Declared Values

The rated parameters of DC machines have been defined in the preceding section. Rated current is the highest permissible current in continuous service that does not cause any damage or failure of DC machine. Rated voltage is the highest voltage which can be maintained permanently without causing breakdown or accelerated aging of electrical insulation. In similar way, the rated levels have been defined for the remaining variables. Rated value of each quantity should be understood as the highest acceptable value to be used in continuous service. Rated quantities are characteristics of the considered electrical machine or its vital parts.

In addition to the rated quantities, the concept of *declared* quantities is frequently encountered as well. Declared values are equal to or lower than the rated values. They are usually written on the plate affixed to the machine and/or presented in catalogue data concerning the machine. Declared quantities are specified by the manufacturer. By specifying a declared quantity, the manufacturer gives a warranty that the machine can bear it during permanent operation without damage. For that reason, they cannot be higher than the rated quantities. A declared quantity can be lower than the rated. Manufacturer can intentionally give declared quantity which is lower than the rated. There could be commercial reasons for such derating. An example to that is the case when manufacturer has large-scale production of 100-kW machines and receives request to deliver only one 90-kW machine. Manufacturing of a single machine is very expensive. Therefore, he would find no economic interest to make and deliver a single machine of 90 kW. Instead, manufacturer takes one machine of 100 kW and affixes a plate declaring it as 90-kW machine. In doing so, the manufacturer gives a warranty that the machine can develop 90 kW during permanent operation and disregards the fact that the actual power can be higher.

13.13 Nameplate Data

Basic data concerning an electrical machine are written on its nameplate. In addition to declared speed and power, declared values are also given for current, voltage, and torque. A plate may contain the following data:

- Declared current of armature winding (I_n)
- Declared current of excitation winding (I_{fn}) (optional)
- Declared voltage of armature winding (U_n)
- Declared voltage of excitation winding (U_{fn}) (optional)
- Declared speed of rotation
- Declared torque
- Declared power
- Declared power factor (for AC machines)
- Method of connecting three-phase stator winding (for AC machines)

Declared speed is usually expressed in revolutions per minute [rpm]. Thus n_n[rpm] $= \Omega_n$[rad/s]$\cdot(30/\pi)$. Declared speed is related to the declared operating conditions. When the declared operating conditions correspond to the rated, declared speed is equal to the rated speed.

The rated speed of DC generators corresponds to the speed when the generator with rated flux operates with rated current and provides rated voltage to electrical loads. Generator current produces voltage drop $R_a I_G$ which is subtracted from the electromotive force. Since $U_n = k_e \Phi_n \Omega_n + R_a I_a = k_e \Phi_n \Omega_n - R_a I_G$, the rated speed of the generator results in electromotive force $E_a = U_n + R_a I_n$. Therefore, $\Omega_n = (U_n + R_a I_n)/k_e \Phi_n$.

The rated speed of DC motors corresponds to the speed when the motor with rated flux ($\Phi = \Phi_n$) operates with rated current. The motor current is directed from the source toward the motor, and therefore $U_n = E_a + R_a I_n$. Therefore, $U_n = k_e \Phi_n \Omega_n + R_a I_n$. At the rated speed, the electromotive force is $E_a = U_n - R_a I_n$. Therefore, $\Omega_n = (U_n - R_a I_n)/k_e \Phi_n$.

Question (13.5): For generator of known parameters $U_n = 220$ V, $I_n = 20$ A, $R_a = 1$ Ω, and $k_e \Phi_n = 1$ Wb, determine the rated speed.

Answer (13.5): Rated speed of the generator is $\Omega_n = (U_n + R_a I_n)/k_e \Phi_n = 240$ rad/s, corresponding to $n_n = 2{,}292$ rpm.

Question (13.6): For motor of known parameters $U_n = 110$ V, $I_n = 10$ A, $R_a = 1$ Ω, and $k_e \Phi_n = 1$ Wb, determine the rated speed.

Answer (13.6): Rated speed of the motor is $\Omega_n = (U_n - R_a I_n)/k_e \Phi_n = 100$ rad/s, corresponding to $n_n = 955$ rpm.

Definition of the rated rotor speed for DC machines may include the voltage drop across the armature resistance. In this case, the rated speed calculated for DC generator is different than the rated speed calculated for DC motor.

Analysis of electrical machines mostly assumes that the rated speed is the ratio of the rated voltage and flux, $\Omega_n \approx U_n/(k_e\Phi_n)$. This definition neglects the voltage drop R_aI_a. The ratio $U_n/(k_e\Phi_n)$ gives the speed that results in electromotive force equal to the rated voltage, provided that DC machine has rated excitation. Disregarding the voltage drop, one obtains the rated speed as $U_n/(k_e\Phi_n)$, while the speed of rotation with rated voltage, rated current, and rated flux will be slightly different. The approximation made is $U_n = R_aI_n + k_e\Phi_n\Omega_n \approx k_e\Phi_n\Omega_n$, and it results in $\Omega_n = U_n/(k_e\Phi_n)$, slightly higher than the speed measured on DC motor running in rated conditions and slightly lower than the speed measured on DC generator running in rated conditions.

In all analyses and calculations where resistance R_a is neglected or it is unknown, it is justifiable to assume that $U_a \approx k_e\Phi\Omega_m$, and that the rated speed is $\Omega_n = U_n/(k_e\Phi_n)$.

In solving the problems where the value of R_a is given, the voltage drop R_aI_a should be taken into account. Then, it is not justified to consider that $\Omega_n \approx U_n/(k_e\Phi_n)$. Expression $\Omega_n = (U_n - R_aI_n)/k_e\Phi_n$ determines the rated speed for motors, while expression $\Omega_n = (U_n + R_aI_n)/k_e\Phi_n$ determines the rated speed for generators.

Chapter 14
Induction Machines

The operating principles of induction machines and basic data concerning constructions of their stator and rotor are presented in this chapter. This chapter includes some basic information regarding construction of induction machines. Discussed and described are the stator windings, the rotor short-circuited cage winding, and slotted and laminated magnetic circuits of both stator and rotor. Fundamentals on creating the revolving magnetic field are reinstated for the three-phase stator winding. Basic operating principles of an induction machine are illustrated on simplified machine with one short-circuited rotor turn. The torque expression is developed and used to predict basic properties of mechanical characteristic. For the purpose of studying the electrical and mechanical properties of induction machines, corresponding mathematical model is developed in Chap. 15 and used within the next chapters. Chapter 16 deals with the steady-state operation, steady-state equivalent circuit and relevant parameters, mechanical characteristics, losses, and power balance. Variable speed operation of induction machines is discussed in Chap. 17, with analysis of constant frequency-supplied induction machines and introduction and analysis of variable frequency-supplied induction machines, fed from PWM-controlled three-phase inverters.

14.1 Construction and Operating Principles

Induction machines have stator comprising three-phase windings. Magnetic axes of the three phases are spatially shifted by $2\pi/3$. If the stator phase windings have sinusoidal currents of the same amplitude and the same angular frequency ω_e, and at the same time their initial phases mutually differ by $2\pi/3$, then the magnetic field within the machine revolves, maintaining the same amplitude. The speed of the field rotation is determined by the angular frequency ω_e of the source voltage. When an induction machine is fed from a network of industrial frequency $f = 50$ Hz, the field rotates at the speed of 100π rad/s. The rotor of an induction machine has a short-circuited cage winding. If the rotor revolves at the same

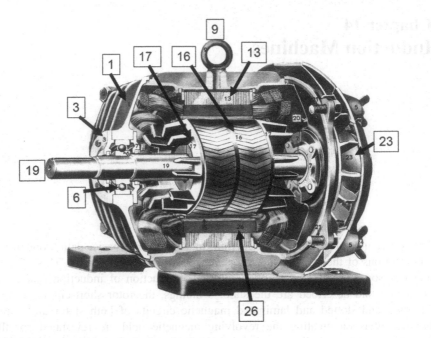

Fig. 14.1 Appearance of a squirrel cage induction motor

speed as the field does, they move synchronously, and there is no relative displacement between the two. In this case, there is no change of the flux in the rotor winding, and no electromotive force is induced. For that reason, there is no current in the rotor short-circuited winding. The speed of rotation of the magnetic field is called *synchronous speed*, and it is denoted by Ω_e. In the case when the difference $\Omega_{slip} = \Omega_e - \Omega_m$ exists between the speed of the field Ω_e and speed of the rotor, there is a change of flux in the rotor. An electromotive force is induced, and the electrical current is established in short-circuited rotor windings, which are usually made as squirrel cage. The frequency of rotor currents ω_{slip} depends on the speed difference Ω_{slip}, also called *slip speed*. The angular frequency ω_{slip} is called *slip frequency*. In machines with two magnetic poles (i.e., with $p = 1$ pair of poles), $\omega_{slip} = \Omega_{slip}$. Joint action of the rotor currents and the stator field results in electromagnetic torque T_{em}. This torque tends to bring the rotor into synchronism with the field. In the case when $\Omega_{slip} = \Omega_e - \Omega_m > 0$, the torque tends to increase the rotor speed and to bring the rotor closer to synchronism with the rotation of the field.

Figure 14.1 gives an insight to construction of an induction machine having rated parameters $U_n = 400$ V, $f_n = 50$ Hz, $P_n = 4$ kW, and $n_n = 1450$ rpm. Number (1) denotes the metal housing that accommodates the machine. The ring denoted by number (9) serves for lifting and transportation. Ball bearings are built-in at the two ends of the shaft (19). The bearings are denoted by numbers (6) and (7). The front bearing is housed in a cartridge (3). The rotor magnetic circuit is denoted by number (16). The rotor conductors are mostly made by casting aluminum into the rotor slots which, in the considered rotor, are not straight but are set obliquely. The rotor conductors are short circuited by aluminum rings (17) at the front and rear sides of the rotor cylinder. The aluminum rings (17) are extended to winglets intended to create an airstream for cooling. The stator conductors (26) are made of copper and covered by electrical insulation. They are laid in the slots of the stator magnetic circuit (13). At the rear side of the motor, the shaft may be equipped with a fan (23) that creates an airstream along the external sides of the housing. This method of assisting the heat transfer is called *self-cooling*. The self-cooling is not suitable for the motors rotating at high speeds, where the fan would create significant losses and acoustic noise.

14.2 Magnetic Circuits

The voltages and currents in the stator windings of an induction machine have angular frequency ω_e. Electromotive forces induced in the rotor have angular frequency ω_{slip}, and they cause electrical currents of the same frequency in short-circuited rotor winding. The flux and magnetic induction of the stator vary at angular frequency ω_e. Induction machines operated from the mains have angular frequency ω_e equal to 100π. The flux and magnetic induction of the rotor have angular frequency ω_{slip}. For induction machines of several kW, the slip frequency is of the order of 1 Hz. Hence, magnetic induction pulsates with respect to the stator magnetic circuit at the line frequency. It also pulsates with respect to the rotor magnetic circuits at lower frequency. In order to reduce iron losses due to eddy currents, both stator and rotor magnetic circuits are laminated, that is, they are made of iron sheets. The shape of these sheets is shown in Fig. 14.2.

By stacking iron sheets, cylindrical magnetic circuits of the stator and rotor are obtained. The stator magnetic circuit is a hollow cylinder. The rotor cylinder is placed axially within the stator. The two parts are separated by the air gap.

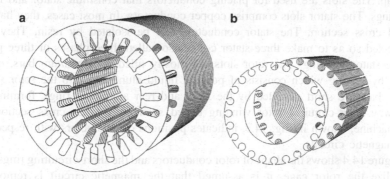

Fig. 14.2 (a) Stator magnetic circuit of an induction machine. (b) Rotor magnetic circuit of an induction machine

Fig. 14.3 Cross section of an induction machine. (**a**) Rotor magnetic circuit. (**b**) Rotor conductors. (**c**) Stator magnetic circuit. (**d**) Stator conductors

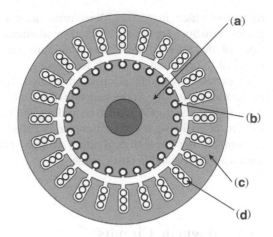

Fig. 14.4 Cage winding

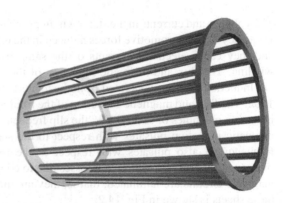

On the internal surface of the stator magnetic circuit, there are axial slots facing the air gap. The same way, the outer surface of the rotor has axial slots facing the air gap. The slots are used for placing conductors that constitute stator and rotor windings. The stator slots comprise copper conductors. In most cases, they have a round cross section. The stator conductors can be isolated by resin. They are connected so as to make three stator coils, also called three phases or three parts of the stator winding. The rotor slots are mostly filled by aluminum bars, often made by casting which consists of pouring liquid aluminum into the rotor slots. These bars are short circuited by the front and rear aluminum rings, forming in this way short-circuited rotor winding called *squirrel cage*. A cross section of the machine, shown in Fig. 14.3, indicates positions of conductors with respect to the magnetic circuits.

Figure 14.4 shows the shape of rotor conductors and the short-circuiting rings. To visualize the rotor cage, it is assumed that the magnetic circuit is removed. Similarity to the cage and the circumstance that the turns are short circuited gave the name *cage rotor*. Sometimes, machines having aluminum-cast short-circuited rotors are called *squirrel cage rotor* machines. In high-power machines, where it is significant to increase the energy efficiency, rotor conductors are made of copper bars. Copper has lower specific resistance compared to aluminum, which reduces specific and total losses in the rotor cage.

Rotor conductors have their electrical insulation made differently. An electrical contact between the rotor bars, cast into the rotor slots, and iron sheets that constitute the rotor magnetic circuit gives a rise to sparse electrical currents that can jeopardize performances of the machine. These currents can be avoided by applying an acid solution to internal surfaces of the rotor slots before casting the aluminum bars or by making the inner surfaces nonconductive in some other way. The insulation layer created in this way separates aluminum bar from the magnetic circuit and prevents any uncontrollable currents. Exceptionally, the rotor of induction machine can have a three-phase winding made of round, insulated copper wire, in the way quite similar to the one used in manufacturing the stator winding. In such cases, the three rotor terminals are made available to the user. Such rotor is also called *wound rotor*, and it is briefly explained within the next paragraphs. It has been used prior to deployment of three-phase, variable frequency static power converters. Wound rotor machines are rarely met nowadays, as a vast majority of induction machines have a cage rotor.

Prescribed method of manufacturing the rotor cage is rather simple, and it does not require high-precision processing nor any special technologies or materials. Manufacturing of the rotor of DC machine is considerably more involved and complicated. It requires mechanical commutator, device that requires rather precise production process and which contains a number of different materials.

Compared to DC machines, induction machines have a number of advantages. They include rather simple manufacturing procedure, robustness, higher specific power $\Delta P/\Delta m$, lower mass and volume, as well as possibility to operate at considerably higher rotor speeds compared to DC machines. Therefore, induction machines are the most widespread machines nowadays. Absence of the brushes and collector eliminates the maintenance and prolongs the lifetime. Robust construction of induction motor results in an improved reliability, which is usually expressed by the mean time between failures (MTBF).

Over the past century, most induction machines were operated from the mains, namely, supplied by AC voltages of fixed amplitude, having the line frequency. These machines were mostly running with constant speed. Speed regulation was possible only with wound rotor.

With recent developments in the area of power converters, semiconductor power switches, digital signal processors (DSP), and digital controls of power converters and drives, it is possible nowadays to design, manufacture, and deploy reliable and affordable systems based on induction motors supplied from static power converters providing variable frequency AC voltages. Variable frequency supply allows for efficient and reliable operation of induction machines over wide range of speeds. Digitally controlled induction motors are frequently used as the torque actuators in motion control systems.

14.3   Cage Rotor and Wound Rotor

In addition to short-circuited rotor, it is possible to encounter *wound rotor* induction machines with *slip rings* used for external access to the rotor winding. Induction motors with wound rotors have been used at times when there was no possibility to change the amplitude and frequency of the stator voltages. The stator used to be fed from the mains, with the amplitude and frequency which could not be changed. At that time, there were no suitable static power converters capable of converting the electrical energy of line-frequency voltages and currents into the energy of variable frequency voltages and currents. Under these conditions, application of induction machines with wound rotors was used to alter the rotor speed of line-frequency-supplied inductions machines. Today, the use of wound rotor motors is declining, and the use of squirrel cage motors is prevailing.

Part (a) in Fig. 14.5 shows a short-circuited cage rotor of induction machine. Part (b) shows a *wound* rotor which has a three-phase winding similar to that of the stator. The three-phase windings are usually star connected, while the remaining three terminals of the rotor winding are connected to metal rings called *slip rings*, mounted at the front end of the machine. When the motor is in service, there are *sliding brushes* pressed against the rings, providing electrical contact and making the rotor terminals accessible to external uses. Sliding brushes are elastic metal-graphite plates which slide, as the rotor revolves, along peripheral surface of the rings. By connecting three external resistors to the rotor circuit, the equivalent resistance of the rotor circuit changes, and this alters the mechanical characteristic of the motor, allowing for desired speed changes.

The need for applying wound rotor machines has disappeared along with the appearance of static power converters which allow continuous change of the supply frequency and, hence, continuous change of the rotor speed.

14.4   Three-Phase Stator Winding

Stator of induction machines has three-phase windings, namely, three separate coils making the system of stator windings. Each phase winding has two terminals. The three-phase windings can be star connected or delta connected. Star connection

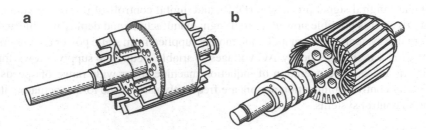

Fig. 14.5 (a) Cage rotor. (b) Wound rotor with slip rings

is denoted by Y and delta connection by Δ. With star connection, one terminal of each phase winding is connected to three-phase AC source, while the other phase terminals are connected to a common node. This common node is called *star point*. With the star point having no other connections (floating), the sum of the three-phase currents must be equal to zero at every instant, $i_a(t) + i_b(t) + i_c(t) = 0$. Most star-connected stator windings have their star point floating. Exceptionally, some large-power, high-voltage[1] induction machines may have their star point connected.[2]

In some cases, phases of the three-phase stator winding are connected to *delta* connection, wherein the three-phase windings are connected into triangle, restricting the phase voltages by $u_a(t) + u_b(t) + u_c(t) = 0$. For the given power rating, delta connection has lower current in stator conductors with respect to star connection. Therefore, delta connection is advantageously used in electrical machines with exceptionally low stator voltages, such as the motors in battery-fed traction drives, where the stator current is very large.

There are advantages of star connection which make it more frequently used.[3] Without the lack of generality, it is assumed throughout this book that the stator phases are star connected.

Magnetic axes of the stator phase windings are spatially shifted by $2\pi/3$, as shown in Fig. 14.6. Desired magnetic field within three-phase induction machines is established by establishing the phase currents of the same amplitude I_m and the same angular frequency ω_e. Their initial phases have to be shifted by $2\pi/3$, the angle that corresponds to the spatial shift between magnetic axes of the three phases. With prescribed currents in the phase windings, magnetic field is established in magnetic circuits and the air gap of the machine. The field revolves at angular speed $\Omega_e = \omega_e$.[4] The phase currents of the same amplitude and frequency, and with the initial phase

---

[1] Voltages in excess to 1 kV are called *high voltages*. The term *medium voltage* is also in use, and it refers to lower end of *high voltages* and corresponds to voltages from 1 up to 10 kV. The upper limit of *medium voltages* is not strictly defined. There are also terms *very high voltages* and *ultrahigh voltage*, both lacking a clear definition.

[2] High-voltage induction machines have an increased insulation stress. Due to transient phenomena, floating star point may have considerable overvoltages. In some cases, star point of high-voltage machines is grounded by means of impedance connected between the star point and the ground.

[3] The sum of the phase voltages of delta-connected phase windings is equal to zero, $u_a(t) + u_b(t) + u_c(t) = 0$. Practical AC machines have imperfect, nonsinusoidal electromotive forces that include harmonics such as the third, which has the same initial phase in all the three-phase windings. This is the property of all $3n$th harmonic, also called *triplian* harmonics. Within the three phases of the stator winding, the waveforms of a triplian harmonics have the same amplitude and phase. Therefore, with star connection, triplian harmonics cannot produce any current due to $i_a(t) + i_b(t) + i_c(t) = 0$. On the other hand, delta connection provides the circular path for triplian harmonics of the stator current. With delta connection, any distortion in electromotive forces that results triplian harmonics contributes to circular currents which compromise the operation of induction machine by increasing losses.

[4] With induction machines having $p > 1$ pairs of magnetic poles, $\Omega_e = \omega_e/p$. Machines with multiple pairs of magnetic poles are explained further on.

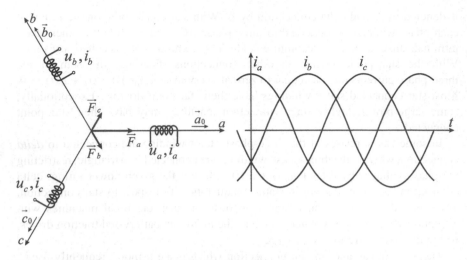

Fig. 14.6 Magnetomotive forces of individual phases

difference of $2\pi/3$, constitute a *symmetrical three-phase system* of electrical currents. Instantaneous values of these currents are given by (14.1), while their change is depicted in the right-hand part of Fig. 14.6.

$$F_a = N_s I_m \cos \omega_e t$$

$$F_b = N_s I_m \cos \left(\omega_e t - \frac{2\pi}{3} \right)$$

$$F_c = N_s I_m \cos \left(\omega_e t - \frac{4\pi}{3} \right) \tag{14.1}$$

In Fig. 14.6, the phase windings are denoted by coils. Each of the three coil signs represents one phase winding. The phase windings have their conductors distributed in a number of stator slots along the machine circumference, next to the inner surface of the stator magnetic circuit. It is understood that conductor density has sinusoidal change along the air-gap circumference, as shown in Fig. 14.7. In this figure, the coil sign denotes the winding and lies on its magnetic axis. Shortened representation of phase windings places a coil sign instead of introducing a number of distributed conductors. These signs are used for clarity. An attempt to represent all the three phases by drawing their individual conductors would result in a drawing which is of little practical value.

Magnetomotive force of phase winding has an amplitude determined by the phase current, while the corresponding vector extends along the magnetic axis of the winding. The winding flux vector has the same direction. Hence, the winding represented by coil symbol has the magnetomotive force and flux vectors directed along the axis indicated by the symbol of coil that represents the winding.

14.5   Rotating Magnetic Field

Each phase winding creates a magnetomotive force along the magnetic axis of the winding. Amplitudes of magnetomotive forces F_a, F_b, and и F_c are dependent on currents $i_a(t)$, $i_b(t)$, and $i_c(t)$. Vector sum of these magnetomotive forces gives the resultant magnetomotive force of the stator $\boldsymbol{F}_s$ (Fig. 14.8). The quotient of the vector $\boldsymbol{F}_s$ and magnetic resistance R_μ gives the stator flux vector. The lines of the stator flux pass through the air gap and encircle the rotor magnetic circuit (Fig. 14.9). Passing through aluminum cast, short-circuited rotor turns, the stator flux contributes to the rotor flux. The coefficient of proportionality is determined by the mutual inductance L_m between the stator winding and the rotor cage. In cases when the rotor revolves in synchronism with the field ($\Omega_m = \Omega_e$), there is no change of the rotor flux. Therefore, maintaining the synchronism, the rotor electromotive force is equal to zero as well as the rotor current.

When the rotor speed is lower than the synchronous speed ($\Omega_m < \Omega_e$), the rotor is lagging with respect to the field. The speed difference $\Omega_{slip} = \Omega_e - \Omega_m > 0$ is called *slip speed*. For the observer residing on the rotor, in the frame of reference of short-circuited rotor cage, the stator flux revolves with relative speed of Ω_{slip}, determined by $\Omega_{slip} = \omega_{slip}$.[5]

Change of flux leads to induction of electromotive force in short-circuited rotor turns. Rotor current is an AC, and it has angular frequency ω_{slip}. It is proportional to the induced electromotive force e and inversely proportional to rotor impedance

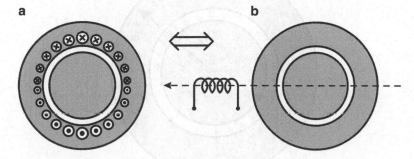

a   **b**

Fig. 14.7 (a) Each phase winding has conductors distributed along machine perimeter. (b) A winding is designated by coil sign whose axis lies along direction of the winding flux

---

[5] In the preceding part of the book, the electrical machines are considered having two-pole magnetic field. They have one north magnetic pole and one south magnetic pole. These machines are called *two-pole* machines, and they have $p = 1$ pair of magnetic poles. Machines with multiple pairs of magnetic poles will be explained as well. As an example, distribution of magnetic field in the air gap may have two north and two south magnetic poles. The number of pairs of poles is denoted by p. It will be shown later that the magnetic field created by AC currents of angular frequency ω_e rotates at angular frequency $\Omega_e = \omega_e/p$. Therefore, for two-pole machines, angular frequency ω_e is equal to the angular speed of rotation Ω_e.

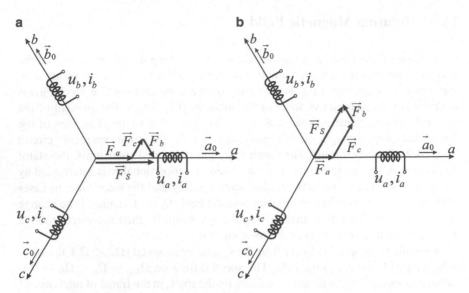

Fig. 14.8 Resultant magnetomotive force of three-phase winding. (**a**) Position of the vector of magnetomotive force at instant $t = 0$. (**b**) Position of the vector of magnetomotive force at instant $t = \pi/3/\omega_e$

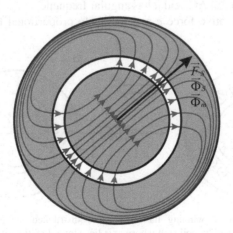

Fig. 14.9 Vector representation of revolving field. (F_s)-vector of the stator magnetomotive force. (Φ_s)-vector of the flux in one turn of the stator. (Φ_m)-vector of mutual flux encircling both the stator and the rotor turns

$R + j\omega_{slip}L$, where R and L are parameters of short-circuited rotor winding. If the slip ω_{slip} is rather small, the rotor current is approximately equal to e/R. Joint action of the magnetic field and currents in rotor conductors creates electromagnetic torque which tends to bring the rotor into synchronism with the field. Namely, in cases where $\Omega_e - \Omega_m > 0$, the torque acts upon the rotor so as to increase the rotor speed Ω_m and bring it closer to the synchronous speed Ω_e.

14.6 Principles of Torque Generation

Principle of operation of an induction machine can be explained by using Fig. 14.10. Represented flux $\boldsymbol{\Phi}_m$ revolves at the synchronous speed Ω_e. It is assumed that the rotor revolves at the speed of $\Omega_m < \Omega_e$, meaning that the rotor is lagging behind the flux by the amount of slip,

$$\omega_{slip} = \Omega_{slip} = \Omega_e - \Omega_m > 0 \tag{14.2}$$

Angle θ_{slip} between the flux vector $\boldsymbol{\Phi}_m$ and the rotor is equal to the integral of the slip; thus, it increases gradually. The figure shows only one short-circuited contour of the rotor in order to explain the principle of the torque generation. The flux within the rotor contour changes with the angle θ_{slip},

$$\theta_{slip}(t) = \theta_{slip_0} + \int_0^t \Omega_{slip}\, d\tau \tag{14.3}$$

Angle between the reference axis of the contour and the flux vector is $\theta_{slip} + \pi/2$. The part of the stator flux which encircles the rotor contour is equal to

$$\Phi_{Rm} = -\Phi_m \cdot \sin \theta_{slip}. \tag{14.4}$$

The total rotor flux includes the effects of rotor current which contribute to the rotor flux in proportion to the coefficient of self-inductance L_R. In the course of gaining an insight into the operating principles, these effects are neglected for the time being. This assumption is justified by the fact that with relatively low slip frequencies, the reactance $L_R\omega_{slip}$ can be neglected. Hence, it is considered instead that the mutual flux $\Phi_{Rm} = -\Phi_m \cdot \sin(\theta_{slip})$ corresponds to the total rotor flux Φ_R.

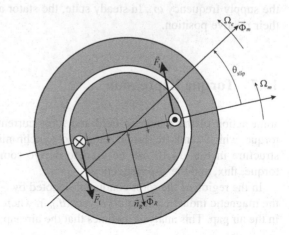

Fig. 14.10 An approximate estimate of the force acting on rotor conductors

Thus, the electromotive force induced in the rotor winding is proportional to the flux Φ_m and to the slip frequency ω_{slip}

$$e = \frac{d\Phi_R}{dt} = -\Phi_m \cdot \Omega_{slip} \cos \theta_{slip} \qquad (14.5)$$

In steady state, $\theta_{slip} = \Omega_{slip}t$. Therefore, rotor current is an AC current. Its amplitude is inversely proportional to the rotor impedance $R + j\omega_{slip}L$. For small values of slip, resistance R is considerably higher than reactance $\omega_{slip}L$. Hence, the rotor current i_R is approximately equal to $e/R \sim \Phi_m\omega_{slip}$

$$u = 0 = R_R i_R + e, \quad i_R = +\frac{\Phi_m \cdot \Omega_{slip}}{R_R} \cos \theta_{slip} \qquad (14.6)$$

In the case under consideration, $i_R > 0$. Direction of the rotor current corresponds to direction shown in Fig. 14.10. Validity of this conclusion is checked by the following reasoning. Since the flux Φ_m advances with respect to the rotor, the flux in the rotor contour in Fig. 14.10 changes its value. It increases in direction opposite to the reference direction n_R.

The induced electromotive force in short-circuited rotor turn and the consequential rotor current have direction opposite to the change of the flux, wherein the flux change is the origin of the electromotive force. Hence, direction of the rotor current pretends to establish the flux which opposes to the original flux change. With reference directions as shown in Fig. 14.10, the rotor current causes the flux change which is directed downward. The rotor current varies in proportion to function cos $(\theta_{slip}) = \cos(\theta_0 + \omega_{slip}t)$. It is an AC current of frequency ω_{slip}. In rotors with multiple turns, AC currents create the flux which revolves with respect to the rotor at the slip speed of $\Omega_{slip} = \omega_{slip}/p$. With the rotor running at the speed of Ω_m, the speed of the rotor flux rotation with respect to the stator is equal to $\Omega_m + \Omega_{slip} = \Omega_m + (\Omega_e - \Omega_m) = \Omega_e$. Hence, the stator and rotor flux vectors of induction machine revolve in synchronism, at the same speed of Ω_e, determined by the supply frequency ω_e. In steady state, the stator and rotor flux vectors maintain their relative position.

14.7 Torque Expression

Joint action of the magnetic field and rotor currents creates the electromagnetic torque which tends to bring rotor into synchronism with the field. Simplified structure in Fig. 14.10 can be used to derive some basic relations between the torque, flux, and the slip frequency.

In the region of the rotor conductor denoted by $\odot$, there is radial component of the magnetic induction equal to $B_m \cdot \cos(\theta_{slip})$, where B_m is the maximum induction in the air gap. This analysis assumes that the air-gap flux is created primarily by the stator currents and that the induced rotor currents have a negligible effect on the magnetic inductance in the air gap. Magnetic induction assumes the maximum value of B_m in the air-gap regions along the flux vector $\boldsymbol{\Phi}_m$. In other directions, its value is smaller, and it changes according to the expression for sinusoidal distribution of the radial component of magnetic induction in the air gap. Hence, for the air-gap region displaced by $\Delta\theta$ from direction of the vector $\boldsymbol{\Phi}_m$, magnetic inductance assumes the value of $B = B_m\cos(\Delta\theta)$. Electrical current of the rotor is equal to $i_R = \Phi_m \cdot \omega_{slip} \cdot \cos\theta_{slip}/R_R$. The force acting on the conductor denoted by $\odot$ is equal to the product $Li_R B$, where L is the axial length of the machine. An equal force is acting on the conductor denoted by $\otimes$. Direction of both forces is positive with respect to the reference tangential direction. Direction of the current in the second conductor is changed. Direction of the magnetic induction is also changed. In the region of conductor $\otimes$, magnetic field comes out of the stator magnetic circuit, passes through the air gap, and enters into the rotor magnetic circuit.

Expression $\Phi_m = (2/\pi)B_m\pi RL = 2B_m RL$ relates flux Φ_m to the maximum value of magnetic induction B_m. In this expression, $R = D/2$ is the radius of the rotor cylinder. Therefore, the expression for magnetic induction obtains the form

$$B = k_1 \cdot \Phi_m \cos\theta_{slip} = B_m \cos\theta_{slip} \tag{14.7}$$

The electromagnetic torque is equal to

$$T_{em} = DL i_R B = DL \frac{\Phi_m \omega_{slip} \cos\theta_{slip}}{R_R} B_m \cos\theta_{slip} \tag{14.8}$$

that is,

$$\begin{aligned} T_{em} &= DL \frac{\Phi_m(p\Omega_{slip})\cos\theta_{slip}}{R_R}(k_1\Phi_m)\cos\theta_{slip} \\ &= k_2\Phi_m^2\Omega_{slip}\cos^2(\theta_{slip}). \end{aligned} \tag{14.9}$$

Therefore, the torque delivered by an induction machine is directly proportional to the slip frequency and to the square of the flux. It is inversely proportional to the rotor resistance.

Question (14.1): In the case considered above, the torque is proportional to $\cos^2(\theta_{slip}) = \cos^2(\theta_0 + \omega_{slip}t)$. Therefore, the torque pulsates from zero up to twice the average value. Is it possible to alter the structure of Fig. 14.10 so as to obtain a constant, ripple-free torque?

Answer (14.1): By adding another short-circuited contour on the rotor, shifted by $\pi/2$ with respect to the existing one, the torque pulsations can be suppressed. The torque acting on conductors of the second contour will be proportional to $\sin^2(\theta_{slip})$. When added to previously obtained torque, proportional to $\cos^2(\theta_{slip})$, the sum of the two becomes $T_{em} = \Phi_m{}^2\omega_{slip}/R_R = \text{const.}$

Chapter 15
Modeling of Induction Machines

This chapter introduces and explains mathematical model of induction machines. This model represents transient and steady-state behavior in electrical and mechanical subsystems of the machine. Analysis and discussion introduces and explains Clarke and Park coordinate transforms. The model includes differential equations that express the voltage balance in stator and rotor windings, inductance matrix which relates flux linkages and currents, Newton differential equation of motion, expression for the air-gap power, and expression for the electromagnetic torque. The model development process starts with replacing the three-phase machine with two-phase equivalent. Namely, the three-phase voltages, currents, and flux linkages are transformed in two-phase variables by appropriate transformation matrix which implements $3\Phi/2\Phi$ transform, also called Clarke coordinate transform. Two-phase model is formulated in stationary coordinate frame. The drawbacks and difficulties in using this model are the rationale for introducing and applying Park coordinate transform, which results in the machine model in synchronous dq coordinate frame. Necessary techniques and procedures of applying and using coordinate transforms are explained in detail, including representation of machine vectors by complex numbers. The operable model of induction machines is obtained in dq coordinate frame which revolves synchronously with the stator field. The merits and practical uses of the model in dq frame are explained at the end of the chapter.

15.1 Modeling Steady State and Transient Phenomena

The work with induction machines requires a sound knowledge of their behavior and principal characteristics. From the electrical access point, it is of interest to find relations between steady-state voltage and currents, so as to obtain an equivalent circuit of the machine, representing the steady-state operation. At the same time, it is important to study the torque–speed relations at the mechanical access of the machine.

S.N. Vukosavic, *Electrical Machines*, Power Electronics and Power Systems,   379
DOI 10.1007/978-1-4614-0400-2_15, © Springer Science+Business Media New York 2013

Analysis of induction machines at steady state is based on *mechanical characteristics* and *steady-state equivalent circuit*. Mechanical characteristic of an induction machine gives relation between steady-state values of the electromagnetic torque and the rotor speed. The mechanical characteristic is dependent on the frequency and amplitude of the stator voltage. Therefore, any change in stator voltage will affect the mechanical characteristic. At steady state, the voltages and currents are sinusoidal quantities of constant amplitude and frequency. For that reason, they can be represented by appropriate phasors.[1] Relations between voltages and currents can be presented by equivalent circuit. A steady-state equivalent circuit is a network consisting of resistances and reactances which serves for calculation of phasors of the stator and rotor currents in conditions with known supply conditions and specified rotor speed.

Working on problems of supplying and controlling induction machines requires a good knowledge of the *dynamic model*. This model comprises differential equations and algebraic expressions relating the machine variables and parameters during transient processes and also in the steady state. Relation between the voltages and currents during transients is given by differential equations describing the voltage equilibrium in the windings, also called *voltage balance equations*. The voltage balance equations describe the *electrical subsystem* of induction machine. The *mechanical subsystem* is described by Newton differential equation of motion. The set of differential equations and expressions describing behavior of the machine is called *mathematical model* or *dynamic model*.

In further considerations, the analyses of electrical and mechanical subsystems of induction machines are presented and explained, resulting in dynamic model. This model includes transforms of the state coordinates, also called *coordinate transforms*. They facilitate the analysis of transient processes in both synchronous and induction machines. Dynamic model is usually mostly used for transient analysis and for solving control problems, but it can also be used to resolve steady states. Starting from dynamic model, one can obtain the steady-state relations; mechanical characteristics; relations between voltages, currents, fluxes, torques, and speed in the steady state; as well as the steady-state equivalent circuit.

The readers with no interest in transient processes in induction machines and with no need to deal with problems of supply and control do not have to study dynamic model of induction machine. Such readers could skip entire Chap. 15 which develops mathematical model and deals with transient processes. The steady-state equivalent circuit can be also determined by using analogy with a transformer, as shown in Sect. 16.6. The analyses of steady-state equivalent circuits and the study of mechanical characteristics of induction machines can be continued in Chap. 16.

---

[1] Phasor is a complex number which represents a sinusoidal AC voltage or current. The absolute value of phasor corresponds to the amplitude, while the phasor argument determines the initial phase of considered voltages and currents. Phasors can be used to represent other quantities that have sinusoidal change in steady state, such as the magnetomotive forces and fluxes.

15.2 The Structure of Mathematical Model

Within the introductory chapters, it is shown that the dynamic model of electrical machines comprises four basic parts. These are:

1. N differential equations of voltage equilibrium
2. Inductance matrix
3. Expression for the torque
4. Newton equation

Differential equations of voltage balance are given by expression

$$\underline{u} = \underline{R} \cdot \underline{i} + \frac{d\underline{\Psi}}{dt}. \tag{15.1}$$

Relation between the fluxes and currents is given by nonstationary inductance matrix

$$\underline{\Psi} = \underline{L}(\theta_m) \cdot \underline{i}. \tag{15.2}$$

The electromagnetic torque is determined by equation

$$T_{em} = \frac{1}{2} \underline{i}^T \frac{d\underline{L}}{d\theta_m} \underline{i} = \frac{1}{2} \sum_{k=1}^{N} \sum_{j=1}^{N} \left(i_k i_j \frac{dL_{jk}}{d\theta_m} \right). \tag{15.3}$$

Transient phenomena in mechanical subsystem are determined by Newton differential equation of motion

$$J \frac{d\Omega_m}{dt} = T_{em} - T_m - k_F \Omega_m. \tag{15.4}$$

The four equations given above define general model applicable to any rotating electrical machine. The model is derived assuming four basic approximations:

1. The effects of distributed parameters are neglected.
2. The energy of electrical field is neglected along with parasitic capacitances.
3. The iron losses are neglected.
4. Magnetic saturation is neglected along with nonlinear $B(H)$ characteristic of ferromagnetic materials.

In the case when a machine has N windings, expression (15.1) contains N differential equations of voltage balance, expression (15.2) provides relation between the winding currents and their fluxes, expression (15.3) gives the electromagnetic torque, and expression (15.4) is in fact Newton differential equation describing variation of the rotor speed. Therefore, in the presented model, there are $N + 1$ differential equations and the same number of *state variables*.

15.3   Three-Phase and Two-Phase Machines

Most induction machines have a three-phase stator winding. Stator AC currents create the vector of magnetomotive force $F_S = F_a + F_b + F_c$. This vector has a radial direction within the machine, and it does not have any axial component. Therefore, it resides in the plane defined by radial and tangential unit vectors of cylindrical coordinate frame. The same plane can be represented by rectangular coordinate system of two orthogonal axes, hereafter denoted by α and β. In order to represent the vector in the α–β coordinate system, directions of axes α and β are defined by their corresponding unit vectors $\boldsymbol{\alpha_0}$ and $\boldsymbol{\beta_0}$.

While the machine has three-phase windings, spatially displaced by $2\pi/3$, relevant vector will be displayed in α–β coordinate system. Therefore, there is a need to express the orientation of the magnetic axes of individual phases in terms of unit vectors $\boldsymbol{\alpha_0}$ and $\boldsymbol{\beta_0}$:

$$\vec{a}_0 = \vec{\alpha}_0,$$

$$\vec{b}_0 = -\frac{\vec{\alpha}_0}{2} + \frac{\sqrt{3}}{2}\vec{\beta}_0,$$

$$\vec{c}_0 = -\frac{\vec{\alpha}_0}{2} - \frac{\sqrt{3}}{2}\vec{\beta}_0. \tag{15.5}$$

With symmetrical set of three-phase voltages, the stator currents can be expressed as

$$i_a = I_m \cos \omega_e t,$$

$$i_b = I_m \cos(\omega_e t - 2\pi/3),$$

$$i_c = I_m \cos(\omega_e t - 4\pi/3), \tag{15.6}$$

and they result in the following magnetomotive forces:

$$\vec{F}_a = N i_a \vec{\alpha}_0,$$

$$\vec{F}_b = N i_b \left(-\frac{1}{2}\vec{\alpha}_0 + \frac{\sqrt{3}}{2}\vec{\beta}_0 \right),$$

$$\vec{F}_c = N i_c \left(-\frac{1}{2}\vec{\alpha}_0 - \frac{\sqrt{3}}{2}\vec{\beta}_0 \right). \tag{15.7}$$

The sum of the three magnetomotive forces results in

$$\vec{F}_s = \vec{F}_a + \vec{F}_b + \vec{F}_c = N\left[\vec{\alpha}_0 \left(i_a - \frac{i_b}{2} - \frac{i_c}{2} \right) + \vec{\beta}_0 \frac{\sqrt{3}}{2}(i_b - i_c) \right],$$

$$\vec{F}_s = \frac{3}{2}N I_m \left[\vec{\alpha}_0 \cos \omega_e t + \vec{\beta}_0 \sin \omega_e t \right], \tag{15.8}$$

the vector which revolves at the speed $\Omega_e = \omega_e$ and maintains the amplitude $F_{Sm} = 3/2\ NI_m$. Modeling three-phase winding encounters certain difficulties. One of them is the fact that the phase currents are not independent variables. They are restrained by relation $i_a + i_b + i_c = 0$, which comes from the circumstance that the windings are star connected. For delta connection, this problem has different nature. Namely, the sum of the phase voltages of delta-connected winding is equal to zero. Considering star-connected winding, conclusion is drawn that only two out of three stator currents are independent variables. Therefore, in all differential equations, the current i_c has to be replaced by $(-i_a - i_b)$, making the equations clumsy and difficult to work with. In addition, angular displacement between the magnetic axis of the phase windings is $2\pi/3$, resulting in nonzero values of mutual inductances. An increased number of nonzero elements in the inductance matrix increases the number of factors in voltage balance equations, making them more involved and less intuitive. On the other hand, hypothetical two-phase machine can be envisaged with only two currents and zero mutual inductance. The mathematical model of an induction machine is more simple if considered machine has two-phase windings on the stator, one of them oriented along unit vector α_0 and the other oriented along unit vector β_0. The stator winding has only two electrical currents, i_α and i_β, and they are independent. Due to orthogonal magnetic axes of the windings, their mutual inductance is zero, simplifying a great deal the voltage balance equations.

In a two-phase machine with stator windings oriented along unit vectors $\boldsymbol{\alpha_0}$ and $\boldsymbol{\beta_0}$, the mathematical model becomes more usable because the winding currents correspond to projections of the magnetomotive force vector on the axes α and β. Namely, the magnetomotive force component along the axis α is $F_\alpha = Ni_\alpha$, which is also projection of the vector F_S on axis α. The magnetomotive force component along the axis β is $F_\beta = Ni_\beta$, equal to projection of the vector F_S on axis β. The same conclusions can be derived for the flux vector. The flux in the phase winding α is equal to the projection of the flux vector $\boldsymbol{\Phi_S}$ on the axis α. Correspondence between the phase quantities and projections of relevant vectors on axes α and β facilitates understanding and using the two-phase model.

One and the same magnetomotive force can be obtained with both three-phase and the two-phase windings. The three-phase system of phase windings of Fig. 15.1 can be replaced by the two-phase system of phase windings, given in Fig. 15.2.

On the basis of (15.8) and assuming that the number of turns is unchanged ($N_{abc} = N_{\alpha\beta}$), the stator magnetomotive force vector F_S retains the same orientation and amplitude provided that the electrical currents in the two-phase system are

$$i_{\alpha s} = i_a - \frac{i_b}{2} - \frac{i_c}{2} = \frac{3}{2}\ i_a = \frac{3}{2}\ I_m \cos \omega_e t$$

$$i_{\beta s} = \frac{\sqrt{3}}{2}(i_b - i_c) = \frac{3}{2}\ I_m \sin \omega_e t. \tag{15.9}$$

Fig. 15.1 Positions of the
phase windings in orthogonal
$\alpha\beta$ coordinate system

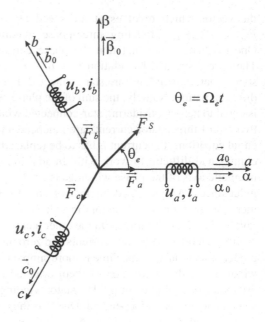

Fig. 15.2 Replacing three-
phase winding by two-phase
equivalent

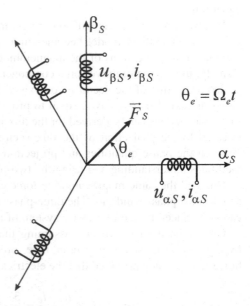

Thought experiment of removing a three-phase winding from induction
machine and replacing them with two orthogonal phase windings results in a
two-phase induction machine. Provided that the electrical currents $i_\alpha(t)$ and $i_\beta(t)$ in
two-phase windings correspond to (15.9), this modified machine would have
the same vector of the stator magnetomotive force $\boldsymbol{F_S}$ as the original

three-phase machine. Consequently, the stator flux vector F_S/R_μ would be the same as well. Moreover, the same flux and speed would result in the same electromagnetic torque. Namely, in addition to the same flux amplitude and speed Ω_e and with the same rotor speed, the rotor would have the same slip frequency, electromotive forces, and currents as the original three-phase machine. This thought experiment results in the following conclusion: *The operation of three-phase induction machine would not change if the three-phase winding is replaced by the two-phase winding, provided that the latter provides the same stator magnetomotive force.* In other words, neither flux nor torque or power of the machine is changed by replacing the three-phase winding by the two-phase equivalent, provided that the magnetomotive force remains invariant. This conclusion will be used further on.

Question (15.1): Starting from the described thought experiment, where three-phase winding system is replaced by two-phase winding system and where the number of turns N_{abc} in each phase of the former is equal to the number of turns $N_{\alpha\beta}$ in each phase of the latter, compare the phase voltages of the two. Is it possible to make a two-phase equivalent of the original machine that would have different voltages and currents? (See Fig. 15.2).

Answer (15.1): One should recall that the maximum value of the electromotive force induced in one turn is $e_1 = \omega_e \Phi_m$, while the maximum value of the winding electromotive force is $e = \omega_e \Psi_m$, while the voltage balance equation for the phase a is $u_a = R_a i_a + d\Psi_a/dt \approx d\Psi_a/dt = -\omega_e \sin(\omega_e t)\Psi_m$. Assumption is that both the original three-phase machine and the equivalent two-phase machine have the same magnetomotive force, flux, torque, and power. Therefore, the stator flux is in both cases of the same amplitude, and it revolves at the same speed. For that reason, the electromotive force induced in one turn is unchanged. Since the number of turns in each phase winding is the same, the voltages u_{abc} and $u_{\alpha\beta}$ are of the same amplitude, and they have the same rms values. Notice that the ratio u/i changes as the original machine is replaced by the equivalent. This ratio has dimension of impedance. Although the voltages are proven to be the same, electrical currents $i_{\alpha\beta}$ of the two-phase machine have their amplitude and rms value larger than currents i_{abc} by 50% (see (15.9)).

Generally, a three-phase stator winding can be replaced by a two-phase stator winding with $N_{\alpha\beta} = mN_{abc}$ turns. In such cases, phase voltages of the two-phase equivalent would be $u_{\alpha\beta} = m\, u_{abc}$. The magnetomotive force F_S would remain unaltered provided that electrical currents of the two-phase equivalent are obtained according to $i_{\alpha\beta} = (3/2) \cdot (i_{abc}/m)$. Hence, the right-hand side of (15.9) should be divided by m.

* * *

Although the two-phase equivalent of induction machine is simple, unambiguous, and intuitive, induction machines are nevertheless manufactured, deployed, and used as three-phase machines with three-phase windings on the stator. Magnetic axes of the stator phases are displaced by $2\pi/3$. There are practical advantages of the three-phase systems over the two-phase systems which resulted in the former being widely used.

Considering the number of conductors required to connect an induction machine to the grid, a three-phase stator winding gets connected to the mains by three lines (wires). The three line-to-line voltages have the same rms value, 0.4 kV for mains supplied low-voltage machines. At steady state, each of the three supply lines has the line current of the same rms values. The number of conductors in the high-voltage transmission lines is also three. Hypothetical two-phase system does not have the same advantages.

Question (15.2): A two-phase induction machine is fed from two voltage sources having the voltages of the same amplitude, phase shifted by $\pi/2$. It is necessary to connect these sources to the machine and use only three supply lines (i.e., three wires). Determine the voltages between the conductors, and compare the rms values of their line currents.

Answer (15.2): A two-phase machine can be fed from the two voltage sources by using four wires to connect each end of the two supplies (and there are two of them) to the winding terminals $\alpha 1$, $\alpha 2$, $\beta 1$, and $\beta 2$ of the two-phase system. It is possible to reduce the number of wires (lines) by using one and the same return path for the two windings. The two return lines, say $\alpha 2$ and $\beta 2$, can be merged and replaced by a single line $\alpha 2 \beta 2$. Then, the number of conductors can be only three. Yet, in this case, the currents in these three lines would not have the same amplitude. The current in the return conductor $\alpha 2 \beta 2$ is $i_\alpha(t) + i_\beta(t) = (3/2) I_m(\cos\omega_e t + \sin\omega_e t)$. It has $2^{0.5}$ times higher amplitude and rms value than the current in remaining two lines. This asymmetry exists in line voltages as well. The voltage between the line $\alpha 1$ and the return conductor $\alpha 2 \beta 2$ corresponds to the phase voltage U_α. The voltage between the line $\beta 1$ and the return conductor $\alpha 2 \beta 2$ corresponds to the phase voltage U_β and has the same amplitude as the previous one. On the other hand, the voltage between conductors $\alpha 1$ and $\beta 1$ is equal to $U_\alpha - U_\beta$, and it has $2^{0.5}$ times higher amplitude.

$$* * *$$

Uneven voltages and currents in three-wired two-phase systems are one of the reasons it never had any wider practical use. The complexity in wiring such system is considerable, since the line conductors cannot be exchanged. On the other hand, equal voltages and currents of the three-phase, three-wire system make the connection process much easier. Connection of the three-phase induction machine to the three-phase mains is much easier since all the wires have the same rms value of electrical current and the same rms value of their line-to-line voltages. The worst consequence of making a random connection is the possibility that the machine would rotate in wrong direction.[2] Nowadays, all the power lines and distribution networks operating with line frequency AC voltages are symmetrical three-phase

---

[2] When this is the case, it is sufficient to exchange any two of the three connections for the machine to change direction and revolve correctly. The two line conductors to be exchanged can be arbitrarily chosen. It is an understatement that this action must be performed in no voltage conditions.

systems. Therefore, even AC machines, such as induction machines and synchronous machines, are made with three-phase stator windings. For the purposes of modeling and analysis, three-phase machines are represented by their two-phase equivalent, so as to achieve clear and usable models, equivalent circuits, and other mathematical representations of machine.

15.4 Clarke Transform

A three-phase machine can be represented by its two-phase equivalent (Fig. 15.3). If the two-phase equivalent produces the same magnetomotive force F_S as the original machine, then the equivalent machine has the same flux, torque, and power as the original three-phase machine. Invariant magnetomotive force is obtained provided that the two-phase equivalent has the number of turns $N_{\alpha\beta}$ and currents $i_\alpha(t)$ and $i_\beta(t)$ that result in the same amplitude and spatial orientation of the vector F_s. In cases where $N_{\alpha\beta} = N_{abc}$, respective stator currents are related by (15.9).

It is not necessary to actually make the two-phase equivalent in order to use the benefits of the two-phase model. Instead, mathematical operation similar to (15.9) can be applied to all the relevant variables. This operation is, as a matter of fact, *coordinate transform* suited to provide the user with a simple, clear, and intuitive model. Relation between the original variables (i_{abc}, u_{abc}, Ψ_{abc}) and their *transformed* counterparts, the two-phase equivalents ($i_{\alpha\beta}$, $u_{\alpha\beta}$, $\Psi_{\alpha\beta}$), is called coordinate transform, and it is expressed by relations similar to (15.9). In the considered case, the three-phase/two-phase transform is applied, named *Clarke transform* after the author.

Generally speaking, the actual state of each system subject to analysis or control is described by *state variables*. The set of state variables uniquely defines the *state* of dynamical system. Putting aside the usual approximations, the set of state variables provides enough information about the system so as to determine its further behavior. A state variable cannot be expressed in terms of other state variables. As an example, only two out of the three-phase currents in a three-phase

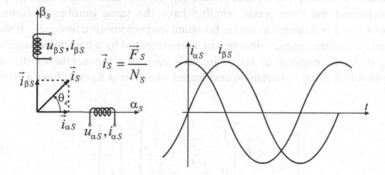

Fig. 15.3 Two-phase equivalent of a three-phase winding

winding are the actual state variables, as the third one is determined from the sum of the other two.

The benefit of coordinate transform can be demonstrated by a simple example. State of an object that moves in three-dimensional space can be described by coordinates x, y, and z in the orthogonal Cartesian coordinate system, as well as by the fist derivative of these coordinates dx/dt, dy/dt, and dz/dt, representing the speed. On the other hand, an observer may have a need to concerning distance r, elevation φ, and azimuth θ in spherical coordinate system. Coordinates x, y, and z can be expressed in terms of spherical coordinate system r, φ, and θ. The function which translates one set of coordinates into another set is called *coordinate transform*. Differential equations of motion can be written by using either the first or the second set of coordinates. The first set of equations would be called *model in x–y–z coordinate system*, and the second *model in spherical coordinate system*. Modeling the system in Cartesian or spherical coordinate system resembles looking into the windowed room through one or the other window. The room remains the same, but the image representing the room changes. Generally, selection of another coordinate system reflects only the observer viewpoint and does not have any impact on the object or system to be modeled. Selection of the appropriate coordinate system and corresponding transform of the state variables may have significant impact on mathematical model. Such model becomes simple, clear, and more intuitive, facilitating decision making regarding control and exploitation of the system. The model in Cartesian coordinate frame is more suitable when modeling an object that moves along the x axis, due to $dy/dt = 0$ and $dz/dt = 0$. An attempt to represent the same motion in spherical coordinate frame results in rather involved changes in coordinates r, φ, and θ. On the other hand, spherical coordinate frame is more suited to describe rotation around the origin or radial motion. Similarly, a three-phase machine can be represented in the original, three-phase domain but also by its equivalent two-phase machine. The latter proves more suitable to study machine properties and characteristics and to specify and design supplies and controls.

Electrical currents of the equivalent two-phase system which represents the three-phase winding are given by (15.10), which is the matrix form of (15.9). The matrix is multiplied by coefficient K_I. In the case when the two-phase equivalent and the three-phase winding have the same number of turns, the value of $K_I = 1$ is required to secure invariant magnetomotive forces F_S. It should be noted that a three-phase winding can be represented by a two-phase equivalent having different number of turns. In such case, for the vector of the stator magnetomotive force F_s to remain unchanged, coefficient K_I must have a different value:

$$\begin{bmatrix} i_{\alpha s} \\ i_{\beta s} \end{bmatrix} = K_I \begin{bmatrix} 1 & -\dfrac{1}{2} & -\dfrac{1}{2} \\ 0 & \dfrac{\sqrt{3}}{2} & -\dfrac{\sqrt{3}}{2} \end{bmatrix} \cdot \begin{bmatrix} i_a \\ i_b \\ i_c \end{bmatrix}. \tag{15.10}$$

The question arises whether the three variables such as i_a, i_b, and i_c can be replaced by only two, $i_{\alpha s}$ and $i_{\beta s}$. Due to $i_a + i_b + i_c = 0$, only the two-phase currents of the original machine are independent state variables, which provides the rational for the transform expressed by (15.10).

15.5   Two-Phase Equivalent

Transform of the state variables of an existing three-phase machine can be understood as a thought experiment which represents a three-phase machine by an imaginary two-phase machine. There is also possibility to actually replace an existing three-phase stator winding by a two-phase stator winding that provides the same magnetomotive force, flux, torque, and power as the original three-phase machine. It is of interest to compare the two induction machines that have the same behavior. One of them is the original three-phase machine, and the other is the two-phase equivalent. It is assumed that two-phase induction machine, called M2, has the same magnetic circuits and the same rotor as the original three-phase machine, M3. If the individual phases of M2 and M3 have the same number of turns, electrical currents in respective stator windings must correspond to the following relation:

$$i_{\alpha s} = i_a - \frac{i_b}{2} - \frac{i_c}{2} = \frac{3}{2}\, i_a$$

$$i_{\beta s} = \frac{\sqrt{3}}{2}(i_b - i_c). \qquad (15.11)$$

so as to provide the same magnetomotive force. With the same magnetomotive force and identical magnetic circuits, both machines have the same flux. Electromotive force in one turn is proportional to the flux and the angular frequency ω_e. Therefore, each turn in machines M2 and M3 has electromotive force of the same amplitude. With $N_{abc} = N_{\alpha\beta}$, electromotive forces induced in phases a, b, c, α, and β have the same peak and rms values. With the assumption that the voltage drop Ri is negligible with respect to the electromotive force, conclusion is drawn that, in the considered case, the phase voltages u_{abc} and $u_{\alpha\beta}$ have the same peak and rms values. Specifically, the voltage across the phase a of machine M3 has the same peak value as the voltage across the phase α of machine M2. On the other hand, considering (15.9), the peak and rms values of the phase currents $i_{\alpha\beta}$ are 3/2 times larger with respect to i_{abc} currents. The above-mentioned considerations show that a three-phase machine can be converted into a two-phase machine by rewinding the stator, yet preserving the same magnetomotive force, flux, torque, and power. Maintaining the same number of turns, the phase voltages remain the same, while the phase currents increase by factor 3/2.

Common practice in applying coordinate transforms to electrical machines includes applying one and the same transformation formula to all the relevant variables, whether voltages, current, or flux linkages. Benefits of this approach will be discussed within subsequent chapters. For the example involving machines M2 and M3, the transformation matrix for electrical currents is given in (15.11). Applying the same formula to voltages, one obtains (15.12) which gives the phase voltage $u_{\alpha s}$ obtained by using the same three-phase to two-phase transform as the one used for currents:

$$u_{\alpha s} = u_a - \frac{u_b}{2} - \frac{u_c}{2} = \frac{3}{2}\,u_a. \qquad (15.12)$$

Apparent problem arises from the fact that the actual phase voltage $u_{\alpha s}$ of the rewound machine M2 does not correspond to the value obtained in (15.12). It seems that there is no way to actually make the equivalent two-phase machine unless the transformation matrices used for voltages are different than those used for currents. Further on, the answer to Question (15.3) proves the opposite. It shows that, with the proper choice of K_I, it is possible to devise a three-phase to two-phase transform that corresponds to two-phase induction machine that can actually be made.

Coordinate transforms do not have to correspond to actual physical systems in order to prove their usefulness in modeling. An example is Park transform, discussed and used in subsequent chapters, which proves very useful in deriving dynamic model and steady-state equivalent circuit and yet results in state variables that correspond to virtual electrical machine that cannot be made.

Considered example includes the three-phase machine M3 which is transformed into two-phase equivalent M2. Machines M2 and M3 have identical magnetic circuit and the same number of turns per phase. The problem that arises with the machine M2 is that it has voltages that do not correspond to those obtained by applying the transform matrix (15.14) on the original voltages u_{abc}. Notice that the currents are transformed according to (15.10), adopting $K_I = 1$. Generally, the voltages can be transformed by using

$$\begin{bmatrix} u_{\alpha s} \\ u_{\beta s} \end{bmatrix} = K_U \begin{bmatrix} 1 & -\dfrac{1}{2} & -\dfrac{1}{2} \\ 0 & \dfrac{\sqrt{3}}{2} & -\dfrac{\sqrt{3}}{2} \end{bmatrix} \cdot \begin{bmatrix} u_a \\ u_b \\ u_c \end{bmatrix}.$$

Therefore, with $K_U = K_I = 1$, the phase voltages and currents obtained by applying the transform on machine M3 do not correspond to the voltages and currents actually measured on machine M2. On the other hand, coefficients K_I and K_U do not have to be equal to 1. Other values can be applied as well. The only practical restriction is $K_I = K_U$, which maintains the ratio between the voltages and currents and secures that all the impedances of the original machine retain their value after the transformation. Not even this restriction is obligatory, yet it is often imposed due to practical reasons. Assuming that the phase current is transformed according to (15.10), with $K_I = 1$, while the voltages are transformed by using $K_U = 2/3$, the voltages and currents derived from the transform will correspond to those measured on two-phase machine M2 which has the same number of turns per phase as three-phase original M3. Drawback of this approach is that the ratio of voltage and current of machines M2 and M3 will not be the same. Thus, proposed transform will affect the impedances. They will not be invariant. Parameter R_S of the three-phase machine would have to be multiplied by 2/3 in order to get the parameter R_S of the two-phase machine. Generally, impedances of the original three-phase machine should be multiplied by K_U/K_I in order to obtain impedances of the two-phase equivalent.

Up to now, discussion was focused on devising a 3-phase to 2-phase transform that corresponds to physical prototypes M2 and M3. In general, transformed quantities can but do not have to correspond to a practical two-phase machine. It is acceptable to adopt $K_I = 1$ and $K_U = 1$ and obtain correct mathematical model. This transform does provide the voltages that can be measured on M2, but it has the advantage of being impedance invariant. On the other hand, transform with $K_I = 1$ and $K_U = 1$ is not invariant in terms of power, namely, $P_{abc} \neq P_{\alpha\beta}$. Nonetheless, such model can be advantageously used. The lack of power invariance has to be kept in mind and taken care of.

Three-phase to two-phase transform with $K_I = K_U = 2/3$ is frequently encountered. It is impedance invariant, but it brings in the relation $P_{abc} = 3/2\ P_{\alpha\beta}$. For better understanding, before listing the properties of Clarke transform, the values of frequently used coefficients K_I and K_U will be described in brief.

15.6 Invariance

If Clarke transform preserves the ratio between voltages and currents of the three-phase original and the two-phase equivalent, it is invariant in terms of impedance. If the ratio between fluxes and currents remains the same, the transform is invariant in terms of inductance. If the expression for power $P_{\alpha\beta}$ of the two-phase equivalent corresponds to the power of the original three-phase machine, then the transform is invariant in terms of power.

It is necessary to point out that transforms which are not power invariant can be advantageously used, provided that the user of mathematical model respects the ratio $P_{abc} = K \cdot P_{\alpha\beta}$.

First-time user of coordinate transforms may nurture doubts whether the mathematical model is correct, considering that it calculates apparently incorrect power due to $P_{abc} \neq P_{\alpha\beta}$. To resolve such doubts, it is important to recall that the state variables obtained by using coordinate transform do not have to correspond to any machine that could be actually made. However, this does not minimize practical values of the mathematical model. As an example, one can start with mathematical model of a simple resistor, $u = Ri$. By performing *coordinate transform* $u_1 = 2u$

and $i_1 = 2i$, one obtains the model having electrical power four times higher than the actual resistor. However, this model is still useful. Given the current, one can calculate the voltage according to $u_1 = Ri_1$. The user should recall the power invariance and calculate the actual power as $u_1 i_1/4$. Representation of u_1, i_1 of a resistor does not represent real resistor, but it stands as a usable model.

Most practical uses of Clarke transform retain the impedance invariance and the inductance invariance, while the lack of power invariance is often acceptable. This means that transforms of the currents, voltages, and fluxes are carried out by using the same transform matrix for all the variables. The transformation matrix is given in (15.13), while the coefficients for voltage and flux transform are given by $K_I = K_U = K_\Psi$:

$$\begin{bmatrix} i_\alpha \\ i_\beta \end{bmatrix} = K_I \begin{bmatrix} 1 & -\dfrac{1}{2} & -\dfrac{1}{2} \\ 0 & \dfrac{\sqrt{3}}{2} & -\dfrac{\sqrt{3}}{2} \end{bmatrix} \begin{bmatrix} i_a \\ i_b \\ i_c \end{bmatrix}. \tag{15.13}$$

It is of interest to use (15.13) with $K_I = 1$ and derive the phase currents of the two-phase equivalent having the same number of turns ($N_{\alpha\beta} = N_{abc}$) as the three-phase original. The three-phase winding with symmetrical set of phase currents $i_a(t) = I_m\cos\omega_e t$, $i_b(t) = I_m\cos(\omega_e t - 2\pi/3)$, and $i_c(t) = I_m \cos(\omega_e t - 4\pi/3)$ can be transformed by using $K_I = 1$ and (15.13). The two-phase equivalent is obtained with $i_\alpha(t) = i_a(t) - i_b(t)/2 - i_c(t)/2 = 3/2 \ I_m\cos\omega_e t$ and $i_\beta(t) = 3^{0.5}/2 \cdot (i_b(t) - i_c(t)) = 3/2 \ I_m\sin\omega_e t$. Hence, the two-phase equivalent has phase currents shifted by $\pi/2$, which corresponds to the spatial shift between magnetic axes of corresponding windings.

In addition to the phase currents, the voltages and fluxes should also be transformed in $\alpha\beta$ coordinate frame. Clarke transform for the voltages and fluxes is given by

$$\begin{bmatrix} u_\alpha \\ u_\beta \end{bmatrix} = K_U \begin{bmatrix} 1 & -\dfrac{1}{2} & -\dfrac{1}{2} \\ 0 & \dfrac{\sqrt{3}}{2} & -\dfrac{\sqrt{3}}{2} \end{bmatrix} \begin{bmatrix} u_a \\ u_b \\ u_c \end{bmatrix}, \tag{15.14}$$

$$\begin{bmatrix} \Psi_\alpha \\ \Psi_\beta \end{bmatrix} = K_\Psi \begin{bmatrix} 1 & -\dfrac{1}{2} & -\dfrac{1}{2} \\ 0 & \dfrac{\sqrt{3}}{2} & -\dfrac{\sqrt{3}}{2} \end{bmatrix} \begin{bmatrix} \Psi_a \\ \Psi_b \\ \Psi_c \end{bmatrix}. \tag{15.15}$$

In general, coefficients K_U and K_Ψ in the above expressions can be arbitrarily selected and do not have to be equal to K_I. The choice $K_I = K_U = K_\Psi$ has the advantages that contribute to legibility and usability of mathematical model that results from transforms.

Selecting $K_I = K_U$, one obtains impedance invariant Clarke transform. Namely, all the resistances, impedances, and other accounts where the ratio u/i appears remain unaltered by the transform. Hence, parameters such as R_S retain their value even in $\alpha\beta$ coordinate frame. Deciding otherwise would create the need to scale all the impedances by the ratio K_U/K_I.

Selection $K_I = K_\Psi$ results in inductance invariant Clarke transform. The self-inductances, mutual inductances, and leakage inductances of all the windings remain unaltered by the transform. Hence, deciding otherwise would create the need to scale all the inductances by the ratio K_Ψ/K_I.

Selection $K_U = K_\Psi$, relation between electromotive forces and fluxes, remains $e = \mathrm{d}\Psi/\mathrm{d}t$. Deciding otherwise would require the two-phase model to include relations such as $e = (K_U/K_\Psi)\,\mathrm{d}\Psi/\mathrm{d}t$.

Throughout this book, it is assumed that $K_I = K_U = K_\Psi$. Other choices are rarely met in reference literature. They result in mathematical models that are correct but more difficult to use than the models obtained by invariant transform.

The freedom of choice is often a problem. An example to that is the attempt to replace the original three-phase machine M3 by actual two-phase prototype M2, already discussed before. Considered is the case where the phase windings of both M3 and M2 have the same number of turns per phase, $N_{\alpha\beta} = N_{abc}$. Clarke transform of the three voltages (u_a, u_b, u_c) calculates the values (u_α, u_β) of the two-phase equivalent. In order to obtain the values of (u_α, u_β) that correspond to voltages actually measured on the machine M2, the coefficient K_U has to be equal to 2/3. This conclusion is already explained, and it relies on the fact that both machines have the same amplitude of electromotive forces induced in single turn. Due to $N_{\alpha\beta} = N_{abc}$, the phase windings of machines M2 and M3 have the same amplitudes of winding electromotive forces. With $e \approx u$, the same holds for the phase voltages as well. Hence, for the Clarke transform to provide the same voltages (u_α, u_β) and currents (i_α, i_β) as the actual prototype M2, with $N_{\alpha\beta} = N_{abc}$, it is necessary to use $K_U = 2/3$ and $K_I = 1$. Correspondence between the obtained two-phase equivalent and the actual machine M2 increases the user confidence. However, there are problems created by the choice $K_I \neq K_U$. The use of different transforms for voltages and currents leads to different ratios u/i in abc and $\alpha\beta$ coordinate systems. In other words, transform is not impedance invariant. Parameters such as resistance R or reactance X have different values in abc and $\alpha\beta$ frames. Any transition from abc to $\alpha\beta$ frame requires impedances to be scaled by 3/2. This does not mean that the model is inaccurate, but it compromises clarity and augments the chances of making errors.

Previous discussion demonstrates that the choice $K_U = 2/3$ and $K_I = 1$ results in Clarke transform that provides two-phase voltage and currents in full correspondence with the actual two-phase machine M2. Yet, such transform is not impedance invariant, and it brings difficulties in using the model. For that reason, decision $K_I = K_U = K_\Psi$ is used throughout this book, although it does not correspond to voltages and currents of the machine M2.

As a rule, while selecting transform of the state variables, it is considered that resistances (R) and inductances (L) should stay invariant. Therefore, invariability of impedances and inductances is set as a prerequisite. In other words, legibility and usability of the model are considered more important than similarity to the actual two-phase prototype M2.

In the analysis and modeling of electrical machines, all the transforms of the state variables are made so as to maintain the ratio of voltages, currents, and fluxes. In this way, transforms are impedance invariant and inductance invariant.

• The same voltage and current transform matrices result impedance invariability.
• The same flux and current transform matrices result in invariable self-inductances, mutual inductances, and leakage inductance.
• The same voltage and flux transform matrices maintain relation $e = \mathrm{d}\Psi/\mathrm{d}t$. With $K_U \neq K_\Psi$, this relation becomes relation $e = (K_U/K_\Psi)\,\mathrm{d}\Psi/\mathrm{d}t$.

Question (15.3): Is it possible to make an actual two-phase machine with $N_{\alpha\beta} = mN_{abc}$ turns which has the same stator magnetomotive force F_s as the original three-phase machine and which has, at the same time, the voltages, currents, and fluxes which correspond to values obtained by Clarke transform performed with $K_U = K_I = K_\Psi$?

Answer (15.3): The actual two-phase equivalent of the original three-phase machine must have the same magnetomotive force, flux, torque, and power as the original machine. A three-phase stator winding can be replaced by a two-phase winding having $N_{\alpha\beta} = mN_{abc}$ turns. Invariability of F_s requires that windings of the two-phase equivalent carry currents $i_{\alpha\beta} = (3/2)\cdot(i_{abc}/m)$. At the same time, the two-phase equivalent has the same electromotive force in a single turn and m times more turns per phase. Therefore, the phase voltages of the two-phase equivalent will be $u_{\alpha\beta} = mu_{abc}$. Finally, with $K_U = K_I$, the ratio of voltages and currents of the original machine (u_{abc}/i_{abc}) is equal to the ratio of voltages and currents of the two-phase equivalent ($u_{\alpha\beta}/i_{\alpha\beta}$). Summarizing the above statements,

$$\frac{u_{\alpha\beta}}{i_{\alpha\beta}} = \frac{m\,u_{abc}}{(3/2)(i_{abc}/m)} = \frac{2m^2}{3}\frac{u_{abc}}{i_{abc}} = \frac{u_{abc}}{i_{abc}}$$

$$\Rightarrow\quad m = \sqrt{\frac{3}{2}}\quad \Rightarrow\quad K_U = K_I = K_\Psi = \sqrt{\frac{2}{3}}.$$

Hence, when three-phase machine is replaced by two-phase equivalent which has $(3/2)^{0.5}$ times more turns per phase, the voltages and currents of the actual two-phase machine correspond to these obtained from the Clarke transform performed with leading coefficient of $(2/3)^{0.5}$. This Clarke transform calculates the voltages, currents, and fluxes in $\alpha\beta$ domain by applying the same transformation matrices to the original voltages, currents, and fluxes in the abc domain. Hence, $K_U = K_I = K_\Psi = (2/3)^{0.5}$. This transform is invariant in terms of impedance, inductance, magnetomotive force, torque, and power. The amplitude of fluxes in windings α and β, the amplitudes and rms values of corresponding voltages, and the amplitudes and rms values of currents are $(3/2)^{0.5}$ times higher compared to the original variables in abc domain. The presence of irrational number $(3/2)^{0.5}$ in calculations is the reason to avoid the Clarke transform with coefficient $K = (2/3)^{0.5}$, notwithstanding its positive sides.

Question (15.4): Prove that Clarke transform with $K_U = K_I = K_\Psi = (2/3)^{0.5}$ is invariant in terms of power.

Answer (15.4): Without lack of generality, it is possible to assume that the machine operates in steady state. Electrical power in each phase winding can be calculated as product of rms values of winding current, voltage, and power factor. For purposes of proving the power invariance, the latter can be considered constant or even equal to 1. Total power of the machine is obtained as the sum of individual phase powers. Consider a three-phase machine with rms values of the phase voltages and currents U_{abc} and I_{abc}; electrical power of the three-phase machine is found to be $3\, U_{abc}I_{abc}$. By applying Clarke transform with coefficients $K_U = K_I = K_\Psi = (2/3)^{0.5}$, one obtains currents and voltages in $\alpha\beta$ domain with amplitudes $(3/2)^{0.5}$ times higher. Therefore, the phase α power is equal to $[(3/2)^{0.5}U_{abc}]\cdot[(3/2)^{0.5}I_{abc}] = 3/2\, U_{abc}I_{abc}$. The same power is obtained in phase β, resulting in total power of $3U_{abc}I_{abc}$, which confirms invariability in terms of power.

Question (15.5): Prove that the application of Clarke transform with $K_U = K_I = K_\Psi = 1$ is not invariant in terms of power but results in $P_{\alpha\beta} = (3/2)P_{abc}$.

Answer (15.5): Assume that voltages and currents of the three-phase machine are known and equal to u_a, u_b, u_c, i_a, i_b, and i_c. By using expression (15.13) for the three-phase/two-phase transform, it is required to determine variables u_α, u_β, i_α, and i_β. By replacing the corresponding variables from the original abc domain in equation $P_{\alpha\beta} = u_\alpha i_\alpha + u_\beta i_\beta$, one obtains that $P_{\alpha\beta} = (3/2)P_{abc}$.

15.6.1 Clarke Transform with K = 1

In the case when $K_U = K_I = K_\Psi = 1$, the two-phase equivalent machine should have the same number of turns in order to provide the same magnetomotive force. Transformed variables include 3/2 times larger amplitudes of voltages and currents. Transform is invariant in terms of impedance and inductance, but it is not power invariant. Hence,

$$N_{\alpha\beta} = N_{abc}, \tag{15.16}$$

$$\left|\vec{i}_{\alpha\beta}\right| = \frac{3}{2} \cdot i_{abc}^{max}, \tag{15.17}$$

$$\left|\vec{u}_{\alpha\beta}\right| = \frac{3}{2} \cdot u_{abc}^{max}. \tag{15.18}$$

The transform is:

- Invariant in terms of impedance
- Invariant in terms of inductance
- Not invariant in terms of power since

$$P_{\alpha\beta} = \frac{3}{2} P_{abc} \tag{15.19}$$

15.6.2 Clarke Transform with K = sqrt(2/3)

In the case when $K_U = K_I = K_\Psi = (2/3)^{0.5}$, the two-phase equivalent machine should have $(3/2)^{0.5}$ times increased number of turns so as to provide the same magnetomotive force. Transformed variables include $(3/2)^{0.5}$ times larger amplitudes of phase voltages and currents. Transform is invariant in terms of impedance, inductance, and power. Hence,

$$N_{\alpha\beta} = \sqrt{\frac{3}{2}} N_{abc} \tag{15.20}$$

$$|\vec{i}_{\alpha\beta}| = \sqrt{\frac{3}{2}} \cdot i_{abc}^{max} \tag{15.21}$$

$$|\vec{u}_{\alpha\beta}| = \sqrt{\frac{3}{2}} \cdot u_{abc}^{max} \tag{15.22}$$

The transform is:

- Invariant in terms of impedance
- Invariant in terms of inductance
- Invariant in terms of power

15.6.3 Clarke Transform with K = 2/3

In the case when $K_U = K_I = K_\Psi = 2/3$, the two-phase equivalent machine should have 3/2 times increased number of turns so as to provide the same magnetomotive force. Transformed variables include the same amplitudes of phase voltages and currents. Transform is invariant in terms of impedance and inductance, but it is not power invariant. Hence,

$$N_{\alpha\beta} = \frac{3}{2}N_{abc} \tag{15.23}$$

$$\left|\vec{i}_{\alpha\beta}\right| = i_{abc}^{max} \tag{15.24}$$

$$\left|\vec{u}_{\alpha\beta}\right| = u_{abc}^{max} \tag{15.25}$$

The transform is:

- Invariant in terms of impedance
- Invariant in terms of inductance
- Not invariant in terms of power since

$$P_{\alpha\beta} = \frac{2}{3}P_{abc} \tag{15.26}$$

15.7 Equivalent Two-Phase Winding

Preceding sections summarize the needs for representing a three-phase machine by its two-phase equivalent. Clarke $3\Phi/2\Phi$ transform is introduced and explained. The choice of transform coefficients is discussed, along with consequences in terms of impedance, inductance, and power invariance of the transform. In further analysis, the following $3\Phi/2\Phi$ transform is adopted and used.

Clarke $3\Phi/2\Phi$ transform of voltages, currents, and fluxes is performed in a unified way, by using the same transform matrix for all the variables having the leading coefficient of $K = 2/3$. The symbol V in the expression (15.27) represents voltage, current, or flux in one phase winding:

$$\begin{bmatrix} V_\alpha \\ V_\beta \end{bmatrix} = \frac{2}{3} \begin{bmatrix} 1 & -\dfrac{1}{2} & -\dfrac{1}{2} \\ 0 & \dfrac{\sqrt{3}}{2} & -\dfrac{\sqrt{3}}{2} \end{bmatrix} \begin{bmatrix} V_a \\ V_b \\ V_c \end{bmatrix}. \tag{15.27}$$

As a consequence, the applied transform is invariant in terms of impedances and inductances. Therefore, parameters such as R_S, R_R, L_m, L_S, and all other inductances and resistances retain their original values.

The peak and rms values of variables in $\alpha\beta$ frame are equal to the peak and rms values of original abc variables. Identities $u_\alpha(t) \equiv u_a(t)$, $i_\alpha(t) \equiv i_a(t)$, and $\Psi_\alpha(t) \equiv \Psi_a(t)$ apply too.

Selected transform with $K = 2/3$ provides $\alpha\beta$ variables that cannot be reproduced by any actual two-phase prototype. Namely, it is not possible to make an actual two-phase stator winding which replaces the three-phase winding, provides the magnetomotive force, and has the stator voltages and currents which correspond to the values obtained by the transform. Invariability of F_s requires $N_{\alpha\beta} = 3/2 \, N_{abc}$, which leads to $u_{\alpha\beta} = 3/2 \, u_{abc}$, owing to equal electromotive forces induced in one turn. On the other hand, selected transform results in $u_{\alpha\beta} = u_{abc}$.

Power of the two-phase equivalent $P_{\alpha\beta} = u_\alpha i_\alpha + u_\beta i_\beta$ is equal to $2/3 P_{abc}$. Namely, since phase quantities in $\alpha\beta$ domain have the same values as the original counterparts in abc domain, the power per phase is equal, resulting in $P_{\alpha\beta} = 2/3 \, P_{abc}$. While using the model, it should be recalled that numerical value $P_{\alpha\beta} = u_\alpha i_\alpha + u_\beta i_\beta$ should be multiplied by $3/2$ in order to get the power of the original three-phase machine. Therefore,

$$P_{abc} = \frac{3}{2} P_{\alpha\beta}. \tag{15.28}$$

The question arises whether the model in $\alpha\beta$ frame with $P_{\alpha\beta} = 2/3 P_{abc}$ stands as an adequate representation of the induction machine having $3/2$ larger power. It has to be recalled that the coordinate transforms result in a mathematical model with variables that do not necessarily correspond to any actual machine. Any attempt to envisage a practical two-phase equivalent and to interpret the variables u_α, i_α, u_β, and i_β as voltages and currents of practical α and β windings may be helpful in understanding and using the model. Yet, the virtual machine in $\alpha\beta$ frame is actually a mathematical fiction, and therefore, relations such as $P_{\alpha\beta} = 2/3 P_{abc}$ do not invalidate the model. Recall that any resistor with voltage u and current i can be represented by mathematical model $u_1 = R i_1$, where the new voltage and current are obtained by coordinate transform $u_1 = 2u$ and $i_1 = 2i$. The model has electrical power 4 times larger than the actual resistor and been used to represent the basic properties of the resistor. Due to lack of the power invariance of the transform $V_1 = 2 \, V$, the user should recall to calculate the actual power as $u_1 i_1 / 4$.

15.8 Model of Stator Windings

By applying Clarke transform, a three-phase machine can be represented by a two-phase equivalent. Axes of virtual phase windings α_S and β_S are still with respect to the stator. The axis α_S is collinear with the magnetic axis of the phase winding a of the original machine. The model where the currents, voltages, and fluxes of the stator are represented by their α_S and β_S components is called *model in stationary coordinate frame*. Given the voltage, current, and flux vectors of the stator winding, their α_S and β_S components can be found as projections of relevant vectors on the axes of α_S-β_S coordinate frame.

Figure 15.4 shows an induction machine represented by a two-phase stator winding and a two-phase rotor winding. Angle θ_m represents the rotor position,

Fig. 15.4 Two-phase
equivalent

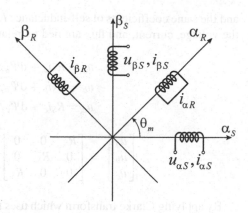

$$\theta_m = \theta_{m0} + \int\limits_0^t \Omega_m d\tau. \qquad (15.29)$$

By reducing the three-phase stator to the two-phase equivalence, the model obtains two virtual stator windings, α_S and β_S, which create the magnetomotive force and flux along axes α_S and β_S of the still coordinate system called *stationary* or *stator* coordinate system. Currents and voltages of these windings are denoted by $u_{\alpha S}$, $u_{\beta S}$, $i_{\alpha S}$, and $i_{\beta S}$, in order to distinguish them from the rotor variables which are introduced later and which make use of subscript R.

Question (15.6): Direction and amplitude of the vector of magnetomotive force of the stator F_s are known. Determine currents $i_{\alpha S}$ and $i_{\beta S}$.

Answer (15.6): The required currents are determined by projections of vector F_s on axes α_S and β_S of the still coordinate system.

15.9 Voltage Balance Equations

Voltage equilibrium in three-phase winding is given by expressions comprising phase voltages u_a, u_b, and u_c; phase currents i_a, i_b, and i_c; and total flux linkages of the phase windings Ψ_a, Ψ_b, and Ψ_c. The flux Ψ_a in the phase winding a has component $L_S i_a$, produced by the current in the same winding. Coefficient L_S is the self-inductance of the stator winding. Other windings on stator and rotor may contribute to the flux Ψ_a. Their contribution is proportional to electrical currents in those windings and also to the coefficient of mutual inductance. For winding denoted by x, the flux contribution is $L_{ax} i_x$, where L_{ax} is mutual inductance between the phase winding a and the winding x, while i_x is the corresponding current. Phase windings have the same number of turns, the same resistance $R_a = R_b = R_c = R_S$, and the same coefficients of self-inductance $L_a = L_b = L_c = L_S$. For any windings, the voltage, current, and flux are tied by relation $u = Ri + \mathrm{d}\Psi/\mathrm{d}t$. Therefore,

$$u_a = R_S i_a + \mathrm{d}\Psi_a/\mathrm{d}t,$$
$$u_b = R_S i_b + \mathrm{d}\Psi_b/\mathrm{d}t, \quad \Rightarrow$$
$$u_c = R_S i_c + \mathrm{d}\Psi_c/\mathrm{d}t.$$

$$\begin{bmatrix} u_a \\ u_b \\ u_c \end{bmatrix} = \begin{bmatrix} R_S & 0 & 0 \\ 0 & R_S & 0 \\ 0 & 0 & R_S \end{bmatrix} \begin{bmatrix} i_a \\ i_c \\ i_c \end{bmatrix} + \frac{\mathrm{d}}{\mathrm{d}t} \begin{bmatrix} \Psi_a \\ \Psi_b \\ \Psi_c \end{bmatrix}. \tag{15.30}$$

By applying Clarke transform which uses the same transformation matrix for the voltages, currents, and fluxes, the voltage balance equations can be transferred to α_S–β_S coordinate frame and expressed in terms of α_S and β_S projections of the voltage, current, and flux vectors. Quantities $\Psi_{\alpha S}$ and $\Psi_{\beta S}$ are projections of the stator flux vector on axes α_S and β_S. They can be calculated by applying the three-phase/two-phase transform to the total fluxes of the phase windings Ψ_a, Ψ_b, and Ψ_c. Moreover, the transformation matrix can be applied to the whole right side of (15.30), obtaining in this way:

$$u_{\alpha S} = R_S i_{\alpha S} + \mathrm{d}\Psi_{\alpha S}/\mathrm{d}t,$$
$$u_{\beta S} = R_S i_{\beta S} + \mathrm{d}\Psi_{\beta S}/\mathrm{d}t. \tag{15.31}$$

The above equation represents the voltage equilibrium in the two-phase equivalent of the stator winding. In addition to modeling the stator, it is required to model the short-circuited rotor cage. Voltage equilibrium equations in the rotor circuit will complete the model of the electrical subsystem of the induction machine.

15.10 Modeling Rotor Cage

The rotor cage contains a relatively large number of conductors which are short circuited by the front and rear rings. An example of the rotor cage separated from the rotor magnetic circuit is shown in Fig. 15.5a. For rotor cage with $N_R = 28$ conductors, it is possible to identify 14 short-circuited turns, each created by one pair of diametrically positioned conductors. Therefore, it is possible to make a model of the rotor comprising 14 short-circuited turns with mutual magnetic coupling, also coupled with α_S and β_S stator windings, as shown in Fig. 15.5b. However, such model would be of little practical value. Its inductance matrix will have dimensions 16×16. Therefore, another approach is needed to model the rotor cage.

As the first step, it is of interest to observe the part (c) in Fig. 15.5 and assume that the rotor flux pulsates along the vertical axis. At this point, it is of interest to

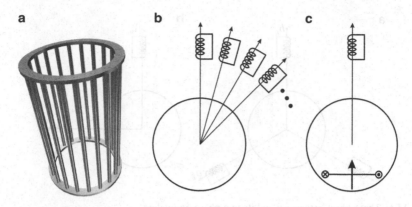

Fig. 15.5 Modeling the rotor cage

derive a model of the rotor cage that would reflect the induction of the rotor electromotive forces and currents in this particular case. The rotor conductors can be considered as a set of turns where the conductors making one turn are positioned symmetrical with respect to the vertical axis, as shown in Fig. 15.5c. With the assumption that pairs of rotor conductors which constitute one turn reside on the same horizontal line, and also assuming that there are no other connections between the turns, electrical currents in such turns would create the rotor flux along vertical axis. Under assumptions, they cannot make any flux in horizontal direction. Therefore, such rotor windings can be denoted by the symbol of coil placed on the vertical axis, as indicated in Fig. 15.5. This symbol represents the rotor turns that can be envisaged as one short-circuited rotor phase with vertical magnetic axis. For the time being, this approach overlooks the circumstance that the front and rear rings make a short circuit for all conductors. Assuming that an external stator flux pulsates along vertical axis, the flux within individual rotor turns would change, resulting in electromotive forces and consequential currents in short-circuited rotor turns. This thought experiment proved that the rotor winding connected according to the part (c) in Fig. 15.5 provides the short-circuiting effect of the rotor cage for vertical pulsations of the external flux. The rotor currents induced due to changes in the external flux act toward suppressing the flux changes. Namely, they contribute to the rotor flux in the direction opposite to the original flux change.

However, the same setup cannot represent the short-circuiting effect of the rotor cage in cases where the changes of the external flux have their horizontal component. For flux pulsations along horizontal direction, there are no induced electromotive forces and no rotor currents since the only phase winding of the rotor has vertical magnetic axis. Hence, it does not react to changes in horizontal flux component. Recall that the setup in Fig. 15.5c reacts only to changes in vertical component of the flux. Therefore, the rotor model comprising only one short-circuited phase winding cannot serve as an accurate representation of phenomena occurring in short-circuited rotor cage.

An actual rotor cage which is short circuited by the front and rare conductive rings exhibits its short-circuiting effects in arbitrary direction. As a matter of fact,

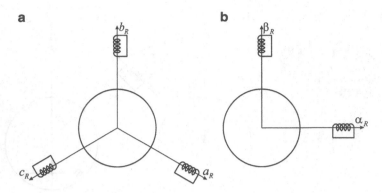

Fig. 15.6 Three-phase rotor cage and its two-phase equivalent

the end rings provide the short circuit between all the rotor bars and make a short-circuited turnout of any pair of rotor bars. Thus, variation of the external flux in any arbitrary orientation induces electromotive forces and currents in rotor turns that have their magnetic axis aligned with the vector of the flux change. The rotor cage is symmetrical, and it has a number of conductors. For this reason, the cage can be modeled as a three-phase winding, with individual rotor phases being short circuited and shifted by $2\pi/3$, as shown in Fig. 15.6a. It is also possible to model the rotor as a two-phase, short-circuited winding, as shown in Fig. 15.6b. Validity of the two-phase model of the rotor cage can be verified by considering an arbitrary variation of the flux and analyzing the rotor reaction. Variation of the external flux with an arbitrary orientation can be represented by two orthogonal flux components lying along horizontal and vertical axis, which correspond to magnetic axis of the representative two-phase winding. Since parameters like resistance and inductance of the phase windings are identical, the short-circuiting effect of the rotor is the same for both flux components. In both horizontal and vertical axis, the rotor reacts to the flux changes by induced electromotive forces and consequential rotor currents. Hence, the two-phase representation of the short-circuited cage provides the model of the rotor reaction which does not depend on spatial orientation of the external flux changes. Therefore, two-phase representation of short-circuited rotor cage is an adequate model of the rotor winding, whatever the number of rotor conductors, provided that all the conductors are the same and that they are equally spaced around the rotor circumference.[3]

In the course of rotor motion, the rotor changes its position θ_m with respect to the stator. Therefore, magnetic axes of the two-phase rotor winding change their relative positions with respect to magnetic axes of the stator winding. In

---

[3] The rotor winding cannot be represented by two-phase equivalent in cases when the rotor cage is damaged. If one or more conductors are broken or disconnected from the short-circuiting rings, the rotor reaction to flux changes will be different in some directions. These cases are out of the scope of this book.

Fig. 15.4, the rotor axes are denoted by α_R and β_R. The voltages in short-circuited phases of the rotor are equal to zero; hence, $u_{\alpha R} = u_{\beta R} = 0$. The rotor currents $i_{\alpha R}$ and $i_{\beta R}$ represent electrical currents induced in the rotor cage, and they create the rotor magnetomotive force $\boldsymbol{F_R}$ whose amplitude and direction depend on currents $i_{\alpha R}$ and $i_{\beta R}$ but also on the rotor position θ_m. In the case when $i_{\alpha R} > 0$ and $i_{\beta R} = 0$, the vector $\boldsymbol{F_R}$ lies along α_R axis. For a given vector $\boldsymbol{F_R}$, currents of the two-phase model of the rotor cage can be determined from projections of this vector on the rotor axes α_R and β_R. The coefficient of proportionality between these currents and magnetomotive force is determined by the number of turns of the two-phase model of the rotor cage. The rotor phases α_R and β_R are virtual phases, that is, they are mathematical fiction that represent the rotor cage. Therefore, the number of turns of such virtual windings can be arbitrarily chosen. It should be noted that the short-circuiting effect of the rotor cage can be modeled by two-phase equivalent with large number of turns made comprising conductors with a lower cross section but also with lower number of turns made of conductors with larger cross section, even with $N_R = 1$. The original cage is aluminum cast, and it has one conductor per slot. For convenience, it is frequently assumed that the two-phase equivalent winding representing the rotor has the same number of turns as the stator phases. In this manner, transformation of rotor variables to the stator side is implied, and all the rotor variables and parameters that appear in the model are already scaled by the appropriate transformation ratio N_S/N_R.

15.11 Voltage Balance Equations in Rotor Winding

Two-phase representation of the stator and rotor windings reduces the mathematical model of the electrical subsystem of an induction machine to a set of four coupled phase windings. One pair of phase winding resides on the stator and the other pair on rotor. Due to rotor motion, the phase windings change their relative position. The voltage balance equation that applies to each of these windings is $u = Ri + d\Psi/dt$, where u, R, i, and Ψ denote the voltage across terminals of the considered phase winding, the winding resistance, electrical current, and total flux, respectively. The rotor winding is short circuited; thus, the voltage balance equations take the form

$$u_{\alpha S} = R_S i_{\alpha S} + \frac{d\Psi_{\alpha S}}{dt},$$

$$u_{\beta S} = R_S i_{\beta S} + \frac{d\Psi_{\beta S}}{dt},$$

$$0 = R_R i_{\alpha R} + \frac{d\Psi_{\alpha R}}{dt},$$

$$0 = R_R i_{\beta R} + \frac{d\Psi_{\beta R}}{dt}. \tag{15.32}$$

15.12 Inductance Matrix

Electrical subsystem of an induction machine is described by 4 differential equations of voltage balance comprising 4 currents and 4 fluxes. Among these 8 variables, there are only 4 state variables. Namely, if electrical currents $i_{\alpha S}$, $i_{\beta S}$, $i_{\alpha R}$, and $i_{\beta R}$ are promoted to the state variables, then the 4 fluxes $\Psi_{\alpha S}$, $\Psi_{\beta S}$, $\Psi_{\alpha R}$, and $\Psi_{\beta R}$ can be expressed in terms of currents. Relation between the fluxes and currents is generally nonlinear, due to nonlinearity of ferromagnetic materials and magnetic saturations. Under assumptions adopted in modeling electrical machines, which include the assumption that ferromagnetic materials have linear B-H characteristic, the flux linkages and electrical currents are in linear relation, defined by the inductance matrix. For the induction machine under consideration, the inductance matrix is given by expression (15.33). Along the main diagonal of the inductance matrix, there are coefficients of self-inductances of the phase windings. Coefficient $L_{11} = L_S$ is self-inductance of the stator winding α_S. Given the magnetic resistance R_μ of the stator flux circuit and the number of turns N_S, self-inductance of the stator can be determined as $L_{11} = L_{22} = L_S = N_S^2/R_\mu$. The stator phases have the same number of turns. At the same time, the air gap does not change along the machine circumference, and therefore, the magnetic resistance is also the same. For that reason, both stator phases have the same self-inductance L_S. Coefficients $L_{33} = L_{44} = L_R$ are self-inductances of rotor phases α_R and β_R. Assuming that the rotor phase windings have the same number of turns as the stator phase windings, the difference in L_S and L_R depends on magnetic resistances encountered by the stator and rotor flux linkages. The field lines of the rotor flux pass through the same air gap as the lines of the stator flux. Therefore, magnetic resistance to the stator flux is approximately equal to the magnetic resistance to the rotor flux. Small difference between L_S and L_R can be seen due to different leakage flux path and different leakage inductances:

$$
\begin{bmatrix} \Psi_{\alpha s} \\ \Psi_{\beta s} \\ \Psi_{\alpha R} \\ \Psi_{\beta R} \end{bmatrix}
=
\begin{bmatrix}
L_{11} & 0 & L_m \cos\theta_m & -L_m \sin\theta_m \\
0 & L_{22} & L_m \sin\theta_m & L_m \cos\theta_m \\
L_m \cos\theta_m & L_m \sin\theta_m & L_{33} & 0 \\
-L_m \sin\theta_m & L_m \cos\theta_m & 0 & L_{44}
\end{bmatrix}
\cdot
\begin{bmatrix} i_{\alpha s} \\ i_{\beta s} \\ i_{\alpha R} \\ i_{\beta R} \end{bmatrix}.
\quad (15.33)
$$

15.13 Leakage Flux and Mutual Flux

If rotor comes to position where magnetic axis of one stator phase coincides with magnetic axis of one rotor phase, then mutual inductance between them assumes maximum value. Figure 15.7 defines the mutual flux and the leakage flux. Flux linkage in one turn of the stator phase and flux linkage in one turn of the rotor phase are given by equations

Fig. 15.7 Mutual flux and leakage flux

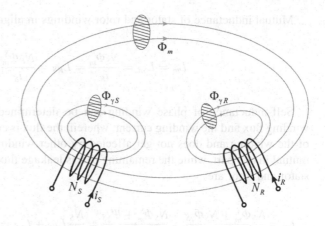

$$\Phi_S = \Phi_{\gamma S} + \Phi_m,$$
$$\Phi_R = \Phi_{\gamma R} + \Phi_m. \tag{15.34}$$

Each flux has mutual component, common for both stator and rotor turns, related to the lines of magnetic field that embrace both windings. In addition, there are leakage flux components. The stator leakage flux is related to magnetic field that encircles only the stator winding. It does not pass through the air gap and does not reach the rotor turns. Both stator and rotor currents contribute to the mutual flux. Mutual flux Φ_m in one turn has a component generated by the stator current (Φ^S_m) and a component generated by the rotor current (Φ^R_m),

$$\Phi_m = \Phi^S_m + \Phi^R_m. \tag{15.35}$$

The flux in phase windings depends on the flux in one turn and on the number of turns per phase. Therefore,

$$\Psi_S = N_S\Phi_S = N_S\Phi_m + N_S\Phi_{\gamma S} = N_S\Phi_m + \Psi_{\gamma S},$$
$$\Psi_R = N_R\Phi_R = N_R\Phi_m + N_R\Phi_{\gamma R} = N_R\Phi_m + \Psi_{\gamma R}. \tag{15.36}$$

In cases where $N_S = N_R$, the mutual flux components in stator and rotor windings are equal. Recall that N_R corresponds to the two-phase equivalent of the rotor winding, a mathematical fiction devised to model the rotor cage, while N_S corresponds to phase windings of the stator. Flux $\Psi_{\gamma S}$ is the *leakage flux* of the stator winding, while $\Psi_{\gamma R}$ is the leakage flux of the rotor winding. Leakage flux in each of the windings is proportional to the winding current. Coefficient of proportionality is *leakage inductance* of the winding. For the windings shown in Fig. 15.7, leakage inductances are given by expression

$$L_{\gamma S} = \frac{\Psi_{\gamma S}}{i_S}, \quad L_{\gamma R} = \frac{\Psi_{\gamma R}}{i_R}. \tag{15.37}$$

Mutual inductance of stator and rotor windings in aligned position equals

$$L_m = L_{SR} = \frac{N_S \Phi_m^R}{i_R} = L_{RS} = \frac{N_R \Phi_m^S}{i_S} . \qquad (15.38)$$

Self-inductance of phase winding can be determined as the quotient of the winding flux and the winding current, wherein the flux is caused only by the current of the winding and does not get affected by other windings currents. This flux is mutual in one part, while the remaining part is leakage flux. Self-inductances of the stator and rotor are

$$L_S = \frac{N_S \Phi_m^S + N_S \Phi_{\gamma S}}{i_S} = \frac{N_S \Phi_m^S + \Psi_{\gamma S}}{i_S} = \frac{N_S}{N_R} L_{RS} + L_{\gamma S} = \frac{N_S}{N_R} L_m + L_{\gamma S},$$

$$L_R = \frac{N_R \Phi_m^R + N_R \Phi_{\gamma R}}{i_R} = \frac{N_R \Phi_m^R + \Psi_{\gamma R}}{i_R} = \frac{N_R}{N_S} L_{SR} + L_{\gamma R} = \frac{N_R}{N_S} L_m + L_{\gamma R}. \qquad (15.39)$$

Therefore, leakage inductances make one part of self-inductances of phase windings. Leakage inductance is higher in the case when magnetic coupling of the two windings is weaker. In the case when the number of turns of the stator and rotor is equal, as well as in the case when the rotor quantities are *transformed* to the stator side, the preceding equation takes the form

$$L_S = L_m + L_{\gamma S},$$
$$L_R = L_m + L_{\gamma R}. \qquad (15.40)$$

15.14 Magnetic Coupling

Leakage flux of the stator and leakage flux of the rotor exist in different magnetic circuits, and they may have different magnetic resistances. For that reason, even the leakage inductances can be different.

Gross part of the stator flux encircles both stator and rotor windings, but there are also some lines of magnetic field that encircle only the stator conductors. They do not cross the air gap and thus do not encircle the rotor conductors. These field lines belong to the leakage flux of the stator. Leakage flux of the stator is a smaller part of the stator flux $\Phi_S = (L_S i_S)/N_S$. Leakage flux of the rotor is defined in similar. Different shapes of the stator and rotor slots as well as differences in the shape and cross section of conductors may result in different magnetic resistances on the path of the stator and rotor leakage fluxes.

Magnetic resistance encountered on the path of the mutual flux is one and the same for both stator and rotor phase windings. In both cases, mutual flux passes through the air gap, where the gross part of the magnetic resistance is encountered. Besides, mutual flux encircles both stator and rotor windings, passing through teeth, yoke, and other parts of stator and rotor magnetic circuits. On the other hand, magnetic resistance encountered along the path of leakage flux component is likely to be different on stator and rotor. Namely, the leakage flux path includes the width of the slots, and these are likely to be different. In general, a narrower slot opening results in a smaller magnetic resistance for the leakage flux and a larger leakage inductance, while a wide slot opening leads to a small leakage inductance.

In electrical machines, the power of electromechanical conversion and the electromagnetic torque depend on the magnetic coupling between the stator and rotor windings. Better coupling leads to more torque and power, hence the intention to keep the leakage as low as possible. Ideally, the coefficient of magnetic coupling of the stator and rotor $k = L_m/(L_S L_R)^{0.5}$ should reach unity. The leakage flux is proportional to difference $1 - k$, and in this case, it reaches zero as well as the leakage inductance coefficients $L_{\gamma S}$ and $L_{\gamma R}$. Practical machines cannot be designed to achieve the coupling coefficient of 1. Such a coupling would require the conductors of the two windings to be next to each other, so as to prevent any leakage flux, and this is not feasible due to practical reasons. The stator and rotor have to be separated by air gap for mechanical and electrical reasons. In machines designed for operation with higher voltages, insulation of individual conductors and windings has to sustain high-voltage stresses. For this reason, insulation layers are thicker, as well as distances between individual conductors. With increased distances between corresponding conductors, the space for the leakage flux is enlarged as well as the leakage flux. Provisional values of the coupling coefficient in low-voltage electrical machines (400 V, 50 Hz) are $k \sim [0.9 .. 0.98]$. In machines designed to operate with high voltages, the values of the coupling coefficient could be considerably lower, even $k < 0.9$.

15.15   Matrix L

Inductance matrix provides the link between the vector column with four total flux linkages and the vector column with four electrical currents. On the main diagonal, inductance matrix has the coefficients of self-inductances. Off the main diagonal, it has the mutual inductances. The mutual inductances describing magnetic coupling between stator and rotor phases are variable. They change in the course of motion.

Neglecting the differences in magnetic resistances for stator and rotor flux linkages, the ratio L_S/L_R depends on the number of stator and rotor turns, $L_{11} = L_{22}$ $= L_S = N_S^2/R_\mu$, $L_{33} = L_{44} = L_R = N_R^2/R_\mu$. Self-inductances are strictly positive, while mutual inductances may assume negative values as well as positive. Mutual inductance L_{jk} determines the flux contribution brought into the phase winding k by the current i_j of the phase winding j. Magnetic coupling between the two windings is reciprocal, $L_{jk} = L_{kj}$; thus, the inductance matrix is symmetrical ($L = L^T$). Mutual inductance of orthogonal windings is equal to zero[4]; thus, $L_{12} = L_{21} = L_{34} = L_{43} = 0$. Coefficient L_{13} of the matrix represents mutual inductance of windings α_S and α_R. Relative position of the considered windings changes as the rotor moves. With $\theta_m = 0$, the windings are placed one against the other, and their magnetic axes coincide. In this position, magnetic coupling peaks, and then current in one winding gives the highest change of flux in the other winding. With $\theta_m = \pi/2$, considered windings are orthogonal, and therefore, $L_{13} = 0$. With $\theta_m = \pi$, positive current in one winding gives negative flux in the other; thus, $L_{13} < 0$. Variation of coefficient L_{13} can be described by function $L_{13}(\theta_m) = L_m \cos(\theta_m)$, where $L_m = k\,(L_S L_R)^{0.5}$ is the maximum value of L_{13}, obtained in position $\theta_m = 0$. Other coefficients of the inductance matrix can be determined in a like manner. It should be noted that the matrix is not stationary. Some coefficients change with the angle $\theta_m = \Omega_m t$. Therefore, there is a nonzero derivative dL/dt. Recall at this point that the electromagnetic torque of electromechanical converter can be obtained as $T_{em} = \frac{1}{2}\,i^T\,(dL/d\theta_m)i$.

$$
\begin{bmatrix} \Psi_{\alpha s} \\ \Psi_{\beta s} \\ \Psi_{\alpha R} \\ \Psi_{\beta R} \end{bmatrix} = \begin{bmatrix} L_s & 0 & L_m \cos\theta_m & -L_m \sin\theta_m \\ 0 & L_s & L_m \sin\theta_m & L_m \cos\theta_m \\ L_m \cos\theta_m & L_m \sin\theta_m & L_R & 0 \\ -L_m \sin\theta_m & L_m \cos\theta_m & 0 & L_R \end{bmatrix} \cdot \begin{bmatrix} i_{\alpha s} \\ i_{\beta s} \\ i_{\alpha R} \\ i_{\beta R} \end{bmatrix}
\tag{15.41}
$$

15.16 Transforming Rotor Variables to Stator Side

Voltage balance equations for rotor windings are given in Sect. 15.9, and they comprise rotor currents $i_{\alpha R}$ and $i_{\beta R}$. The rotor currents are not directly accessible. They cannot be measured by accessing the rotor cage and inserting measurement devices. Moreover, the two-phase model replaces the rotor cage by an equivalent two-phase winding brought into the short circuit. Hence, the rotor currents $i_{\alpha R}$ and $i_{\beta R}$ are not the currents flowing through the rotor bars but the currents of the two-phase equivalent which has replaced the cage. The short-circuiting effect of the cage can be modeled by the two-phase equivalent having large number of conductors of small cross-sectional area or small number of conductors of large cross-sectional area. Therefore, the number of turns in the two phases depicting the rotor can be arbitrarily selected, as explained in the following example.

---

[4] This assumption is not valid for nonlinear magnetic circuits. Magnetic saturation contributes to so-called *cross saturation* in orthogonal windings, phenomenon where the flux in one of the windings changes the saturation level in common magnetic circuit and, hence, changes the flux of the other winding.

The magnetomotive force created by electrical currents of all the conductors placed in one rotor slot is equal to the sum of their currents. The slot may have only one conductor with current of 100 A or 100 conductors each carrying 1 A, and in both cases, the magnetomotive force will be 100 ampere-turns. This value corresponds to circular integral of magnetic field H along closed contour encircling the slot. Hence, the system with two-phase windings that models the rotor cage could have an arbitrary number of turns, as long as the product Ni of the rotor current and the number of turns equals the value created by the original short-circuited cage.

The freedom in choosing the number of turns of the two-phase rotor equivalent is most frequently used to introduce $N_R = N_S$. Assuming that the rotor has the same number of turns as the stator results in $L_S = N_S{}^2/R_\mu = N_R{}^2/R_\mu = L_R$ and gives $L_m = k\,(L_S L_R)^{0.5} = kL_S = kL_R$, while the leakage inductances of the stator and rotor become $L_{\gamma S} = L_S - L_m = (1-k)L_S = (1-k)L_R = L_{\gamma R}$. The obtained expressions are based on the assumption that differences in magnetic resistances for the stator and rotor fluxes are negligible. This assumption is valid in most cases.

The inductance matrix allows that each of the four flux linkages is expressed in terms of electrical currents. For example, the flux in phase α of the stator is

$$\Psi_{\alpha S} = L_S i_{\alpha S} + L_m \cos\theta_m\, i_{\alpha R} - L_m \sin\theta_m i_{\beta R}. \tag{15.42}$$

Question (15.7): Stator currents of an induction machine are $i_{\alpha S} = I_{mS}\cos\omega_e t$ and $i_{\beta S} = I_{mS}\sin\omega_e t$, where $\omega_e > 0$ and rotor currents are $i_{\alpha R} = I_{mR}\sin\omega_x t$ and $i_{\beta R} = I_{mR}\cos\omega_x t$ having the angular frequency $0 < \omega_x < {<}\omega_e$. The machine operates at steady state. By using (15.42) for flux $\Psi_{\alpha S}$, determine the rotor speed.

Answer (15.7): Currents $i_{\alpha S}$ and $i_{\beta S}$ produce the stator magnetomotive force and flux rotating in positive direction. Phase sequence of the given rotor currents is such that they create magnetic field which rotates relative to the rotor at the speed of $\omega_x = \Omega_x$ in negative direction. In steady state, the rotor and stator fields revolve synchronously. Therefore, it is concluded that $\omega_m = \omega_x + \omega_e$. The same conclusion can be obtained from the expression for flux, $\Psi_{\alpha S} = L_S I_{mS}\cos\omega_e t + L_m I_{mR}(\cos\omega_m t\,\sin\omega_x t - \sin\omega_m t\,\cos\omega_x t) = L_S I_{mS}\cos\omega_e t - L_m I_{mR}\sin(\omega_m t - \omega_x t)$. The elements of this must have the same frequency in steady-state conditions since the stator and rotor variables rotate at the same speed ω_e, maintaining the relative positions unchanged. This condition is met in cases $\omega_m - \omega_x = +\omega_e$ as well as $\omega_m - \omega_x = -\omega_e$, that is, for the speeds of rotation $\omega_m = \omega_x + \omega_e$ or $\omega_m = -\omega_e + \omega_x$. According to the assumed conditions, $0 < \omega_x \ll \omega_e$, and the solution is $\omega_m = \omega_x + \omega_e$. In this solution, the slip $\omega_{slip} = \omega_e - \omega_m = -\omega_x$ is negative; hence, the rotor is rotating faster than the field. The machine operates in generator mode.

Question (15.8): Starting from the inductance matrix of the system of windings α_S, β_S, α_R, and β_R, prove that the torque is equal to $T_{em} = (3/2)\,(\Psi_{\alpha S}\,i_{\beta S} - \Psi_{\beta S}\,i_{\alpha S})$.

Answer (15.8): The electromagnetic torque is given by expression $T_{em} = \frac{1}{2} i^T [dL$ $(\theta_m)/d\theta_m] i$ where $L(\theta_m)$ is the inductance matrix whose elements are dependent on position θ_m of the rotor with respect to the stator. Variable elements of the inductance matrix are $L_{13} = L_{31}$, $L_{14} = L_{41}$, $L_{23} = L_{32}$, and $L_{24} = L_{42}$, while the remaining coefficients are constant and result in $dL_{jk}/d\theta_m = 0$. The calculation can be simplified because $L^T = L$; thus, the result can be obtained by doubling the contributions of coefficients L_{13}, L_{14}, L_{23}, and L_{24}. Finally, one obtains $\frac{1}{2} i^T [dL(\theta_m)/$ $d\theta_m] i = -L_m \sin\theta_m\, i_{\alpha S}\, i_{\alpha R} - L_m \cos(\theta_m) i_{\alpha S}\, i_{\beta R} + L_m \cos(\theta_m) i_{\alpha R}\, i_{\beta S} - L_m \sin(\theta_m) i_{\beta R}$ $i_{\beta S}$. The same result is obtained by starting from expression $(\Psi_{\alpha S}\, i_{\beta S} - \Psi_{\beta S}\, i_{\alpha S})$ and introducing the replacement where the fluxes are expressed from the first and second row of the inductance matrix. In expression $T_{em} = (3/2)\,(\Psi_{\alpha S}\, i_{\beta S} - \Psi_{\beta S}$ $i_{\alpha S})$, coefficient $3/2$ is the consequence of adopting the $3\Phi/2\Phi$ transform with $K_U = K_I = K_\Psi = 2/3$.

15.17 Mathematical Model

In subsequent considerations, mathematical model of induction machine is presented in terms of coordinates α_S, β_S, α_R, and β_R. The voltage balance equations and inductance matrix were defined already within previous sections. The model is completed by adding Newton equation and the torque expression $T_{em} = (3/2)(\Psi_{\alpha S}$ $i_{\beta S} - \Psi_{\beta S}\, i_{\alpha S})$. This set of differential equations and algebraic expressions constitutes mathematical model of induction machine, based on previously adopted approximations. The model is summarized in (15.43), (15.44), (15.45), and (15.46). It can be used in its present form to predict dynamic behavior and steady-state properties of induction machines. For that to be achieved, it is sufficient to enter (15.43), (15.44), (15.45), and (15.46) into program for computer simulation of dynamic systems. Hence, developed model is the correct representation of behavior of induction machines. Yet, it has drawbacks that hinder further analytical considerations and introduce difficulties in drawing conclusions and deriving the steady-state characteristics.

There could be specific situations where the given model cannot serve as an accurate representation of the induction machine. In cases where the iron losses cannot be neglected, or the magnetic saturation is emphasized, as well as in cases where the remaining two approximations do not hold, the model may give erroneous results. In such cases, the model has to be modified and upgraded so as to include the effects that were neglected in the first place. The four approximations that were adopted in modeling electrical machines are listed and explained in introductory chapters.

The model contains α_S and β_S components of the stator variables, as well as α_R and β_R components of the rotor variables. Equations (15.43), (15.44), and (15.45) remain unaltered whatever the choice of the leading coefficient K of $3\Phi/2\Phi$ transform. In (15.46), it is assumed that $K = 2/3$ is used. This choice is used throughout the book, and it requires the $\alpha\beta$ power and torque to be multiplied by $3/2$ in order to obtain the power and torque of the original.

It is of interest to recall that the choice of K has to do with selecting the number of turns $N_{\alpha\beta}$ of the equivalent two-phase machine. As already shown, the choice $N_{\alpha\beta} = (3/2)^{0.5} N_{abc}$ and $K = (2/3)^{0.5}$ results in a two-phase equivalent which has power, impedance, and inductance invariance. The two-phase equivalent is a mathematical fiction, and it does not have to be actually made. Yet, envisaging the variables $i_{\alpha S}$, $i_{\beta S}$, $i_{\alpha R}$, and $i_{\beta R}$ as electrical currents of actual phase windings helps understanding the basic voltage, current, and flux vectors of the machine, and it helps using the model.

Components of the voltage, current, and flux in the model are projections of the relevant vectors of the voltage, current, and flux on axes of coordinate systems α_S–β_S and α_R–β_R. Stator vectors are projected on axes α_S and β_S of stationary coordinate system, while rotor vectors are projected on axes α_R and β_R of coordinate system that revolves with the rotor.

Complete model is summarized by (15.43), (15.44), (15.45), and (15.46). The symbol p in (15.46) represents the number of pairs of magnetic poles, discussed in Chap. 16. Preceding considerations assumed that $p = 1$, namely, that magnetic field has one north pole and one south pole:

$$u_{\alpha s} = R_s i_{\alpha s} + \frac{\mathrm{d}\Psi_{\alpha s}}{\mathrm{d}t}, \quad u_{\beta s} = R_s i_{\beta s} + \frac{\mathrm{d}\Psi_{\beta s}}{\mathrm{d}t}, \tag{15.43}$$

$$0 = R_R i_{\alpha R} + \frac{\mathrm{d}\Psi_{\alpha R}}{\mathrm{d}t}, \quad 0 = R_R i_{\beta R} + \frac{\mathrm{d}\Psi_{\beta R}}{\mathrm{d}t}, \tag{15.44}$$

$$\begin{bmatrix} \Psi_{\alpha s} \\ \Psi_{\beta s} \\ \Psi_{\alpha R} \\ \Psi_{\beta R} \end{bmatrix} = \begin{bmatrix} L_s & 0 & L_m \cos\theta_m & -L_m \sin\theta_m \\ 0 & L_s & L_m \sin\theta_m & L_m \cos\theta_m \\ L_m \cos\theta_m & L_m \sin\theta_m & L_R & 0 \\ -L_m \sin\theta_m & L_m \cos\theta_m & 0 & L_R \end{bmatrix} \cdot \begin{bmatrix} i_{\alpha s} \\ i_{\beta s} \\ i_{\alpha R} \\ i_{\beta R} \end{bmatrix}, \tag{15.45}$$

$$T_{em} = \frac{3}{2} p \left(\Psi_{\alpha S} i_{\beta S} - \Psi_{\beta S} i_{\alpha S} \right). \tag{15.46}$$

15.18 Drawbacks

The above model is an adequate representation of dynamic and steady-state behavior of induction machines, but it has drawbacks that make further uses more difficult. Such uses are the steady-state analysis, deriving equivalent circuits, and conceiving and designing control algorithms. The key problems with the model are:

1. The presence of trigonometric functions in differential equations
2. The state variables exhibiting sinusoidal change even in steady state

The consequences of the above issues on clarity and usability of the model can be seen by considering the voltage balance equation. Further on, discussion will close by proposing further steps that should be taken to obtain a more clear and more intuitive model.

By expressing the flux in terms of electrical currents and introducing the resulting expression in voltage balance equations for the stator phase windings, the following expressions are obtained:

$$u_{\alpha s} = R_s i_{\alpha s} + L_s \frac{di_{\alpha s}}{dt} + L_m \cos \theta_m \frac{di_{\alpha R}}{dt} - \omega_m L_m \sin \theta_m i_{\alpha R}$$
$$- L_m \sin \theta_m \frac{di_{\beta R}}{dt} - \omega_m L_m \cos \theta_m i_{\beta R},$$

$$u_{\beta s} = R_s i_{\beta s} + L_s \frac{di_{\beta s}}{dt} + L_m \sin \theta_m \frac{di_{\alpha R}}{dt} + \omega_m L_m \cos \theta_m i_{\alpha R}$$
$$+ L_m \cos \theta_m \frac{di_{\beta R}}{dt} - \omega_m L_m \sin \theta_m i_{\beta R}.$$

Presence of trigonometric functions in differential equations makes the steady-state analysis more difficult. An attempt to derive a steady-state equivalent circuit becomes more involved. Hypothetic removal of trigonometric functions from the voltage balance equations results in

$$u_{\alpha S} = R_S i_{\alpha S} + L_S \frac{di_{\alpha S}}{dt} + L_m \frac{di_{\alpha R}}{dt}$$

which simplifies greatly the steady-state relations and makes it more obvious. Applying Laplace transform to the previous equations, an algebraic expression is obtained which relates the complex images of voltages and currents,

$$U_{\alpha S}(s) = R_S I_{\alpha S}(s) + s L_S I_{\alpha S}(s) + s L_m I_{\alpha R}(s)$$

At steady state, electrical current $i_{\alpha s}$ exhibits sinusoidal change. Assuming that, for the sake of an example, all the electrical currents have the angular frequency ω, the steady-state analysis implies that the operator s becomes $j\omega$, resulting in the following expression:

$$U_{\alpha S}(s) = R_S I_{\alpha S}(s) + j\omega L_S I_{\alpha S}(s) + j\omega L_m I_{\alpha R}(s)$$

which represents the voltage balance equation in a contour comprising the voltage source $U_{\alpha S}$, resistance R_S, an inductance $(L_S - L_m)$ carrying current $I_{\alpha S}$, and an inductance L_m carrying current $(I_{\alpha R} + I_{\alpha S})$. Considered example demonstrates that a set of differential equations with constant coefficients provides the grounds for deriving an equivalent circuit that represent the machine in the steady state. The presence of trigonometric functions in voltage balance equations makes this impossible.

Sinusoidal change of state variables even in steady-state conditions brings in more difficulties in steady-state analysis. At the same time, similar difficulties are encountered in designing control structures for the current, flux, and torque regulation. The models with state variables that remain unaltered in the steady state are simple to grasp. In their steady-state equations, all the time derivatives of the state variables disappear, and the remaining expressions are easy to understand and use. Yet, the present model does not have this possibility. The state variables of the model, such as $i_{\alpha s}$, exhibit sinusoidal changes even in the steady state.

Most mathematical models are usually formulated in such way that their state variables remain constant in the steady state. The time derivatives of these state variables are then equal to zero. In such cases, steady-state relations are obtained from differential equations by removing the time derivatives.

Perpetual changes of state variables even in steady state make the control problems more difficult to solve. An example to that is the applications of electrical motors in industrial robots or autonomous vehicles, where the electrical motors are used for controlling the motion. For that to be achieved, it is necessary to *regulate* some relevant motor variables, such as the flux, torque, speed, and current. The term *regulation* implies:

- Definition of a desired *reference* value for the controlled variable (such as the phase current) that should be reached and maintained
- Measurement of the controlled variable (current) and calculation of the *error*, the deviation of the controlled variable from the desired value
- Performing *control algorithm*, calculation procedure or formula which receives the *error* and calculates the *control*, the output of the control algorithm[5]
- Bringing the control variable to the system under control through an *executive organ* or *actuator*[6]

When the reference value does not change, the steady-state error can be reduced to zero by adding integral control action into the control algorithm. Corresponding control action is proportional to the integral of the error. Yet, this maintains the error at zero only for constant references. In cases when the reference value has perpetual changes, control actions would act upon the controlled variable to track these changes. This is achieved at the cost of an error which cannot be removed.

---

[5] By using devices called *actuators* or *amplifiers*, control output can affect the system and change the controlled variable. Control algorithm is suited to reduce error and bring the controlled variable toward the reference. Well-suited control algorithm ensures progressive reduction of the error which, after a while, reaches zero. Hence, the steady-state value of errors is expected to be equal to zero.

[6] Control of phase currents of most electrical machines often includes switching power converters employing power transistors and PWM techniques, as well as digital signal controllers. The control algorithm is usually implemented by programming digital signal controllers. Control variables are obtained in numerical form, as binary-coded digital words that reside within processor registers. On the other side, the actual control variable that has the potential to change the phase current is the voltage across the phase winding which may change within the range of ± 600 V. The executive organ is often a switching transistor which receives gating signals from the digital signal controller.

Namely, controlled variable would track the reference with a tracking error that depends on the rate of change of the reference. The presence of error can jeopardize the operation of the whole system.

Induction machines have alternating currents in their phase windings. Hence, even in the steady state, the variables $i_{\alpha s}$ and $i_{\beta s}$ exhibit sinusoidal changes. Whenever there is a need to regulate stator currents of induction machine, controlled variables are variable even at steady state. This creates the need to transform the mathematical model in such way that the steady-state values of the state variables are constant and to use this model to formulate the control algorithm.

Question (15.9): Is there an operating mode of an induction machine where the frequency of stator currents is zero?

Answer (15.9): Considering induction machines supplied from constant frequency voltage sources, the frequency of the stator electrical currents is determined by the frequency of voltages fed to the stator winding. If induction machine is connected to the mains, the frequency of the stator current will be 50 Hz, irrespective of the rotor speed, torque, or power. For that reason, the stator currents may have different frequency only in cases when the induction machine is supplied from a variable frequency, variable voltage source such as the static power converter which uses semiconductor power switches and operates on pulse-width modulation principles. In this way, the stator winding can be supplied with a symmetrical system of three-phase voltages of variable amplitude and variable frequency.

In further considerations, it is assumed that the supply voltage has variable frequency ω_e which can be adjusted to achieve desired operating mode. For two-pole machine, $\omega_e = \Omega_m + \omega_{slip}$, where Ω_m is the rotor speed while ω_{slip} is the slip frequency. Operating mode where DC current flows through the stator is the one where $\Omega_m = -\omega_{slip}$, resulting in $\omega_e = 0$. It should be noted that the slip frequency is proportional to the developed electromagnetic torque. Therefore, the operating mode with $\omega_e = 0$ is reached when the torque and speed of rotation are of different signs. The rotor rotates at the speed which has the same magnitude as the slip frequency, but it has opposite sign. In one of such cases, the motor is stopped and develops the torque $T_{em} = 0$. In another example, the rotor revolves at -300 rpm, it develops positive torque, and it has the rotor currents of 5 Hz.

15.19 Model in Synchronous Coordinate Frame

The problems of analysis and control based on mathematical model in α–β coordinate frame arise due to the fact that the state variables exhibit sinusoidal change even in the steady state. Namely, projections of the vectors of currents, voltages, and flux linkages on the axes of α–β coordinate frame change as sinusoidal functions even in steady state, with constant amplitude of magnetomotive force F_S, constant flux, and constant rotor speed. This deficiency of the model can be removed by applying another transform of the state coordinates and replacing the existing state variables by the new ones. The new variables should be selected so that their values at steady state do not vary. The new coordinate transform should maintain the invariability in terms of impedance, inductance, and power. For that to be achieved, the same transformation matrix should be applied to all the variables, whether the currents, voltages, or flux linkages. The question arises about how to devise this new coordinate transform.

15.20 Park Transform

In the model of the machine which is formulated in stationary α–β frame, the stator currents $i_{\alpha S}$ and $i_{\beta S}$ are projections of the vector of magnetomotive force $F_S = N_S i_S$ on axes of the α_S–β_S coordinate system. The problem arises due to the fact that F_S is a rotating vector. Its rotation with respect to α–β frame makes the projections $i_{\alpha S}$ and $i_{\beta S}$ variable. In steady state, they become sinusoidal function. In order to solve the problem, it is necessary to envisage a new coordinate system which is rotating at the same speed as the magnetomotive force F_S. In this case, projections of this vector on new pair of axes do not change in the steady state, as the relative position between the revolving vector and the new frame does not change. The same conclusion applies for the voltage and flux vectors. Therefore, the new transform of the state coordinates should formulate the model of induction machine in a new coordinate system which revolves synchronously with the field.

By adopting a synchronously rotating coordinate system having axes d and q, projections i_d and i_q of the stator current vector $i_S = F_S/N_S$ on these new axes have constant steady-state values. Therefore, by transforming all the stator quantities from the stationary α_S–β_S coordinate system to the synchronously rotating d–q coordinate system, a model of the stator winding is obtained where the relevant quantities have constant steady-state values.

Transform implies a relation that expresses the variables of the d–q system (i_d and i_q) in terms of the variables of the α_S–β_S coordinate system ($i_{\alpha S}$ and $i_{\beta S}$). Applying a coordinate transform does not introduce any change to the considered induction machine nor to its flux or torque. The transform merely represents a *different point of view* and represents one and the same actual systems by means of another mathematical representation. Therefore, no matter whether the mathematical model is formulated in the stationary or in the synchronous coordinate frame, it must describe the same vectors of the magnetomotive force, flux, voltage, and current as well as the same torque, speed, and power of electromechanical conversion. Consequently, both the model in α–β frame and the model in d–q frame must have one and the same vector of the stator magnetomotive force F_S. In other words, the vector F_S created by currents $i_{\alpha S}$ and $i_{\beta S}$ must have the same amplitude and spatial orientation as the vector F_S created by i_d and i_q, the stator currents transformed into d–q coordinate frame.

Coordinate transform is essentially a mathematical operation, and it does not have to result in state variables that correspond to an actual, physical machine.

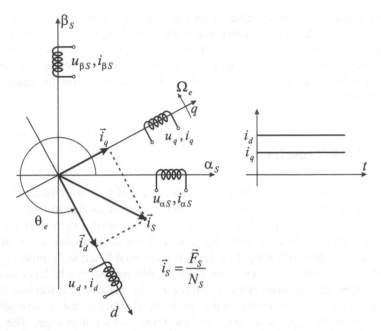

Fig. 15.8 Position of d–q coordinate frame and corresponding steady-state currents in virtual phases d and q

In many cases, it is not even possible to construct an induction machine that would have the voltages and currents that correspond to the ones obtained from coordinate transform. Yet, an attempt to represent the variables i_d and i_q as electrical currents of a new, virtual stator winding may help understanding and using the new model. The new transform can be represented by *removing* the stator phase windings α_S and β_S and *installing* new stator phases d and q, with their magnetic axes lying along the d and q of the new coordinate frame, as shown in Fig. 15.8. These d–q windings cannot be actually made, so it is correct to call them *virtual*. Electrical currents i_d and i_q of these new virtual windings must result in the same vector F_S as the previous, created by electrical currents in phases α_S and β_S.

In the following figures and illustrations that involve the new d–q coordinate frame, new notation is introduced for the phase windings and electrical currents in d–q frame. The *virtual* stator windings are denoted by d and q, while the *virtual* rotor windings are denoted by D and Q. The stator currents in new coordinate frame are denoted by i_d and i_q, while the rotor currents are i_D and i_Q. Components of electrical currents in d–q frame are equal to projections of current vectors on axes d and q. Should virtual windings d and q actually exist, the currents i_d and i_q would result into the same vector of the stator magnetomotive force F_S which actually exists in the original machine. Neither the stator nor the rotor could have their actual phase windings residing in the d–q frame. The only purpose of showing them in illustrations is to help visualizing the state variables which are obtained by rotational transform, such as i_d and i_q, and also to facilitate the model analysis and use.

15.21 Transform Matrix

Relations of the rotational transform can be derived from the condition of invariability of the vector F_S. In the preceding figure, the angle θ_e denotes the angular shift of the revolving axis d with respect to the still axis α_S. Projection of the current component $i_{\alpha S}$ on axis d is $i_{\alpha S}\cos\theta_e$, while projection of current component $i_{\beta S}$ on the same axis is $i_{\beta S}\sin\theta_e$. Hence, projection of the vector F_S/N_S on axis d is equal to $i_{\alpha S}\cos\theta_e + i_{\beta S}\sin\theta_e$. Considering d–q frame variables, the current i_q produces the component of F_S in direction of the axis q, and it cannot contribute to d component of F_S. Hence, the current i_d has to be equal to $i_{\alpha S}\cos\theta_e + i_{\beta S}\sin\theta_e$ in order to achieve invariability of vector F_S (Fig. 15.9). By summing projections of currents $i_{\alpha S}$ and $i_{\beta S}$ on axis q, one obtains that current i_q in virtual phase winding q must be equal to $-i_{\alpha S}\sin\theta_e + i_{\beta S}\cos\theta_e$. Park rotational transform is summarized by using the matrix given by (15.47). Determinant of the transform matrix is equal to 1. By transforming any vector from stationary frame to synchronously rotating frame, one obtains the vector of the same amplitude as the original. By applying the same transform matrix to voltages, currents, and fluxes, one obtains the relevant variables in d–q coordinate system (u_d, u_q, i_d, i_q, Ψ_d, Ψ_q). Proposed transform is invariable in terms of impedance, inductance, and power (det $T = 1$). Variables Ψ_d and Ψ_q denote total flux linkages of the virtual stator windings in d and q axes.

$$\begin{bmatrix} i_d \\ i_q \end{bmatrix} = \begin{bmatrix} \cos\theta_s & \sin\theta_s \\ -\sin\theta_s & \cos\theta_s \end{bmatrix} \cdot \begin{bmatrix} i_{\alpha s} \\ i_{\beta s} \end{bmatrix} = \underline{T} \cdot \begin{bmatrix} i_{\alpha s} \\ i_{\beta s} \end{bmatrix} \tag{15.47}$$

Question (15.10): Starting from expression $u_{\alpha S} = R_S i_{\alpha S} + d\Psi_{\alpha S}/dt$, is it possible to state that $u_d = R_S i_d + d\Psi_d/dt$?

Answer (15.10): The stator phase winding d is a virtual winding. Even though, it is possible to formulate the voltage balance equation which comprises the variables u_d, i_d

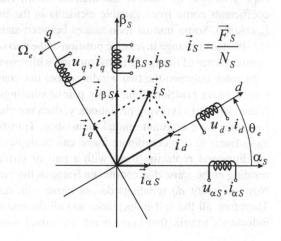

Fig. 15.9 Projections of F_S/N_S on stationary and rotating frame

and other variables of the d–q frame. The presence of unchanged stator resistance R_S is expected since the transform is invariant in terms of impedance. However, it cannot be stated that $u_d = R_S i_d + \mathrm{d}\Psi_d/\mathrm{d}t$. Namely, the voltage balance equation $u = Ri + \mathrm{d}\Psi/\mathrm{d}t$ applies to any winding that actually exists or the winding that could actually be made. Relation $u = Ri + \mathrm{d}\Psi/\mathrm{d}t$ was used to model the voltage balance in α_S–β_S coordinate frame. Although the original induction machine has three phases, the two-phase stator winding in α_S–β_S frame can be actually made. Therefore, the stator phase windings α_S and β_S can be considered as real windings. Thus, the voltage balance equation in α_S–β_S frame is $u = Ri + \mathrm{d}\Psi/\mathrm{d}t$. On the other hand, the phase windings in dq frame cannot be made. The analysis performed later on proves that the voltage u_d is not equal to $R_S i_d + \mathrm{d}\Psi_d/\mathrm{d}t$. The actual voltage balance equations in dq frame, describing the virtual voltages, will be derived by applying the transform matrix to equations $u_{\alpha S} = R_S i_{\alpha S} + \mathrm{d}\Psi_{\alpha S}/\mathrm{d}t$ and $u_{\beta S} = R_S i_{\beta S} + \mathrm{d}\Psi_{\beta S}/\mathrm{d}t$.

15.22   Transforming Rotor Variables

It is beneficial to have both the stator and the rotor variables residing in the same frame of coordinates. For that to be achieved, it is necessary to apply rotational transform to the rotor variables and find their DQ equivalents.

Park transform of the stator phase windings α_S and β_S results in two virtual stator phases in dq frame. In the same way, it is necessary to replace the short-circuited rotor cage by two virtual rotor phases residing in d–q coordinate system. For clarity, notation adopted hereafter implies that the stator variables have lower case subscripts dq, while the rotor variables have upper case subscripts DQ.

It is necessary to explain the need for transforming the stator and rotor variables to the same, synchronously rotating, coordinate system. First of all, it should be recalled that the voltage balance equations in stationary α–β coordinate frame are difficult to cope with due to variable coefficients of differential equations. These variable coefficients come from variable elements in the inductance matrix, such as $L_{13} = L_m \cos(\theta_m)$. Some mutual inductances between stator and rotor phases are variable (15.48) due to change in relative position of the two windings. In (15.45), this comes as a consequence of the α_R–β_R frame moving with respect to the α_S–β_S frame. The mutual inductance between the two windings does not change as long as they maintain the same relative position. Hence, any pair of windings residing in the same coordinate frame, whether revolving or stationary, does not change relative position, and therefore, they have constant mutual inductance. Transformation of rotor windings from $\alpha_R \beta_R$ frame to d–q coordinate frame can be depicted as removing the original $\alpha_R \beta_R$ windings and replacing them with a pair of virtual rotor phases denoted by DQ, residing in the same d–q coordinate frame as the virtual stator phases denoted by dq. Now the stator dq windings do not move with respect to the rotor DQ windings. Therefore, all the self-inductances and all the mutual inductances are constant. The inductance matrix that corresponds to virtual windings in dq frame has constant

Fig. 15.10 Rotor coordinate
system and *dq* system

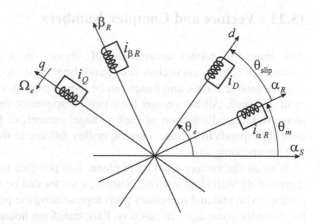

elements. It will be shown later that, as a consequence, the voltage balance equations
in *dq* frame do not have variable coefficients.

In addition to obtaining a constant inductance matrix, transformation of the rotor
variables to d–q system is also required in order to obtain a mathematical model
where the rotor variables have constant steady-state values. The actual rotor cage
has AC currents of angular frequency $\omega_{slip} = \omega_e - \omega_m = \omega_e - p\Omega_m$. These
currents create magnetomotive force and flux of the rotor which revolve at
the speed Ω_{slip} with respect to the rotor. With $p = 1$, their speed with respect to the
stator is $\Omega_{slip} + \Omega_m = \omega_{slip} + \omega_m = \omega_e$. In other words, the rotor magnetomotive
force and flux vectors revolve at the same speed (in synchronism) with the stator
vectors. Some phase and/or spatial shift among these variables may exist at steady
state. Therefore, projections of the rotor variables on d–q do not vary at steady state.
Hence, the actual rotor variables should be transformed to d–q coordinate frame.

Rotational transform of the rotor variables is illustrated in Fig. 15.10. It should
be noted that the product of the rotor currents and the number of rotor turns $N_R\, i_{\alpha R}$
and $N_R\, i_{\beta R}$ are equal to projections of the rotor magnetomotive force vector $\boldsymbol{F}_R$ on
axes α_R–β_R of the rotor coordinate system. The angular displacement between axes
α_R and α_S is determined by the rotor position θ_m. Synchronously rotating d–q
coordinate system leads by θ_{slip} with respect to the rotor; thus, its advance with
respect to the stator is $\theta_e = \theta_{slip} + \theta_m$.

Transformation matrix for the rotor variables is derived starting from invariability of the rotor magnetomotive force. Projection of the rotor phase current $i_{\alpha R}$ on
axis d is equal to $i_{\alpha R} \cos\theta_{slip}$, while projection of the rotor phase current $i_{\beta R}$ on the
same axis is equal to $i_{\beta R} \sin\theta_{slip}$. Therefore, the current of the virtual rotor winding
D is equal to $i_D = i_{\alpha R} \cos\theta_{slip} + i_{\beta R} \sin\theta_{slip}$. Similarly, $i_Q = -i_{\alpha R} \sin\theta_{slip} + i_{\beta R}$
$\cos\theta_{slip}$. These relations can be written in matrix form, as shown in the (15.48). The
same transformation matrix is applied to all rotor variables, resulting in voltages of
virtual rotor windings $u_D = u_Q = 0$ and providing total flux linkages Ψ_D and Ψ_Q:

$$\begin{bmatrix} i_D \\ i_Q \end{bmatrix} = \begin{bmatrix} \cos\theta_{slip} & \sin\theta_{slip} \\ -\sin\theta_{slip} & \cos\theta_{slip} \end{bmatrix} \cdot \begin{bmatrix} i_{\alpha R} \\ i_{\beta R} \end{bmatrix}. \tag{15.48}$$

15.23   Vectors and Complex Numbers

Park transform relates coordinates of vectors in α–β coordinate system to coordinates of the same vectors in d–q coordinate system. All the voltages, currents, magnetomotive forces, and fluxes can be represented by vectors, either in $\alpha\beta$ frame or in dq frame. All the vectors have two components, being associated with two-phase windings. Projection of each voltage, current, or flux vector on one of the axes corresponds to voltage, current, or flux linkage in the phase winding residing on corresponding axis.

With all the vectors residing in plane, it is possible to use complex notation in representing individual vectors. Namely, a vector can be represented by a complex number, with real and imaginary parts representing the projections of the vector on the two orthogonal axes. In this way, Park transform notation can be simplified, and the transformation matrix reduced to a complex number. As an example, the vector of the stator current can be expressed in terms of its α_S component, directed along α_S axis, and its β_S component, directed along β_S axis. If α_S axis is given attributes of the real axis and axis β_S attributes of the imaginary axis, the α_S–β_S plane is interpreted as the complex plane, where the current vector is represented by complex number,

$$\vec{i}_{\alpha\beta S} = \vec{\alpha}_0 i_{\alpha S} + \vec{\beta}_0 i_{\beta S} \Rightarrow \underline{i}_{\alpha\beta S} = i_{\alpha S} + j\, i_{\beta S}.$$

Similarly, considering the current vector in d–q system, axis d can be given attributes of the real axis, while axis q can be treated as imaginary axis, converting in this way d–q space into a complex plane, where the current is represented by the complex number $\underline{i}_{dq} = i_d + j i_q$. Complex notation of current vectors is not unique. With axis d as real axis, the number $\underline{i}_{dq}$ is obtained, different than the complex number $\underline{i}_{\alpha\beta}$ obtained with axis α_S as real axis.

15.23.1   Simplified Record of the Rotational Transform

By using the complex notation, Park transform can be written as

$$\underline{i}_{dq} = i_d + j i_q = \underline{i}_{\alpha\beta S} e^{-j\theta_e},$$

$$\underline{i}_{\alpha\beta S} e^{-j\theta_e} = \left(i_{\alpha S} + j i_{\beta S} \right)\left(\cos(\theta_e) - j \sin(\theta_e) \right)$$
$$= \left(i_{\alpha S} \cos(\theta_e) + i_{\beta S} \sin(\theta_e) \right) + j \left(-i_{\alpha S} \sin(\theta_e) + i_{\beta S} \cos(\theta_e) \right).$$

The inverse transform can be written as

$$\underline{i}_{\alpha\beta S} = \underline{i}_{dq} e^{+j\theta_e}.$$

Complex number $\underline{i}_{dq} = i_d + ji_q$ is a compact way of representing the stator current vector by representing the two state variables, i_d and i_q, as a single complex number. The choice of real axis (d) and complex axis (q) is in accordance with the fact that axis q leads by $\pi/2$ and that imaginary unit $j = \exp(j\pi/2)$ represents the phase shift of $\pi/2$. Complex notation $\underline{i}_{dq}$ acquires another significance in steady state. With i_d and i_q remaining constant, the steady-state value of $\underline{i}_{dq}$ becomes a complex constant, *phasor*. The amplitude and argument of this phasor determine the amplitude and the initial phase of the stator current.

15.24 Inductance Matrix in dq Frame

By application of Park transform, the rotor and stator windings are transferred to synchronously rotating d–q system. The virtual stator windings are denoted by subscripts d and q, while virtual rotor windings are denoted by subscripts D and Q. Rotation of dq frame does not change relative position of stator and rotor virtual phases; thus, all coefficients of relevant inductance matrix remain constant. The flux of the stator winding d is determined by the first row of matrix, $\Psi_d = L_s i_d + L_m i_D$. The mutual inductance is constant since windings d and D do not change their relative positions.

Revolving frame has an angular speed Ω_e which is the same as the speed of the revolving field. The speed Ω_e is determined by the angular frequency of stator voltages and currents, ω_e. For two-pole machines with $p = 1$, $\Omega_e = \omega_e$. The angle between axes d and α_S is equal to

$$\theta_e = \theta_e(0) + \int_0^t \Omega_e d\tau. \tag{15.49}$$

Stator currents are

$$\begin{bmatrix} i_d \\ i_q \end{bmatrix} = \begin{bmatrix} \cos\theta_e & \sin\theta_e \\ -\sin\theta_e & \cos\theta_e \end{bmatrix} \cdot \begin{bmatrix} i_{\alpha s} \\ i_{\beta s} \end{bmatrix}. \tag{15.50}$$

Complex notation of the stator current is

$$\underline{i}_{dq} = i_d + ji_q = e^{-j\theta_e}\underline{i}_{\alpha\beta}. \tag{15.51}$$

Rotor currents are

$$\begin{bmatrix} i_D \\ i_Q \end{bmatrix} = \begin{bmatrix} \cos\theta_{slip} & \sin\theta_{slip} \\ -\sin\theta_{slip} & \cos\theta_{slip} \end{bmatrix} \cdot \begin{bmatrix} i_{\alpha R} \\ i_{\beta R} \end{bmatrix}. \tag{15.52}$$

Fig. 15.11 Stator and rotor
windings in *dq* coordinate
frame

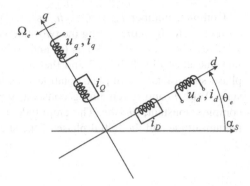

The stator and rotor windings are represented by virtual *dq* and *DQ* windings
residing in *dq* frame. Their magnetic axes coincide, and they do not move relative to
one another. Therefore, their mutual inductances do not change. The mutual
inductance between virtual windings *d* and *D* is L_m, as well as the mutual inductance between windings *q* and *Q*. The mutual inductance between windings in
orthogonal axes, such as *d* and *Q*, is equal to zero. The inductance matrix is

$$
\begin{bmatrix} \Psi_d \\ \Psi_q \\ \Psi_D \\ \Psi_Q \end{bmatrix} = \begin{bmatrix} L_s & 0 & L_m & 0 \\ 0 & L_s & 0 & L_m \\ L_m & 0 & L_R & 0 \\ 0 & L_m & 0 & L_R \end{bmatrix} \cdot \begin{bmatrix} i_d \\ i_q \\ i_D \\ i_Q \end{bmatrix} = \underline{L} \cdot \begin{bmatrix} i_d \\ i_q \\ i_D \\ i_Q \end{bmatrix}. \tag{15.53}
$$

Question (15.11): The electromagnetic torque acting on moving part of the system
which comprises several magnetically coupled contours is determined from the
expression $T_{em} = \frac{1}{2}\, i^T(dL/d\theta_m)i$, where L is the inductance matrix. Taking into
account the matrix given in (15.53), it is concluded that $dL/d\theta_m = 0$ which,
introduced into the torque expression, gives $T_{em} = 0$. Is this consideration correct?

Answer (15.11): Expression $T_{em} = \frac{1}{2} i^T(dL/d\theta_m)i$ has been derived starting from the
expression for the field energy and using the voltage equilibrium equations for actual
physical windings. There is no proof that the same expression applies for the inductance matrix and electrical currents of virtual, inexistent windings. The torque expression $\frac{1}{2}\, i^T(dL/d\theta_m)i$ can be used only with the inductance matrix L and electrical
currents that correspond to actual physical windings. Hence, substitution of (15.53)
into the torque expression is erroneous. Notice that the torque expression can be used
in conjunction with $\alpha\beta$ variables. This is due to the fact that one can actually build a
two-phase machine with $\alpha\beta$ windings which represents the original three-phase
machine. Virtual windings such as *d, q, D*, and *Q* cannot exist in an actual machine.
Yet, they are introduced as a means to understand and use Park rotational transform.

Question (15.12): The self-inductance L_S of the stator phase windings of a three-
phase machine and coefficient of magnetic coupling between the stator and rotor *k*
are known. Determine the coefficients of inductance matrix for the winding system
in Fig. 15.11.

Answer (15.12): The inductance matrix in (15.53) is obtained by applying Clarke $3\Phi/2\Phi$ transform to the original three-phase machine and then applying Park rotational transform to the two-phase $\alpha\beta$ equivalent. The applied transforms are invariant in terms of inductance. Therefore, elements L_{11} and L_{22} of the matrix are equal to L_S. Considering the rotor model, it is usually assumed that the short-circuited rotor winding has the same number of turns as the stator and that magnetic resistance along paths of the two fluxes is approximately equal; thus, $L_S = L_R$. The coefficients of mutual inductance are equal to $L_m = k(L_S L_R)^{0.5} = kL_S$.

Successive application of Clarke and Park transforms results in the transform known as Blondel transform.

15.25 Voltage Balance Equations in *dq* Frame

Mathematical model of the electrical subsystem comprises the voltage balance equations, differential equations that express the equilibrium of the supply voltage, the voltage drops, and electromotive forces in actual phase windings. These equations can be transformed from $\alpha\beta$ frame to dq frame, synchronous with the revolving field. In $\alpha\beta$ stationary coordinate system, voltage equilibrium in the stator phase windings is given by

$$u_{\alpha S} = R_S i_{\alpha S} + \mathrm{d}\Psi_{\alpha S}/\mathrm{d}t, \qquad u_{\beta S} = R_S i_{\beta S} + \mathrm{d}\Psi_{\beta S}/\mathrm{d}t. \qquad (15.54)$$

Multiplication of the second equation by j and summing the two equations result in a single voltage balance equation with voltages, fluxes, and currents expressed as complex numbers,

$$\underline{u}_{\alpha\beta S} = (u_{\alpha S} + ju_{\beta S}) = R_S(i_{\alpha S} + ji_{\beta S}) + \mathrm{d}(\Psi_{\alpha S} + j\Psi_{\beta S})/\mathrm{d}t$$
$$= R_S \underline{i}_{\alpha\beta S} + \mathrm{d}\underline{\Psi}_{\alpha\beta S}/\mathrm{d}t. \qquad (15.55)$$

Voltages of virtual d-phase and q-phase windings are obtained from $\alpha\beta$ voltages by applying Park transform,

$$\underline{u}_{dq} = u_d + ju_q = \underline{u}_{\alpha\beta S}e^{-j\theta_e} = \left(R_S \underline{i}_{\alpha\beta S} + \mathrm{d}\underline{\Psi}_{\alpha\beta S}/\mathrm{d}t\right)e^{-j\theta_e}. \qquad (15.56)$$

Variables $\underline{i}_{\alpha\beta S}$ and $\Psi_{\alpha\beta S}$ of the stationary coordinate system can be represented in terms of dq variables by applying inverse Park transform, $\underline{i}_{\alpha\beta S} = \underline{i}_{dq}\exp(-j\theta_e)$, to obtain

$$\underline{u}_{dq} = \left(R_S \underline{i}_{dq}e^{+j\theta_e} + \mathrm{d}(\underline{\Psi}_{dq}e^{+j\theta_e})/\mathrm{d}t\right)e^{-j\theta_e}$$
$$= R_S \underline{i}_{dq} + \mathrm{d}\underline{\Psi}_{dq}/\mathrm{d}t + j\omega_e\underline{\Psi}_{dq}. \qquad (15.57)$$

Therefore, voltage balance equations of virtual stator phases in dq frame do not have the form $u = Ri + \mathrm{d}\Psi/\mathrm{d}t$. They contain an additional member which comes as a consequence of Park transform. The above equation with complex numbers can be separated into real and imaginary parts, resulting in two scalar equations. The same procedure can be applied to the voltage balance equations of the rotor windings, but this time the angle θ_e is replaced by angle θ_{slip}. This is due to the fact that the angle between the original rotor coordinate system (α_R–β_R) and the target dq coordinate system is $\theta_{slip} = \theta_e - \theta_m$. For the virtual rotor windings (DQ), the following voltage balance equations are obtained:

$$
\begin{aligned}
\underline{u}_{DQ} &= \left(R_R \underline{i}_{DQ} e^{+j\theta_{slip}} + \mathrm{d}\left(\underline{\Psi}_{DQ} e^{+j\theta_{slip}}\right)/\mathrm{d}t \right) e^{-j\theta_{slip}} \\
&= R_R \underline{i}_{DQ} + \mathrm{d}\underline{\Psi}_{DQ}/\mathrm{d}t + j\omega_{slip}\underline{\Psi}_{DQ}.
\end{aligned}
\tag{15.58}
$$

15.26 Electrical Subsystem

This section summarizes the model of the electrical subsystem of induction machine expressed in synchronous dq frame, where the state variables are constant in the steady state and where the inductance matrix of virtual windings has constant elements. This model is based on the four approximations adopted in modeling induction machines. They are detailed in introductory chapters, and they include:

1. Neglected effects of distributed parameters.
2. The energy of electrical field is neglected along with parasitic capacitances.
3. Neglected are the iron losses.
4. Magnetic saturation is neglected along with nonlinear $B(H)$ characteristic of ferromagnetic materials.

Moreover, the leading coefficient of Clarke transform, used to replace the three-phase original with the two-phase equivalent, is $K = 2/3$. For simplicity, further considerations assume that the induction machine under consideration is two-pole machine where $p = 1$ and that electrical angular frequency ω corresponds to mechanical angular speed Ω, resulting in $\omega = p\Omega$.

The following equations give complete mathematical model of the electrical subsystem of an induction machine in the synchronously rotating dq coordinate system. The two voltage balance equations with complex variables are split into four voltage balance equations with scalars:

$$
u_{\mathrm{d}} = R_s i_{\mathrm{d}} + \frac{\mathrm{d}\Psi_{\mathrm{d}}}{\mathrm{d}t} - \omega_e \Psi_q,
\tag{15.59}
$$

$$u_q = R_s i_q + \frac{\mathrm{d}\Psi_q}{\mathrm{d}t} + \omega_e \Psi_d, \tag{15.60}$$

$$0 = R_R i_D + \frac{\mathrm{d}\Psi_D}{\mathrm{d}t} - \omega_{slip} \Psi_Q, \tag{15.61}$$

$$0 = R_R i_Q + \frac{\mathrm{d}\Psi_Q}{\mathrm{d}t} + \omega_{slip} \Psi_D. \tag{15.62}$$

The inductance matrix relates the flux linkages and currents of virtual windings:

$$\begin{bmatrix} \Psi_d \\ \Psi_q \\ \Psi_D \\ \Psi_Q \end{bmatrix} = \begin{bmatrix} L_s & 0 & L_m & 0 \\ 0 & L_s & 0 & L_m \\ L_m & 0 & L_R & 0 \\ 0 & L_m & 0 & L_R \end{bmatrix} \cdot \begin{bmatrix} i_d \\ i_q \\ i_D \\ i_Q \end{bmatrix}. \tag{15.63}$$

Question (15.13): Starting from expression for the torque $T_{em} = (3/2)(\Psi_{\alpha S} i_{\beta S} - \Psi_{\beta S} i_{\alpha S})$, express the torque T_{em} as function of the fluxes and currents in dq coordinate frame.

Answer (15.13): By using complex notation in vector representation, where $\Psi_{\alpha\beta S} = \Psi_{\alpha S} + j\Psi_{\beta S}$ and $\underline{i}_{\alpha\beta S} = i_{\alpha S} + j i_{\beta S}$, the torque can be written as $T_{em} = (3/2) \,\mathrm{Im}\, (\Psi^*_{\alpha\beta S} \underline{i}_{\alpha\beta S})$, where Im denotes the function taking imaginary part of the complex number, while $\Psi^*_{\alpha\beta S}$ is conjugate value of the complex number, the number with the imaginary part having the opposite sign. On the basis of Park transform expression $\Psi_{dq} = \Psi_d + j\Psi_q = \exp(-j\theta_e)\, \Psi_{\alpha\beta S}$ and $\underline{i}_{dq} = i_d + j\, i_q = \exp(-j\theta_e)\, \underline{i}_{\alpha\beta S}$, one obtains

$$\begin{aligned} T_{em} &= \frac{3}{2}\mathrm{Im}\left[\left(\underline{\Psi}_{dq} e^{j\theta_e} \right)^* \left(\underline{i}_{dq} e^{j\theta_e} \right) \right] \\ &= \frac{3}{2}\mathrm{Im}\left[\underline{\Psi}^*_{dq} e^{-j\theta_e} \underline{i}_{dq} e^{j\theta_e} \right] \\ &= \frac{3}{2}\mathrm{Im}\left[\underline{\Psi}^*_{dq} \underline{i}_{dq} \right] = \frac{3}{2}\mathrm{Im}\left[\Psi_d i_q - \Psi_q i_d \right]. \end{aligned}$$

Chapter 16
Induction Machines at Steady State

In this chapter, steady state operation of induction machines is studied with the aim to derive the equivalent circuit and mechanical characteristic of induction machine. The steady state model is derived from the dynamic model, developed and explained in the previous chapter. The voltage balance equations at steady state are used to develop the steady state equivalent circuit of the machine with squirrel cage rotor. At the same time, the concept of the equivalent transformer is introduced to derive the same steady state equivalent circuit. The equivalent circuit is used to determine the steady state currents, torque, power, losses, and flux linkages. Typical resistances and inductances of the equivalent circuit are explained and discussed, along with typical experimental procedures for their measurement and estimation. The system of relative units is introduced and explained, along with benefits that come from its use. Characteristic examples are studied to develop skills in working with relative units and selecting the base quantities used in scaling the absolute values into relative values. The functions that approximate the mechanical characteristic of induction machine and typical mechanical loads are introduced and explained. Natural mechanical characteristic is analyzed along with the start-up mode, rated operation, and no load operation. Breakdown torque is studied and explained in both motor and generator modes. Stable and unstable equilibrium points on mechanical characteristic are discussed and explained. The influence of machine resistances and reactances on the start-up torque, breakdown torque, breakdown slip, and coefficient of efficiency is analyzed and explained. This chapter proceeds by summarizing energy losses in windings, magnetic circuits, and mechanical losses due to rotation. Calculation of steady state losses is explained on the basis of the steady state equivalent circuit. The losses are presented in the form of power balance chart drawn for induction machine that operates in motoring mode. Simplified power balance is derived by splitting the air-gap power into rotor losses and mechanical power, according to the relative slip s. This chapter ends with deriving power balance chart for induction generator and discussing generator operating mode of induction machines.

S.N. Vukosavic, *Electrical Machines*, Power Electronics and Power Systems, 427
DOI 10.1007/978-1-4614-0400-2_16, © Springer Science+Business Media New York 2013

16.1   Input Power

Induction machines have short-circuited cage winding on the rotor, and they have three-phase stator winding. The stator winding is supplied from the source of three-phase voltages: u_a, u_b and u_c. No electrical power is supplied through the rotor winding. Electrical power supplied through the stator windings is $p_e = u_a i_a + u_b i_b + u_c i_c$. Using Clarke transform, three-phase machine is replaced by two-phase equivalent. With Clarke transform coefficient $K = 2/3$, the input power to the machine is equal to

$$
p_e = \frac{3}{2}\left(u_{\alpha S} i_{\alpha S} + u_{\beta S} i_{\beta S}\right)
$$

$$
= \frac{3}{2}\text{Re}\left[\left(u_{\alpha S} + j u_{\beta S}\right)\left(i_{\alpha S} - j i_{\beta S}\right)\right] = \frac{3}{2}\text{Re}\left[\left(\underline{u}_{\alpha\beta S}\right)\left(\underline{i}_{\alpha\beta S}\right)^{*}\right].
$$

By using Park transform, α_S and β_S phase windings are replaced by virtual d and q windings in synchronous coordinate frame. With $\underline{i}_{dq} = \exp(-j\theta_e)\, \underline{i}_{\alpha\beta S}$ and $\underline{u}_{dq} = \exp(-j\theta_e)\, \underline{u}_{\alpha\beta S}$, the input power can be expressed in terms of dq voltages and currents:

$$
p_e = \frac{3}{2}\text{Re}\left[\left(\underline{u}_{\alpha\beta S}\right)\left(\underline{i}_{\alpha\beta S}\right)^{*}\right] = \frac{3}{2}\text{Re}\left[\left(\underline{u}_{dq}e^{j\theta_e}\right)\left(\underline{i}_{dq}e^{j\theta_e}\right)^{*}\right]
$$

$$
= \frac{3}{2}\text{Re}\left[\underline{u}_{dq}e^{j\theta_e}\left(\underline{i}_{dq}\right)^{*}e^{-j\theta_e}\right] = \frac{3}{2}\text{Re}\left[\underline{u}_{dq}\left(\underline{i}_{dq}\right)^{*}\right] = \frac{3}{2}\left(u_d i_d + u_q i_q\right).
$$

Starting from expression $p_e = (3/2)(u_d i_d + u_q i_q)$ for power delivered to the machine from the electrical source, it is possible to calculate components of the input power and determine the electromagnetic torque. By using voltage balance equations for the stator winding, one obtains the source power p_e as

$$
p_e = \frac{3}{2}\left(u_d i_d + u_q i_q\right)
$$

$$
= \frac{3}{2}R_s\left(i_d^2 + i_q^2\right) + \frac{3}{2}\left(\frac{\mathrm{d}\Psi_d}{\mathrm{d}t}i_d + \frac{\mathrm{d}\Psi_q}{\mathrm{d}t}i_q\right) + \frac{3}{2}\omega_e\left(\Psi_d i_q - \Psi_q i_d\right)
$$

$$
= p_{Cu1} + \frac{\mathrm{d}W_m}{\mathrm{d}t} + p_\delta.
$$

Component p_{cu1} represents losses in copper of the stator winding. Component $\mathrm{d}W_m/\mathrm{d}t$ has dimension of power and represents the rate of change of the energy accumulated in magnetic field, the first derivative of the field energy. During the operation of induction machine, the air-gap flux may change its value. Therefore, the instantaneous value of $\mathrm{d}W_m/\mathrm{d}t$ is positive when the flux increases and negative when the flux decreases. The average value of $\mathrm{d}W_m/\mathrm{d}t$ must be equal to zero, since the accumulated energy cannot increase indefinitely nor can it decrease indefinitely.

In the case when iron losses p_{Fe} in the stator magnetic circuit are significant, they are accounted for along with dW_m/dt and subtracted from the source power. The remaining power coming from the source, denoted by p_δ and called *air-gap power* or *power of rotating field*, is transferred to the rotor. It is of interest to investigate the nature of the air-gap power and determine the mechanical power and torque which are, via shaft, transferred to mechanical load, namely, to work machine.

16.2 Torque Expression

The source power calculated in dq coordinate frame is

$$u_d i_d + u_q i_q = R_S\left(i_d^2 + i_q^2\right) + \left(\frac{d\Psi_d}{dt}i_d + \frac{d\Psi_q}{dt}i_q\right) + \omega_e\left(\Psi_d i_q - \Psi_q i_d\right). \quad (16.1)$$

With Clarke transform using $K = 2/3$, this power has to be multiplied by $3/2$ in order to get the input power to the original machine. The first factor on the right-hand side of equation is p_{cu1}, representing the copper losses in the stator winding. The second factor is dW_m/dt, while the remaining power is passed to the rotor through the air gap. The air-gap power of the original machine is

$$p_\delta = \frac{3}{2}\omega_e\left(\Psi_d i_q - \Psi_q i_d\right). \quad (16.2)$$

The air-gap power can also be obtained by calculating the surface integral of the pointing vector $E \times H$ through the cylindrical surface enveloping the rotor and passing through the air gap. Actually, the air-gap power represents the flux of the vector $E \times H$ through the air gap, toward the rotor. The air-gap power is positive when the vector $E \times H$ is directed from the stator toward the rotor. The air-gap power is equal to the product of the electromagnetic torque and the angular speed $\Omega_e = \omega_e/p$ of the revolving field, also called synchronous speed. Therefore, the torque can be calculated by dividing p_δ and the synchronous speed. Hence, $T_{em} = p_\delta/\Omega_e$, and this results in

$$T_{em} = \frac{3}{2}p\left(\Psi_d i_q - \Psi_q i_d\right). \quad (16.3)$$

Newton equation of motion determines variation of the speed:

$$J\frac{d\Omega_m}{dt} = T_{em} - \sum T_L. \quad (16.4)$$

16.3 Relative Slip

If the rotor of an induction machine rotates at the speed of $\Omega_m \neq \Omega_e$, different than the synchronous speed Ω_e, which determines the angular speed of the magnetic field, then the slip speed is $\Omega_{slip} = \Omega_e - \Omega_m$. The rotor speed changes during transients and represents one of the state variables. At steady state, the rotor revolves at a constant speed. The slip speed Ω_{slip} determines the angular frequency of the rotor currents, $\omega_{slip} = p\Omega_{slip}$. The ratio $s = \omega_{slip}/\omega_e = \Omega_{slip}/\Omega_e$ is called *relative slip*. For a two-pole motor fed from the mains with the line frequency of 50 Hz, the synchronous speed is 50 rev/s or 3,000 rpm (revolutions per minute). If the rotor revolves at 2,700 rpm, the relative slip is $s = 300/3,000 = 0.1$. For induction motor with locked rotor (the rotor which does not rotate), relative slip is $s = 1$.

16.4 Losses and Mechanical Power

Electromagnetic torque T_{em} acts on the rotor which revolves at the speed Ω_m. Related mechanical power $p_{mR} = T_{em}\Omega_m$ is also called internal mechanical power. It is slightly different than the mechanical power $p_m = T_m\Omega_m$ transferred to the external mechanical load via shaft. The difference between $p_{mR} = T_{em}\Omega_m$ and $p_m = T_m\Omega_m$ appears due to internal mechanical losses within induction machines. These losses are the air friction, friction in the bearings, and other losses in mechanical subsystem of the machine. Therefore, electromagnetic torque T_{em} is different than the torque T_m which is passed to the work machine through the shaft. Detailed analysis of losses in induction machine is given in Sect. 16.24, along with the balance of power which is shown in Fig. 16.17.

Mechanical power p_{mR} differs from the air-gap power p_δ by the amount of losses in rotor winding and rotor magnetic circuit. Electrical currents in rotor bars create losses proportional to the square of the current, Ri^2. Although the rotor bars are usually made of aluminum, losses in the cage are frequently called *copper losses* and denoted as P_{Cu2}. In addition to losses in the rotor windings, there are losses in the rotor magnetic circuit, denoted by P_{FeR}. Rotor magnetic circuit is, like that of the stator, made of iron sheets. It contains magnetic field with sinusoidal change at angular frequency ω_{slip}, creating losses due to hysteresis and eddy currents. The slip frequency ω_{slip} is much lower than the stator frequency ω_e; thus, it is justified to neglect iron losses in the rotor. This approximation is not justified for operating modes where the assumption $\omega_{slip} \ll \omega_e$ does not hold, namely, where the relative slip does not satisfy relation $s \ll 1$.

One example where the rotor iron losses cannot be neglected is the case of where the induction motor with its rotor at standstill gets connected to the mains. This is the most common way of starting mains supplied induction motors. With locked rotor, $\Omega_m = 0$ and $s = 1$. The frequency of electrical currents in the rotor cage corresponds to the line frequency as well as the frequency of changes of magnetic induction B within the rotor magnetic circuit.

Fig. 16.1 Components of the air-gap power

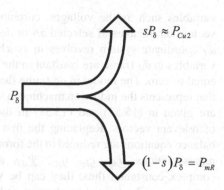

In steady state operation, the slip frequency is much lower than the line frequency. Therefore, iron losses in the rotor magnetic circuit are rather low. Total losses within the rotor are obtained by adding the rotor cage Joule losses to the iron loses in the rotor magnetic circuit. At steady state, mechanical power P_{mR} is equal to $T_{em}\Omega_m = T_{em}\Omega_e(1 - s) = (1 - s)P_\delta$. Therefore, remaining part sP_δ is the power of losses in the rotor:

$$P_{Cu2} + P_{FeR} = sP_\delta. \tag{16.5}$$

Question (16.1): A two-pole induction motor is fed from the mains and operates at steady state. It is known that the iron losses in stator magnetic circuit cannot be neglected. The losses in copper of the stator windings P_{cu1}, apparent power S which the motor takes from the mains, and phase delay φ of the current with respect to the voltage are known. Determine the air gap and electromagnetic torque.

Answer (16.1): In steady state, the average value of the rate of change dW_m/dt of the field energy is equal to zero. The air-gap power is $P_\delta = S \cos(\varphi) - P_{Fe} - P_{cu1}$. The electromagnetic torque is equal to $T_{em} = pP_\delta/\omega_e = P_\delta/\Omega_e$, where ω_e is the angular frequency of the mains.

Question (16.2): A mains-fed induction machine rotates at the speed of 2,700 rpm. Estimate the coefficient of efficiency of the machine.

Answer (16.2): Relative slip is $s = (3,000-2,700)/3,000 = 0.1$. Neglecting losses in the stator, one obtains that the useful power is close to 90% of input power, while the rotor losses account for 10% of the input power.

16.5 Steady State Operation

Park transform of state variables has been introduced with the aim to obtain mathematical model of induction machine where the steady state values of state variables are constant. With all two-phase representations of induction machine, variables such as the voltages, currents, or fluxes are projections of respective vectors on the axes of selected $\alpha\beta$ or dq coordinate system. With Park transform, dq coordinate system revolves in synchronism with the field. Therefore, all the variables in dq frame are constant in the steady state, and their first derivatives are equal to zero. The model in dq frame facilitates derivation of the equivalent circuit that represents the induction machine in steady state. The voltage balance equations are given in (15.57) and (15.58) in the form that uses complex representation of relevant vectors. Replacing the first time derivatives with zeros, the voltage balance equations are reduced to the form $\underline{U} = R\underline{I} + j\omega\underline{\Psi}$. At steady state, complex representations $\underline{u}_{dq}$, $\underline{i}_{dq}$, $\underline{i}_{DQ}$, $\underline{\Psi}_{dq}$, and $\underline{\Psi}_{DQ}$ of relevant vectors become complex constants; thus, they can be written as $\underline{U}_s$, $\underline{I}_s$, $\underline{I}_R$, $\underline{\Psi}_s$, and $\underline{\Psi}_R$. These constants can be treated as phasors, as both quantity represent the amplitude and phase of variables that have sinusoidal change at steady state. They differ from common phasors[1] by representing the vectors associated to three-phase winding system. At the same time, common phasors have the amplitude which corresponds to rms value of relevant sinusoidal variable, while the absolute value of complex numbers $\underline{U}_s$, $\underline{I}_s$, $\underline{I}_R$, $\underline{\Psi}_s$, and $\underline{\Psi}_R$ corresponds to maximum value of relevant sinusoidal variable. Namely, the absolute value of the phasor $\underline{I}_s$ corresponds to maximum value of electrical currents $i_{\alpha s}(t)$ and $i_{\beta s}(t)$. On the other hand, these relations change if the leading coefficient K of Clarke transform is not 2/3. In general, the ratio between the rms value of the stator voltage and the absolute value $U_s = (U_d^2 + U_q^2)^{0.5}$ of complex constant $\underline{U}_s$ is determined by coefficient K used in $3\Phi/2\Phi$ transform. Considerations throughout this book assume that $K = 2/3$, which results in the module of U_S equal to the peak value of phase voltages:

$$\underline{u}_{dq} = R_s\underline{i}_{dq} + \frac{d}{dt}\underline{\Psi}_{dq} + j\omega_e\underline{\Psi}_{dq},$$

$$0 = R_R\underline{i}_{DQ} + \frac{d}{dt}\underline{\Psi}_{DQ} + j\omega_{slip}\underline{\Psi}_{DQ},$$

$$\underline{u}_{dq} = \underline{U}_S, \ \underline{i}_{dq} = \underline{I}_S, \ \underline{i}_{DQ} = \underline{I}_R, \ \underline{\Psi}_{dq} = \underline{\Psi}_S, \ \underline{\Psi}_{DQ} = \underline{\Psi}_R.$$

In order to facilitate the analysis of steady state operation of induction machines, it is desirable to represent the voltage balance equations by the steady state equivalent circuit. There is, however, a problem in doing that. The stator equations comprise the angular frequency ω_e while the rotor equations have the frequency ω_{slip}. If both equations are to be represented by a unique circuit which makes the use of phasors to represent the voltage balance equations in the steady state, such circuit

---

[1] Common use of phasor has to do with representing the variables with sinusoidal change by complex numbers, where the phasor amplitude represents the rms value of corresponding AC variable, while the phasor argument represents the initial phase. An example is the complex number $\underline{V} = V \cdot \cos(\varphi) + j \cdot V \cdot \sin(\varphi)$ which represents the AC variable $v(t) = V \cdot \text{sqrt}(2) \cdot \cos(\omega t + \varphi)$.

must make the use of the same angular frequency. Namely, the phasor concept is applicable to current circuits where all the voltages and currents have the same angular frequency. Stator voltages and currents have the angular frequency ω_e:

$$\underline{U}_S = R_S \underline{I}_S + j\omega_e \underline{\Psi}_S.$$

Hence, the same frequency should be used in voltage balance equations describing the rotor winding. At steady state operating conditions, relative slip $s = \omega_{slip}/\omega_e$ is constant:

$$s = \frac{\omega_{slip}}{\omega_e} = \frac{\Omega_{slip}}{\Omega_e}.$$

In cases where the relative slip is equal to zero, there are no electromotive forces induced in short-circuited turns of the rotor cage, and the rotor current is equal to zero. This condition is called *no load condition*, wherein the rotor revolves in synchronism with the magnetic field. In all other cases, relative slip s is different than zero. With $s \neq 0$, the rotor equation can be divided by relative slip s, and the voltage balance of the rotor circuit becomes

$$0 = \frac{R_R}{s} \underline{I}_R + j\omega_e \underline{\Psi}_R \tag{16.6}$$

while the voltage balance equation for the stator remains

$$\underline{U}_S = R_S \underline{I}_S + j\omega_e \underline{\Psi}_S. \tag{16.7}$$

The fluxes can be expressed as functions of electrical currents in the windings: $\underline{\Psi}_R = L_R \underline{I}_R + L_m \underline{I}_S$, $\underline{\Psi}_S = L_S \underline{I}_S + L_m \underline{I}_R$. Recall at this point that the two-phase equivalent of the rotor cage, whether α_R-β_R or D-Q, may have an arbitrary number of turns N_R; the present analysis assumes that $N_R = N_S$. Much like with power transformers, the sum of stator and rotor currents $\underline{i}_m = \underline{i}_s + \underline{i}_R$ can be called and treated as *magnetizing current*. Now, one can write $\underline{\Psi}_R = L_{\gamma R} \underline{I}_R + L_m \underline{I}_m$, $\underline{\Psi}_S = L_{\gamma s} \underline{I}_s + L_m \underline{I}_m$, where $L_{\gamma R} = L_R - L_m$ and $L_{\gamma s} = L_S - L_m$ are the leakage inductances of the respective windings, while $\underline{\Psi}_m = L_m \underline{I}_m = \Psi_{md} + j\Psi_{mq}$ is *magnetizing flux*. This flux passes through the air gap and encircles both the stator and rotor windings. Magnetizing flux is also called *air-gap flux*. Equations (16.6) and (16.7) provide the steady state relations between voltages, currents, and fluxes of an induction machine. In order to determine the steady state equivalent circuit, it is necessary to express the flux linkages as functions of winding currents and inductance coefficients:

$$\underline{\Psi}_s = L_{\gamma s} \underline{I}_s + L_m \underline{I}_m, \tag{16.8}$$

$$\underline{\Psi}_R = L_{\gamma R} \underline{I}_R + L_m \underline{I}_m.$$

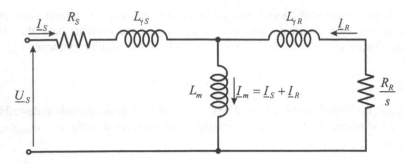

Fig. 16.2 Steady state equivalent circuit

Steady state voltage balance equations for stator and rotor windings can be represented by the equivalent circuit shown in Fig. 16.2. Voltage $\underline{U}_s$ fed from the left-hand side of the circuit is equal to $\underline{U}_s = \underline{U}_{dq} = U_d + jU_q$, where U_d and U_q are the values obtained by applying the $3\Phi/2\Phi$ transform to the phase voltages u_a, u_b, and u_c and then Park transform to obtain dq components. Phasors $\underline{I}_s = I_d + jI_q$ and $\underline{I}_R = I_D + jI_Q$ determine the amplitudes and phases of the stator and rotor currents at steady state. Resistance R_s and inductance $L_{\gamma s}$ represent the resistance and leakage inductance of the stator winding. Resistance R_R and inductance $L_{\gamma R}$ represent the resistance and leakage inductance of the equivalent two-phase rotor winding that represent the rotor cage. Parameters R_R and $L_{\gamma R}$ are *referred to the stator side*. That means that the values of R_R and $L_{\gamma R}$ correspond to the equivalent two-phase rotor winding which has the same number of turns per phase as the stator windings, $N_S = N_R$. All further developments start with an assumption that the rotor variables are referred to the stator side, namely, that short-circuited rotor cage is modeled by a pair of short-circuited windings α_R and β_R which have the same number of turns as the stator windings.

Referring the rotor parameters to the stator side is similar to scaling the secondary circuit impedance of a power transformer to the primary side. With impedance Z_2 of the secondary circuit, and with the transformation ratio of the power transformer $m = N_1/N_2$, the primary side equivalent of the secondary impedance becomes $m^2 Z_2$. With an induction motor, secondary winding of *transformer* is short-circuited rotor cage, while the stator phases represent the primary winding. Two diametrically positioned rotor bars can be considered as one-phase winding of the rotor. This phase winding has $N_R = 1$ turns. Assume that resistance of this short-circuited turn is $R_2 = 1\,m\Omega$ and that the stator winding has $N_S = 40$ turns; the value R_R of the rotor resistance referred to the stator side becomes $R_R = 1\,m\Omega \cdot (N_S/N_R)^2 = 1.6\,\Omega$. The assumption adopted in all the subsequent considerations is that the short-circuited cage is represented by a two-phase winding with $N_R = N_S$ turns; thus, all rotor variables are implicitly referred to the stator side.

16.6 Analogy with Transformer

The steady state equivalent circuit of induction machine can be determined by using the analogy with transformer, where the stator stands for primary winding of transformer, while the rotor represents short-circuited secondary. The stator of an induction machine has three-phase winding. Therefore, the analogy can be made with a three-phase transformer having short-circuited secondary.

The difference in operation of an induction machine and a transformer is that the stator and rotor currents do not have the same angular frequency. The reason for that is the rotor motion. It is of interest to consider a two-pole induction machine supplied from symmetrical three-phase mains of the line frequency ω_e. Magnetic field of the machine revolves at the speed of $\Omega_e = \omega_e$. With locked rotor ($\Omega_m = 0$), the field revolves at the same speed Ω_e with respect to both stator and rotor. Therefore, the angular frequency of electromotive forces induced in rotor cage is equal to the line frequency ω_e. The angular frequency of rotor electromotive forces and currents is also called slip frequency, and it is calculated as $\omega_{slip} = s\omega_e$, where $s = (\Omega_e - \Omega_m)/\Omega_e$ is relative slip, which is equal to 1 in locked rotor conditions. Hence, in locked rotor conditions, an induction machine corresponds in full to a three-phase transformer with short-circuited secondary windings. This situation changes when the rotor is set to motion and revolves at the speed $\Omega_m > 0$. Since the rotor revolves in the same direction as the field, the difference $\Omega_e - \Omega_m$ between the two speeds gets smaller. Therefore, rotation of the field with respect to the rotor cage and consequential electromotive forces and currents has the frequency $\omega_{slip} = s\omega_e$ which is smaller than the line frequency due to $s < 1$. Due to rotor motion, induction machine operates as three-phase transformer with short-circuited secondary which revolves with respect to the primary and, therefore, has the electrical currents of reduced angular frequency $\omega_{slip} = s\omega_e$. The presence of different frequencies in stator and rotor circuits is an obstacle to deriving an equivalent circuit that would represent the whole machine. Further considerations are directed to this aim.

The resultant magnetomotive force and flux Ψ_m in the air gap arise due to electrical currents of both the stator and rotor windings. The sum of the stator and rotor currents $\underline{I}_m = \underline{I}_S + \underline{I}_R$ represents the magnetizing current, the same way as the sum $\underline{I}_m = \underline{I}_1 + \underline{I}_2'$ of the primary current $\underline{I}_1$ and secondary current $\underline{I}_2' = (N_2/N_1)\underline{I}_2$ represents the magnetizing current of the transformer. In transformers, $\underline{I}_2' = (N_2/N_1)\underline{I}_2$ represents the primary side equivalent of the secondary current, or the secondary current referred to the primary side. In induction machine model, the rotor cage is replaced by equivalent short-circuited winding with the same number of turns as the stator phases, thus $N_S = N_R$, which permits the magnetizing current to be written as $\underline{I}_m = \underline{I}_S + \underline{I}_R$. Multiplying the magnetizing current by the number of turns gives the resultant magnetomotive force in the machine, $\underline{F}_m = \underline{F}_S + \underline{F}_R$, which is the sum of the magnetomotive forces of the stator and rotor windings. Mutual flux is also called air-gap flux, it encircles both stator and rotor windings, it passes through the air gap, and it is calculated from expression $\Phi_m = \underline{F}_m/R_\mu$, where R_μ represents magnetic resistance and Φ_m is the mutual flux in one turn.

Fig. 16.3 Voltage balance
in stator winding

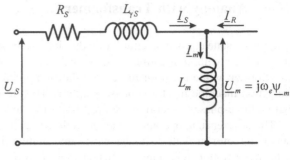

Fig. 16.4 Voltage balance
in rotor winding

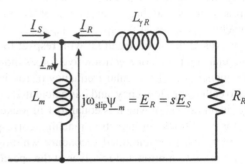

Mutual flux of the winding $\underline{\Psi}_m$ is obtained by adding up the mutual flux linkages of individual turns. With concentrated windings, where all the turns reside in the same place and, hence, have the same flux Φ_m, the winding flux is obtained as $\underline{\Psi}_m = N_N \Phi_m$, where N_N represents the number of turns. For windings with sinusoidal distribution of conductors, individual turns do not reside at the same place and do not have the same flux. The winding flux is obtained by integration, as explained in introductory chapters, resulting in $\underline{\Psi}_m = (\pi/4) N_N \Phi_m$. In the stator phase windings of an induction machine, the flux has sinusoidal change at the angular frequency of the supply voltages. Mains supplied machines have the line frequency ω_e. The electromotive force induced in the stator windings due to rotation of the mutual flux $\underline{\Psi}_m$ is $\underline{E}_S = \underline{U}_m = j\omega_e \underline{\Psi}_m$. The voltage $\underline{U}_m$ is called *magnetizing branch* voltage, as it appears across the element L_m in Fig. 16.3. The voltage balance equation of the stator winding is $\underline{U}_S = R_S \underline{I}_S + j\omega_e L_{\gamma S} \underline{I}_S + \underline{E}_S$, and it is illustrated in Fig. 16.3.

Since $N_S = N_R$, the same mutual flux $\underline{\Psi}_m$ exists in short-circuited rotor turns. The mutual flux in rotor turns has sinusoidal change. The flux revolves with respect to the rotor at slip speed Ω_{slip}. Therefore, the frequency of the flux changes is ω_{slip}. The flux changes result in the rotor electromotive force $\underline{E}_R = s\underline{U}_m = j\omega_{slip} \underline{\Psi}_m$. The voltage balance equation of the rotor cage is $\underline{U}_R = 0 = R_R \underline{I}_R + j\omega_{slip} L_{\gamma R} \underline{I}_R + \underline{E}_R$, and it is shown in the Fig. 16.4.

All the impedances of the rotor equivalent circuit in Fig. 16.4 can be divided by the relative slip $s = \omega_{slip}/\omega_e$, while maintaining the circuit topology and leaving all the currents unchanged. Modified circuit will have the voltages divided by s as well as

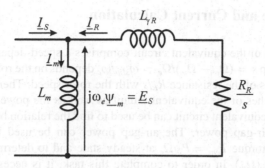

Fig. 16.5 Rotor circuit after division of impedances by s

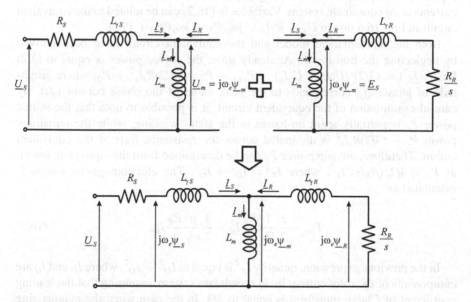

Fig. 16.6 Steady state equivalent circuit

the impedances. Resistance of the rotor branch takes the value R_R/s. Reactances $\omega_{slip}L$ obtain new values $\omega_e L$ due to $\omega_{slip}/s = \omega_e$. In this manner, a new numerical value of the angular frequency becomes ω_e, which facilitates connecting the rotor and stator circuits (Figs. 16.3, 16.4). The voltages in the rotor circuit are increased $1/s$ times; thus, the electromotive force $\underline{E}_R = s\underline{U}_m$ becomes equal to $\underline{U}_m$. After division of impedances by the relative slip s, the rotor equivalent circuit becomes as shown in Fig. 16.5.

By connecting the equivalent stator circuit to the equivalent rotor circuit where the impedances are multiplied by $1/s$, one obtains the equivalent circuit shown in Fig. 16.6. It represents the steady-state relations of currents, voltages, and flux linkages of an induction machine.

16.7   Torque and Current Calculation

The rotor branch of the equivalent circuit comprises a speed-dependent resistance R_R/s. Relative slip $s = (\Omega_e - \Omega_m)/\Omega_e = \omega_{slip}/\omega_e$ depends on the rotor speed, which results in changes of the resistance R_R/s with the rotor speed. Therefore, electrical currents in branches of the equivalent circuit as well as the power depend on the rotor speed. The equivalent circuit can be used to find the relation between the rotor speed and the air-gap power. The air-gap power can be used to calculate the electromagnetic torque $T_{em} = P_\delta/\Omega_e$ at steady state and to determine mechanical characteristic $T_{em}(\Omega_m)$. In order to complete this task, it is necessary to analyze the equivalent circuit and calculate the steady state value of the air-gap power. Equation (16.2) expresses the air-gap power as function of flux linkages and currents in dq coordinate system. Variables in (16.2) can be related to the equivalent circuit in Fig. 16.6 due to $\underline{U}_S - R_S\underline{I}_S = j\omega_e\underline{\Psi}_S = j\omega_e(\Psi_d + j\Psi_q)$.

Both the mathematical model and the equivalent circuit have been derived by neglecting the iron losses. At steady state, the source power is equal to $(3/2)$ $\mathrm{Re}(\underline{U}_S\,\underline{I}_S^{*}) = (3/2)\,(U_dI_d + U_qI_q) = P_{cu1} + P_\delta = (3/2)R_SI_S^2 + P_\delta$, where amplitude of phasor $|\underline{I}_s|$ corresponds to the peak value of the phase current $i_a(t)$. By careful examination of the equivalent circuit, it is possible to note that the source power P_e is partially spent on losses in the stator winding, while the remaining power $P_e - (3/2)R_sI_s^2$ is dissipated across the resistance R_R/s of the equivalent circuit. Therefore, *air-gap power* P_δ can be determined from the equivalent circuit as $P_\delta = 3/2\,(R_R/s)\,I_R^2$, where $I_R^2 = I_D^2 + I_Q^2$. The electromagnetic torque is calculated as

$$T_{em} = \frac{3}{2}\frac{1}{\Omega_e}\frac{R_R}{s}I_R^2 = \frac{3}{2}\frac{p}{\omega_e}\frac{R_R}{s}I_R^2. \tag{16.9}$$

In the previous expression, quantity I_R^2 is equal to $I_D^2 + I_Q^2$, where I_D and I_Q are components of the rotor current in dq coordinate system, assuming that the leading coefficient of Clarke transform is equal to 2/3. In the case when the magnetizing current is significantly smaller than the stator current, $|\underline{I}_m| \ll |\underline{I}_s|$, it is justified to make the assumption $I_R^2 \approx I_S^2$. The rotor current and magnetizing current can be expressed in terms of the stator current:

$$\underline{I}_{R1} = -\underline{I}_R = \frac{j\omega_eL_m}{j\omega_eL_m + j\omega_eL_{\gamma R} + R_R/s}\,\underline{I}_S,$$

$$\underline{I}_m = \frac{j\omega_eL_{\gamma R} + R_R/s}{j\omega_eL_m + j\omega_eL_{\gamma R} + R_R/s}\,\underline{I}_S.$$

When the machine has a relatively small slip s, the active part of the rotor impedance is significantly larger than the reactive part of the same impedance, $R_R/s \gg \omega_eL_{\gamma R}$. In this case, the magnetizing current and the rotor current can be approximated as

$$\underline{I}_m \approx \frac{R_R/s}{j\omega_e L_m + R_R/s} \underline{i}_S, \qquad \underline{I}_{R1} \approx \frac{j\omega_e L_m}{j\omega_e L_m + R_R/s} \underline{I}_S.$$

Under assumptions, the rotor current $\underline{I}_R$ has phase advance of $\pi/2$ with respect to the magnetizing current $\underline{I}_m$. Due to this phase shift, the impact of relatively small I_m on amplitude of the stator current is significantly reduced. With $I_S{}^2 = I_m{}^2 + I_R{}^2$ and $I_m{}^2 \ll I_R{}^2$, the amplitude of the stator current is approximately equal to the amplitude of the rotor current, $I_R{}^2 \approx I_S{}^2$. Therefore, approximate expression for the electromagnetic torque is $T_{em} = 3R_R I_S{}^2/(2s\Omega_e)$, while approximate value of the air-gap power is $P_\delta = 3/2\ (R_R/s)\ I_S{}^2$.

Question (16.3): The leading coefficient K of Clarke $3\Phi/2\Phi$ transform is equal to $2/3$. Discuss the relation between the stator current $\underline{I}_S$ and the phase currents $i_{\alpha S}(t)$ and $i_{\beta S}(t)$. Relate the expression for the air-gap power $P_\delta = 3/2\ (R_R/s)\ I_S{}^2$ to the expression $u_a i_a + u_b i_b + u_c i_c$ for the power of the three-phase winding system.

Answer (16.3): One of the properties of Clarke transform with $K = 2/3$ is that all the variables have the same peak values in the original abc domain and in the two-phase α_S–β_S coordinate system. Therefore, if the rms value I_{rms} of stator currents i_a, i_b, and i_c is known, then the peak value of phase currents $i_{\alpha S}(t)$ and $i_{\beta S}(t)$ of the two-phase equivalent is equal to $I_{rms}\ 2^{0.5}$. Phase currents $i_{\alpha S}(t)$ and $i_{\beta S}(t)$ have mutual phase shift of $\pi/2$; thus, $i_{\alpha S}{}^2 + i_{\beta S}{}^2 = 2I_{rms}{}^2$. Park coordinate transform of currents $i_{\alpha S}$ and $i_{\beta S}$ does not alter the amplitude of the stator current vector; thus, $I_S{}^2 = i_d{}^2 + i_q{}^2 = 2I_{rms}{}^2$. The air-gap power $P_\delta = 3/2\ (R_R/s)\ I_S{}^2$ can be written as $P_\delta = 3(R_R/s)\ I_{rms}{}^2$, which corresponds to the power of a three-phase star-connected symmetrical resistive load R_R/s having the current of rms value equal to the rms value of the motor phase currents.

16.8 Steady State Torque

Electromagnetic torque developed by an induction machine at steady state depends on the supply voltage, on the rotor speed Ω_m, and on the machine parameters. It can be determined by the following procedure:

- Determine relative slip $s = (\Omega_e - \Omega_m)/\Omega_e$.
- Introduce the supply voltage $\underline{U}_S$ and the resistance R_R/s in the equivalent circuit and calculate the steady state rotor current $\underline{I}_R$.
- Determine the air-gap power $P_\delta = 3/2\ (R_R/s)\ i_R{}^2$.
- Determine the torque by dividing the power by synchronous speed, $T_{em} = P_\delta/\Omega_e$.

With the assumption that $I_R{}^2 \approx I_S{}^2$, that is, that the magnetizing current is relatively small $|\underline{I}_m| \ll |\underline{I}_s|$, the stator and rotor currents are equal to

$$\underline{I}_S \approx \frac{U_S}{(R_S + R_R/s) + j\omega_e(L_{\gamma R} + L_{\gamma S})} \approx \underline{I}_R,$$

$$I_S^2 \approx \frac{U_S^2}{(R_S + R_R/s)^2 + \omega_e^2(L_{\gamma R} + L_{\gamma S})^2} \approx I_R^2. \qquad (16.10)$$

Value of electromagnetic torque is equal to the quotient of the air-gap power and the revolution speed of the magnetic field, also called *synchronous speed* and denoted by Ω_e. Considered so far are the two-pole induction machines, where the phase windings a, b, and c are spatially shifted by $2\pi/3$. With phase currents i_a, i_b, and i_c of the same amplitude and the same angular frequency ω_e and with initial phases displaced by $2\pi/3$, two-pole induction machine has magnetic field which revolves at the speed of $\Omega_e = \omega_e$ and which has two diametrical magnetic poles, displaced by π. Every set of windings that creates magnetic field having one north pole and one south pole is called *two-pole* winding. There are ways to construct the winding system which creates magnetic field with multiple pairs of magnetic poles. With phase windings made of two or more part distributions along the machine circumference, it is possible to create magnetic field with two or more pole pairs. In general, stator winding could have $2p$ magnetic poles, where p determines the number of pole pairs. Induction machines analyzed so far have $p = 1$ pole pairs, resulting in $\Omega_e = \omega_e$.

Magnetic field of machines with multiple-pole pairs revolves slower than the field of two-pole machines. Synchronous speed in an induction machine with $p > 1$ pairs of poles is equal to $\Omega_e = \omega_e/p$. A more detailed analysis of the multipole machines and distribution of their magnetic fields will be carried out further on. Unless otherwise stated, it is assumed that induction machine under the scope is a two-pole machine ($p = 1$) with $\Omega_e = \omega_e$, where the symbol ω_e represents electrical frequency of currents and voltages while Ω_e stands for the angular speed of the magnetic field. Notwithstanding the number of magnetic poles, the electromagnetic torque is the quotient of the air-gap power and the angular speed of the magnetic field, also called synchronous speed:

$$T_{em} = \frac{1}{\Omega_e} P_\delta = \frac{1}{\Omega_e} \frac{3}{2} \omega_e \left(\Psi_d I_q - \Psi_q I_d\right) = \frac{3p}{2} \left(\Psi_d I_q - \Psi_q I_d\right). \qquad (16.11)$$

The rotor mechanical speed is denoted by Ω_m. The electrical equivalent of the rotor speed is $\omega_m = p\Omega_m$. For two-pole machines, $\Omega_m = \omega_m$, since $p = 1$. Separate notation of the mechanical speed and electrical frequency is also introduced for the slip speed. The electrical equivalent of the slip speed is the angular frequency of rotor currents, and it is denoted by ω_{slip}, while mechanical speed of the rotor lagging with respect to the synchronous speed is denoted by Ω_{slip}. For two-pole machines with $p = 1$, the angular frequency of the rotor currents and the slip speed have the same value, $\Omega_{slip} = \omega_{slip}$.

With the assumption $|\underline{I}_m| \ll |\underline{I}_s|$, the electromagnetic torque is

$$T_{em} = \frac{P_\delta}{\Omega_e} = \frac{3}{2\Omega_e} \frac{R_R}{s} \frac{U_S^2}{(R_S + R_R/s)^2 + \omega_e^2 (L_{\gamma R} + L_{\gamma S})^2}$$

$$= \frac{3pR_R}{2\omega_e s} \frac{U_S^2}{(R_S + R_R/s)^2 + \omega_e^2 (L_{\gamma R} + L_{\gamma S})^2}. \tag{16.12}$$

In the above expression, U_S is the peak value of the phase voltages, p is the number of pole pairs, ω_e is the frequency of stator voltages, and currents R_S and $L_{\gamma S}$ are parameters of the stator winding, while R_R and $L_{\gamma R}$ are parameters of the rotor winding referred to the stator side. In order to determine the mechanical characteristic, it is required to calculate relative slip $s = (\Omega_e - \Omega_m)/\Omega_e = (\omega_e - \omega_m)/\omega_e$, to introduce s in the above expression, and to calculate the torque.

Question (16.4): Determine the expression for electromagnetic torque of an induction machine starting from the steady state equivalent circuit. The assumption $|\underline{I}_m| \ll |\underline{I}_s|$ cannot be made. The product $R_S R_R/s$ can be neglected compared to $X_{\gamma S} X_{\gamma R}$.

Answer (16.4): The electromagnetic torque is determined as the ratio of the air-gap power and the synchronous speed Ω_e. The air-gap power P_δ is equal to the power across the resistance R_R/s, residing on the right-hand side of the equivalent circuit. In order to calculate the air-gap power without customary approximation ($|\underline{I}_m| \ll |\underline{I}_s|$), it is necessary to find the rotor current and calculate $P_\delta = (3/2)\,(R_R/s)\,I_R^2$. By adopting notation $\underline{Z}_S = R_S + j\omega_e L_{\gamma S} = R_S + jX_{\gamma S}$, $\underline{Z}_R = R_R/s + j\omega_e\,L_{\gamma R} = R_R/s + jX_{\gamma R}$, and $\underline{Z}_m = j\omega_e\,L_m = jX_m$, the stator current can be expressed as

$$\underline{I}_S = \frac{U_S}{\underline{Z}_S + \frac{\underline{Z}_R \underline{Z}_m}{\underline{Z}_R + \underline{Z}_m}} = \frac{U_S(\underline{Z}_R + \underline{Z}_m)}{\underline{Z}_S \underline{Z}_m + \underline{Z}_R \underline{Z}_m + \underline{Z}_R \underline{Z}_S},$$

while the rotor current becomes

$$\underline{I}_R = \frac{-U_S \underline{Z}_m}{\underline{Z}_S \underline{Z}_m + \underline{Z}_R \underline{Z}_m + \underline{Z}_R \underline{Z}_S}$$

$$= \frac{-U_S jL_m \omega_e}{(R_S + jL_{\gamma S}\omega_e)jL_m\omega_e + (R_R/s + jL_{\gamma R}\omega_e)jL_m\omega_e + (R_R/s + jL_{\gamma R}\omega_e)(R_S + jL_{\gamma S}\omega_e)},$$

adopting the rotor current reference direction from right to left. Dividing the numerator and denominator by impedance $j\omega_e L_m$,

$$\underline{I}_R = \frac{-U_S}{(R_S + jL_{\gamma S}\omega_e) + \left(\frac{R_R}{s} + jL_{\gamma R}\omega_e\right) + \left(\frac{L_{\gamma S}}{L_m}\frac{R_R}{s} + \frac{L_{\gamma R}}{L_m}R_S\right) + j\left(\frac{L_{\gamma R}L_{\gamma S}\omega_e}{L_m} - \frac{R_R R_S}{s\omega_e L_m}\right)}.$$

By neglecting $R_S R_R/(s\omega_e)$ compared to the product $L_{\gamma S}L_{\gamma R}\omega_e$, the product $R_S R_R/s$ becomes negligible with respect to $X_{\gamma S} X_{\gamma R}$, resulting in

$$
\underline{I}_R = \frac{-\underline{U}_S}{R_S\left(1 + \frac{L_{\gamma R}}{L_m}\right) + \frac{R_R}{s}\left(1 + \frac{L_{\gamma S}}{L_m}\right) + j\left(L_{\gamma S}\omega_e + L_{\gamma R}\omega_e\left(1 + \frac{L_{\gamma S}}{L_m}\right)\right)}
$$

$$
= \frac{-\underline{U}_S}{R_S\frac{L_R}{L_m} + \frac{R_R}{s}\frac{L_S}{L_m} + j\left(L_{\gamma S}\omega_e + L_{\gamma R}\omega_e\frac{L_S}{L_m}\right)}.
$$

Coefficients $v^S = L_S/L_m > 1$ and $v^R = L_R/L_m > 1$ are introduced, and their values are close to one. In the case when $L_S = L_R$, mutual inductance is equal to $L_m = k(L_S L_R)^{0.5} = kL_S$; thus, $v^S = v^R = 1/k$. Coupling coefficient of the windings k is close to one; thus, $v^S = v^R \approx 1$. With this in mind, the electromagnetic torque can be determined by using the following expression:

$$
T_{em} = \frac{P_{ob}}{\Omega_e} = \frac{3}{2\Omega_e}\frac{R_R}{s}I_R^2 = \frac{3}{2\Omega_e}\frac{R_R}{s}\frac{U_S^2}{\left(v^R R_S + v^S R_R/s\right)^2 + \omega_e^2\left(v^S L_{\gamma R} + L_{\gamma S}\right)^2}
$$

$$
= \frac{3pR_R}{2\omega_e s}\frac{U_S^2}{\left(v^R R_S + v^S R_R/s\right)^2 + \omega_e^2\left(v^S L_{\gamma R} + L_{\gamma S}\right)^2}.
$$

In the case when $v^S = v^R \approx 1$, the expression for the torque takes the form (16.12)

$$
T_{em} = \frac{3pR_R}{2\omega_e s}\frac{U_S^2}{\left(R_S + R_R/s\right)^2 + \omega_e^2\left(L_{\gamma R} + L_{\gamma S}\right)^2}.
$$

This form will be used in most subsequent considerations. Higher values of coefficients v^S and v^R are encountered in induction machines with increased leakage inductance and reduced coefficient of coupling k.

16.9   Relative Values

Absolute values of winding resistances and reactances $X = L\omega$ are expressed in ohms. Parameter $R_S = 1\ \Omega$ alone does not provide the grounds to estimate the voltage drop across the stator resistance, nor it helps judging on the stator copper losses. For that to achieve, other information, such as the rated voltage and the rated current are needed as well. Relative value of the stator resistance provides more information in this regard. The *relative* or *normalized* value of variables and parameters are dimensionless numbers obtained by dividing the absolute value by the *base* value. Hence, the base value is considered to be 100% or 1 *per unit*, also designated by 1 [p.u.]. Assuming that the base value for the machine impedances is $Z_B = 2\ \Omega$, while the stator resistance is $R_S = 1\ \Omega$, it is possible to find the relative value of the stator resistance as $r_S = R_S/Z_B = 0.5$ p.u. or 50%.

To facilitate estimation of copper losses of the stator winding and the voltage drop across the stator resistance, the base value for the machine impedances is often selected as $Z_B = Z_n = U_n/I_n$, wherein the rated values of the voltage U_n and current I_n of the electrical machine should be known. For a motor having $U_n = 220$V, $I_n = 2.2$A, and resistance $R_S = 1\Omega$, the relative value of the stator resistance is $r_S = 0.01$. In rated operating conditions, the voltage drop $R_S I_n$ corresponds to 1% of the rated voltage, while the copper losses correspond to 1% of the rated power. Whatever the size and type of electrical machine, the information on r_S provides the grounds to estimate the relative value of the voltage drop and the relative value of the copper losses in the winding. The absolute value of R_S cannot be used for the same purpose unless additional information on the machine is made available. As an example, the motor with the same resistance $R_S = 1\Omega$ but with $U_n = 110$V and $I_n = 22$A has a quite large voltage drop $R_S I_n$, corresponding to 20% of the rated voltage.

The voltage drop and the copper losses in the stator winding are determined in a direct manner by the relative value of the stator resistance. This value is denoted either by r_S or by R_S^{rel}. If the base value of the impedance is determined from rated voltages and currents, $Z_B[\Omega] = Z_n[\Omega] = U_n[V]/I_n[A]$, relative value of the stator resistance is equal to $r_S = R_S^{rel} = R_S[\Omega]/Z_n[\Omega]$. Since $r_S = R_S I_n/U_n = R_S I_n^2/(U_n I_n)$ $= R_S I_n^2/S_n$, it can be concluded that relative value of the stator resistance $r_S = 0.01$ indicates that the rated current causes a voltage drop across the stator resistance of 1% of the rated voltage and that the copper losses of the stator amount 1% of the rated apparent power S_n. The base value of impedance is usually determined as the ratio of the rms values of the phase voltages and currents:

$$Z_B[\Omega] = Z_n[\Omega] = \frac{U_{n,rms.}^{phase}[V]}{I_{n,rms.}^{phase}[A]} \quad \Rightarrow \quad \frac{R_S I_n}{U_n} = R_S^{rel} = r_S. \quad (16.13)$$

With a star-connected induction motor with line-to-line voltages of $220 \cdot$ sqrt(3) V, with phase voltages of 220V, and with rated current of 10A, the base impedance is $Z_B = Z_n = 22\Omega$. Rotor resistance $R_R[\Omega]$ is implicitly referred to the stator side. Its relative value is determined in the same way:

$$r_R = R_R^{rel} = \frac{R_R[\Omega]}{Z_B[\Omega]} = \frac{R_R[\Omega]}{Z_n[\Omega]}. \quad (16.14)$$

Currents and voltages of the stator and rotor are normalized (made relative) on the basis of the rated values; thus, relative value of 1%, or 100%, corresponds to the rated current or voltage:

$$i_S = I_S^{rel} = \frac{I_S[A]}{I_n[A]}, \quad u_S = U_S^{rel} = \frac{U_S[V]}{U_n[V]}. \quad (16.15)$$

In doing so, it is understood that the rms value of the voltage U_S gets divided by the rated rms value U_n. It is also possible to take the peak value of the stator voltage, but in this case, it should be divided by the peak value of the rated voltage. Thus, the rms value of 110V can be divided by the rated rms value of 220V, obtaining $u_S = 0.5$. The same result is obtained by dividing the peak values of corresponding voltages, $155.56/311.12 = 0.5$.

Relative values of reactances and corresponding inductances are obtained by dividing their voltage drop with the rated voltage, assuming that they carry rated current with rated angular frequency ω_n. Hence, the value $X[\Omega]$ is divided by the base impedance. Relative value of an inductance has the same value as the relative value of reactance:

$$l_S = L_S^{rel} = \frac{L_S \omega_n I_n}{U_n} = \frac{L_S[\mathrm{H}]\omega_n[\mathrm{rad/s}]}{Z_n[\Omega]}, \quad x_S = L_S^{rel} = X_S^{rel}. \tag{16.16}$$

Equation (16.16) can be rewritten in the form which comprises the base value of the inductance L_B:

$$l_S = L_S^{rel} = \frac{L_S[\mathrm{H}]}{L_B[\mathrm{H}]}, \quad L_B = \frac{Z_n}{\omega_n}.$$

Relative value of the flux is obtained by dividing the amplitude of the flux vector by the flux base value Ψ_B. The amplitude of the flux vector is the flux of the corresponding winding. The base value Ψ_B is obtained by dividing the peak value of the rated voltage $U_{n(\max)}$ and the rated angular frequency ω_n. The flux base value is obtained from the voltage balance equation in the phase winding of the stator, $u_a = R_a i_a + d\Psi_a/dt$. Neglecting the voltage drop across the stator resistance, one obtains $u_a = d\Psi_a/dt$. Assuming that the flux has the peak value of Ψ_S and that it has sinusoidal change of the frequency of ω_S, the peak voltage across the phase winding is $U_{S(\max)} = \Psi_S \omega_S$. Therefore, the base value Ψ_B is derived as $U_{n(\max)}/\omega_n$.

While the steady state voltages and currents are normalized to their rms values, the flux is referred to its peak, maximum value. The change of the flux in each phase winding is sinusoidal at steady state. Therefore, it is possible to define and use the rms value of the flux. Yet, this approach is seldom used. Further considerations assume that the stator and rotor fluxes Ψ_S and Ψ_R correspond to their maximum values. The peak value of the stator flux corresponds to the amplitude of the stator flux vector, and it is obtained as $\Psi_S^2 = \Psi_{\alpha S}^2 + \Psi_{\beta S}^2$. The flux normalization is performed by dividing the absolute value with the base value Ψ_B, which corresponds to the maximum value of the flux in each phase winding of the machine that operates with rated voltages and with the rated frequency. The base value Ψ_B is very close to the rated value Ψ_n of the flux, which is, at the same time, the knee point of the magnetizing curve. There is a very small difference between the two, caused by the voltage drops that make the product $\omega_n \Psi_n$ slightly lower than the peak phase voltage $U_{n(\max)}$. This is due to a finite value of the voltage drop across the stator resistance $R_S I_S$ which makes the electromotive force in the stator winding different than the winding voltage. As a consequence, the relative value of the rated flux Ψ_n is slightly lower than 1. Hence, the peak stator flux of an induction machine operating in steady state under rated conditions is slightly lower than Ψ_B, resulting in $\Psi_n/\Psi_B < 1$.

Steady state peak values of flux linkages can be obtained from the steady state equivalent circuit in Fig. 16.6. If the voltage phasor $\underline{U}_S$ has the absolute value which corresponds to the maximum of the stator phase voltages, the complex numbers $\underline{\Psi}_S$, $\underline{\Psi}_R$, and $\underline{\Psi}_m$, obtained from the equivalent circuit, have the absolute values corresponding to the peak values of the stator flux, rotor flux, and the mutual (air-gap) flux. Solving the equivalent circuit for the stator and rotor currents, the mutual flux Ψ_m can be calculated by $|\underline{\Psi}_m| = |L_m \underline{I}_m| = |L_m \underline{I}_S + L_m \underline{I}_R|$. Similarly, one can calculate the stator and rotor flux amplitude. Their relative values are obtained by dividing the absolute with the base value Ψ_B, derived as $U_{n(max)}/\omega_n$:

$$\psi_S^{rel} = \frac{\Psi_S[\text{Wb}]}{\Psi_B[\text{Wb}]}, \quad \psi_R^{rel} = \frac{\Psi_R[\text{Wb}]}{\Psi_B[\text{Wb}]}, \quad \psi_{\gamma S}^{rel} = \frac{\Psi_{\gamma S}[\text{Wb}]}{\Psi_B[\text{Wb}]}, \quad \psi_{\gamma R}^{rel} = \frac{\Psi_{\gamma R}[\text{Wb}]}{\Psi_B[\text{Wb}]},$$

$$\psi_m^{rel} = \frac{\Psi_m[\text{Wb}]}{\Psi_B[\text{Wb}]}, \quad \Psi_B[\text{Wb}] = \frac{U_n^{rms}\sqrt{2}[\text{V}]}{\omega_n[\text{rad/s}]}. \tag{16.17}$$

Relative value of the rotor speed is usually determined as $\Omega_m^{rel} = \omega_m^{rel} = \Omega_m/\Omega_{en} = \omega_m/\omega_{en}$, where ω_{en} is the rated supply frequency while Ω_{en} is the rated synchronous speed. This choice of the speed base value allows the slip s to be calculated as $s = 1 - \Omega_m^{rel}$. On the other hand, with rated operating conditions, normalized rotor speed is lower than 1 due to $\Omega_n < \Omega_e$. The value $\Omega_n/\Omega_S = 1 - s_n$ is slightly lower than 1, and the difference is the rated relative slip, s_n. Instead of selecting the rated synchronous speed Ω_{en} as the base speed, it is also possible to choose the rated speed Ω_n, slightly lower than the rated synchronous speed. In this case, the relative speed of rotation is determined as $\Omega_m^{rel} = \Omega_m/\Omega_n$. The rated synchronous speed would be greater than one. This approach is rarely encountered, and the base speed is mostly selected to be the rated synchronous speed.

Relative value of the electromagnetic torque is determined by dividing the absolute value (the value expressed in [Nm]) by the selected base value T_B [Nm]. For the base value of the torque, it is possible to select the rated torque T_n. This results in the rated torque having the relative value of 1 (100%). Taking the rated torque for the base value has certain shortcomings. The product $P_n = T_n \Omega_n$ provides the rated output power of the machine. At the same time, $P_n = S_n \cdot \cos\varphi_n \cdot \eta_n$, where S_n is the rated apparent power, $\cos\varphi_n$ is the power factor in rated conditions, while η_n is the rated efficiency. Hence, the choice $T_B = T_n$ requires the rated efficiency and the rated power factor to be known and taken into account. Therefore, in most cases, the base value of the torque is chosen as $T_B = S_n[\text{VA}]/\Omega_n[\text{rad/s}] > T_n$, where $S_n = 3U_{(f)n}I_{(f)n}$ is the rated value of the apparent power of induction machine, while $\Omega_n = \omega_n/p$ is the rated synchronous speed.

Relation $T_B > T_n$ comes from the relation between the rated apparent power and rated active power of a machine. The apparent power S_n is larger than the rated power. The rated power delivered by the induction motor through the shaft is equal to $P_{en}\eta_n$, where $P_{en} = S_n \cdot \cos(\varphi_n)$ is the power of the electrical source, in rated operating conditions, while $\eta_{nom} = P_n/P_{en} < 1$ is the coefficient of efficiency. Rated power is equal to the product of the rated torque, obtained at the shaft, and rated rotor speed $\Omega_n = \Omega_{en}(1 - s_n)$, where s_n is rated relative slip while Ω_{en} is the rated synchronous speed. By equating $P_n = T_n\Omega_{en}(1 - s_n)$ and $P_n = S_n\cos(\varphi_n)\,\eta_n$, one obtains the relation of the torque base value $T_B = S_n/\Omega_{en}$ and the rated value T_n:

$$\frac{T_n}{T_B} = \frac{T_n}{\left(\frac{S_n}{\Omega_{en}}\right)} = \frac{\cos(\varphi_n)\eta_n}{1 - s_n} < 1. \tag{16.18}$$

Question (16.5): Prove that the relation $\cos(\varphi_n)\,\eta_n/(1 - s_n) < 1$ holds for any three-phase induction machine.

Answer (16.5): For an induction machine that operates in rated conditions, the air-gap power P_δ is obtained by subtracting the stator copper losses and the stator iron losses from the source power, $P_\delta = P_{en} - P_{cu1} - P_{Fe1}$. The air-gap power is divided in two parts: the losses in the rotor windings $s_n P_\delta$ and the internal mechanical power $P_{mR} = (1 - s_n)P_\delta$ which is converted from electrical to mechanical form. This power is equal to $P_{mR} = T_{em}\,\Omega_n$. The torque $T_{em} > T_n$ is slightly higher than the rated torque due to losses caused by friction and ventilation. Assuming that the stator copper losses P_{cu1}, the stator iron losses P_{Fe1}, and the losses due to friction and ventilation are negligible, relation between the input and output power reduces to $P_n = P_{en}(1 - s_n)$, while the efficiency becomes $\eta_n = (1 - s_n)$. Taking into account all the losses mentioned as negligible, it is concluded that $\eta_n < (1 - s_n)$. At the same time, $\cos(\varphi_n) < 1$; thus, the original statement has been proved. As a consequence, relative value of the rated torque obtained by normalizing T_n with the base value of $T_B = S_n/\Omega_{en}$ is less than one.

16.10   Relative Value of Dynamic Torque

The choice of the base value for the electromagnetic torque $T_B = S_n/\Omega_{en}$ has the consequence of $T_n^{rel} < 1$, making the relative value of the rated torque inferior to one. As an example, the relative value of the torque at rated operating conditions may be equal to 0.9% or 90%. Advantages of choosing $T_B = S_n/\Omega_{en}$ instead of $T_B = T_n$ are more obvious from the expression which determines the relative torque in terms of relative currents and relative fluxes. Starting from the expression for electromagnetic torque during transients

$$T_{em} = \frac{3p}{2}\left(\Psi_d i_q - \Psi_q i_d\right),$$

by normalizing the torque using the base value $T_B = T_n$ and adopting the notation Ψ_{dn}, Ψ_{qn}, i_{dn}, i_{qn} for dq values of stator currents and fluxes at rated operating conditions, one obtains

$$T_{em}^{rel} = \frac{T_{em}}{T_{nom}} = \frac{(\Psi_d i_q - \Psi_q i_d)}{(\Psi_{dn} i_{qn} - \Psi_{qn} i_{dn})}. \qquad (16.19)$$

By dividing the numerator and denominator of this expression by the base flux and current values, $\Psi_n = \Psi_B$ and $I_n = I_B$, one obtains the expression for relative torque T_{em}^{rel} in terms of relative fluxes and currents. Denominator of expression (16.20) contains relative values of the stator flux and currents that exist in rated operating conditions,

$$T_{em}^{rel} = \frac{T_{em}}{T_n} = \frac{\left(\Psi_d^{rel} i_q^{rel} - \Psi_q^{rel} i_d^{rel}\right)}{\left(\Psi_{dn}^{rel} i_{qn}^{rel} - \Psi_{qn}^{rel} i_{dn}^{rel}\right)} = K_{RM}\left(\Psi_d^{rel} i_q^{rel} - \Psi_q^{rel} i_d^{rel}\right), \qquad (16.20)$$

where

$$K_{RM} = \frac{1}{\left(\Psi_{dn}^{rel} i_{qn}^{rel} - \Psi_{qn}^{rel} i_{dn}^{rel}\right)}.$$

Constant K_{RM} in the previous expression is not equal to one. The value $K_{RM} > 1$ depends on the rated slip, the rated power factor, and the rated efficiency. Each machine has its own value of coefficient K_{RM}, and it has to be made available in order to perform normalization of the torque. The value $1/K_{RM}$ is equal to the vector product of the relative values of the flux and the stator current at rated conditions. At rated conditions, the stator current has its rated value $I_n = I_B$; thus, its relative value is equal to one. Therefore, relative values i_{dn}^{rel} and i_{qn}^{rel} meet the equation $i_{dn}^{rel} \cdot i_{dn}^{rel} + i_{qn}^{rel} \cdot i_{qn}^{rel} = 1$. With rated supply voltages, relative value of the stator flux Ψ_S^2 is slightly lower than one due to voltage drop $R_S i_S$. Therefore, relative d and q components of the flux meet the equation $\Psi_{dn}^{rel} \cdot \Psi_{dn}^{rel} + \Psi_{qn}^{rel} \cdot \Psi_{qn}^{rel} \approx 1$. The vector product of the stator current and flux vectors, both with amplitude close to one, depends on the sine of the angle between the two vectors. Therefore, $K_{RM} = 1$ would require this angle to be $\pi/2$, which is not feasible in steady state operation with rated supply and rated speed. The angle between the stator flux vector and the stator current vector can be estimated from the equivalent circuit. The steady-state angle between the two vectors is determined by the phase difference between corresponding phasors, represented within the equivalent circuit. The phase of the stator flux is close to the phase of the magnetizing flux and, hence, the phase of the magnetizing current. Neglecting the rotor leakage inductance, the rotor current is leading by $\pi/2$ with respect to the magnetizing current. The stator current is the sum of the two, $I_S = I_m + I_{R1}$. Therefore, it leads with respect to the flux by an angle which cannot be $\pi/2$ due to $I_m \neq 0$.

Question (16.6): Estimate the angle between the stator flux and the stator current of an induction machine at rated operating conditions. Use the equivalent circuit to find the phase difference between the flux phasor and the current phasor. In doing so, consider that the difference between the stator voltage and the voltage across magnetizing branch is negligible.

Answer (16.6): Starting from steady state equivalent circuit, the estimate of the angle can be obtained by neglecting the voltage drop across the stator series impedance $(R_S + j\omega_e L_{\gamma S})\underline{I}_S$. The voltage across magnetizing branch $\underline{U}_m = j\omega_e \underline{\Psi}_m \approx \underline{U}_S = j\omega_e \underline{\Psi}_S$ is then approximately equal to the supply voltage; thus, the magnetizing flux is considered approximately equal to the stator flux $\underline{\Psi}_m \approx \underline{\Psi}_S$. Stator current is equal to the sum of the magnetizing current $\underline{I}_m$ and the rotor current $\underline{I}_{R1} = -\underline{I}_R$. The initial phase and spatial orientation of the flux $\underline{\Psi}_m \approx \underline{\Psi}_S$ is determined by the initial phase of magnetizing current $\underline{I}_m$. The rotor current $\underline{I}_{R1}$ is equal to the voltage $\underline{U}_m = j\omega_e \underline{\Psi}_m$ divided by the rotor impedance $(R_R/s + j\omega_e L_{\gamma R})$. The slip $s = s_n$ is significantly smaller than 1; thus, $R_R/s \gg \omega_e L_{\gamma R}$. Therefore, the rated value of the rotor impedance is mainly resistive, and the phase lagging of the current $\underline{I}_{R1}$ with respect to the voltage $\underline{U}_m$ is relatively small. With $tg(\theta) \approx \theta$, this phase lag is approximately equal to $s_n \omega_e L_{\gamma R}/R_R$ rad. It can be concluded that the phase lead of the rotor current $\underline{I}_{R1}$ with respect to $\underline{I}_m$ is somewhat smaller than $\pi/2$. With the stator and rotor currents significantly larger than the magnetizing current, the stator current $\underline{I}_S = \underline{I}_{R1} + \underline{I}_m$ will also lead with respect to $\underline{I}_m$ and with respect to the stator flux $\underline{\Psi}_m \approx \underline{\Psi}_S$. This phase lead is close to $\pi/2$. It is larger in cases where the magnetizing current is smaller, but it never reaches $\pi/2$.

* * *

Preceding analysis shows that during rated operating mode of an induction machine, the angle between the stator flux and current cannot reach the value of $\pi/2$. Therefore, the coefficient K_{RM} of (16.20) is greater than 1. This coefficient can be expressed as function of φ_n, η_n, and s_n. Coefficient K_{RM} is required for the torque normalization according to expression $T_{em}^{rel} = K_{RM} (\Psi_d^{rel} \cdot i_q^{rel} - \Psi_d^{rel} \cdot i_d^{rel})$, where the flux linkages and electrical currents are expressed in relative units. Calculation of K_{RM} presents a difficulty as the values φ_n, η_n, and s_n are machine specific and they change from one machine to another. For this reason, preferred torque normalization does not use the base value of $T_B = T_n$. Instead, the value of $T_B = S_n/\Omega_{en}$ is used.

As a consequence of selecting $T_B = S_n/\Omega_{en}$, the formula that calculates the relative torque in terms of relative flux linkages and relative currents becomes rather simple. This formula is developed in (16.21), where U_n and I_n represent the rms values of the rated voltages and currents. The torque expression makes the use of dq variables Ψ_d, i_q, Ψ_q, and i_d. Therefore, it is necessary to introduce relative values for the flux and current components in dq frame. With Clarke transform having the leading coefficient of 2/3, the peak values of the phase currents correspond to the amplitude of the stator current in dq frame, $(i_d^2 + i_q^2)^{0.5}$. At the same time, the maximum value of the flux linkage in one phase corresponds to the amplitude of the flux in dq frame. Therefore, relative values of i_d and i_q are determined by dividing the absolute currents with the peak of the rated current, $I_n \cdot \text{sqrt}(2)$. At the same time, the flux linkages Ψ_d and Ψ_q are normalized by dividing their absolute values with $U_n \cdot \text{sqrt}(2)/\omega_{en}$, where ω_{en} is the rated supply frequency:

$$T_{em}^{rel} = \frac{T_{em}}{T_B} = \frac{\frac{3}{2}p(\Psi_d i_q - \Psi_q i_d)}{\frac{S_n}{\Omega_{en}}} = \frac{\frac{3}{2}p(\Psi_d i_q - \Psi_q i_d)}{\frac{p}{\omega_{en}}3U_n I_n}$$

$$= \frac{(\Psi_d i_q - \Psi_q i_d)}{\left(\frac{U_n \sqrt{2}}{\omega_{en}}\right)(I_n \sqrt{2})} = \left(\Psi_d^{rel} i_q^{rel} - \Psi_q^{rel} i_d^{rel}\right). \tag{16.21}$$

The formula (16.21) does not make use of any coefficients such as K_{RM}, 3/2, or p. The relative torque is obtained as the vector product of normalized stator current and normalized stator flux, both expressed in dq coordinate frame. Whenever the need appears to express the absolute torque, expressed in [Nm], the relative torque should be multiplied by $T_B = S_n/\Omega_{en}$, which makes all the analysis with $T_B = S_n/\Omega_{en}$ rather simple. The only disadvantage is the circumstance that the relative value of the rated torque is inferior to one, $T_n^{rel} < 1$, that is, lower than 100%.

The problem of selecting the base value for the torque of AC machines is different than the problem of selecting T_B with DC machines. The input power to the armature winding of DC machines is equal to the product of the armature current and the armature voltage. This product does not have to be multiplied with the power factor $\cos\varphi$, which is the case in AC machines. The absence of power factor $\cos\varphi$ in DC machines makes the calculations of power and torque less involved.

16.11 Parameters of Equivalent Circuit

The analysis of steady state equivalent circuit requires the knowledge of parameters R_S, R_R, $L_{\gamma S}$, $L_{\gamma R}$, and L_m. It is of interest to have a basic idea of their range and to learn about practical ways to measure relevant parameters and/or to calculate parameters from data declared on the machine nameplate.

Stator resistance of a star-connected winding can be determined by connecting two of the three stator terminals to a DC power supply. By measuring DC voltage between the terminals, one obtains the sum of voltage drops in two-phase windings, $2R_S I_{DC} = U_{DC}/I_{DC}$. Relative value of stator resistance R_S is usually larger with small machines, and it drops down as the machine power increases. With induction machines rated a couple of hundreds of Watts, relative value of R_S is close to 10%. For induction machine in excess to 300 kW, relative value of the stator resistance can be as low as 0.1%.

Question (16.7): What is the reason that makes the relative value of winding resistances smaller for large power machines?

Answer (16.7): In electrical machines, there are losses in magnetic circuits and in current circuits, also called iron losses and copper losses. Specific power of iron losses is the iron loss per unit volume or unit mass of the magnetic circuit. It depends on magnetic induction B_m and frequency of the field changes in considered magnetic circuit. Considering two electrical machines with the same flux density (induction B) and the same frequency, then the specific power of iron losses has the same values in both machines. Therefore, the ratio of P_{Fe1} and P_{Fe2} is determined by the ratio of volumes of respective magnetic circuits V_{1f}/V_{2f}, or by the ratio of their masses m_{1f}/m_{2f}.

Specific power of losses in copper is the amount of loss power per unit volume or unit mass, and it is dependent on current density $[A/mm^2]$ and specific conductivity σ of the metal (copper) used to make the conductors. Considering two electrical machines with the same current density and with conductors made of the same material (copper), their specific power of copper losses is equal. The ratio of P_{Cu1} and P_{Cu2} is dependent on the quantity of the material used to make the windings, namely, the ratio P_{Cu1}/P_{Cu2} is determined by the ratio of volumes V_{1Cu}/V_{2Cu} or mass m_{1Cu}/m_{2Cu} of the copper used in making respective windings.

If one of the two machines has diameter which is two times larger and it has twice the length of the other, smaller machine, its volume is eight times larger ($V \sim l^3$) than the volume of the smaller machine. At the same time, assumption is made that the two machines are similar and that both magnetic circuit and the windings of the larger machine are obtained by starting from relevant parts of the smaller machine and doubling their linear dimensions. Assuming that both machines have the same current density, the same induction B_m, and the same operating frequency f, then the power of losses of the larger machine is eight times higher.

Losses of energy in the magnetic and current circuits of electrical machines generate heat which increases temperatures of the vital parts of the machine. Maintaining integrity and functionality of machine parts such as the electrical insulation of the windings, ferromagnetic materials, bearings, and other parts requires containing the machine temperature within safe limits. Excessive temperatures can cause permanent damage to vital parts of the machine. Therefore, it is necessary to provide cooling that removes excess heat and keeps the machine temperature within safe limits. At steady state, when the machine temperature sets to a constant value, the heat created by the conversion losses within the machine is equal to the heat removed by cooling. Different cooling methods such as convection, conduction, and radiation remove the amount of heat which depends on the temperature difference between the machine and the surrounding environment. The heat transfer is also proportional to the surface area of the machine parts getting in touch with the environment and also the contact area with the cooling fluids. The surface is proportional to $S \sim l^2$. Therefore, the larger machine with doubled diameter and twice the axial length can be cooled with four times larger surface area than the smaller machine.

It can be concluded that machines with l times larger linear dimensions have l^3 times higher losses, while their ability to remove heat is increased l^2 times. Under circumstances, the temperature of the machine with l times larger linear dimensions is going to increase l times. This statement relies on the assumption that the specific power of losses remains unaltered. The above consideration shows that larger machines require more efficient cooling methods and reduction of specific power of losses. For this reason, large machines are usually designed and made to have smaller values of magnetic induction B_m and smaller current densities in their windings. As a consequence, the windings are built of conductors with larger cross section, which results in smaller winding resistances. Therefore, relative value of the stator resistance is smaller for machines of larger power. It is necessary to note that the value $R_S^{rel} = 0.001$, corresponding to machine of 1 MW, actually means that the copper losses in the stator winding have the power of 1 kW. Namely, the energy of 1,000 J is converted into heat in each second. Heat removal at rates of 1,000 J per second may require forced cooling. Any further increase in R_S would bring additional copper losses and result in an increase in the winding temperature that may damage electrical insulation.

Ratio of the power of losses and cooling surface of smaller machines is much more favorable. Therefore, the problem of cooling is less emphasized, and there is no need to use more copper by increasing the cross section of conductors. For very small machines, relative values of R_S in excess of 10% can be tolerated. This means that 10% of the rated power accounts for the copper losses, which introduces the question of energy efficiency. It is of interest to notice that many small electrical machines operate intermittently, such as the motors in hand dryers, and remain disconnected for most of the time. In such cases, the energy that has to be used to manufacture electrolytic copper required for the motor windings may be more significant than the energy lost in the copper losses during the machine lifetime.

* * *

Magnetizing inductance L_m determines the magnetizing current $\underline{I}_m = \underline{I}_S + \underline{I}_R$ which is required to achieve desired flux $\underline{\Psi}_m = L_m \underline{I}_m$. With most induction machines, magnetizing current $\underline{I}_m$ is much smaller than the rated current. By analogy with power transformer, considering the stator as the primary winding and the rotor as the secondary winding, magnetizing current corresponds to *no load current* of the transformer, the current that circulates in primary winding when the secondary winding does not have any current. In the case when an induction machine is not coupled to mechanical load, the rotor revolves with no external resistance. The load torque T_L is equal to zero. Internal mechanical losses such as the bearing friction and the air resistance are negligible in most cases. Therefore, the rotor revolves at the speed which is very close to the synchronous speed and the relative slip is $s \approx 0$. Impedance R_R/s takes a very high value; thus, rotor current can be neglected, and the rotor circuit can be considered open. The equivalent circuit of the induction machine reduces to a series connection of the stator resistance R_S, the stator leakage inductance $L_{\gamma S}$, and the magnetizing inductance L_m.

The experiment with an induction machine connected to the mains having the rated voltage and the rated frequency applies the rated supply to the stator windings. With shaft disconnected from any mechanical load or work machine, the rotor revolves at synchronous speed. Such experiment is called *no load test*. At steady state, the stator current is equal to $\underline{I}_0 = \underline{U}_{Sn}/(R_S + j\omega_{en}L_S)$. Neglecting the stator resistance and the leakage inductance, one obtains $L_S \approx L_m \approx U_{Sn}/(I_0\,\omega_{en})$.

No load current of induction machines of small power is within range of 50–70% of the rated current. For high-power machines, no load current can be lower than 20% of the rated current.

Question (16.8): What is the reason that makes the relative no load current smaller for high-power machines?

Answer (16.8): High-power machines are designed to have smaller value of magnetic induction B_m in their magnetic circuits. Cooling of high-power machines is more difficult. Therefore, they are designed to have smaller values of magnetic induction in iron parts and smaller current density in copper conductors. The slope $\Delta B/\Delta H$ of the magnetizing characteristic of the ferromagnetic material (B–H curve) is higher in regions that are closer to the origin of the B–H plane. Consequently, magnetic permeability is higher, which reduces magnetic resistance $R_{\mu Fe}$ of the iron parts of the magnetic circuit. Moreover, high-power machines have larger dimensions and more favorable ratio of the air gap δ and machine diameter D. Due to smaller δ/D ratio, the air-gap part of magnetic resistance is smaller. Smaller values of magnetic resistance result in higher values of inductance, which is inversely proportional to magnetic resistance due to $L = N^2/R_\mu$. Self-inductance of the stator winding L_S is proportional to the ratio N^2/R_μ, while magnetizing inductance is slightly smaller due to $L_m = kL_S$, where coupling coefficient k is somewhat lower than one. It can be concluded that a lower magnetic resistance and improved δ/D ratio of high-power machines result in higher relative values of their magnetizing inductance L_m and smaller relative values of their no load currents.

* * *

Leakage inductances of the stator and rotor are coefficients of proportionality between the respective leakage flux and winding current. With $N_S = N_R$ and $L_S = L_R$, $L_{\gamma S} = (1 - k)L_S$, $L_{\gamma R} = (1 - k)L_S$, and $L_m = kL_S$. The number of rotor turns N_R is related to virtual, equivalent rotor winding which replaces the rotor cage. Therefore, in most cases, the stator and rotor leakage inductances can be considered equal,[2] $L_{\gamma S} \approx L_{\gamma R}$. In cases where short-circuited rotor cage is represented by equivalent rotor winding having the same number of turns as the stator, $N_S = N_R$, all the rotor parameters are referred to the stator side without scaling due to the transfer ratio $m = N_S/N_R = 1$. Inductances of the stator and rotor are proportional to squared number of turns and inversely proportional to magnetic resistance. The mutual and leakage fluxes of the stator and rotor windings exist in magnetic circuits

---

[2] There is no unique way to determine the mutual flux (i.e., the air-gap flux). The surface used to calculate the surface integral of the magnetic induction and find the air-gap flux passes through air gap, but it can be placed closer to the stator or closer to the rotor. Therefore, the same machine can be modeled by using different sets of values ($L_{\gamma S}$, $L_{\gamma R}$, L_m). It is of interest to note that all of these sets provide correct mathematical model of the machine behavior.

with approximately equal lengths and cross sections. They share the same air gap, and they are made of the same iron sheets. For this reason, it is justified to assume that $L_S \approx L_R$ and $L_{\gamma S} \approx L_{\gamma R}$. The assumption is valid for majority of induction machines. Exceptions to this appear with induction machines with large differences in dimensions and shapes of the stator and rotor slots.

An approximate value of leakage inductances can be determined by the *short-circuit test*. During this test, the rotor of the induction machine is *locked*. It cannot move and its speed is equal to zero. During the test, the rotor motion is prevented by mechanical means. In locked rotor conditions, the relative slip is equal to one, $s = 1$. Impedance R_R/s obtains relatively small value R_R. By neglecting the magnetizing current, the equivalent impedance of the motor becomes $\underline{U}_S/\underline{I}_S = R_S + R_R + j\omega_e L_{\gamma S} + j\omega_e L_{\gamma R}$. For machines with rated power in excess to 10 kW, relative values of resistances R_S and R_R are very low, and they can be neglected. The leakage inductance is then determined as $L_{\gamma S} \approx L_{\gamma R} = \frac{1}{2} U_S/(I_S\omega_e)$, where U_S and I_S are the rms values of the stator voltages and currents measured during the short-circuit test. With small machines, the winding resistance cannot be neglected. During the short-circuit test, it is necessary to determine the phase shift between the voltages and currents in order to distinguish between the real and imaginary parts of the short-circuit impedance $\underline{Z}_K = R_S + R_R + j\omega_e L_{\gamma S} + j\omega_e L_{\gamma R}$.

The short-circuit reactance $\omega_{en}(L_{\gamma R} + L_{\gamma S})$ at rated conditions has the relative value that ranges from 0.05 up to 0.3. Smaller values are met in lower power low-voltage (400 V) machines, while higher values correspond to induction machines for medium voltages (6 kV) and powers of the order of 1 MW. In high-power and high-voltage machines, distances between conductors of stator and rotor windings are larger. Larger distances are necessary for the purposes of a more efficient cooling. At the same time, insulation between conductors has to withstand higher voltages, which contributes to an increased distance between conductors. As a consequence, the coupling coefficient k between stator and rotor windings is smaller, which increases leakage inductances due to $L_{\gamma S} \approx L_{\gamma R} \approx (1 - k)L_S$:

$$L_{\gamma S} = L_S - L_m, \quad L_{\gamma R} = L_R - L_m. \tag{16.22}$$

The short-circuit reactance of an induction machine is approximately equal to the sum of the stator and rotor leakage reactances $\omega_{en}(L_{\gamma R} + L_{\gamma S}) = X_{\gamma S} + X_{\gamma R}$. A small difference between the actual short-circuit reactance $L_{\gamma e}\omega_{en}$ and the sum of the stator and rotor leakage reactances exists due to a finite value of $L_m\omega_{en}$.

The actual value of the short-circuit reactance can be determined by considering equivalent transformer. The stator and rotor windings of an induction machine are coupled by magnetic field in the same way as the primary and secondary windings of a transformer. A short-circuited induction motor with $R_S, R_R \ll \omega_{en}(L_{\gamma R} + L_{\gamma S})$ is equivalent to a transformer with short-circuited secondary and with negligible winding resistances. The input (equivalent) inductance $L_{\gamma e} = X_{\gamma e}/\omega_{en}$ of a short-circuited induction machine, also called *the equivalent leakage inductance*, can be calculated from the circuit in Fig. 16.7:

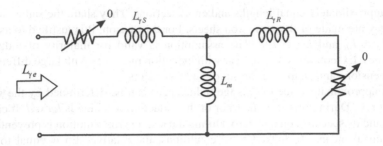

Fig. 16.7 Equivalent leakage inductance

$$L_{\gamma e} = L_{\gamma S} + \frac{L_m L_{\gamma R}}{L_m + L_{\gamma R}} = \frac{L_{\gamma S} L_R + L_m L_{\gamma R}}{L_R}$$

$$= \frac{L_{\gamma S} L_R + L_m L_R - L_m L_R + L_m L_{\gamma R}}{L_R} = \frac{L_S L_R - L_m^2}{L_R}. \tag{16.23}$$

The obtained expression is similar to the one for the input inductance L_u of a short-circuited transformer with known self-inductance of the primary winding (L_1), self-inductance of the secondary winding (L_2), and with mutual inductance M. This expression reads $L_u = (L_1 L_2 - M^2)/L_2$. In the case when $L_{\gamma R} \ll L_m$, it can be considered that $(L_{\gamma R} + L_{\gamma S}) \approx L_{\gamma e}$.

Relative value of reactance $L_{\gamma e} \omega_{en}$ is relatively small, within the range from 0.05 up to 0.3. Relative value of the current is approximately equal to the ratio of the relative value of voltage and relative value of the short-circuit reactance. Therefore, by connecting the rated voltage to a short-circuited induction machine, one obtains currents which are much higher than the rated current. Due to high losses, operation of the machine with the locked rotor must be very short, in order to avoid undesirable increase of temperature and damage of insulation. For this reason, the short-circuit test is usually performed with lower stator voltage, so as to achieve an acceptable stator current that would not damage the machine under the test. In the case when the short-circuit test is performed with the rated current ($I_S = 1$ [p.u.]), then the relative voltage across the winding is required to be $U_S = X_{\gamma e} \ll 1$. In the case when the short-circuit experiment is performed with the rated voltage, the relative stator current is $I_S \approx 1/X_{\gamma e}$ and is much higher than the rated current. Even the reactance obtained by the short-circuit experiment with rated voltage may be different from the reactance obtained by the short-circuit experiment performed with rated current. This difference can appear due to magnetic saturation which alters the magnetic resistance as well as the values of inductances. With very high stator currents, the leakage flux is high, and the magnetic induction B in the stator teeth may reach the saturation level. The teeth belong to the magnetic circuit of the leakage flux. Therefore, magnetic saturation of stator teeth reduces the leakage inductance and the leakage reactance. Similar conditions are met in the rotor circuit. During short-circuit test, the stator current is approximately equal to the rotor current I_{R1}. Therefore, with large stator currents, the magnetic material on the path of the rotor leakage flux reaches saturation, which leads to an increase of magnetic resistance and a decrease of the rotor leakage inductance. The short-circuit test with elevated currents demonstrates that the induction machine is a nonlinear system. Mathematical model and the equivalent circuit developed since describe the machine behavior provided that the modeling approximations are valid and that the machine operates with all the variables remaining reasonably close to rated conditions. In cases with very large currents, very large speeds, excessive slip frequencies, or excessive flux amplitudes, parameters of the steady state equivalent circuit cannot be considered constant. Moreover, even the steady state equivalent circuit may not be the adequate way to describe the machine behavior. Nonlinear phenomena in induction machines are intentionally kept out of this book.

Resistance of the rotor winding referred to the stator side is denoted by R_R. Relative value of parameter R_R is usually close to the relative value of R_S. In low-power machines, relative R_R is close to 10%, while for induction machines in excess to 1 MW, it could be as low as $R_R \approx 0.1\%$. In essence, the actual resistance of the rotor bars can hardly be measured. Yet, parameter R_R in the equivalent circuit could be determined from the equivalent circuit. With known $\underline{U}_S$ and $\underline{I}_S$, and with all the parameters of the equivalent circuit known, except for R_R/s, it is possible to calculate the value of R_R/s expressed in $[\Omega]$. Provided that the rotor speed is available as well, one can calculate the relative slip and obtain precise value of the rotor resistance. The following considerations provide the means for finding a quick estimate of the rotor resistance.

16.11.1 Rotor Resistance Estimation

The rotor resistance R_R can be calculated from the impedance obtained by the locked rotor test. This impedance is $\underline{Z}_K \approx R_S + R_R + j\omega_e L_{\gamma S} + j\omega_e L_{\gamma R}$. Taking the real part of the locked rotor test and subtracting the stator resistance, it is possible to obtain the parameter R_R which makes part of the short-circuit impedance. This value may be different than the resistance encountered by the rotor currents having the rated slip frequency. In the locked rotor test, the rotor currents have the line frequency, the same as the stator currents. At rated operating conditions, the frequency of rotor currents ranges from $f_{slip} \sim 0.5$ to $f_{slip} \sim 5$ Hz. Due to relatively large cross section of rotor conductors, the skin effect is emphasized even at the line frequency. The skin effect in rotor bars consists in pushing the rotor currents toward the air gap and reducing the currents deep in the rotor slots. This uneven distribution of currents leads to an increase in the rotor equivalent resistance at elevated frequencies. Therefore, the rotor resistance at $f_{slip} = 50$ Hz, measured in the short-circuit test, is considerably higher than the rotor resistance at rated slip frequency.

Another way of estimating the parameter R_R is based on determining the time constant of the rotor circuit, $\tau_R = L_R/R_R$. The rotor time constant τ_R can be determined from the voltage change between the two terminals of the stator winding. This voltage change has to be measured after disconnecting the machine from the mains. In this case, the voltage between the stator terminals reflects the changes of the induced electromotive force, proportional to the rotor flux. In the prescribed conditions, the flux decays with the time constant τ_R, which provides the grounds for R_R estimation. Before commencing with measurement, induction machine should be running with no load ($T_L = 0$), connected to the mains which provides the rated voltage of the rated frequency. It is assumed that the machine accelerated up to the synchronous speed, that the steady state has been established, and that the torque, power, and slip are close to zero, while the magnetic circuits has the rated flux. At this point, a safe, isolated voltage probe[3] should be connected between the two terminals of the stator windings, bringing the line voltage to a digital storage oscilloscope. Upon disconnecting the machine from the mains, the stator currents drop to zero. Considering the equivalent circuit, the stator contour remains open, while the remaining rotor current circulates through the magnetizing branch. Hence, the residual rotor current provides the magnetizing current and supports the flux. The short-circuited rotor cage opposes to any change of the flux, attempting to preserve the flux found at the instant of disconnecting the mains. The rotor counter electromotive forces and currents are induced to oppose to the flux change. It is of interest to study the change of the rotor current at the instant of disconnecting the mains. Considering the equivalent circuit prior to disconnection, the rotor current was close to zero, while the stator current was equal to the magnetizing current I_m. Disconnecting the stator winding from the mains cuts the stator current down to zero,[4] while the magnetizing current shifts into the rotor circuit. The rotor circuit is established due to short-circuited cage attempting to maintain the flux unaltered. Hence, although the machine is disconnected from the mains, the magnetizing flux is maintained by the rotor cage. Therefore, residual flux revolves within the machine at the rotor speed, which remains close to the synchronous speed. At first, there are no significant changes of the flux amplitude. Electromotive force $\Psi_m \omega_e$ is induced in the stator windings. It has the amplitude close to the rated voltage, the frequency close to the supply frequency, and it can be measured between the open stator terminals. It is assumed that, in the absence of mechanical load, the rotor speed remains constant throughout the experiment.

---

[3] The instant of opening the switch which connects the stator winding to the mains may result in transient overvoltages across the contacts being opened and across the voltage probe. Therefore, it is advisable to use the probe that withstands the voltages in excess to the rated voltages. With 0.4 kV machines, a 10 kV probe would do. The probe has to be passed through the insulator which secures galvanic insulation between the motor terminals and the oscilloscope, where the later has to be connected to protective ground for safety.

[4] It may take up to a couple of milliseconds to open the switch contacts and extinguish the electric arc between the opened contacts, which eventually brings the stator current to zero. The arcing cannot be avoided, and it is more emphasized when the leakage inductance is larger. As a matter of fact, the arcing *burns* the energy of the leakage flux, $W_{\gamma e} \sim L_{\gamma e} i_s^2$. The switch opening time is very short compared to the machine dynamics, and it is neglected in the present discussion.

By time, the amplitude of this electromotive force gradually decays. Since the stator circuit is open, the rotor current and, hence, the magnetizing current exist in the contour comprising three elements: L_m, $L_{\gamma R}$, and R_R. Therefore, the current decays exponentially with the time constant determined by the ratio of inductance and resistance of the circuit, $\tau_R = L_R/R_R$. As a consequence, the envelope of the induced electromotive force drops according to the law $\exp(-t/\tau_R)$. Therefore, by determining the envelope of the line voltage, the rotor time constant can be estimated as the time required for the amplitude to drop e times, that is, to reduce to some 37% of the initial value.

There is also possibility to calculate an approximate value of R_R from data available on the nameplate of the machine. Parameter R_R can be estimated from the equivalent circuit by using the rated voltages and currents. It is necessary to make a series of assumptions regarding the equivalent circuit in rated conditions. By neglecting the stator resistance and magnetizing inductance, the equivalent circuit reduces to series connection of the equivalent reactance $L_{\gamma e}\omega_{en}$ and rotor resistance R_R/s. In rated operating conditions, the voltage across the resistance R_R/s of the equivalent circuit is equal to $(U_n^2 - (I_n L_{\gamma e}\omega_{en})^2)^{1/2}$. If relative values are used, then $U_n = 1$ and $I_n = 1$. Reactance $L_{\gamma e}\omega_{en}$ has relative value ranging from 0.05 to 0.25. The phasor of the voltage drop across this reactance is orthogonal to the phasor $I_n R_R/s$; thus, the voltage drop $I_n L_{\gamma e}\omega_{en}$ has no significant impact on the voltage across the resistance R_R/s. With $L_{\gamma e}\omega_{sn} = 0.25$ and $U_n = 1$, the relative value of the rotor voltage $I_n R_R/s$ is 97%. Therefore, it is justifiable to assume that the rotor voltage in rated conditions is close to 1 relative unit. With rated current in both stator and rotor windings, impedance R_R/s_n is equal to $U_n/I_n = Z_B$. On the basis of introduced approximations, it follows that $R_R \approx s_n Z_B$. In other words, relative value of rotor resistance $r_R = R_R/Z_B$ is close to the rated value of relative slip s_n. Hence, from nameplate of an induction motor that comprise data $I_n = 22$ A, $U_{fn} = 220$ V ($U_{ln} = 380$ V), $f_n = 50$ Hz, and $n_n = 2{,}700$ rpm, an estimate of the motor parameter R_R is $s_n Z_B = 1$ Ω.

16.12 Analysis of Mechanical Characteristic

The electromagnetic torque of an induction machine is equal to the ratio of the air-gap power P_δ, which is transferred from the stator to the rotor, and synchronous speed $\Omega_e = \omega_e/p$, which is the revolving speed of the magnetic field. From the analysis of the steady state equivalent circuit, the torque can be calculated from expression

$$T_{em} = \frac{P_\delta}{\Omega_e} = \frac{3}{2\Omega_e} \frac{R_R}{s} I_R^2 .$$

By solving the equivalent circuit, shown in Fig. 16.8, one can obtain the stator and rotor currents. Rotor current is equal to

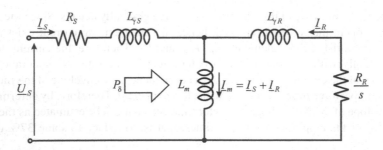

Fig. 16.8 Equivalent circuit of induction machine

$$\underline{I}_{R1} = -\underline{I}_R = \frac{\underline{Z}_m \, \underline{U}_S}{\underline{Z}_m \underline{Z}_S + \underline{Z}_m \underline{Z}_R + \underline{Z}_S \underline{Z}_R} \tag{16.24}$$

where

$$\underline{Z}_m = j\omega_e L_m, \quad \underline{Z}_S = R_s + j\omega_e L_{\gamma s}, \quad \underline{Z}_R = R_R/s + j\omega_e L_{\gamma s}.$$

At steady state, the torque is equal to

$$T_{em} = \frac{P_\delta}{\Omega_e} = \frac{3}{2\Omega_e} \frac{R_R}{s} I_R^2$$

$$= \frac{3pR_R}{2\omega_e s} \frac{U_S^2}{\left(v^R R_S + v^S R_R/s\right)^2 + \omega_e^2\left(v^S L_{\gamma R} + L_{\gamma S}\right)^2}, \tag{16.25}$$

where $v^S = L_S/L_m$ and $v^R = L_R/L_m$. U_S and I_R denote stator voltage and rotor current, with absolute values corresponding to peak values of the phase variables. The same expression can be used if the absolute values of phasors $\underline{U}_S$ and $\underline{I}_R$ correspond to respective rms values, but then, the value of the above expression should be doubled, whereby coefficient 3/2 becomes 3. Coefficients v^S and v^R are close to one, and they are dependent on the leakage inductance of the machine. Magnetizing inductance L_m is several tens of times higher than the leakage inductance $L_{\gamma S}$, which is in denominator of the torque expression. With the introduced approximations, the stator and rotor currents are equal to

$$\underline{I}_S \approx \underline{I}_{R1} \approx \frac{U_s}{\left(R_s + \frac{R_R}{s}\right) + j\omega_e\left(L_{\gamma s} + L_{\gamma R}\right)}, \tag{16.26}$$

and the torque is equal to

$$T_{em} = \frac{3pR_R}{2\omega_e s} \frac{U_S^2}{\left(R_S + R_R/s\right)^2 + \left(\omega_e L_{\gamma e}\right)^2}$$

$$= 3\frac{1}{\Omega_e} \frac{R_R}{s} \frac{U_{S(eff)}^2}{\left(R_S + R_R/s\right)^2 + \left(\omega_e L_{\gamma e}\right)^2}. \tag{16.27}$$

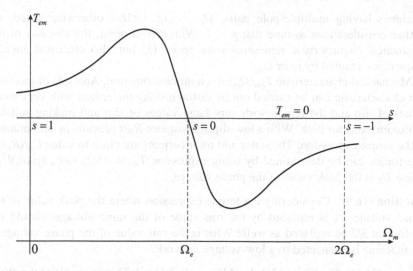

Fig. 16.9 Mechanical characteristic

The obtained expression can be used for deriving the mechanical characteristic of induction machine $T_{em}(\Omega_m)$, which represents the steady state relation of the torque and the rotor speed. In order to determine $T_{em}(\Omega_m)$, it is required to:

- Determine relative slip $s = (\Omega_e - \Omega_m)/\Omega_e$.
- Determine the rms value of the phase voltage $U_{S(rms)}$.
- Introduce this value (alternatively, the peak value $U_{S(rms)} \cdot$ sqrt(2), the synchronous speed $\Omega_e = 2\pi f_S/p$, the resistance R_R/s, and other motor parameters in the expression for the electromagnetic torque (16.27).
- Calculate the torque according to the formula comprising coefficient 3 if the rms value of the voltage is used or 3/2 if U_S denotes the peak value of the phase voltage.

Later on, in (16.37), it will be demonstrated that the torque in (16.27) can be expressed in a more compact way. Expression such as (16.28) will be obtained by making reasonable approximations:

$$T_{em}(s) \approx \frac{K_1}{s + \frac{K_2}{s}}. \tag{16.28}$$

Mechanical characteristic is shown in Fig. 16.9. The torque is on the ordinate, while the abscissa represents the rotor mechanical speed. With two-pole induction machines ($p = 1$), angular frequency ω_e of stator currents and voltages corresponds to the speed Ω_e of the revolving field, also called synchronous speed. Hence, $\Omega_e = \omega_e/p = \omega_e$. At the same time, $\Omega_{slip} = \omega_{slip}$ and $\Omega_m = \omega_m$. With induction machines having multiple-pole pairs, $\Omega_e = \omega_e/p$. Unless otherwise stated, all further considerations assume that $p = 1$. With this in mind, the abscissa of the mechanical characteristic represents rotor speed Ω_m but also electrical angular frequencies denoted by ω or ω_m.

Mechanical characteristic $T_{em}(\Omega_m)$ is a nonlinear function. Analysis of mechanical characteristic can be carried out by distinguishing the region with very small values of slip and the region with very large values of slip and making suitable approximations for both. With a low slip s, resistance R_R/s prevails in denominator of the torque expression. The stator and rotor currents are close to value $U_S/(R_R/s)$. The torque can be determined by using expression $T_{em} = (3/2)\ (s/R_R)(p/\omega_e)U_S^2$, where U_S is the peak value of the phase voltage.

Question (16.9): Considering the torque expression where the peak value of the phase voltage U_S is replaced by the rms value of the same voltage, should the coefficient 3/2 be replaced as well? What is the rms value of the phase voltage if the machine is connected to a low-voltage network?

Answer (16.9): Coefficient 3/2 should be replaced by 3. The rms value of the phase voltage is 220 V.

16.13 Operation with Slip

Introductory considerations discussing principles of operation of induction machines demonstrated that, at relatively small values of relative slip, the torque is proportional to the slip (14.9, Question 14.1). With two-pole machine where $p = 1$, the angular speed of the rotating field Ω_e, also called synchronous speed, is equal to the angular frequency of the supply voltages ω_e. If the slip is small, the rotor speed is close to the synchronous speed. In such case, impedance R_R/s is the largest of all impedances in the denominator of the torque expression (16.27). Therefore, the expression reduces to

$$T_{em} \approx \frac{3}{2}\frac{1}{\Omega_e}\frac{R_R}{s}\frac{U_S^2}{\left(\frac{R_R}{s}\right)^2} \Rightarrow T_{em} \approx \frac{3p}{2}\frac{s}{R_R}\frac{1}{\omega_e}U_S^2. \tag{16.29}$$

Therefore, the torque is proportional to relative slip. Multiplication of the numerator and denominator by angular frequency results in expression

$$T_{em} \approx \frac{3p}{2}\frac{\omega_{slip}}{R_R}\frac{1}{\omega_e^2}U_S^2.$$

With the assumption that the voltage drop across the stator resistance is negligible, the ratio of the peak value of the phase voltage U_S and the stator frequency

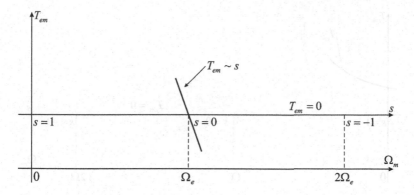

Fig. 16.10 Mechanical characteristic small slip region

ω_e represents the peak value of the flux in the stator phase windings, also called the amplitude of the stator flux:

$$\Psi_S \approx \frac{U_S}{\omega_e} \Rightarrow T_{em} \approx \frac{3p}{2} \frac{\Psi_S^2}{R_R} \cdot \omega_{slip}. \tag{16.30}$$

This corresponds to equation (15.3) of the introductory section discussing the operating principles of induction machines. Therefore, with low slip, the torque $T_{em} = 3p\omega_{slip}\Psi_S^2/(2R_R)$ is proportional to the slip, proportional to the square of the stator flux, and inversely proportional to the rotor resistance. It can be concluded that the mechanical characteristic of induction motor is linear in the region of small slip frequencies and that the rate of change (slope) is proportional to the ratio Ψ_S^2/R_R.

16.14   Operation with Large Slip

Abscissa of the mechanical characteristic shows the rotor speed $\omega_m = \Omega_m$. The same axis can be used to show relative slip, which is equal to zero for $\Omega_m = \Omega_e$. At the origin of the mechanical characteristic, where $\Omega_m = 0$, relative slip is equal to 1.

In the region of high slips, prevailing part of the impedance in denominator of the torque expression is the equivalent leakage reactance $L_{\gamma e}\omega_e$, which depends on the supply frequency. At the rated supply frequency, reactance $X_{\gamma e} = L_{\gamma e}\omega_{en}$ has relative value ranging 10% up to 25%. With $\Omega_m = 0$ and $s = 1$, the stator current is approximately equal to $I_{ST} = U_s/X_{\gamma e}$. It is called *start-up current*, and it reaches $4I_n$ to $10I_n$. As the machine accelerates, the speed Ω_m increases and the slip s reduces. The stator current remains high and almost constant until the slip s reduces enough to make the impedance R_R/s prevail over $X_{\gamma e}$. Hence, for a wide range of slip values, the stator current is close to $U_s/X_{\gamma e}$, and it does not depend on the rotor speed.

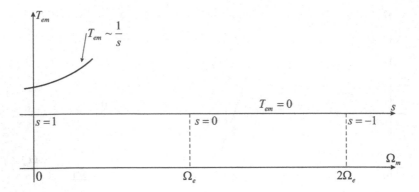

Fig. 16.11 Mechanical characteristic in high-slip region

The torque is proportional to the square of the current and to the resistance R_R/s. For this reason, in the region with high slip, the torque is inversely proportional to relative slip. By reducing the speed of rotation, relative slip increases while the torque decreases following hyperbolic law:

$$\underline{I}_s \approx \frac{U_s}{j\omega_e L_{\gamma e}},$$

$$T_{em} \approx \frac{3}{2} \frac{R_R}{s} \frac{1}{\omega_e} \frac{U_s^2}{L_{\gamma e}^2 \omega_e^2}. \tag{16.31}$$

The torque developed at speed of $\Omega_m = 0$ with slip $s = 1$ is called the start-up torque. By applying the usual approximations, the start-up torque is obtained as $T_{ST} = (3p/2)\,(R_R/\omega_e)\,(U_S/X_{\gamma e})^2$, where U_S denotes the peak value of the phase voltage, while ω_e is the angular frequency of the stator voltages.

16.15   Starting Mains Supplied Induction Machine

When an induction motor at standstill gets connected to the three-phase mains with line voltages of 400 V and line frequency of $f_e = 50$ Hz, the start-up current $I_{STrms} \approx U_{Srms}/X_{\gamma e}$ appears, exceeding by far the rated current. The start-up torque $T_{ST} = (3p/2)\,(R_R/\omega_e)\,(U_S/X_{\gamma e})^2$ is developed, accelerating the rotor and increasing the speed Ω_m. As the speed reaches the synchronous speed Ω_e, the relative slip gradually decreases as well as the stator current. Reduction of relative slip increases R_R/s and reduces the rotor and stator currents. In the absence of load torque T_L, the acceleration ends with the speed reaching the synchronous speed, while the stator current reduces to no load current $I_{0rms} \approx U_{Srms}/X_m$, which circulates through the magnetizing branch of the equivalent circuit.

If the starting torque is not sufficient to prevail over the load torque and friction, and the rotor does not start to move, the stator current $I_{STrms} \approx U_{Srms}/X_{\gamma e} \approx 5I_n$ is maintained. In this regime, the losses are significantly larger than the losses at rated condition. The copper losses in the stator winding are some 25 times larger than the copper losses at rated conditions. Therefore, the rise of the machine temperature is very fast. If such a state is not discontinued quickly, the insulation and other vital parts get overheated and permanently damaged.

16.16 Breakdown Torque and Breakdown Slip

Mechanical characteristic of induction machine is linear in the region where the rotor speed is close to the synchronous speed and the relative slip is small. With the rotor speed close to zero, and with the relative slip close to 1, the torque is inversely proportional to the slip. Between the two zones, mechanical characteristic exhibits a maximum, which corresponds to the highest torque obtained with given stator supply. The maximum value of the torque is called *breakdown torque*. The breakdown torque T_b can be determined from the torque expression, by finding extremum of the function $f(s) = T_{em}(s)$. The slip value s_b which results in the maximum torque is called *breakdown slip*. Assuming that the stator resistance R_S is negligible, the breakdown slip is determined from expression

$$\frac{dT_{em}}{ds} = \frac{d}{ds}\left[\frac{3}{2} \cdot \frac{R_R}{\omega_e \cdot s} \frac{U_s^2}{\frac{R_R^2}{s^2} + \omega_e^2 L_{\gamma e}^2}\right] = 0,$$

$$s_b = \pm \frac{R_R}{\omega_s L_{\gamma e}}. \tag{16.32}$$

First derivative of function $T_{em}(s)$ is equal to zero for $s = +s_b = +R_R/X_{\gamma e}$ and for $s = -s_b = -R_R/X_{\gamma e}$. At the speed of $\Omega_m = \Omega_e(1 - R_R/X_{\gamma e})$, there is the maximum torque $T_{em} = +T_b$ in motoring mode, where the torque is positive and the speed is somewhat lower that the synchronous speed. For negative value of s, at the speed of $\Omega_m = \Omega_e(1 + R_R/X_{\gamma e})$, there is the torque extremum $T_{em} = -T_b$ in generator mode, when the torque is negative and the rotor speed is somewhat higher than the synchronous speed. In Fig. 16.12, the breakdown slip s_b is somewhat higher than the value encountered with most standard induction machines.

The breakdown frequency $\omega_b = \omega_e s_b = R_R/L_{\gamma e}$ does not depend on the supply frequency, and it is equal to the quotient of the rotor resistance and the equivalent leakage inductance. By introducing the substitution $s = s_b = R_R/X_{\gamma e}$ and approximation $R_S \approx 0$ in the torque expression, it is possible to calculate the breakdown torque. Breakdown torque in the generator mode has the same amplitude but the opposite sign:

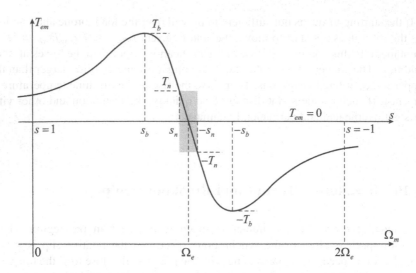

Fig. 16.12 Breakdown torque and breakdown slip on mechanical characteristic

$$T_b = \frac{3}{2}\frac{1}{\Omega_e}\frac{R_R}{s_b}\frac{U_S^2}{(R_R/s_b)^2 + (\omega_e L_{\gamma e})^2} = \frac{3}{2}\frac{1}{\Omega_e}\frac{U_S^2}{2\omega_e L_{\gamma e}}$$

$$= \frac{3p}{4\omega_e}\frac{U_S^2}{\omega_e L_{\gamma e}} = \frac{3p}{4L_{\gamma e}}\frac{U_S^2}{\omega_e^2} = \frac{3p}{4L_{\gamma e}}\Psi_S^2. \tag{16.33}$$

The breakdown torque can be calculated in terms of the stator flux $\Psi_S \approx U_S/\omega_e$. The above expression can be written in the form $T_b = (3p/4)\ \Psi_S^2/L_{\gamma e}$:

$$T_b = \frac{3p}{2L_{\gamma e}}\frac{U_{Seff}^2}{\omega_e^2}. \tag{16.34}$$

Therefore, the breakdown torque is proportional to the square of the stator flux and inversely proportional to the equivalent leakage inductance. For this reason, the breakdown torque is higher for induction machines having higher coupling coefficient k between the stator and rotor windings. In the process of designing induction motors which are started by connection to the mains, the choice of the coupling coefficient and equivalent leakage inductance is the result of compromise. Namely, smaller values of $L_{\gamma e}$ give higher breakdown torque, which is desirable, but also higher starting current $I_{STrms} \approx U_{Srms}/X_{\gamma e}$, which is not desirable.

16.17 Kloss Formula

In the case when the breakdown slip s_b and breakdown torque T_b are known, function $T(s)$ can be reduced to the form where parameters of the machine do not appear. Starting from expressions for the breakdown torque and breakdown slip

$$T_b = 3 \frac{1}{\Omega_e} \frac{U_{Srms}^2}{2\omega_e L_{\gamma e}}, \quad s_b = \frac{R_R}{\omega_b L_{\gamma e}}, \tag{16.35}$$

the expression for the electromagnetic torque can be reduced to the following form:

$$
\begin{aligned}
T(s) &= \frac{3}{\Omega_e} \frac{R_R}{s} \frac{U_{Srms}^2}{(R_R/s)^2 + (\omega_e L_{\gamma e})^2} \\
&= T_b \left(\frac{\Omega_e}{3} \frac{2\omega_e L_{\gamma e}}{U_{Srms}^2} \right) \left(\frac{3}{\Omega_e} \frac{R_R}{s} \frac{U_{Srms}^2}{(R_R/s)^2 + (\omega_e L_{\gamma e})^2} \right) \\
&= T_b \frac{R_R}{s} \frac{2\omega_e L_{\gamma e}}{(R_R/s)^2 + (\omega_e L_{\gamma e})^2} \\
&= T_b \frac{2}{\frac{R_R/s}{\omega_e L_{\gamma e}} + \frac{\omega_e L_{\gamma e}}{R_R/s}} = \frac{2T_b}{\frac{s_b}{s} + \frac{s}{s_b}}.
\end{aligned}
\tag{16.36}
$$

The obtained expression, also known as *Kloss formula*, is based on assumption that the stator resistance R_S is negligible. In addition, it is assumed that the mutual inductance L_m is so large that the magnetizing current $|I_m|$ can be neglected compared to the current in the stator and rotor windings. Calculation of the torque on the basis of expression

$$T(s) = \frac{2T_b}{\frac{s_b}{s} + \frac{s}{s_b}} \tag{16.37}$$

is relatively simple because it requires the knowledge of only two parameters: the breakdown torque and the breakdown slip.

Question (16.10): Compare the breakdown torque in motor and generator modes.

Answer (16.10): According to Kloss formula, and also the formula for electromagnetic torque $T_{em}(s)$ where the influence of the stator resistance R_S is neglected, the breakdown torque in motor mode is equal to the breakdown torque in generator mode. The difference between the two appears only in the case when R_S is not negligible, and it can be determined by solving the equivalent circuit while taking into account the resistance R_S, obtaining the corresponding expression for the torque $T_{em}(s)$, and finding the extremum of the function $T_{em}(s)$. The same conclusion can be reached considering the impact of the voltage drop $R_S I_S$ on the flux amplitude. Starting from the expression for the breakdown torque $T_b = (3p/4)\,\Psi_S^2/L_{\gamma e}$, which shows that T_b depends on the square of the stator flux, it is possible to conclude that the breakdown torque in motoring mode has a lower absolute value than the breakdown torque in generator mode. The phasor of the stator flux can be determined on the basis of expression $\underline{\Psi}_S = (\underline{U}_S - R_S\underline{I}_S)$. In motor mode, the active component of the current is directed from the source to the machine; thus, the voltage drop $R_S\underline{I}_S$ makes the stator flux somewhat lower than U_S/ω_e. In generator mode, direction of the current is altered, thus increasing the flux above the value of U_S/ω_e. Therefore, the absolute value of the breakdown torque in generator mode is higher than that in motoring mode.

The maximum torque of the mechanical characteristic $T_{em}(s)$ is called *breakdown* torque since the operation with $T_L = T_{em} = T_b$ at the rotor speed of $\Omega_m = \Omega_e(1 - s_b)$ is prone to transition to the zone $\Omega_m < \Omega_e(1 - s_b)$ where the slope of the mechanical characteristic $\Delta T/\Delta\omega$ changes sign. Transition to the zone where $\Omega_m < \Omega_e(1 - s_b)$ and $s > s_b$ leads to a progressive reduction of the rotor speed and eventually brings the rotor down to zero speed.

16.18   Stable and Unstable Equilibrium

If the torque of work machine T_L is constant, the breakdown torque and breakdown slip separate the mechanical characteristic in two parts. The part where $s < s_b$ results in stable equilibrium $T_{em} = T_L$, while the part with $s > s_b$ results in unstable equilibrium $T_{em} = T_L$. The analysis starts from Newton equation of motion:

$$J\frac{d\Omega_m}{dt} = T_{em}(\Omega_m) - T_L. \tag{16.38}$$

Figure 16.13 shows the regions of stable and unstable equilibrium. The initial assumption is that the load torque is constant. Point UE represents an operating point in the region of unstable equilibrium. If the rotor speed drops by $\Delta\Omega_m$, as

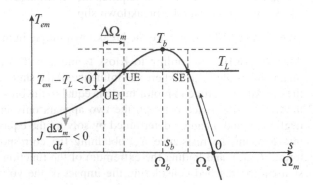

Fig. 16.13 Regions of the stable and unstable equilibrium

shown in the figure, the operating point moves from right to left and arrives at the point denoted by UE1, where the load torque T_L remains unchanged, while the electromagnetic torque $T_{em}(\Omega_m)$ decreases. It is of interest to notice that the electromagnetic torque T_{em} reduces as the operating point moves to left from the point UE. The first derivative of the speed is $J d\Omega_m/dt = T_{em}(\Omega_m) - T_L$, and it becomes negative at UE1, thus leading to further decelerations. The operating point moves progressively to the left and goes toward the origin. Hence, once disturbed, the motor will not return to point UE.

In the same figure, point SE is the operating point in the region of stable equilibrium. If the speed of rotation reduced due to an external disturbance, the operating point moves from right to left, the load torque T_L remains unchanged, while the electromagnetic torque $T_{em}(\Omega_m)$ increases. Derivative of the speed is positive; thus, the operating point returns to the equilibrium state, which is the point SE.

The breakdown torque is the highest achievable torque. Starting from no load point with $s = 0$, where $T_L = 0$ and $\Omega_m = \Omega_e$, gradual increase of the load torque results in gradual decrease of the rotor speed. At the same time, the relative slip and the electromagnetic torque are increased until the electromagnetic torque does not reach the load torque. When the value T_{em} reaches the load torque T_L, the point SE of stable equilibrium is reached. When the load torque reaches the value of the breaking torque T_b, the electromagnetic torque T_{em} increases as well, and the operating point in T_{em}–Ω_m diagram (Fig. 16.13) reaches the *breakdown* point (T_b, Ω_b). Any further load torque disturbance leads to crossing the breakdown point of $T_{em}(\Omega_m)$ characteristic. The operating point passes toward the left and enters the region of unstable equilibrium. In turn, this leads to progressive reduction of the rotor speed. Assuming that the load torque does not change with the speed and remains constant, the rotor decelerates toward zero speed; it changes direction of rotation and proceeds accelerating in the opposite direction. With reactive load torque which resists the motion in both direction, the rotor would decelerate to zero and eventually stop.

16.19   Region Suitable for Continuous Operation

Permanent operation of an induction machine is possible in the operating regions where the stator current does not exceed the rated current. In the cases where $I_S > I_n$, losses in the machine exceed the permissible level. In a prolonged operation with increased losses, the temperature of the machine exceeds the safe limits and causes permanent damages of the vital parts. The region where continued operation is permitted is shown in Fig. 16.14.

At no load, the slip is close to zero and the stator windings have no load current $\underline{I}_0 = \underline{U}_{Sn}/(R_S + j\omega_{en}L_S)$ which is considerably lower than the rated current. Described operating point corresponds to the crossing of the mechanical characteristic and abscissa of the diagram in Fig. 16.14. Increasing the load torque T_L slows

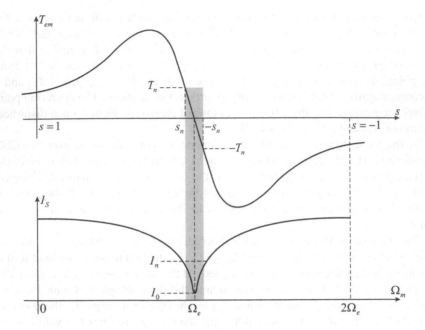

Fig. 16.14 Electromagnetic torque and stator current in the steady state

down the rotor, increases the slip, decreases impedance R_R/s, and causes the stator and rotor currents to increase. At rated slip $s = s_n$, induction machine develops the rated torque T_n, it rotates at the rated speed $\Omega_n = \Omega_{en}(1 - s_n)$, and the stator windings have rated current I_n.

In the case when the load torque T_L becomes negative, it tends to accelerate the rotor, increasing the rotor speed above synchronous speed. The slip becomes negative, as well as the impedance R_R/s, resulting in negative air-gap power and negative torque. The machine operates in generator mode and converts mechanical energy to electrical energy. At the slip which is equal to the negative rated slip, $s = -s_n$, induction machine develops the torque $-T_n$; it rotates at the rotor speed $\Omega_{en}(1 + s_n)$ which is higher than the synchronous speed, while the stator windings have rated current (I_n).

At rated supply conditions, stator current is maintained within the rated limits for slips $|s| \leq s_n$, that is, for speeds $\Omega_{en}(1 - s_n) \leq \Omega_m \leq \Omega_{en}(1 + s_n)$. For small power induction machines, the speed may deviate from the synchronous speed by $\pm 10\%$ maintaining at the same time the stator current within the rated limit. For medium- and high-power machines, permissible deviation of the rotor speed is less than 1%. Continued operation of induction machines is possible with stator currents that remain within the rated limits, $|I_S| \leq I_n$. According to Fig. 16.14, condition $|I_S| \leq I_n$ is met for the rotor speeds that remain in close vicinity of the synchronous speed. Therefore, in order to accomplish continuous variation of the rotor speed, it is necessary to change its synchronous speed. This is achieved by varying the frequency of the stator voltages and currents, which is achieved by supplying the induction machine from a three-phase source providing variable frequency AC voltages.

16.20 Losses and Power Balance

One of the assumptions taken in modeling of electrical machines has been neglecting the losses in magnetic circuit, also called iron losses. These losses have one part proportional to $B^2 f^2$, caused by eddy currents, and the other part proportional to $B^2 f$, caused by hysteresis of the $B(H)$ curve. The power balance discussed hereafter takes into account the iron losses as well.

Within the rotor of an induction machine, there are variations of the magnetic induction at relatively low-slip frequency. Thus, the power of losses within the rotor magnetic circuit is relatively small. Power of iron losses in the stator magnetic circuit is considerably higher since the magnetic field in the stator core varies at the frequency of the supply voltages ω_e. Besides iron losses and copper losses, the process of electromechanical conversion has the losses in mechanical subsystem. Mechanical losses include the motion resistances such as the air resistance and friction in bearings.

16.21 Copper, Iron, and Mechanical Losses

The most significant losses in an induction machine are:

- Copper losses in stator windings, $R_S I_S^2$
- Losses in rotor cage winding, $R_R I_R^2$
- Iron losses in stator magnetic circuit, $\sigma_H f_e B_m^2 + \sigma_V f_e^2 B_m^2$
- Iron losses in rotor magnetic circuit, $\sigma_H f_k B_m^2 + \sigma_V f_k^2 B_m^2$
- Mechanical losses due to rotation, $k_F \Omega_m^2$

Losses in copper of the stator winding are proportional to the square of the rms value of the stator current, $P_{cu1} = 3R_S I_{Srms}^2$.

The losses P_{Fe1} in stator magnetic circuit are proportional to square of the magnetic induction, and they are dependent on stator frequency f_e. If the eddy current losses prevail, the iron losses are proportional to the square of the stator frequency.

Currents in rotor bars create losses which are proportional to the square of the rotor current, $P_{Cu2} = 3R_R I_{Rrms}^2$. In addition to losses in the rotor winding, there are also iron losses in the rotor magnetic circuit, where the magnetic field varies at the slip frequency ω_{slip}. The slip frequency ω_{slip} is much lower than the stator frequency. Therefore, the rotor iron losses are often neglected. Yet, there are operating modes where the rotor iron losses are considerable and cannot be neglected. One of them is the start-up mode, when the machine with the rotor at standstill gets connected to the mains and has the slip frequency of $\omega_{slip} = \omega_e$, contributing to significant iron losses in the rotor magnetic circuit.

Losses due to rotation are caused by friction in bearings, air resistance, and other phenomena in mechanical subsystem of the induction machine which oppose to the rotor motion. The torque due to the air resistance is proportional to the square of the rotor speed, while the corresponding power depends on the third power of the rotor speed. On the other hand, the sum of all the losses due to rotation is often modeled by an approximate formula $P_F \approx k_F \Omega_m{}^2$, with the corresponding torque being $T_F = k_F \Omega_m$. The friction torque T_F is subtracted from the electromagnetic torque T_{em} which is generated by electromagnetic forces. Thus, the torque available at the output shaft becomes $T_m = T_{em} - T_F$, and it is somewhat lower than T_{em}.

16.22   Internal Mechanical Power

Electrical source connected to the stator windings supplies the input power P_e to the induction machine. At steady state, the input electrical power can be determined on the basis of expression $P_e = 3U_{Srms} I_{Srms} \cos(\varphi)$, where U_{Srms} is the rms value of the phase voltages and I_{Srms} is the rms value of the stator currents, while $\cos(\varphi)$ is the power factor. The power of losses $P_{cu1} = 3R_S I_{Srms}{}^2$ represents the copper losses in the stator windings, and it depends on the squared rms value of the stator currents. The losses in the stator magnetic circuit, also called the stator iron losses, are denoted by P_{Fe1}, and they depend on the stator frequency ω_e and the magnetic induction B. Considering that the stator flux depends on the magnetic induction in the stator iron, the stator iron losses can be expressed in terms of the squared amplitude of the stator flux. The air-gap power $P_\delta = P_e - P_{cu1} - P_{Fe1}$ is transferred from the stator to the rotor, and it passes through the air gap. Electromagnetic torque $T_{em} = P_\delta/\Omega_e$ is obtained by dividing the air-gap power by the speed of rotation Ω_e of the magnetic field, also called synchronous speed:

$$P_\delta = P_e - 3R_S i_{S_{rms}}^2 - P_{Fe1} = T_{em}\Omega_e. \tag{16.39}$$

The product of the electromagnetic torque $T_{em} = P_\delta/\Omega_e$ and the rotor speed Ω_m results in the internal mechanical power P_{mR}. This power is slightly lower than the air-gap power, as the rotor does not revolve at the synchronous speed. The difference between the air-gap power and the internal mechanical power covers the rotor losses, namely, the copper losses in the rotor and relatively small iron losses in the rotor:

$$P_{mR} = T_{em}\Omega_m \;\Rightarrow\; P_\delta - P_{Cu2} - P_{Fe2} \approx P_\delta - P_{Cu}^{rot}. \tag{16.40}$$

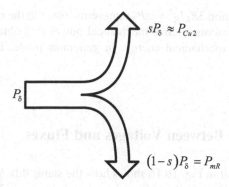

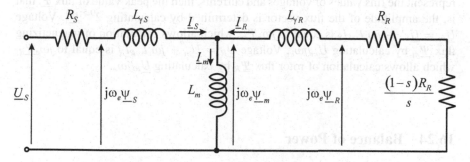

Fig. 16.15 Air-gap power split into rotor losses and internal mechanical power

Fig. 16.16 Equivalent circuit and relation between voltages and fluxes

Losses in the rotor are

$$P_{\gamma R} = P_\delta - P_{mR} = T_{em}(\Omega_e - \Omega_m) = sP_\delta = P_{Cu2} + P_{Fe2}. \qquad (16.41)$$

The rotor speed is lower than the synchronous speed by the amount of slip. Mechanical power P_{mR} created inside the machine is equal to $P_{mR} = T_{em}\Omega_m$. At steady state, mechanical power P_{mR} is equal to $T_{em}\Omega_m = T_{em}\Omega_e(1-s) = (1-s)P_\delta$. Therefore, the balance sP_δ is equal to the power of losses in the rotor, and these are the losses in the rotor windings P_{cu2} and the losses in the rotor magnetic circuit P_{Fe2}. Since rotor frequency is relatively small, it is most often justified to assume that $P_{Fe2} \approx 0$ and use the expression $sP_\delta = P_{Cu2}$.

Useful mechanical power $P_m = T_m\Omega_m$ is different from internal mechanical power $P_{mR} = T_{em}\Omega_m$ by the amount of losses due to rotation. These are the losses in mechanical subsystem, including the air resistance and friction in bearings.

The equivalent circuit of induction machine shown in Fig. 16.16 is modified in order to separate the rotor losses $P_{Cu2} = sP_\delta$ from the internal mechanical power P_{mR}. For this to achieve, resistance R_R/s in the rotor branch of the equivalent circuit is split in two parts: R_R and $R_R(1-s)/s$. Assuming that I_R is the rms value of the rotor current, dissipation $3R_R I_R^2 = sP_\delta$ represents losses in the rotor winding, while $3R_R I_R^2 (1-s)/s$ represents internal mechanical power P_{mR} obtained by converting electrical energy to mechanical energy. In generator mode, internal mechanical power is negative.

16.23 Relation Between Voltages and Fluxes

The equivalent circuit in Fig. 16.16 shows how the stator flux Ψ_S, mutual flux Ψ_m, and rotor flux Ψ_R can be calculated from the equivalent circuit. Voltage $\underline{U}_S^1$ is equal to $\underline{U}_S - R_S \underline{I}_S$, and it is determined by product $j\omega_e \Psi_S$, which allows calculation of the stator flux Ψ_S by finding $\underline{U}_S^1/j\omega_e$. In the case when variables $|\underline{U}_S|$, $|\underline{I}_S|$, and $|\underline{I}_R|$ represent the rms values of voltages and currents, then the peak value of flux Ψ, that is, the amplitude of the flux vector is determined by calculating $2^{0.5}|\Psi|$. Voltage $\underline{U}_m = \underline{U}_S^1 - j\omega_e L_{\gamma S} \underline{I}_S$ is equal to $j\omega_e \Psi_m$, which allows calculation of magnetizing flux Ψ_m by calculating $\underline{U}_m/j\omega_e$. Voltage $\underline{U}_R = \underline{U}_m + j\omega_e L_{\gamma R} \underline{I}_R$ is equal to $j\omega_e \Psi_R$, which allows calculation of rotor flux Ψ_R by calculating $\underline{U}_R/j\omega_e$.

16.24 Balance of Power

Balance of power of an induction machine operating as a motor is presented in Fig. 16.17. The indicated powers P_e, P_δ, P_{mR}, and P_m are positive and related as $P_e > P_\delta > P_{mR} > P_m$. Coefficient of efficiency in motor operating mode is $\eta = P_m/P_e$.

Useful mechanical power obtained at the output shaft is

$$P_m = P_{mR} - k_F \Omega_m^2. \tag{16.42}$$

If the stator copper losses, the stator iron losses, and the losses due to rotation are neglected, only the rotor winding losses are taken into account. The electrical power taken from the source, that is, from the mains is

$$P_e \approx P_\delta \approx P_{Cu2} + P_m = sP_\delta + (1-s)P_\delta. \tag{16.43}$$

In generator mode, the machine receives mechanical power from the shaft. Therefore, the *output* power $P_m = T_m \Omega_m$ is negative, considering adopted reference directions. In the above figure, it is assumed that positive power depicts the energy transfer from left to right. Indicated power amounts P_e, P_δ, P_{mR}, and P_m are negative in generator mode, and they are related by $P_e > P_\delta > P_{mR} > P_m$, that is, $|P_e| < |P_\delta| < |P_{mR}| < |P_m|$.

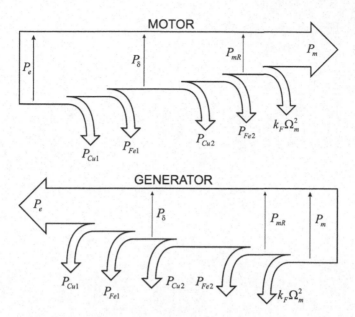

Fig. 16.17 Balance of power of an induction machine

Considering generator mode and adopting the reference power flow from right to left, negative values are omitted and the balance of power of the machine is more obvious. A generator is, via shaft, supplied by the input mechanical power P_m. By subtracting the losses of the mechanical subsystem, which are the losses due to rotation, one obtains the internal mechanical power which is converted to electrical power. The losses in the rotor windings and magnetic circuit are subtracted from the obtained electrical power ($-P_{mR} > 0$) to obtain the air-gap power ($-P_{ob} > 0$) which is transferred to the stator. After subtracting the losses in stator copper and iron, one obtains the electrical power ($-P_e > 0$) provided by the generator, which can be transferred to consumers via three-phase network connected to the stator windings of the induction generator. Coefficient of efficiency in generator mode is equal to $\eta = P_e/P_m$. Induction generators are often used in small hydro-electric power plant stations and wind power plants. When connected to the mains, induction generators have magnetic field that rotates at the synchronous speed of $\Omega_e = 2\pi \cdot 50/p$. Mechanical power obtained from the water flow or from the wind accelerates the rotor and makes it revolve at speeds which are higher than the synchronous speed. In this regime, relative slip is negative, as well as the equivalent resistance R_R/s of the equivalent circuit. For this reason, electrical power absorbed by the machine from the network is negative, meaning that the machine actually supplies the network with electrical energy. This energy is obtained by converting the mechanical work into electrical energy.

Chapter 17
Variable Speed Induction Machines

This chapter discusses the means for the speed change of induction machines. The speed regulation is required in both generators and motors. Induction machines that serve as generators in wind power stations revolve at variable speed. Therefore, the machine and the associated equipment must ensure conversion of mechanical work in electrical energy at variable speed. The machines used as motors often serve in motion control applications, where the speed changes in continuous manner.

In the first part of this chapter, the means for altering the rotor speed of mains-supplied induction machines are discussed and explained. Possibilities are considered to adjust the rotor speed of induction machines operating with constant frequency of stator voltages. The analysis considers changes in mechanical characteristic and the rotor speed due to variations of the voltage amplitude. Variation of the rotor resistance is studied as the means of changing the rotor speed of wound rotor induction machines. The impact of the number of magnetic pole pairs p on synchronous speed Ω_e and rotor speed Ω_m is reinstated, and the change of the number of poles is looked upon as the means of changing the rotor speed. An introduction to electrical machines with multiple pole pairs is given by studying distribution of the magnetic field of an electrical machine with $2p = 4$ magnetic poles. The possibility of designing the stator winding so as to achieve the magnetic field with multiple pole pairs is studied on a sample winding that can be switched to produce either $2p = 2$ poles or $2p = 4$ poles. The expressions for synchronous speed, rotor speed, and slip frequency are given as functions of the number of magnetic pole pairs. Discussion on constant frequency-supplied induction machine closes with establishing deficiencies, limitations, energy losses, and design problems arising in mains-supplied induction machines.

The second part of this chapter deals with induction machines supplied from variable frequency sources such as the three-phase inverters with switching power transistors and pulse width modulation control. This chapter introduces basic aspects and problems of variable frequency supply. Operation of induction machines fed from variable frequency static power converter is introduced and studied. A short review of power converter topologies used for supplying induction machines is presented, along with methods for continuous change in the stator

S.N. Vukosavic, *Electrical Machines*, Power Electronics and Power Systems,   475
DOI 10.1007/978-1-4614-0400-2_17, © Springer Science+Business Media New York 2013

voltage amplitude and frequency, suited to accomplish desired rotor speed and desired flux. The effects of changing the supply frequency on mechanical characteristic are analyzed in both the constant flux region and in the field weakening region. The basic approaches the torque, flux, and power control are outlined for an induction machine fed from a variable frequency static power converter. Family of mechanical characteristics obtained by frequency variation is presented and explained. Based upon the study of operating limits of the machine and operating limits of associated three-phase inverter, steady state operating area and transient operating area are derived in T-Ω plane and studied for variable frequency operation of induction machines. The limits of constant power operation in field weakening mode are determined, explained, and expressed in terms of the machine leakage inductance. Finally, this chapter discusses the differences in construction and parameters of mains-supplied induction machines and inverter-supplied induction machines.

17.1   Speed Changes in Mains-Supplied Machines

In majority of applications of electrical machines, it is required to accomplish a continuous variation of the rotor speed. Some of examples are motion control tasks in production machines and industrial robots and propulsion tasks in electrical vehicles, fans, pumps, and similar.

The rotor speed of induction machines is different from the synchronous speed by the amount of the slip. When an induction motor is supplied from the mains, the stator current is maintained within rated limits under condition that the slip remains relatively low, $|s| \leq s_n$, namely, for the speed range $\Omega_{en}(1 - s_n) \leq \Omega_m \leq \Omega_{en}(1 + s_n)$. Hence, continuous operation of mains-supplied induction machine is restricted to a rather narrow range of speeds. For medium- and high-power machines, the rated slip is lower than 1%; thus, the condition $I_S \leq I_n$ is maintained within the range of speeds from 99% up to 101% of the synchronous speed. Operation at higher slip frequencies involves high losses and high currents in stator and rotor windings. The use of induction machine outside the zone where $|s| \leq s_n$ and $I_S \leq I_n$ leads to an increase of the machine temperature. For this reason, continued service of induction machines is possible only with relatively small values of slip. Therefore, it is justified to conclude that the speed of rotation Ω_m is close to the synchronous speed Ω_e. Synchronous speed $\Omega_e = \omega_e/p$ is determined by the angular frequency of the stator voltages ω_e, that is, by the frequency of electrical currents in stator windings. For two-pole machines, where the number of pole pairs p is equal to 1, the synchronous speed is equal to the angular frequency of the supply. For this reason, variation of the rotor speed of an induction machine requires a variable frequency of the stator voltages and currents. This can be accomplished by supplying the machine from a three-phase source of AC voltages having variable frequency. Most common solution to this is the three-phase inverter, a power converter which employs semiconductor power switches. Inverters are mostly used in conjunction with three-phase diode rectifiers. Three-phase diode rectifiers are power converters supplied from the mains with line voltages of 400 V and line frequency of $f = 50$ or 60 Hz. The rectifier converts the AC line voltages into DC voltage. This DC voltage is fed to the three-phase inverter, which converts the DC voltage into a set of three-phase AC voltage of variable frequency and variable amplitude. Finally, at the output terminals of the inverter, a three-phase system is available with frequency and voltage amplitude that can be changed to suit the needs of the induction machine. Contemporary inverters apply pulse width modulation control and make use of semiconductor switches such as bipolar transistors (BJT), MOSFET transistors, and IGBT transistors. Industrial use of such devices started in last decades of the twentieth century. At present, power converters using power transistors are standard industrial units for supplying induction machines and changing their speed (Fig. 17.10a).

Induction machines have been in use for more than 100 years. During the first century of industrial use of induction machines, there were no semiconductor switches suitable for designing the three-phase inverters. Therefore, another kind of components, devices, and techniques was being devised and used to achieve continuous variation of the rotor speed of induction machines. Induction machines were supplied from the mains providing the voltages of industrial frequency, whether $f_e = 50$ Hz or $f_e = 60$ Hz. There were no ways of altering the supply frequency and providing continuous change of the synchronous speed. Therefore, most induction machines were primarily used in constant speed applications. Particular procedures, methods, and devices were used in applications requiring variable speed operation, all of them conditioned and restricted by technology limits of the times. Traditional approaches to speed variations include:

1. Variation of stator voltage
2. Variation of rotor resistance
3. Variation of the number of poles

Besides, in some cases, induction motors were connected to mechanical load by means of transmission mechanism with gears or some other mechanical transducers. Using particularly suited transmission with variable transmission ratio, it was possible to change the load speed while the induction motor speed remained constant, close to the synchronous speed.

In further text, the effects of the traditional approaches (1), (2), and (3) will be reviewed.

17.2 Voltage Change

With constant frequency supply, mechanical characteristic of an induction machine crosses the abscissa Ω_m of $T_{em}-\Omega_m$ plane at $\Omega_m = \Omega_e = 2\pi f_e/p$, where $T_{em} = 0$ (Fig. 17.1). Upon loading, the slip increases and the speed decreases. In generator mode, where $T_{em} < 0$ and $\omega_{slip} < 0$, the torque amplitude $|T_{em}|$ increases as the

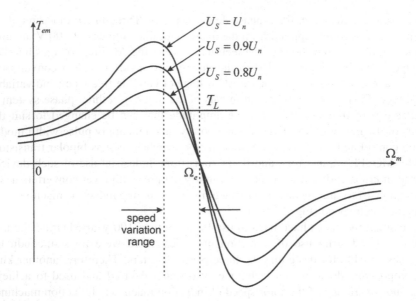

Fig. 17.1 Effects of voltage changes on mechanical characteristic

rotor speed goes beyond Ω_e. The electromagnetic torque is proportional to the slip and also proportional to the square of the stator flux Ψ_S. In turn, the flux amplitude Ψ_S is proportional to the quotient of the stator voltage and the angular frequency, $\Psi_S \approx U_S/\omega_e$.

Reduction of the stator voltage results in reduction of the flux. At the same time, the slope of the mechanical characteristic $S = \Delta T/\Delta\omega$ is reduced as well. As a consequence, the speed would exhibit larger reduction for the same torque. In the region of small slip, the electromagnetic torque varies according to the law $T_{em} = k\omega_{slip}\Psi_S^2/R_R$. Therefore, reduction of the flux Ψ_S causes an increase in the slip speed ω_{slip} and reduction of the rotor speed.

Speed can be changed by the voltage variations in a limited range. Namely, the speed can be varied within the range that does not exceed the breakdown slip s_b. A consequence of the speed reduction by the slip increase is an increase of the rotor losses $P_{Cu2} \approx sP_\delta$, which may lead to increase of the machine temperature.

Variation of the rotor speed of an induction machine obtained by changing the amplitude of the supply voltage is seldom used. One possible use of the voltage control is in fan drives, where mechanical power is proportional to the third power of the rotor speed. It takes a relatively small change of the rotor speed to produce significant change of the torque and power. Hence, the speed regulation based on reduction of the stator voltage does not produce significant losses $P_{Cu2} \approx sP_\delta$ since the air-gap power $P_\delta \sim \Omega_m^3$ reduces considerably even with relatively small drop in the rotor speed. For small fan drives with induction motors, voltage can be reduced by inserting variable resistors in series with the stator winding. The series connection of the stator windings and variable resistors is fed from the mains with constant voltages. An increase in series resistance reduces the voltage across the stator windings, increases the slip, and reduces the rotor speed. Disadvantage of this approach is the presence of Joule losses in series resistors.

17.3 Wound Rotor Machines

Most induction machines have the rotor winding made of cast aluminum, having the form of a cage with two short-circuiting rings. On the other hand, the rotor winding can be made in the same way as the stator winding by placing insulated copper conductors in the rotor slots and connecting these conductors in series, so as to make a star-connected three-phase winding. The three end terminals of such winding can be made available from the stator side. In most cases, the ends of the rotor winding are connected to three conductive rings that are fastened to the shaft and mutually isolated. These rings rotate with the shaft, but they are electrically isolated from the shaft too. From the stator side, there are three conductive brushes fastened to the stator and pressed against revolving rings. They touch the external surface of the rings and provide electrical contacts. The brushes slide along the circumference of the rings, which are also called *slip rings*. Induction machine with slip rings is also called *wound rotor* machine. The three brushes make the rotor winding ends available for stator side connections. A variable three-phase resistor can be connected to the brushes, providing the means for changing the equivalent resistance of the rotor circuit. The effects of inserting an external resistor into the rotor circuit are the same as the effects of hypothetical changes of the rotor resistance of the rotor with short-circuited cage. Variation of an externally connected three-phase resistor changes mechanical characteristic of the machine in the way shown in Fig. 17.2. Sample wound rotor machine is given in Fig. 17.3.

When the brushes are brought into short circuit, the wound rotor is short-circuited. In such case, behavior of the machine and its mechanical characteristic are the same as with a squirrel cage induction machine. By way of slip rings and brushes, additional resistance R_{ext} can be inserted in the rotor circuit, increasing in this way the value of equivalent rotor resistance $R_{Re} = R_R + R_{ext}$ which should substitute R_R in the steady state equivalent circuit.

Considering the expression for the breakdown torque, it is concluded that $T_b \sim \Psi_S^2/L_{\gamma e}$. Hence, an increase in the rotor resistance R_{Re} does not affect the value of the breakdown torque. On the other hand, the slope $\Delta T/\Delta\omega$ of the mechanical characteristic is inversely proportional to the rotor resistance, $T_{em} \sim k\omega_{slip}\Psi_S^2/R_{Re}$; thus, increasing the total rotor resistance R_{Re} decreases the slope of the mechanical characteristic $|\Delta T/\Delta\omega|$. Breakdown slip $s_b = R_{Re}/X_{\gamma e}$ is proportional to the rotor resistance; thus, increasing the external resistor R_{ext} increases the breakdown slip in both motor and generator modes. Total effects of variation of the rotor resistance on mechanical characteristic are shown in Fig. 17.2. While breakdown torque remains unchanged, breakdown slip increases, while slope of the mechanical characteristic gradually decreases.

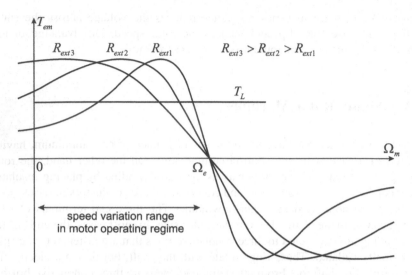

Fig. 17.2 Influence of rotor resistance on mechanical characteristic

The effects of the rotor resistance changes on the rotor speed are investigated by assuming that the load torque is constant as well as the supply frequency and the synchronous speed. In Fig. 17.2, the load characteristic becomes a horizontal line. The operating speed is obtained at the crossing of the load characteristic and the mechanical characteristic. It is observed in the figure that the intersection of the two characteristics occurs at lower speeds if the equivalent rotor resistance is larger. Hence, the rotor speed can be changed by changing the external resistor connected in the rotor circuit via slip rings and brushes. Continuous variation of the resistance enables continuous variation of the rotor speed. Unlike squirrel cage motors, wound rotor motors can operate with very large slips. When operating with slips in excess to the rated slip, the stator and rotor currents of the squirrel cage motors exceed the rated currents and result in overheating. An induction motor with wound rotor may operate with higher slip values. With elevated slip, resistance R_{Re}/s of the rotor branch in the equivalent circuit is reduced, but the equivalent resistance $R_{Re} = R_R + R_{ext}$ is increased due to external resistance, keeping the stator and rotor currents within acceptable limits.

A shortcoming of the described approach is poor efficiency caused by additional losses in the external resistor. The speed $\Omega_m = \Omega_e(1-s)$ is controlled in the way which keeps synchronous speed Ω_e constant. The rotor speed Ω_m is lowered on account of an increased slip s. With an increase in slip, the rotor losses $sP_\delta = P_{Cu2}$ are increased as well. Increased losses P_{Cu2} do not cause overheating of the induction machine, because significant part of these losses is dissipated in the external three-phase resistor. Nevertheless, the efficiency of the system is significantly reduced. Mains-supplied two-pole induction motor has the synchronous speed of $n_e = 3,000$ rpm. In cases when the rotor speed is reduced to 1,500 rpm,

Fig. 17.3 Wound rotor with
slip rings and external
resistor. (**a**) Three-phase rotor
winding. (**b**) Slip rings. (**c**)
Stator. (**d**) Rotor. (**e**) External
resistor

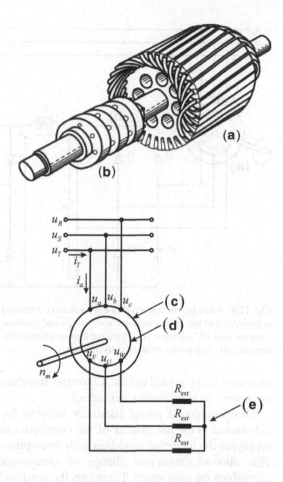

one half of the air-gap power is dissipated in the rotor circuit, while the other half is
converted into mechanical power. The efficiency of this induction motor is
lower than 50%. Poor efficiency is the consequence of a high value of sP_δ, also
called *slip power*.

Efficiency of an induction machine with wound rotor and external resistor is
poor due to large slip power sP_δ which is converted into heat. Efficiency can be
increased by recovering the slip power back to the mains. In the middle of the
twentieth century, semiconductor diodes and thyristors suitable for industrial
applications have been developed and put to use. Static power converters have
been designed comprising diodes and thyristors. Connecting a static power converter to the rotor circuit, the slip power sP_δ is transferred to the diode rectifier,
shown in Fig. 17.4, which converts AC rotor currents to DC currents in the choke
denoted by L_d. Further on, thyristor converter (C) converts DC currents to AC
currents, the latter having the line frequency and being directed back into the mains
through the transformer (D). In this way, the slip power sP_δ is returned to the mains

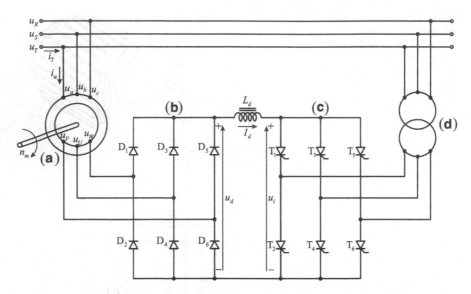

Fig. 17.4 Static power converter in the rotor circuit recuperates the slip power. (**a**) The converter is connected to the rotor winding via slip rings and brushes. (**b**) Diode rectifier converts AC rotor currents into DC currents. (**c**) Thyristor converter converts DC currents into line frequency AC currents. (**d**) Slip power recovered to the mains

instead of being wasted in heat. Converter structures connected into the rotor circuit are known as *synchronous cascades*.[1]

Development of power transistor suitable for building high-power inverters culminated in the last quarter of the twentieth century. It provided the means for supplying the induction machines with three-phase voltages of variable frequency. This allowed continuous change of synchronous speed as a viable way of controlling the rotor speed. Therefore, the need for wound rotor induction machines and cascade static power converters gradually declined.

---

[1] Slip rings' access to the rotor winding can be used to take the slip power out of the rotor circuit, as shown in Fig. 17.4. It is also possible to use the static power converter of different topology and to use it to supply the power to the rotor circuit. In this case, slip power and slip speed are negative, and the rotor revolves at the speed $\Omega_m > \Omega_e$. The two considered topologies are called *subsynchronous cascade* and *supersynchronous cascade*. Over the past century, there were also applications of wound rotor induction machines with slip rings and a four-quadrant (reversible) static power converter in the rotor circuit. With four-quadrant rotor converter, wound rotor machine can operate with $\Omega_m > \Omega_e$ as well as with $\Omega_e > \Omega_m$. Some early wind power solutions were conceived with wound rotor induction generators and static power converter in the rotor circuit. The advantage of this approach is relatively low slip power which results in relatively low voltage and current ratings of semiconductor power switches. More recent wind power generators are based on squirrel cage induction machines and full-power transistor-based static power converters that provide the interface between the constant frequency mains and variable frequency stator voltages.

17.4 Changing Pole Pairs

The rotor speed of a squirrel cage induction machine is close to the synchronous speed $\Omega_e = \omega_e/p$, where ω_e is the angular frequency of the power supply, while $2p$ is the number of magnetic poles of the machine. The mains-supplied machines operate with constant stator frequency, and their synchronous speed cannot be varied.[2] There is, however, a possibility to vary the number of poles $2p$. For a two-pole machine with $p = 1$ and with $f = 50$ Hz, the synchronous speed is $n_e = 3,000$ rpm. By increasing the number of poles to $2p = 4, 6$, or 8, one obtains synchronous speeds of 1,500, 1,000, and 750 rpm, respectively. Hence, the synchronous speed as well as the rotor speed can be changed in discrete steps by altering the number of magnetic poles.

It is of interest to discuss the number of magnetic poles in AC and DC machines. In DC machines, magnetic field is created by permanent magnets or by the excitation winding. The number of magnetic poles is determined by design of the machine, and it is equal to the number of main stator poles. For that reason, it cannot be changed during the operation. The number of magnetic poles in DC machines affects design and the construction of mechanical commutator, machine windings, and the form of magnetic circuits. For that reason, it is not possible to change the number of magnetic poles unless the whole construction of the machine is changed. In AC machines, magnetic field is created by electrical current in the stator windings. With suitable design of the stator windings, appropriate reconnection of the stator phases can be used to change the number of magnetic poles. Hence, the synchronous speed of an induction motor as well as the rotor speed can be changed by altering the electrical connections of the stator phases.

In many applications of electric drives, it is not required to accomplish a continuous variation of the rotor speed. Instead, it is sufficient to have two or three discrete values of the speed which can be selected as required. Then, the problem of regulation of the rotor speed can be solved by using an induction machine fed from a constant frequency source, under conditions that the number of poles can be varied. In the case of an induction machine, magnetic poles of the rotating field are not related to any particular part of the magnetic circuit. Instead, they rotate with respect to the stator and rotor magnetic circuits. The number of poles is dependent on distribution of electrical current in the stator slots. This distribution depends on the supply currents and on the method used to make the stator winding. So far, induction machines have been considered with two-pole magnetic field which has diametrically positioned north and south magnetic poles, as shown in the upper part of Fig. 17.5. The same figure also shows distribution of the stator currents which create magnetic field with two north and two south poles.

In the lower part of Fig. 17.5, electrical currents in diametrically positioned conductors have the same direction. In one pair of diametrically placed conductors,

---

[2] When the machine is supplied from an inverter with power transistors, the frequency of the supply can be varied. Consequently, the rotor speed can be varied too.

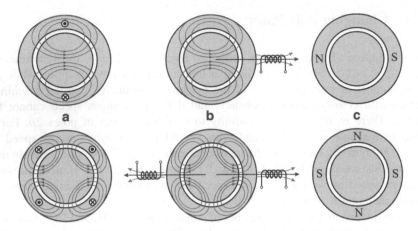

Fig. 17.5 Two-pole and four-pole magnetic fields. (**a**) Windings. (**b**) Magnetic axes. (**c**) Magnetic poles

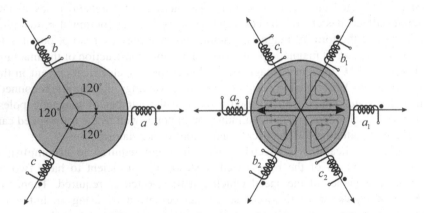

Fig. 17.6 Three-phase four-pole stator winding

direction is $\odot$. The other pair of diametrically positioned conductors is displaced by $\pi/2$ with respect to the first pair, and it has electrical currents of the opposite direction $\otimes$. The magnetomotive force created by the system of four conductors comprises two zones where the lines of magnetic field pass from rotor to stator and two zones where the lines of magnetic field pass from stator to rotor. Therefore, stator currents in this example create magnetic field having two north poles and two south poles. This field is called a four-pole field. Magnetic field under consideration has four magnetic poles or two-pole pairs; hence, $p = 2$.

Example given in Fig. 17.5 shows that the three-phase windings with appropriate distribution of conductors can create a four-pole rotating magnetic field. On the left-hand side in Fig. 17.6, there is a three-phase stator winding with spatial displacement of $2\pi/3$ between the phases, which creates a two-pole magnetic field in the air gap. A stator winding which creates a two-pole field is often called *two-pole winding*. The right-hand side of the same figure shows the method of forming a three-phase stator winding with magnetomotive force that creates a four-pole magnetic field in the air gap. Each of the phases a, b, and c is split into two sections, each with the same number of turns. The sections of the phase windings are spatially displaced by π, while the angle between the neighboring sections is $\pi/3$, which is one half of the spatial shift of a two-pole winding.

Phase a of the four-pole stator winding consists of two diametrically positioned sections. The sections are connected in series; thus, the conductors of both sections carry the same current $i_a(t)$. The reference terminals of each of the sections are marked by dots. Convention of marking the reference terminal is the following: when the current enters the dot-marked terminal of the considered section, then the section creates the magnetomotive force and field which start from the rotor magnetic circuit and propagate through the air gap toward the considered section. In the case when the two sections, actually the two halves a_1 and a_2 of the phase winding a, are connected in series and in such way that the phase current $i_a(t)$ enters the dot-marked terminals of both halves, magnetic field is created with two north and two south poles. Namely, at positions $\theta = 0$ and $\theta = \pi$, it creates the field with lines that pass from the rotor to the stator. Due to flux conservation law (div $\boldsymbol{B} = 0$), corresponding field lines must pass from the stator to the rotor at positions $\theta = \pi/2$ and $\theta = 3\pi/2$. In this way, a four-pole field is created, the field with two pairs of poles ($p = 2$), having two north magnetic poles and two south magnetic poles. With phase currents $i_a(t)$, $i_b(t)$, and $i_c(t)$ that have sinusoidal change of the same amplitude and frequency, and which are phase shifted by $2\pi/3$, consequential magnetic field revolves at the speed determined by the supply frequency, and it does not change the amplitude. It can be shown that the speed of rotation Ω_e of the four-pole field is equal to one half of the supply frequency ω_e. From Fig. 17.6, it should be noted that one of the magnetic poles is against the section a_1 at $t = 0$, when the phase current $i_a(t) = I_m \cos \omega_e t$ has the value $i_a(t) = +I_m$. Corresponding magnetic pole is marked by an arrow $\rightarrow$ which is directed toward the section a_1. The phase current $i_b(t) = I_m \cos(\omega_e t\ 2\pi/3)$ is delayed, and it reached its maximum positive value at $t_1 = 2\pi/(3\omega_e)$. Over the interval $0 < t < t_1$, the considered magnetic pole is shifting from the section a_1 toward the section b_1, where it arrives at the instant t_1. At $t_2 = 4\pi/(3\omega_e)$, the phase current $i_c(t)$ reaches its maximum value. Then the considered magnetic pole is placed against the section c_1. At the end of one full period, at $t_3 = 2\pi/\omega_e$, the maximum of $i_a(t) = +I_m$ is repeated in the phase current $i_a(t)$. At the same time, the magnetic pole being tracked reaches position against the section a_2. During one period of voltages and currents (*electrical period*) $T = 2\pi/\omega_e$, the phase change of electrical currents is $\Delta\theta_e = 2\pi$, while the spatial displacement of the magnetic pole is $\Delta\theta_m = \pi$. Therefore, the speed of rotation of the field is one half of the supply frequency, $\Omega_e = \omega_e/2$.

It should be noted that the four-pole field shown in Fig. 17.6 has two diametrically positioned poles of the same polarity. During one period of variation of electrical variables, both poles shift by π; hence, they switch their places.

In general, when stator windings are arranged to make a rotating magnetic field with $2p$ poles, the synchronous speed of the field rotation is $\Omega_e = \omega_e/p$. The torque of multipole induction machines is determined on the basis of expression $T_{em} = P_\delta/\Omega_e = pP_\delta/\omega_e$, according to (17.3).

By changing the number of magnetic poles, windings obtain a different number of slots per phase, different winding factors such as belt factor and chord factor, and different magnetic induction, winding resistance, leakage inductance, and self-inductance.

17.4.1   Speed and Torque of Multipole Machines

The number of magnetic pole pairs of an electrical machine is denoted by p. Synchronous speed of an induction motor is equal to

$$\Omega_e = \frac{\omega_e}{p}, \tag{17.1}$$

while the speed of rotor rotation is equal to

$$\Omega_m = \frac{\omega_m}{p} = \frac{\omega_e - \omega_{slip}}{p} = \Omega_e - \Omega_{slip}, \tag{17.2}$$

where ω_{slip} is the angular frequency of the rotor currents. The electromagnetic torque is

$$T_{em} = \frac{P_\delta}{\Omega_e} = p\frac{P_\delta}{\omega_e}. \tag{17.3}$$

17.5   Characteristics of Multipole Machines

Multipole machines ($p > 1$) make better use of the copper and iron compared to two-pole machines. There are applications of induction machines where selection of higher number of poles offers higher specific torque and power compared to the solutions involving two-pole machines. A rationale of these statements is presented here.

It is known that one turn of the phase winding of two-pole machine is positioned diametrically at angular distance $\Delta\theta \approx \pi$. When one of these conductors is under the north magnetic pole of the revolving field, the other is below the south pole. At the front and rear of the machine, the two conductors are connected by end turns. In two-pole machines, end turns are relatively long. Their length is one half of the machine circumference. Longer conductors contribute to increased consumption of copper, higher resistance of the winding, and higher losses in copper. In multipole machines, conductors making one turn are placed at angular distance $\Delta\theta \approx \pi/p$, which corresponds to the distance between the two magnetic poles. Length of the end turns in multipole machines is much shorter, which reflects favorably to the total mass of consumed copper, reduces winding resistance, and reduces power losses.

Moreover, multipole machines make a more efficient use of ferromagnetic materials. Magnetic field lines pass from the zone of the north magnetic pole of the stator, pass through the air gap, enter the rotor magnetic circuit, pass through the air gap for the second time, and reach the zone of the south magnetic pole of the stator. Following that, the field lines pass through the stator yoke and return to the north pole. Passing tangentially along the stator perimeter, the field lines in a two-pole machine cover the angular distance of $\Delta\theta \approx \pi$. Passing tangentially, the field lines do not pass through the zone where the stator slots are located. Instead, they pass through the outer part of the stator magnetic circuit called *yoke*. The yoke is required to reduce magnetic resistance on the flux path. On the other hand, the field passing through the yoke does not contribute to electromechanical conversion. Instead, it increases the iron weight and the total mass of the machine. In multipole machines, the path covered by the field lines between the two magnetic poles is shorter, and it is equal to $\Delta\theta \approx \pi/p$. Hence, the flux path through the yoke is shorter, and the iron usage is improved.

17.5.1 Mains-Supplied Multipole Machines

Synchronous speed of mains-supplied machines is determined by the line frequency f_e, and it is equal to $\Omega_e = 2\pi f_e/p$. Their specific power depends on the number of pole pairs. It can be determined by considering the relation between the electromagnetic torque and the size of the machine. It has been shown in the preceding sections that the available torque of the machine depends on its volume, resulting in proportion[3] $T_{em} \sim V \sim D^2 L$, where D and L are diameter and axial length of the machine. With $T_{em} = P_\delta/\Omega_e = pP_\delta/(2\pi f_e)$, one obtains $V \sim pP_\delta$, namely, the size of mains-supplied multipole induction machine increases with the number of poles. Given the air-gap power P_δ, dimensions of the machine are proportional to $V \sim D^2 L \sim pP_\delta$. In case where the power of the induction machine is predefined, its size is proportional to p. As an example, one can compare masses of standard induction motors with rated power of 1.1 kW, designed for the line voltage of $U_S = 400$ V and for the rated frequency of $f = 50$ Hz. While two-pole motor has a mass of $m \approx 8$ kg, four-pole motor has $m \approx 13$ kg, six-pole motor has $m \approx 16$ kg, and eight-pole motor has $m \approx 23$ kg. Practical values of motor masses are different from prediction $m \approx k \cdot p$ owing to the effects that were neglected in the preceding analysis.

---

[3] Electromagnetic torque depends on the fourth power of linear dimensions. Hence, $T_{em} \sim V^{4/3}$.

17.5.2   Multipole Machines Fed from Static Power Converters

In applications of induction machines fed from static power converters, it is possible to adjust the supply frequency and the synchronous speed according to needs. Static power converters are mostly transistorized inverters which use the pulse width modulation (PWM) to provide symmetric, three-phase system of voltages of variable frequency and variable amplitude. With the possibility of changing the stator supply frequency, the same synchronous speed $\Omega_e = \omega_e/p$ can be obtained with machines having different number of poles. Hence, there is a choice to select the number of poles in order to make a better use of the copper and iron. Two-pole machines have lower specific torque and lower specific power than equivalent four-pole and six-pole machines. In sample design where the induction machine is expected to run with the rotor speed Ω_m, it is necessary to supply the stator voltages which create magnetic field which revolves at the speed of $\Omega_e \approx \Omega_m$. One way to accomplish that is by selecting a two-pole machine and setting the supply frequency to $\omega_e = \Omega_e$ or by selecting a multipole machine ($p > 1$) and setting the supply frequency to $\omega_e = p\Omega_e$. In both cases, the machine has the same synchronous speed, develops the same electromagnetic torque, and gives the same power. With multipole machine ($p = 2$, 3, or 4), specific power is higher due to improved usage of iron and copper. Therefore, multipole induction machine is smaller and lighter. It should be noted that these advantages of multipole machines are lost in the case when the number of poles is extremely high. In such cases, stator frequencies $\omega_e = p\Omega_e$ are exceptionally high, and this leads to a significant increase in iron losses. In designing magnetic circuits of induction machines operating with high frequencies, magnetic circuit cannot be made of iron sheets. Instead, ferrites or other special ferromagnetic materials have to be used.

17.5.3   Shortcomings of Multipole Machines

Angular frequency of the stator currents and voltages required for the given rotor speed depends on the number of poles, $\omega_e = p\Omega_e$. With the target speed of $\Omega_m \approx \Omega_e$, the stator frequency is approximately equal to $p\Omega_m$. Hence, with inverter-supplied machines that have variable supply frequency, the same rotor speed can be achieved with lower number of poles $2p$ and lower supply frequency or with multiple pole pairs and higher supply frequency. It is known that losses in magnetic circuit due to eddy currents are proportional to the square of the frequency, while hysteresis losses grow linearly with the frequency. Therefore, specific iron losses in multipole machines are larger than specific losses in two-pole machine. In the process of the machine design, it is necessary to envisage adequate cooling or to reduce the peak value of the magnetic induction B_m in order to keep the losses within permissible limits.

With multipole machines, it is more difficult to achieve quasisinusoidal distribution of stator conductors. Stator winding is formed by placing conductors of the same reference direction $\otimes$ under one magnetic pole and conductors of the opposite reference direction $\odot$ under the opposite magnetic pole. In two-pole machines, the width of magnetic poles is close to $\Delta\theta \approx \pi$. In multipole machines, the width of magnetic poles is $\Delta\theta \approx \pi/p$. Therefore, conductors of the same reference direction are placed within the angular interval $\Delta\theta \approx \pi/p$, p times narrower than in the case of a two-pole machine. Conductors of the stator winding are placed in the stator slots. For a machine having N_Z slots, there are a finite number of discrete locations where the conductors could be placed. Therefore, the conductors cannot have an ideal sinusoidal distribution. Instead, they are distributed into a finite number of slots in a way that creates an approximate, quasisinusoidal distribution. For multipole machines, the range of $\Delta\theta \approx \pi/p$ comprises p times less slots than the range $\Delta\theta \approx \pi$ of two-pole machines. Hence, with $p > 1$, it is even more difficult to accomplish distribution of conductors which is close to sinusoidal. As a consequence, induced electromotive forces in multipole machines could have an increased amount of higher harmonics, increased cogging torque, and increased losses due to higher harmonics.

Question (17.1): The torque of multipole induction machines is determined by $T_{em} = P_\delta/\Omega_e = pP_\delta/\omega_e$. Considered is an induction machine of axial length L and diameter D. The stator windings can be made so as to create magnetic field with an arbitrary number of pole pairs p. Is it possible to increase the available torque by increasing the number of poles?

Answer (17.1): For a given value of maximum induction B_{max} and given value of permissible current density in conductors, the torque available from an electrical machine is proportional to l^4, the fourth power of linear dimensions, or $V^{4/3}$, where V is the machine volume. Therefore, for the machine of given dimensions, the way of making the stator winding cannot have major impact on the available torque. In other words, the available torque does not depend on the number of magnetic poles. This conclusion can also be derived by representing the torque generation process as the interaction of electrical current in rotor bars with magnetic field created by the stator. Due to a limited current density, electrical current in rotor bars cannot be increased, and their limit is unaffected by the number of poles. Magnetic induction B_m is determined by the characteristics of iron sheets. Electromagnetic torque depends on electrical currents in rotor bars, magnetic induction, the rotor length, and radius. By neglecting the secondary effects, it can be concluded that the torque of an induction machine of given dimensions does not depend on the number of poles, namely, it does not depend on the method of making the stator windings.

Question (17.2): A four-pole induction machine designed for mains supply of 3×400 V, 50 Hz, has the rated power P_n. The stator winding is removed from the stator slots. New stator winding is made, designed for the same power supply conditions, producing two magnetic poles. Make an estimate of the rated power of the new machine.

Answer (17.2): The power is product of the rotor speed and the electromagnetic torque. Rewinding a four-pole machine with a two-pole winding doubles the synchronous speed and, hence, the rotor speed. The torque depends on magnetic induction B and the sum of electrical currents in individual slots. Assuming that the flux density B in the air gap does not change much and that the current density remains the same, the two-pole machine will generate the torque comparable to the torque delivered by the original, four-pole machine. Hence, the two-pole machine has the potential of delivering $2P_n$.

Question (17.3): It has been shown that stator winding of an induction machine can be wound so that it creates a rotating magnetic field with four or more magnetic poles. Does the number of poles influence the rotor construction?

Answer (17.3): Rotor of a squirrel cage induction machine consists of a relatively large number of bars which are short-circuited by conducting rings at both rotor ends, at the front and the rear. The electromotive forces and currents in short-circuited rotor contours depend on the speed of the rotor relative to the field, that is, on the slip speed $\Omega_{slip} = \omega_{slip}/p$ and on magnetic induction in the air gap. Under the north magnetic pole of the rotating field, induced rotor currents have one direction, while under the south magnetic pole of the rotating field, they are of the opposite direction. Therefore, the number of poles of the consequential rotor field is determined by the number of poles of the stator field. In other words, the same rotor can be used within a two-pole machine as well as in a multipole machine.

17.6   Two-Speed Stator Winding

Change in the synchronous speed of an induction machines and change in the rotor speed can be accomplished by changing the number of poles. In order to accomplish that, it is necessary to have the possibility of changing the stator winding so as to change the number of magnetic poles of the stator magnetomotive force. In Fig. 17.7, stator winding of an induction motor is shown with each of the three-phase windings made of two sections. Sections a_1 and a_2 of phase winding a are made to create magnetomotive forces of the same course but of opposite directions. With connection shown on the right-hand side of the figure, the stator winding creates a four-pole magnetic field. In the left side of the figure, direction of phase currents in sections a_1, b_1, and c_1 is maintained, while direction of phase currents in sections a_2, b_2, and c_2 is changed. In each phase, both sections create magnetomotive forces in the same direction, and they create two-pole magnetic field.

The motion of the stator magnetomotive force vector during one third of the supply voltage period $T = 1/f_e$ is shown in Fig. 17.8 for two-pole and four-pole configurations.

Change of the rotor speed by means of changing the number of poles requires commutation of internal connections between the stator sections. This change is to

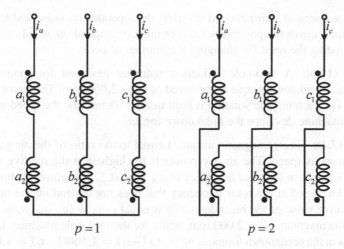

Fig. 17.7 A two-speed stator winding. By changing connections of the halves of the phase windings, two-pole (*left*) or four-pole (*right*) structures are realized

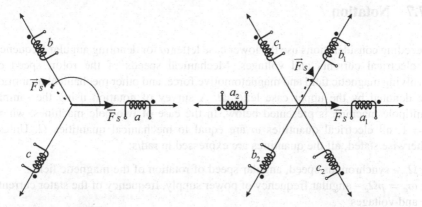

Fig. 17.8 Rotation of magnetomotive force vector in 2-pole and 4-pole configuration

be made in the course of the machine operation. In order to perform these changes, it is necessary to operate dedicated switches that make or break connections between individual sections and to achieve the winding configuration that results in desired number of poles and desired synchronous speed. The speed is usually changed automatically, in the absence of operator. Therefore, the state of the switches is controlled from a digital controller which issues command voltages. The need for a number of controlled switches makes the application relatively complex. In addition to that, additional shortcoming of the speed variation by changing the number of poles is discontinuous nature of this kind of control. Namely, the rotor speed cannot exhibit continuous change. Instead, one can select two or three discrete values of synchronous speed, and this is accomplished by connecting the stator windings in two or three configurations.

The appearance of transistorized inverters that operated on pulse width modulation principles opens the possibility for continuous change of the supply frequency, thus eliminating the need for changing the number of poles.

Question (17.4): A two-pole induction machine designed for mains supply develops the breakdown torque at the speed of $n_b = 2{,}000$ rpm. The stator winding is removed and a four-pole winding is built instead. Determine the speed where the four-pole machine develops the breakdown torque.

Answer (17.4): Electromagnetic torque is equal to the ratio of the air-gap power and synchronous speed. The air-gap power is the highest at the relative slip $s_b = R_R/(L_{\gamma e}\omega_e)$, that is, at the rotor frequency of $\omega_b = R_R/L_{\gamma e}$. Therefore, the breakdown torque is developed at the rotor frequency that does not depend on the number of poles. Relative value of the breakdown slip is equal to $s_b = (n_e-n_b)/n_e = 1/3$. For the two-pole machine, $n_b = 2{,}000$ rpm, while for the four-pole machine, the speed that results in the breakdown torque is $n_b = n_e(1-s_b) = 1{,}500(1-s_b) = 1{,}000$ rpm.

17.7   Notation

Preceding considerations use the lower case letter ω for denoting angular frequency of electrical currents and voltages. Mechanical speeds of the rotor, speed of revolving magnetic field and magnetomotive force, and other mechanical quantities are denoted by the upper case letter Ω. A survey of notation using the sample multipole machine is presented below. In the case of two-pole machines, where $p = 1$, all electrical quantities ω are equal to mechanical quantities Ω. Unless otherwise stated, all the quantities are expressed in rad/s:

- Ω_e – synchronous speed, angular speed of rotation of the magnetic field
- $\omega_e = p\Omega_e$ – angular frequency of power supply, frequency of the stator currents and voltages
- Ω_m – the rotor speed
- $\omega_m = p\Omega_m$ – electrical representation of the rotor speed
- $n = 9.54\Omega_m$ – the rotor speed expressed in rpm (revolutions per minute)
- $\omega_{slip} = \omega_e - p\Omega_m$ – angular frequency of the rotor currents
- $\Omega_{slip} = \Omega_e - \Omega_m$ – the slip speed of the rotor, the speed of lagging behind the synchronous speed

For a four-pole induction machine ($p = 2$) with rated supply frequency $f_e = 50$ Hz and rated speed of $n_n = 1{,}350$ rpm, characteristic angular frequencies and speeds in rated operating conditions are the following:

- Angular frequency of stator voltages – $\omega_e = 100\pi$
- Synchronous speed – $\Omega_e = 50\pi$, $n_e = 1{,}500$ rpm
- Angular frequency of rotor currents – $\omega_{slip} = 10\pi$
- Slip speed – $\Omega_{slip} = 5\pi$, $n_{slip} = 150$ rpm
- Rotor mechanical speed – $\Omega_m = 45\pi$, $n_n = 1{,}350$ rpm

17.8 Supplying from a Source of Variable Frequency

Preceding sections discussed traditional approaches for changing the rotor speed of induction machines, specifically:

* Variation of the stator voltage
* Variation of the rotor resistance
* Variation of the number of poles

Their drawbacks are:

* Difficult implementation
* High energy losses
* No possibility to achieve continuous speed change over a wide range

Continuous speed variation over a wide range relies on power converters which make use of transistor switches. They operate on the basis of pulse width modulation and provide the possibility for continuous variation of the supply frequency, which results in continuous variation of the synchronous speed and, hence, the rotor speed. In this way, variable speed is obtained without the need to operate the induction machine with increased slip frequencies. Therefore, there is no increase in conversion losses due to the speed change. Moreover, there is no need to use wound rotor, slip rings, or any special design of induction machine. Subsequent sections provide a brief introduction to variable frequency supply of induction machines.

17.9 Variable Frequency Supply

It is of interest to investigate the nature of the stator voltages in variable speed induction machines. In phase a of the stator winding, the current is $i_a(t)$, and the flux is $\Psi_a(t)$. Assuming that the leakage flux is small, the flux of the phase winding a comes as a consequence of the rotating magnetic field. At the instant when the vector of the rotating field is aligned with the axis of the phase a, the flux $\Psi_a(t)$ reaches its maximum value Ψ_m. With revolving field, the flux in phase a varies according to sinusoidal law. Phase voltage is equal to $u_a(t) = R_S\, i_a(t) + d\Psi_a(t)/dt$. The change of the phase voltage u_a is shown in Fig. 17.9. Neglecting the voltage drop across the stator resistance, one obtains

$$u_\alpha = u_a \approx \frac{d\Psi_{\alpha s}}{dt} = \omega_e \Psi_m \sin(\omega_e t - \varphi). \tag{17.4}$$

Slip of induction machines is relatively small; thus, the frequency of power supply is determined from the rotor speed, $\omega_e = p\Omega_e = p(\Omega_m + \Omega_{slip}) \approx p\Omega_m$. Therefore, variation of the rotor speed requires the power supply for the stator

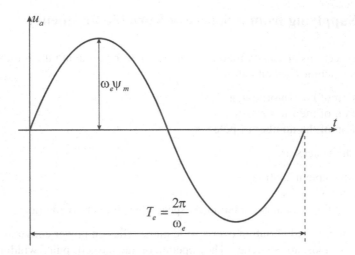

Fig. 17.9 Desired shape of the phase voltage

winding which provides a three-phase system of voltages with variable frequency and variable amplitude ($\Psi_m \omega_e$). Continuous change in the rotor speed requires continuous change of the voltage amplitude and frequency. Therefore, PWM controlled three-phase inverters are required to provide continuous change of both the voltage amplitude $\Psi_m \omega_e$ and the period $T_e = 2\pi/\omega_e$.

17.10   Power Converter Topology

Simplified schematic diagram of the power converter intended for supplying induction machines is given in Fig. 17.10a, along with the shape of the line voltage obtained at the output terminals (Fig. 17.10b). The change of the voltage across the output terminals comprises a train of voltage pulses. Averaged value of this pulse-shaped waveform has sinusoidal change with adjustable amplitude and frequency. Such waveforms are fed to the stator terminals of induction machines and used for supplying three-phase machines by voltages of variable frequency and amplitude. Converter topology includes a three-phase diode rectifier with six diodes, shown in the left side of the figure. It converts AC voltages and currents, provided from the three-phase AC mains, into DC voltages and currents across the parts L_{DC} and C_{DC}. These parts are placed in the middle of the converter, between the rectifier and the inverter, and they are called *intermediate DC circuit*, *DC link*, or *DC bus*. DC voltage E across the capacitor C_{DC} is fed to the three-phase inverter, the switching structure which makes the use of six power transistors. Each transistor is used in *switching mode*, namely, it is either *opened* (off, $i_{CE} \approx 0$) or *closed* (on, $u_{CE} \approx 0$). Transistor power switches are organized in three groups, called *inverter phases* or *inverter arms*. Each arm has two transistor switches, connected in series and

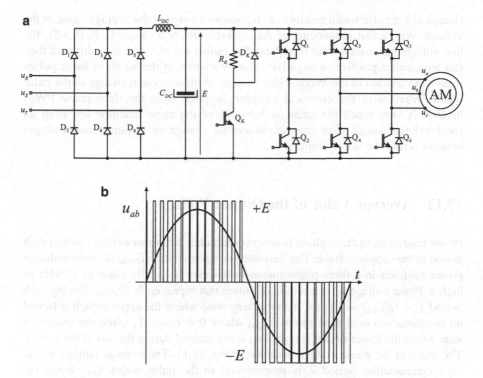

Fig. 17.10 (a) Three-phase PWM inverter with power transistors. (b) Typical waveform of line-to-line voltages

attached between the plus rail and the minus rail of the DC bus. At each instant, only one switch in each arm is turned on. Turning on both switches would result in a short circuit across the DC bus circuit. Turning the upper switch on brings the output phase to the potential of positive DC bus rail. Turning the lower switch on brings the output phase to the potential of negative DC bus rail.

17.11 Pulse Width Modulation

By taking the negative rail of the DC link circuit for the reference potential, turning on of the upper switch results in phase voltage $u_a = +E$, while turning on the lower switch results in $u_a = 0$. The same applies for the phase voltages u_b and u_c. Hence, the phase voltages take discrete values $u \in \{0, +E\}$. At the same time, line-to-line voltages such as $u_{ab} = u_a - u_b$ may take the values $u \in \{-E, 0, +E\}$. Hence, the instantaneous value of the line voltage cannot be changed in continuous manner. Instead, it takes one of the three discrete values. However, a fast exchange of the switching states results in a train of pulses of variable width. The width of the voltage pulses affects the average value of the voltage waveform. A continuous change of the pulse width results in a continuous change of the average value of the voltage. With a fast sequencing of the available discrete values $\{-E, 0, +E\}$, the line voltage becomes a train of pulses. The pulses are of variable width, and they can have either positive or negative value. Variation of the width of these pulses results in variation of the average line voltage. With sinusoidal change of the pulse width, behavior of the electrical machine supplied from the three-phase PWM inverter is very much the same as behavior of the same machine fed from an ideal voltage source with smooth, sinusoidal change of instantaneous voltages brought across the stator terminals.

17.12 Average Value of the Output Voltage

Power transistors in three-phase inverters commutate a number of times within each period of the supply voltage. The frequency of commutation f_{PWM} of semiconductor power switches in a three-phase transistor inverter is usually close to 10 kHz or higher. Phase voltage $u_a(t)$ is a train of pulses that repeat each $1/f_{PWM}$. During each period $T = 1/f_{PWM} = 100$ μs, the switching state where the upper switch is turned on is maintained over time interval t_{ON}, where $0 < t_{ON} < T$, while the switching state where the lower switch is turned on is maintained during the rest of the period. The shape of the phase voltage is shown in Fig. 17.11. The average voltage within each commutation period T is proportional to the pulse width t_{ON}. When the potential of the negative rail of the DC link is taken as the reference potential, the phase voltage over the interval $0 < t < t_{ON}$ is equal to $+ E$, while the voltage during the remaining part of the period T is $u_a = 0$. Continuous variation of t_{ON} over the range $0 < t_{ON} < T$ results in average value $u_a{}^{av} = E(t_{ON}/T)$ change from 0 to $+ E$.

$$U_{av} = \frac{1}{T} \cdot \int_{NT}^{(N+1)T} u_a dt = \frac{t_{ON}}{T} E. \qquad (17.5)$$

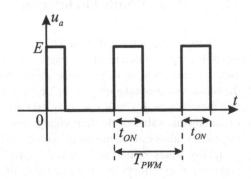

Fig. 17.11 Pulse width
modulation: upper switch is
on during interval t_{ON}

17.13 Sinusoidal Output Voltages

The width of the voltage pulses that constitute the phase voltage can be varied or *modulated*. Variation of the pulse width is called *pulse width modulation* (PWM). Starting from the expression $u_a^{av} = E(t_{ON}/T)$, desired average voltage u_a^{av} can be used to calculate the time t_{ON}, which determines the average value of the phase voltage within one switching period $T = 1/f_{PWM}$.

In order to achieve variation of the average voltage value $u_a^{av}(t)$, that is, to obtain the phase voltage that changes the average value in successive switching periods T, the pulse width t_{ON} should change as $t_{ON}(t) = T u_a^{av}(t)/E$. As a matter of fact, it is not correct to write $t_{ON}(t)$, as the pulse width assumes one discrete value in each switching period T. Namely, the pulse width over the period $[nT..(n + 1)T]$ is determined by discrete value $t_{ON}(n) = T u_a^{av}(n)/E$. If $f_e = 50$ Hz is the desired frequency of the phase voltage, while φ is the initial phase, and the number $0 < A < 1$ determines the desired amplitude, the pulse width should be varied according to

$$t_{ON}(n) = \frac{T}{2} + \frac{T}{2}A\sin(2\pi f_e \cdot nT - \varphi). \tag{17.6}$$

With this change of the pulse width, the phase voltage u_a is obtained with average value over successive intervals T that change according to expression

$$u_a^{av}(n) \approx \frac{E}{2} + \frac{E}{2}A\sin(2\pi f_e \cdot nT - \varphi). \tag{17.7}$$

The frequency component f_e of the phase voltage has the amplitude which can be varied by changing parameter A, while the frequency and phase are determined by parameters f_e and φ. Commutation frequency f_{PWM} has to be considerably higher than the desired frequency of the phase voltage, $f_e << f_{PWM}$. In the expression for $u_a^{av}(n)$, there is a DC component of the voltage which is equal to $E/2$. This is the consequence of selecting the negative rail of the DC bus for the reference potential V_0. By turning on the upper switch Q_1 in Fig. 17.10a, the phase voltage u_a is equal to E. With Q_2 turned on, $u_a = 0$. Other choices for reference potential result in different values of the phase voltage u_a.

DC component in the phase voltages is equal in all three phases. It has no impact on the operation of the induction machine. Namely, the stator winding is connected to three-phase inverter by three conductors, and the operation of the machine is determined by line-to-line voltages. Line-to-line voltage u_{ab} is equal to u_a-u_b. Subtracting the two-phase voltages removes the DC components. It is of interest to point out that the choice of the reference potential is arbitrary one. Therefore, it cannot have an impact on the operation of the electrical machine. In order to confirm this statement, one can calculate the line voltage $u_{ab}^{av}(t)$ as the difference between the phase voltages $u_a^{av}(t)$ and $u_b^{av}(t)$. The voltage $u_a^{av}(t)$ is given in (17.7) while the voltage $u_b^{av}(t)$ is equal to

$$u_b^{av}(t) \approx \frac{E}{2} + \frac{E}{2} A \sin(2\pi f_e \cdot nT - \varphi - 2\pi/3),$$

and it lags behind u_a by $2\pi/3$. By calculating $u_{ab}^{av}(t) = u_a^{av}(t) - u_b^{av}(t)$, DC components $E/2$ are canceled, resulting in line-to-line voltage u_{ab}^{av} which does not have a DC component:

$$u_{ab}^{av}(t) \approx \frac{AE\sqrt{3}}{2} \sin(2\pi f_e \cdot nT - \varphi + \pi/6).$$

Digital implementation of the pulse width modulation implies that parameters A, f_e, and φ are the numbers represented by binary record in RAM memory. They can be adjusted to the needs of the actual operating regime of the induction machine.

Within each period $T_e = 1/f_e$ of phase voltages, there are a finite number of pulses. The width of these pulses is modulated (changed) in the way to obtain sinusoidal change of the average value $u^{av}(t)$.

The process of approximating sinusoidal change of the phase voltage by means of a finite train of pulse width-modulated impulses has similarities with the procedure of making distributed windings with quasisinusoidal distribution of conductors placed in a finite number of slots.

17.14   Spectrum of PWM Waveforms

Sinusoidal PWM is used to obtain a sequence of variable width pulses. Averaged pulses provide the phase voltages of the desired frequency f_e. When the pulse width has sinusoidal change, according to expression $t_{ON}(n) = (T/2)[1 + A\sin(2\pi f_e nT - \varphi)]$, the average values with each switching period $T = 1/f_{PWM}$ change according to expression $u^{av}(n) \approx (E/2) + (E/2) A \sin(2\pi f_e nT - \varphi)$.

The commutation frequency f_{PWM} has to be considerably higher than the desired frequency of the phase voltage, $f_e \ll f_{PWM}$, so as to obtain a smooth, gradual change of average voltage between the successive switching periods $T = 1/f_{PWM}$ and to obtain the operation similar to feeding the machine from an ideal source. The frequency f_e of the phase voltages determines the synchronous speed, and it is called *basic* or *fundamental*. It ranges from several tens to several hundreds of cycles per second. The frequency f_{PWM} is called *commutation* or *switching* frequency, and it ranges from 5 to 20 kHz.

The spectrum of a pulse width-modulated sequence of phase voltage pulses contains:

- DC component $E/2$
- Slowly varying AC component of frequency f_e, created by sinusoidal variation of the pulse width, called the *basic* or *fundamental* frequency component

- Frequency component at the commutation frequency $f_{PWM} = 1/T$, created by the train of variable width voltage pulses that keep repeating each T
- A series of frequency components with smaller amplitudes and with frequencies $m{\cdot}f_{PWM}$ that are integer multiples of the switching frequency f_{PWM}
- A series of frequency components at frequencies $m{\cdot}f_{PWM} \pm n{\cdot}f_e$, produced by interaction between the switching frequency f_{PWM} and the basic frequency f_e

DC component of the phase voltage is the same in all phases; thus, it has no influence on line voltages. The fundamental AC component at the basic frequency f_e is the desired result, the voltage required across the stator terminals. It has adjustable amplitude and adjustable frequency. Assuming that the high-frequency content of the spectrum can be neglected, the three-phase inverter with PWM control can be regarded as the source of sinusoidal voltages with adjustable amplitude and adjustable fundamental frequency f_e.

Spectral component at the switching frequency $f_{PWM} = 1/T \approx 10$ kHz is the consequence of the pulsating nature of the three-phase inverter. Component of the voltage at the switching frequency has the amplitude determined by the DC link voltage E. Due to considerable amplitude, the effects of this frequency component cannot be neglected. The effects of the pulsed supply on an induction machine should be analyzed in order to establish the impact of pulsating voltage on the machine and verify whether this way of supplying the stator winding is acceptable. Subsequent discussion proves that the stator current, the electromagnetic torque, and the rotor speed of an induction machine, supplied from PWM controlled three-phase inverter, resemble the current, torque, and speed obtained with an equivalent source of smooth, sinusoidal phase voltages. Therefore, the operation with PWM power supply can be considered equivalent to the operation with an ideal source of sinusoidal waveforms providing the voltages of instantaneous value such as $u(t) = (E/2) + (E/2){\cdot}A\sin(2\pi f_e t - \varphi)$.

17.15 Current Ripple

The voltage balance in the phase winding a is given by the equation $u_a(t) = R_S i_a(t) + \mathrm{d}\Psi_a/\mathrm{d}t$. The flux Ψ_a can be represented as sum of the mutual flux Ψ_{ma}, which passes through the air gap and encircles both stator and rotor windings, and the leakage flux of the stator winding, which is proportional to the leakage inductance L_γ. The voltage balance equation assumes the form $u_a(t) = R_S i_a(t) + L_\gamma \mathrm{d}i_a(t)/\mathrm{d}t + \mathrm{d}\Psi_{ma}/\mathrm{d}t$, where $L_\gamma i_a(t)$ is the leakage flux in the phase a of the stator winding. Rotating magnetic field changes its relative position with respect to the phase winding a; thus, the flux Ψ_{ma} exhibits sinusoidal change of the frequency determined by the synchronous speed. For this reason, the electromotive force $\mathrm{d}\Psi_{ma}/\mathrm{d}t$ is sinusoidal, and it has the frequency $\omega_e \approx \omega_m$ and the amplitude $\Psi_m \omega_e$, where Ψ_m represents the maximum value of the mutual flux. With $f_e \ll f_{PWM}$, the change in the considered electromotive force is slow compared to the switching phenomena.

While considering the effects of switching power supply on the machine behavior, slow variations of the electromotive force can be neglected; thus, the voltage balance equation reduces to $u_a(t) = R_S i_a(t) + L_\gamma di_a(t)/dt$.

By applying Laplace transform on previous differential equation, one obtains complex image of the stator current, $I_a(s) = U_a(s)/(R_S + sL_\gamma)$. Function $W(s) = 1/(R_S + sL_\gamma)$ represents *transfer function* of the stator winding, which is the ratio of the complex image of the stator current and the complex image of the stator voltage. Transfer function $W(s)$ provides the means to calculate the stator current response to excitation by the stator voltage of known amplitude and frequency. In the case of interest, the voltage excitation has the switching frequency f_{PWM}. The function is obtained by neglecting slow-varying electromotive force; thus, it is applicable for calculating the response at frequencies as high as f_{PWM}, but it cannot be used for the analysis of the machine response to lower excitation frequencies, such as the fundamental frequency f_e.

The ratio of the current and voltage at the frequency $\omega = 2\pi/T = 2\pi f_{PWM}$ is obtained by introducing $s = j\omega$ in function $W(s)$; thus, the function assumes the form $W(j\omega) = 1/(R_S + j\omega L_\gamma)$. At switching frequencies of the order of several kHz, it is justified to introduce the assumption $R_S << \omega L_\gamma$ and to obtain relation $|I(-j\omega)/U(j\omega)| \approx 1/(L_\gamma\omega)$. This expression shows that electrical machines behave as low-pass filters. When exposed to high-frequency voltages, the stator current response to such excitation is smaller as the excitation frequency increases. In other words, low-frequency excitation has much larger impact on the stator currents than high-frequency excitation. At frequencies close to $f_{PWM} \approx 10$ kHz, reactance $L_\gamma\omega$ is so large that the voltage pulses have a very small influence on the stator currents. Typical change of the phase current of an induction machine supplied from three-phase PWM inverter is shown in Fig. 17.12.

With pulsed supply, the phase voltages exceed the desired sinusoidal waveform during t_{ON} and then fall below during the reminder of the PWM switching period. Therefore, the stator currents oscillate around their mean values. These oscillations have the frequency of the switching bridge $f_{PWM} = 1/T = \omega/(2\pi)$. When the switching frequency is sufficiently high, the amplitude of these oscillations is rather small, and their effect on the operation of the induction machine can be neglected. An estimate of the amplitude of oscillations of the stator current can be obtained by using expression $|I(j\omega)/U(j\omega)| \approx 1/(L_\gamma\omega)$. This expression is applicable for excitation by sinusoidal voltages of the frequency $\omega = 2\pi\ f_{PWM}$. The three-phase switching inverter does not output sinusoidal waveforms at the switching frequency. Instead, it provides rectangular voltage pulses of the period $1/f_{PWM}$. Nonetheless, the formula can be used to obtain a rough estimate of the stator current oscillations, also called *ripple*. In most cases, the current ripple amounts from 1% to 5% of the rated current.

Question (17.5): Induction machine of rated frequency $f_{en} = 50$ Hz has an equivalent leakage reactance $x_{\gamma e} = 20\%$. The machine is supplied from a three-phase transistor inverter of the switching frequency $f_{PWM} = 10$ kHz. Provide an estimate of the stator current ripple that appears due to pulsed power supply.

Fig. 17.12 Stator current
with current ripple

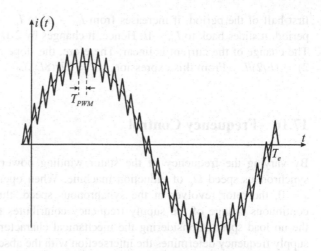

$i(t)$

T_{PWM}

T_e

t

Answer (17.5): The amplitude of the pulses generated by the switching source and
supplied to the stator winding terminals is equal to the voltage E of the DC link
circuit. Voltage E must be sufficient to provide the rated AC voltage at the inverter
output. Hence, for the purpose of making an estimate, the DC link voltage can be
assumed to have the relative value of $E \approx 1$. The leakage reactance $X_{\gamma e}$ at the rated
frequency f_{en} has relative value of $x_{\gamma e} = 0.2$. The switching frequency is 200 times
higher than the rated frequency. Reactance is proportional to the frequency. At the
switching frequency, the relative value of the reactance is $x_{\gamma e(PWM)} = x_{\gamma e}(f_{PWM}/f_e)$
$= 40$, that is, 4,000%. The ripple current comes as the quotient of the voltage,
having the relative value of 1, and the leakage reactance at the switching frequency.
Therefore, the relative value of the current ripple due to pulsed supply is estimated
as $\Delta I \approx 1/x_{\gamma e(PWM)}$ =2.5%.

Question (17.6): Induction motor is supplied from a three-phase transistor inverter
with DC link voltage E and with the switching frequency of $f_{PWM} = 1/T$. The speed
of rotation is equal to zero; thus, the electromotive force induced in the stator
winding can be neglected. Resistance of the stator winding is also negligible. The
leakage inductance $L_{\gamma e}$ of the motor is known. It can be assumed that potential of
the star point node is in between the positive and negative DC bus rails and it is
chosen for the reference potential. Determine the shape and amplitude of
oscillations of the stator current in the case when $t_{ON} = T/2$.

Answer (17.6): With reference potential point in between the DC bus rails, the
output phase voltage can be either $+ E/2$ or $-E/2$. Having neglected the
electromotive force and the voltage drop across the stator resistance, the voltage
balance equation reduces to $u_a = L_{\gamma e} \, di_a/dt$. During the first half of the period T,
that is, during the interval t_{ON}, the upper switch of the inverter phase a is turned on,
thus $L_{\gamma e} \, di_a/dt = E/2$. Therefore, the change of current is linear. Similar conclusion
applies for the second half of the period, when the voltage is $u_a = - E/2$.
The current oscillates around its average value I_{av} with an amplitude of ΔI. During

first half of the period, it increases from $I_{av} - \Delta I$ to $I_{av} + \Delta I$. In the second half period, it slides back to $I_{av} - \Delta I$. Hence, it changes by $2\Delta I$ within each half period. The change of the current is linear. Therefore, the slope di_a/dt is equal to $2\Delta I/(T/2) = (E/2)/L_{\gamma e}$. From this expression, $\Delta I = ET/(8L_{\gamma e})$.

17.16   Frequency Control

By varying the frequency of the stator winding power supply, one varies the synchronous speed Ω_e of induction machine. When operating with $T_{em} = 0$ and $s = 0$, the rotor revolves at the synchronous speed, thus $\Omega_m = \Omega_e$. Therefore, continuous change of the supply frequency contributes to continuous change of the no load speed. Considering the mechanical characteristic $T_{em}(\Omega_m)$, the stator supply frequency determines the intersection with the abscissa. For that reason, the frequency changes would affect as well the rotor speed of loaded induction machines. A family of mechanical characteristics obtained by varying the stator frequency is shown in Fig. 17.13.

In addition to the no load speed, mechanical characteristic of an induction machine is also characterized by the breakdown torque, breakdown slip, and stiffness $S = |\Delta T_{em}/\Delta \Omega_m|$. Breakdown slip $\Omega_b = p\omega_b = pR_R/L_{\gamma e}$ is determined by the machine parameters, and it does not depend on the power supply frequency. The breakdown torque $T_b = (3p/4)\Psi_s^2/L_{\gamma e}$ and the stiffness of the mechanical characteristic $S = k\Psi_s^2/R_R$ both depend on the square of the stator flux Ψ_s^2. The amplitude of the stator flux depends on the ratio of the power supply voltage and frequency. While operating with the rotor speeds lower than the rated speed Ω_n, it is desirable to maintain the rated flux, that is, the maximum flux that can be achieved within the machine. At high speeds, it is necessary to perform the field weakening and to operate with the flux inversely proportional to the speed. With $\Omega_m > \Omega_n$, the flux has to be reduced in order to maintain the stator voltage U_S within the limits of the rated voltage U_n.

To achieve flux control, it is necessary to perform simultaneous change of the supply voltage and the supply frequency. Relation between the voltage, frequency, and flux is derived from the equivalent circuit. The voltage balance equation of the stator winding of an induction machine which operates at steady state is

$$\underline{U}_s = R_s\underline{I}_s + j\omega_e(L_{\gamma s}\underline{I}_s + \underline{\Psi}_m). \tag{17.8}$$

By neglecting the voltage drop across the stator resistance, voltage balance equation of the stator winding at steady state becomes $\underline{U}_S \approx j\omega_e\Psi_S = j\omega_eL_{\gamma S} + j\omega_e\Psi_m = j\omega_eL_{\gamma S}\,\underline{i}_S + j\omega_eL_m\underline{I}_m$. The flux amplitude is determined by the ratio of the maximum voltage and the supply frequency, $\Psi_S \approx U_S/\omega_e$. When the stator voltages are obtained from a three-phase transistor inverter, the frequency ω_e of the basic (fundamental) component determines the synchronous speed, while the quotient of the voltage amplitude and the angular frequency ω_e determines the amplitude of the

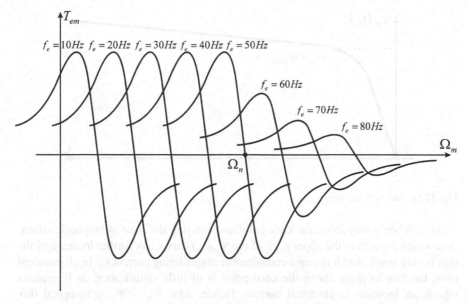

Fig. 17.13 Family of mechanical characteristics obtained with variable frequency supply

stator flux. In DC machines, the excitation flux depends on electrical currents in a separate excitation winding, while the electrical power subject to electromechanical conversion is supplied through the armature winding. Hence, DC machines have two electrical ports, and these are the excitation winding terminals and the armature winding terminals. Induction machines are supplied from the stator side only. Hence, both the machine excitation and the electrical power subject to electromechanical conversion pass from the three-phase inverter into the stator winding terminals. The flux, torque, and power of induction machines all depend on the voltages supplied to the stator terminals.

It is of interest to recall criteria for selecting the flux amplitude in various operating conditions. The torque developed in any electrical machine can be calculated as vector product of the flux and the current. Given the target torque T_{em}, electrical current required for the torque generation is proportional to the ratio T_{em}/Ψ_S, that is, it is inversely proportional to the flux. Lower currents are preferred as they lead to lower losses. Thus, it is beneficial to use higher values of the flux, whenever possible. With three-phase inverter supply, the flux is determined by the ratio of the voltage and the frequency, $\Psi_S \approx U_S/\omega_e$. The flux can be increased up to the value $\Psi_{max} \approx \Psi_n$ which marks the knee of the magnetizing characteristic shown in Fig. 17.14.

The flux values in excess to Ψ_{max} result in saturation of the magnetic circuit. The difference between the air-gap flux (i.e., mutual flux) and the stator flux is considered negligible for the discussion in course. Therefore, with $\Psi_m \approx \Psi_S$, the flux is determined by the magnetizing current i_m which circulates in the magnetizing branch of the equivalent circuit. This current is on the abscissa of the magnetizing characteristic

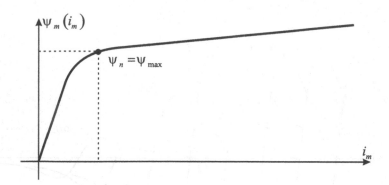

Fig. 17.14 Magnetizing curve

$\Psi_m(i_m)$. When going above the knee point and entering the zone of magnetic saturation, which extends in the upper right of the $\Psi_m(i_m)$ curve, any further increase of the flux is very small, and it requires considerable magnetizing current i_m. In all practical uses, the flux increase above the knee point is of little significance, as it requires significant increase in electrical current. Hence, with $\Psi_m > \Psi_n$, a marginal flux increment would require very high currents, accompanied by consequential copper losses. For this reason, the flux is maintained at the rated value Ψ_n by keeping the ratio U_S/ω_e equal to Ψ_n, unless other reasons and specific circumstances require the operation with reduced flux:

$$\underline{\Psi}_m \approx \underline{\Psi}_s \approx \frac{U_s}{j\omega_e}. \tag{17.9}$$

17.17   Field Weakening

Operation with $U_S/\omega_e = \Psi_n$ is not always possible. At higher rotor speeds, it is necessary to reduce the flux. In operation with $\Omega_m > \Omega_n$, it is necessary to increase the stator frequency above the rated value. In order to keep the flux at its rated value, it is necessary to have the stator voltage of $U_S = \omega_e\Psi_n$. With $U_S/\omega_e = \Psi_n$ and $\omega_e > \omega_n$, it is necessary to increase the stator voltage above the rated level.

At steady state, the stator voltage must not exceed the rated value. Otherwise, electrical insulation of windings will be damaged. For this reason, three-phase inverters designed for the stator winding power supply are made to produce voltages within the range $0 < U_S < U_n$. It would not make sense to make the power supply capable of providing the voltages that can cause damage to the machine. Therefore, three-phase inverters such as the one shown in Fig. 17.10 cannot produce the output voltage which exceeds the rated voltage of the motor. Considering the operation above the rated speed, where $U_S \leq U_n$ and $\omega_e > \omega_n$, the flux cannot be maintained at the rated level, and it has to be reduced. With the stator voltage equal to the rated value and, hence, constant, and with the operating frequency of the stator power supply in progressive rise, the flux of the machine is inversely proportional to the rotor speed. The expression describing the flux change at high speeds can be derived from the following discussion.

With rated flux, the electromotive force in the machine is equal to $\omega_e \Psi_n$. At the rated speed $\Omega_m \approx \Omega_n$, the electromotive force reaches the rated voltage $U_n \approx \omega_n \Psi_n$. This discussion neglects the slip and overlooks the difference between the air-gap flux and the stator flux. With limited voltage, the rated flux cannot be maintained in operation above the rated speed. The highest sustainable voltage that is applicable to the stator terminals is the stator rated voltage U_n. At speeds $\Omega_m \approx \Omega_e > \Omega_n$, the electromotive force must not exceed the rated voltage. For that to achieve, the flux must not exceed $\Psi_s(\omega_e) = \Psi_n(\omega_n/\omega_e)$. In this case, the electromotive force would be equal to the rated voltage. At higher speeds, the flux is inversely proportional to the rotor speed and, hence, inversely proportional to the supply frequency, $\Psi \sim 1/\omega$.

For an induction machine supplied from the three-phase inverter, the stator voltage and frequency have to be changed in order to obtain desired rotor speed. Calculation of the stator voltage amplitude and frequency in terms of the rotor speed is described below.

In operation below the rated speed, the stator frequency is $\omega_e < \omega_n$, the stator voltage $U_S \approx \omega_e \Psi_n$ is proportional to frequency and lower than the rated voltage, while the flux in the machine is constant and it has rated value:

$$\Psi_m = \frac{U_s}{\omega_e} = \frac{U_n}{\omega_n} = const. \quad \Rightarrow \quad \frac{U}{f} = const. \tag{17.10}$$

During operation at the speeds above the rated speed, stator voltage is maintained at rated value, which is the highest voltage available from the three-phase inverter. The stator frequency increases with the speed, and the flux decreases according to the law $\Psi \sim 1/\omega$. The machine operates in the field weakening regime:

$$\Psi(\omega)|_{|\omega|>\omega_n} = \frac{\omega_n}{\omega} \cdot \Psi_n. \tag{17.11}$$

Finally, variation of the flux is determined by

$$\Psi(\omega) = \begin{cases} \omega \leq \omega_n & \Rightarrow & \Psi_n \\ \omega > \omega_n & \Rightarrow & \frac{\omega_n}{\omega} \cdot \Psi_n \end{cases}. \tag{17.12}$$

With that in mind, it is possible to envisage the family of mechanical characteristics obtained by frequency variation. In the following diagrams (Fig. 17.15) and expressions, legibility is helped by assuming that $\Omega_m \approx \Omega_e$, as well as $\omega_e \approx \omega_m = p\Omega_m$. Hence, the subscript may be omitted due to assumption $\omega \approx \omega_e \approx \omega_m$.

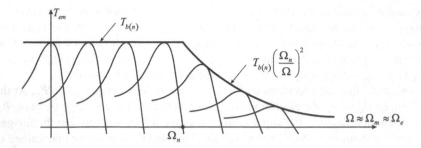

Fig. 17.15 The envelope of mechanical characteristics obtained with variable frequency

At speeds below the rated speed, the ratio U_S/ω_e does not change, and the flux is constant. Consequently, all mechanical characteristics obtained for the supply frequencies $\omega_e < \omega_n$ have the same value of the breakdown torque and the same slope. By changing the supply frequency over the range of $0 < \omega_e < \omega_n$, a family of mechanical characteristics is obtained having the same slope and the same breakdown torque. The characteristics can be drawn by starting with natural characteristic, obtained with the rated frequency, and performing translation toward the origin of the $T(\Omega)$ diagram. At speeds above the rated speed, induction machine operates in field weakening regime. The stator voltage amplitude is maintained at the rated value, while the flux decreases according to the law $\Psi_S(\omega) = \Psi_n(\omega_n/\omega_e)$. No load speed of the resulting mechanical characteristics is determined by the stator supply frequency, $\Omega_e = \omega_e/p$. Since the breakdown torque T_b and the slope S are proportional to the square of the flux, they decrease proportionally to the square of the speed, $T_b \sim 1/\omega^2$. The breakdown torque obtained at operation with the rated flux is denoted by $T_{b(n)}$. In field weakening region, the breakdown torque is $T_b(\omega_e) = T_{b(n)}(\omega_n/\omega_e)^2$. Therefore, the envelope of mechanical characteristics obtained with variable frequency supply in field weakening regime decreases with the square of the speed and frequency, $T_b \sim 1/\omega^2$.

17.17.1 Reversal of Frequency-Controlled Induction Machines

The rotor speed Ω_m of induction machines is close to the synchronous speed Ω_e. In order to change direction of the rotor speed, it is necessary to change direction of the revolving field and invert the synchronous speed. In mains-supplied machines, changing the phase sequence results in $\Omega_e = -\omega_e/p$. Inverter-supplied machines can be reversed without rewiring the phases.

The power supply frequency ω_e may take a negative value. The number ω_e resides in RAM memory of digital controller, and it is used to calculate t_{ON} intervals according to expressions similar to (17.6). Entering a negative value for ω_e leads to generation of three-phase system of stator voltages which create magnetic field that revolves in the opposite direction. In such cases, the synchronous speed and the rotor speed change direction.

Speed reversal by means of negative supply frequency does not require any change of the wiring. It is not necessary to exchange the two-phase conductors for the induction machine to change direction of rotation. It is sufficient to insert a negative value of ω_e in the expression that calculates the pulse width $t_{ON}(n) = (T/2)$ $[1 + A\sin(\omega_e nT - \varphi)]$ of the voltage pulses created by the switching action of the three-phase inverter. During the operation with $\omega_e < 0$, mechanical characteristic is transferred to the second and third quadrant of the T_{em}-Ω_m plane.

17.18 Steady State and Transient Operating Area

By varying the frequency and amplitude of the stator voltages, it is possible to achieve operation in all four quadrants of the T_{em}-Ω_m plane. It is of interest to establish the steady state operating area, that is, the region of T_{em}-Ω_m plane that encircles all of the operating points where the machine can operate permanently and with no damage. In a like manner, transient operating area contains the operating points that can be reached within short time intervals. Steady state operating limits of induction machine operating in the first quadrant are given in Fig. 17.16 and explained henceforth.

Continuous operation of an induction machine at certain operating mode requires that the voltages and currents are within the rated limits. At the same time, the sum of the losses should be within the rated limits in order to avoid overheating and damage to the machine.

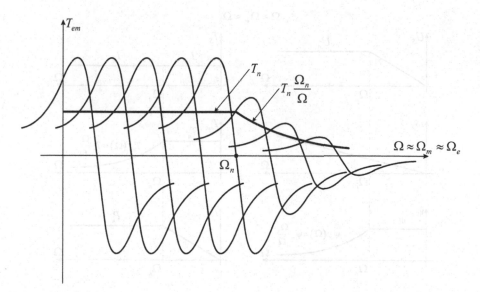

Fig. 17.16 Steady state operating limits in the first quadrant

During operation at speeds below the rated speed, the flux is maintained at the rated value. With rated currents in stator windings and with the rated flux, induction machine provides the rated torque T_n at the shaft. The rated torque is available in continuous service at all speeds where the stator frequency remains within the rated limits, $|\omega_e| \leq \omega_n$.

In the field weakening region, the stator frequency exceeds the rated value, $|\omega_e| > \omega_n$. The flux varies according to the law $\Psi_S(\omega) = \Psi_n(\omega_n/\omega)$. For this reason, the torque available in continuous service in the field weakening regime is inversely proportional to the rotor speed, that is, inversely proportional to the stator frequency. The torque available in the field weakening operation can be represented by expression

$$T_n \frac{\Omega_n}{\Omega} \tag{17.13}$$

which defines boundaries of the steady state operating area in the zone of higher speeds (Fig. 17.16).

17.19   Steady State Operating Limits

Steady state operating limits of relevant variables are shown in Fig. 17.17. Any value that does not exceed the limits is sustainable in continuous operation. The limits are given for the voltage, current, stator frequency, torque, flux, and power of an

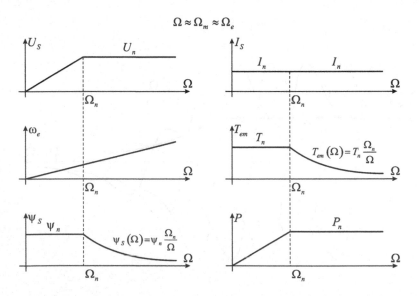

Fig. 17.17 Steady state operating limits for the voltage, current, stator frequency, torque, flux, and power. The region $\Omega_m < \Omega_n$ is with constant flux and torque, while the field weakening region $\Omega_m < \Omega_n$ is with constant power

induction machine supplied from variable frequency, variable voltage source. For clarity of diagrams, it is assumed that $\Omega \approx \Omega_e \approx \Omega_m$. The region $\Omega_m < \Omega_n$ below the rated speed is called *constant flux* or *constant torque* region. The region $\Omega_m > \Omega_n$ above the rated speed is called *flux weakening region* or *constant power region*.

Diagrams in Fig. 17.17 represent the change of steady state operating limits for U_S, I_S, T_{em}, P, and Ψ_S. Therefore, they comprise only the first quadrant with $\Omega > 0$. However, the limits such as $T_{em}(\Omega_m)$ apply to all the four quadrants of T_{em}-Ω_m plane. Namely, the same limits of the continuous service apply for both directions of the speed, and they are equally valid in motor operating mode as well as in generator operating mode. It is shown in the figure that the stator frequency ω_e increases proportionally to the rotor speed. By neglecting the slip, relation between the rotor speed and the stator frequency is $\omega_e \approx p\Omega_m$. In constant flux zone, at speeds lower than the rated speed, the ratio U_S/ω_e is kept constant. Upon reaching the rated speed, the voltage is maintained at the rated value. Further increase of the rotor speed gets the machine in the field weakening mode, where the voltage remains constant while the angular frequency of the power supply keeps increasing. This results in flux decrease. In the field weakening regime, the flux changes according to $\Psi_S(\omega) = \Psi_n(\omega_n/\omega_e)$.

In constant flux region, the available torque is constant, while in field weakening region, the torque is limited by $T_{em}(\Omega) \approx T_n(\Omega_n/\Omega_e)$. The power available in constant flux region ($|\Omega_e| \leq \Omega_n$) increases linearly with the speed. In field weakening, the available torque drops according to $T_{em} \sim 1/\Omega$. Therefore, the power available in field weakening has a constant value, $P(\Omega) \approx \Omega T_n(\Omega_n/\Omega) = P_n$. For this reason, the field weakening region is called *constant power region*. By recognizing the secondary effects, which have been neglected in the first approximation, it can be shown that the power available in the field weakening regime is somewhat higher than the rated power.[4]

17.19.1 RI Compensation

In the process of calculating the stator voltage, the voltage drop across the stator resistance has been neglected. For clarity, approximation $\underline{U}_S = R_S \underline{I}_S + j\omega_e \Psi_S \approx j\omega_e \Psi_S$ has been made. The stator resistance has a very low relative value. Therefore, this approximation does not introduce any significant error in calculations, provided that the rotor speeds are sufficiently high and that the electromotive force $\omega_e \Psi_S$ has the value significantly higher than the voltage drop $R_S I_S$. At very small speeds, where the electromotive force is comparable to the voltage drop $R_S I_S$, this approximation cannot be justified.

---

[4] Due to flux decrease in the field weakening regime, the magnetizing current I_m is lower than the rated magnetizing current. This allows for a slight increase in the rotor current liable for the torque generation.

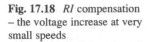

Fig. 17.18 *RI* compensation
– the voltage increase at very
small speeds

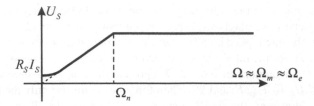

If an induction machine operates at very small speed, the angular frequency is very small as well. Maintaining the rule that the stator voltage U_S is proportional to the supply frequency, the flux of the machine is obtained below the rated flux. Consider the case where the ratio U/f is retained even at very low speeds, notwithstanding the resistive voltage drop. The stator voltage is equal to $U_S = \omega_e \Psi_n$. Assuming that the mechanical load T_L is close to zero, the machine operates with the slip of $s = 0$ and with $I_R = 0$. With $\underline{U}_S = U_S$, the stator current is equal to $\underline{I}_S = U_S/(R_S + j\omega_e L_S)$, and the stator flux is $\Psi_S = L_S\underline{I}_S = L_S(U_S/(R_S + j\omega_e L_S)) = \Psi_n \cdot (jL_S\omega_e/(R_S + j\omega_e L_S))$. In cases where $\omega_e L_S >> R_S$, the stator flux amplitude is equal to Ψ_n, and it does not depend on the parameter R_S. At very low speeds, the flux amplitude decreases due to the voltage drop $R_S I_S$.

According to diagram $U_S(\Omega)$ in Fig. 17.17, the operating point $\Omega_e = \omega_e/p = 0$ results in the stator voltage $U_S = \omega_e \Psi_n = 0$. With $U_S = 0$ and $R_S \neq 0$, the stator flux is equal to zero as well. With low supply frequencies and with $U_S = \omega_e \Psi_n$, the actual stator flux is lower than the rated value. A low flux amplitude at very low frequencies reduces the start-up torque and has adverse effect on the operation of induction machines at low speeds. These effects can be reduced by changing the control law $U_S = \omega_e \Psi_n$ and increasing the supply voltage amplitude at low speeds in the manner shown in Fig. 17.18.

17.19.2   Critical Speed

According to Fig. 17.17, induction machine operating in the field weakening region is capable of providing a constant rated power. This figure has been derived based upon certain assumptions. One of them is neglecting the difference between the stator flux and the air-gap flux, namely, neglecting the voltage drop across the leakage inductance. The leakage reactance is proportional to the supply frequency. At very high speeds, the operating frequency increases up to the levels where the leakage reactance cannot be neglected. Therefore, there is a limit to the constant power operation. The speed Ω_{cr} is the maximum speed where the machine is still capable of delivering the rated power. The operation above this speed is feasible but with a power lower than the rated power. The speed Ω_{cr} is called *critical speed*. It denotes the intersection of functions $T_s(\omega) = T_n(\omega_n/\omega)$ and $T_b(\omega) = T_{b(n)}(\omega_n/\omega)^2$. An approximate value of the critical speed will be determined in the subsequent analysis. To keep the discussion simple, it is assumed that $p = 1$ and that electrical frequencies ω have the same values as the relevant speeds Ω. At the same time, the slip frequency and the slip speed are considered negligible ($\omega_{slip} \ll \omega_e$, $\Omega_{slip} \ll \Omega_e$) allowing the rotor speed Ω_m to be replaced with the synchronous speed Ω_e.

The operation at a constant, rated power in the zone of field weakening requires the torque $T_s(\omega) \approx T_n(\omega_n/\omega)$. The function $T_s(\omega)$ delimits the steady state operating limit for the torque. Namely, the curve $T_s(\omega)$ expresses the maximum steady state torque at the given speed. The limit torque $T_s(\omega)$ is obtained with rated stator current I_n. On the other hand, the envelope of breakdown torques varies according to the law $T_b(\omega) \approx T_{b(n)}(\omega_n/\omega)^2$, where $T_{b(n)}$ is the breakdown torque obtained at the rated power supply conditions, with the rated flux. The function $T_b(\omega)$ represents the maximum transient torque available at the given speed. This transient torque T_b can be maintained only for a short interval of time, as it requires the stator currents $I_S > I_n$, and therefore cause increased losses and temperature rise. The function $T_s(\omega)$ crosses the function $T_b(\omega)$ at the speed $\omega_{cr} = \omega_n(T_{b(n)}/T_n)$.

For speeds above ω_{cr}, the available breakdown torque $T_b(\omega)$ is smaller than the torque $T_s(\omega) \approx T_n(\omega_n/\omega)$ which is permissible in continuous service. Hence, the transient torque limit falls below the torque limit in continuous service, which appears a contradiction. For the proper understanding, it is of interest to understand the difference between the functions $T_s(\omega)$ and $T_b(\omega)$. The curve $T_s(\omega)$ represents the torque T_s which is available at the given speed of ω provided that the current in the stator windings is $I_S = I_n$. Hence, in a way, the curve $T_s(\omega)$ represents the limit $I_S < I_n$ expressed in $T(\omega)$ plane. It is of interest to notice that the torque values $T_s(\omega)$ are feasible only in cases where the stator current can actually reach the rated current. On the other hand, the curve $T_b(\omega)$ is the actual limit for the instantaneous torque. At the given speed ω, the function $T_b(\omega)$ provides the peak torque available from the induction machines of the given parameters. Above critical speeds, the curve $T_b(\omega)$ provides the values of the breakdown torque available for $\omega > \omega_{cr}$. At the same time, the values indicated by $T_s(\omega)$ cannot be reached for speeds $\omega > \omega_{cr}$. At elevated supply frequencies, the leakage reactance increases. With the stator voltage limited to the rated value and with an increased leakage reactance, the stator current cannot reach the rated values. Hence, for the speeds $\omega > \omega_{cr}$, the stator current falls below I_n in both transient and steady state service. This leads to situation where $T_s(\omega) > T_b(\omega)$.

Induction machines with variable frequency supply can operate above critical speed, but their power will fall below the rated power. In this range of speeds, the available torque will drop proportionally to the square of the rotor speed, while the available power will drop proportionally the speed, $P \sim 1/\omega$. Transient and steady state operating limits of an induction machine are given in Fig. 17.19.

With the assumed approximations, the critical speed is

$$\Omega_{cr} = \Omega_n \frac{1}{2x_{\gamma e}} = \frac{T_{b(n)}}{T_n}, \tag{17.14}$$

where

$$x_{\gamma e} = \frac{X_{\gamma e}}{Z_n} = \frac{L_{\gamma e}\omega_n}{Z_n} = \frac{L_{\gamma e}\omega_n I_n}{U_n}.$$

Fig. 17.19 Transient and
steady state operating limits

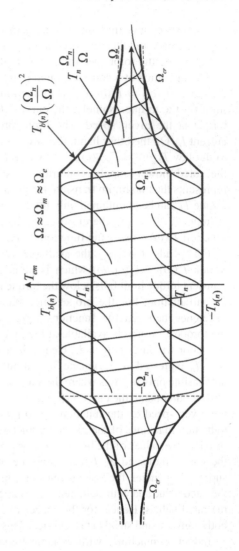

Question (17.7): Induction motor connected to the voltage source with rated
frequency and rated voltage amplitude develops the stator current $I_P = 5I_n$ in
locked rotor conditions ($\Omega_m = 0$). Provide an estimate of relative values of the
breakdown torque and the critical speed.

Answer (17.7): The breakdown torque obtained with the rated power supply is
determined by expression

$$T_{b(n)} = \frac{3p}{\Omega_e} \frac{U_{Sn}^2}{2X_{\gamma en}},$$

where U_{Sn} is the rated rms value of the phase voltage, while $X_{\gamma en}$ is the leakage inductance at the rated stator frequency. By introducing approximations $R_S = 0$, $L_m \gg L_{\gamma e}$, and $(R_R/s_n) \approx U_n/I_n \gg X_{\gamma en}$, the rated torque can be represented by the following expression:

$$T_{nom} = \frac{3p}{\Omega_e} \frac{R_R}{s_n} \frac{U_{Sn}^2}{\left(\frac{R_R}{s_n}\right)^2 + X_{\gamma en}^2} \approx \frac{3p}{\Omega_e} \frac{R_R}{s_n} \frac{U_{Sn}^2}{\left(\frac{R_R}{s_n}\right)^2} \approx \frac{3p}{\Omega_e} \frac{U_{Sn}^2}{\left(\frac{R_R}{s_n}\right)} \approx \frac{3p}{\Omega_e} U_{Sn} I_{Sn},$$

where s_n represents the rated value of the relative slip.

Relative value of the breakdown torque is equal to the ratio of the two preceding expressions,

$$m_{b(n)} = \frac{T_{b(n)}}{T_n} = \frac{\frac{U_{Sn}^2}{2X_{\gamma en}}}{U_{Sn} I_{Sn}} = \frac{1}{2x_{\gamma en}},$$

where $x_{\gamma en}$ is the relative value of the leakage reactance. An approximate value of the leakage reactance can be determined from the start-up current, $x_{\gamma en} \approx 1/I_P = 0.2$. Relative value of the initial torque is equal to $m_{b(n)} = 2.5$. Critical speed $\Omega_{cr} = \Omega_n(T_{b(n)}/T_n)$ is the highest speed at which the rated power can still be obtained. Relative value of the critical speed (Ω_{cr}/Ω_n) is equal to the relative value of the breakdown torque, thus $\omega_{cr} = 2.5\omega_n$.

17.20 Construction of Induction Machines

Induction machines have been in use since the end of the nineteenth century. During the first hundred years of their application, the switching power transistors and other components required for variable frequency supply were not available. For that reason, induction machines were supplied from the mains, with the voltages having a constant, line frequency. Therefore, all the induction machines used in this period were designed and optimized for constant frequency operation. Starting up of the induction motors was performed by connecting them to a three-phase network of industrial frequency 50/60 Hz.

17.20.1 Mains-Supplied Machines

At start-up time, a mains-supplied induction motor has the rotor speed $\Omega_m = 0$ and the stator voltages of the rated amplitude and frequency. The start-up current in the stator windings is $I_P \approx U_{Sn}/X_{\gamma e}$, where $X_{\gamma e} \approx X_{\gamma S} + X_{\gamma R}$ is the equivalent leakage reactance. Small values of the leakage reactance would result in high start-up

Fig. 17.20 (a) Semi-closed
slot. (b) Open slot

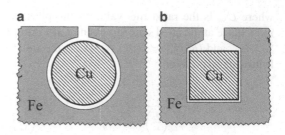

currents. A reactance of $x_{\gamma e} \approx 10\%$ gives the start-up current which is 10 times higher than the rated current. Such current results in the stator copper losses that exceed the losses under rated condition by 100 times. At the same time, large start-up currents result in considerable drop in the mains voltage and affect other electrical loads that are connected to the same line. The start-up mode of an induction motor lasts until the rotor speed comes close to the synchronous speed, where the relative slip s comes down and the impedance R_R/s in the equivalent circuit obtains the value $R_R/s >> X_{\gamma e}$, therefore causing the stator current to decrease and come down to acceptable levels. The acceleration time depends on the load inertia J, and it can last from several hundreds of milliseconds up to several seconds. Losses in the windings during the start-up are proportional to the square of the initial current, and they cause a steep rise of the motor temperature. Therefore, the acceleration cannot last long. Unless the machine approaches the synchronous speed in a short time, it has to be disconnected from the mains in order to avoid dangerous temperatures and preserve the machine integrity. The start-up current is much higher than the rated current, and this may pose a problem for the installations, fuses, and cabling. Mains-supplied induction machines have to be designed to sustain large start-up currents without damage.

In order to reduce the start-up current of mains-supplied induction machines, it is necessary to design such machines so as to provide higher leakage inductances. The leakage inductance is proportional to the ratio N^2/R_μ, where N is the number of turns of the relevant winding, while R_μ is magnetic resistance along the path of the leakage flux. By reducing the magnetic resistance R_μ, it is possible to increase the leakage inductance and leakage reactance. This would contain the start-up current of an induction machine. One of the ways to achieve this is the use of the semi-closed and closed slots in the rotor magnetic circuit (Fig. 17.20).

Making the rotor slot opening toward the air-gap narrower reduces the magnetic resistance along the path of the leakage flux. With narrow top of the rotor slot, the leakage flux path through the air is made shorter, which reduces the magnetic resistance. An increase in the leakage inductance and reactance would decrease the start-up current. There are also side effects. An increase in leakage inductance reduces the breakdown torque, which is inversely proportional to the leakage reactance. In the process of designing an induction machine intended for constant frequency operation, the choice of leakage reactance is the result of a compromise. The final value should result in acceptable start-up currents, but it should not make an unacceptable reduction of the breakdown torque.

Fig. 17.21 Double cage of mains-supplied induction machines. (**a**) Brass cage is positioned closer to the air gap. (**b**) Copper or aluminum cage is deeper in the magnetic circuit

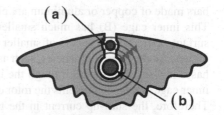

At start-up of an induction motor supplied from the mains, it is necessary to develop the start-up torque T_P as high as possible. This would prevail over the motion resistances T_L and provide for acceleration. At start-up, the acceleration is equal to $d\Omega_m/dt = (T_P - T_L)/J$, and it strongly depends on the start-up torque. Higher values of T_P result in higher acceleration, resulting in short acceleration times, lower amount of heat caused by losses, and lower increase in temperature. Shortening the start-up reduces the thermal stress and extends the lifetime of the machine. The start-up torque $T_P = (3p/\Omega_{en})R_R I_P^2$ is dependent on the square of the start-up current, and it is proportional to the rotor resistance R_R. In order to increase the start-up torque, it is necessary to increase the rotor resistance R_R. This can be accomplished by making the rotor bars with a smaller cross section or making them of materials with higher specific resistance, such as brass. However, an increase in the rotor resistance affects the steady state losses in the rotor windings. In rated operating condition, the copper losses in the rotor would be higher due to elevated resistance R_R. This would reduce the coefficient of efficiency of the machine, increase the temperature, and eventually reduce the rated power. High efficiency during steady state regimes requires the rotor resistance to be as small as possible. At the same time, the need to maximize the start-up torque requires the use of rotor resistances as high as possible. This contradiction was resolved by making the rotor cage so as to obtain frequency dependence of the rotor resistance.

The rotor winding can be designed so as to have a frequency-dependent resistance. Parameter R_R can be made dependent on the frequency of the rotor currents, namely, on the slip frequency. In start-up mode, the slip frequency is high. It is equal to the line frequency due to the relative slip being equal to one. At steady state, the machine operates with the speeds that are close to the synchronous speed, and the slip frequency is very low, of the order to 1 Hz. The change in the slip frequency can be used to obtain variable rotor resistance that would suit the needs of mains-supplied induction machines.

At start-up, it is necessary to have high values of R_R in order to have a high start-up torque. Then, the frequency of the rotor currents is equal to that of the stator currents, $f_{slip} = f_e = 50$ Hz. Hence, it is necessary to make the rotor cage so that it pays relatively high resistance to electrical currents of the line frequency.

At steady state, the speeds are close to the rated value and to the synchronous speed. The frequency of rotor currents is much lower, and it is close to $f_{slip} \sim 1$ Hz. Equivalent resistance of the rotor cage at low frequencies should be as low as possible.

By building a double cage, like the one shown in Fig. 17.21, the rotor winding can be made with frequency-dependent resistance. Low-resistance, large cross-sectional bars made of copper or aluminum are placed deeper into the rotor magnetic circuit. This inner cage (B) has much smaller resistance to DC currents. Closer to the surface, there are brass bars of smaller cross section (A). Their resistance is much higher. At very low frequencies, such as the rated slip frequencies, the rotor current has a low-resistance path through the inner cage. At line frequency of 50 Hz, the inner cage reactance prevents the rotor circuits from passing through the inner cage. Therefore, the start-up current in the rotor circuit passes through the brass cage, which is closer to the air gap, which has a much higher resistance and therefore provides a higher start-up torque.

It is of interest to notice that the leakage reactance of the inner cage is much higher than the leakage reactance of the brass cage. The figure shows the lines of the magnetic field of the leakage flux. The copper bars are encircled by a large number of field lines. For this reason, the leakage flux, leakage inductance, and leakage reactance are relatively high. The brass cage, placed much closer to the air gap, is encircled by a smaller number of field lines. Its leakage flux and leakage reactance are considerably smaller. At start-up, the frequency of rotor currents is $f_{slip} = f_e$ =50 Hz, which increases the values of rotor reactance $X_\gamma = L_\gamma \omega_{slip} = L_\gamma \omega_e$. Due to relatively high frequency, reactances of both cages prevail over resistances, and the impedance of each of the cages is mainly reactive, $X_\gamma >> R_R$. Since reactance of the brass cage is considerably smaller, the rotor start-up current passes mainly through the brass cage, the cage with higher resistance. When the motor enters the steady state, the frequency of the rotor currents is considerably smaller, and it is close to $f_{slip} \sim 1$ Hz. Therefore, the impedance of both cages is mainly resistive, as the resistances prevail over reactances, $X_\gamma = L_\gamma \omega_{slip} << R_R$. Since the resistance of the lower (copper) cage is considerably smaller, the rotor current at steady state passes mainly through the low-resistance copper bars.

It can be concluded that the rotor currents in a double-cage rotor pass through the upper, brass cage during start-up, while they get shifted to the lower, copper cage during operation at steady state, where the speed is close to the synchronous speed and the slip frequency is low. In this way, the rotor resistance is made frequency dependent. At steady states close to rated operating conditions, equivalent rotor resistance is low, while during the start-up, equivalent rotor resistance is high. Manufacturing double cage increases complexity of the production process and increases the costs. Therefore, it is used mainly for machines with larger rated power and/or larger start-up torque requirements.

The effects similar to those created by double cage can be obtained by designing rotor slots with an increased depth and decreased width. An example of such deep slot is shown in Fig. 17.22. The slot contains a rotor bar of rectangular cross section. With very low rotor frequencies, where the leakage reactances are of no importance, the rotor current is distributed equally across the cross section of the rotor bar. The currents that pass at the bottom of the slot are encircled by a number of lines of the magnetic field, that is, by relatively large leakage flux. On the other hand, there are also currents next to the surface, facing the air gap. They are encircled by considerably lower number of field lines and have much smaller leakage flux.

Fig. 17.22 A deep rotor slot and distribution of rotor currents

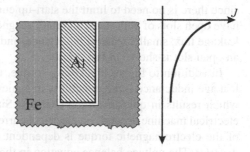

When the rotor bars have current of relatively high slip frequency, variable leakage flux creates electromotive force which opposes to electrical currents. The current passing through the upper part of the slot is encircled by a small leakage flux. Therefore, the reactive electromotive force for this current is smaller. On the other hand, the current passing through the lower part of the slot is buried into the rotor magnetic circuit, and it is encircled by a larger leakage flux. Thus, the reactive electromotive force for this current is much higher. It impedes the current flow and pushes the rotor current toward the air gap. An example of an uneven distribution of the rotor current is given in the right-hand side of the figure. Since the currents of relatively high slip frequency pass through a smaller part of the rotor bar cross section, the equivalent resistance of the rotor is increased. On the other hand, the current distribution at low slip frequencies is even, and the equivalent rotor resistance is much lower.

17.20.2   Variable Frequency Induction Machines

In previous section, different approaches to designing mains-supplied induction machines have been outlined. They were focus on resolving the problems of constant frequency induction machines. Most important problems include limiting the start-up current, providing sufficient start-up torque, and providing a satisfactory efficiency in steady state conditions.

Modern induction machines are supplied from three-phase switching inverters which make use of power transistors. They produce three-phase voltage system of variable frequency and variable amplitude. Parameters of the power supply are suited to serve the target operating modes. The start-up of frequency-controlled induction machine does not imply a large start-up current. Instead, the stator frequency is reduced to the value close to the rated slip frequency, and the voltage amplitude is determined so as to produce the rated flux. In this way, development of the start-up torque does not require the stator currents that exceed the rated current, unless the motion resistances do exceed the rated torque. Hence, frequency-controlled induction machines are never exposed to rated voltage in locked rotor condition. Therefore, they do not need to have an increased leakage inductance, since there is no need to limit the start-up current. Instead, they can be designed to have open slots of both stator and rotor magnetic circuits, which results in a smaller leakage flux, smaller leakage inductance, and larger breakdown torque. Example of an open slot is shown in Fig. 17.20b.

In addition to higher breakdown torque, the advantage obtained by decreasing leakage inductance is the possibility to achieve faster changes of the stator current, which results in quicker torque changes. Since the electromagnetic torque of an electrical machine depends on electrical currents in the windings, the rate of change of the electromagnetic torque is dependent on the first derivative of the current, $di_S(t)/dt$. The voltage balance equation in the stator winding can be represented by $u_S = R_S i_S + L_{\gamma e} di_S/dt + e$. Therefore, the first derivative of the stator current $di_S/dt = (u_S - R_S i_S - e)/L_{\gamma e}$ is inversely proportional to the leakage inductance. With lower leakage inductances, it is possible to achieve larger rate of change of the electromagnetic torque.

Reduction of leakage inductance can also have negative consequences. Due to a finite number of slots carved into magnetic circuits and due to non-sinusoidal distribution of conductors, the windings of an induction machine contain electromotive forces that have higher harmonics. These harmonics cause electrical currents of the same frequency. The amplitude of such currents in rotor bars is directly proportional to the amplitudes of the relevant frequency component of the electromotive force, and inversely proportional to the winding impedance. The winding impedance at higher frequencies is determined primarily by the leakage reactance. For this reason, reduction of the leakage inductance leads to increased amplitudes of winding currents caused by higher harmonics, and increases the current ripple caused by the PWM supply. The adverse effects can be avoided by careful design of the rotor slots by shaping the stator and rotor magnetic circuits so as to reduce non-sinusoidal distribution of the field and to design the stator and rotor windings so as to reduce the electromotive forces induced due to distortions and higher harmonics.

Question (17.8): Rotor bars are placed in slots which are separated by teeth. The lines of the magnetic field are directed along the path of smaller magnetic resistance. Therefore, magnetic induction is high in rotor teeth and significantly lower in rotor slots. Conductors carrying rotor currents are placed in slots, where magnetic induction B is close to zero. Considering the force exerted on conductors, it depends on magnetic induction and electrical current. Apparently, this force is going to be very low. Explain the fact that, notwithstanding the abovementioned, induction machine does generate considerable torque.

Answer (17.8): The torque generation can be represented as the result of forces acting on conductors. With conductors placed in relatively deep slots, magnetic induction within the iron teeth exceeds by far the magnetic induction within aluminum-filled slots. Ratio of magnetic induction in slots and magnetic induction in teeth is close to μ_0/μ_{Fe}. Therefore, only a very small force is acting on conductors. Instead, forces act on rotor surfaces that separate ferromagnetic materials, such as iron in magnetic circuits, from nonmagnetic materials, such as aluminum conductors or the air gap. They act on the surface walls between slots and teeth. The forces acting on the rotor teeth can be explained by using the concept called *equivalent magnetic pressure*.[5] Calculation of spatial distribution of the magnetic field in relatively complex, three-dimensional structure such as the slotted rotor is rather involved. In most cases, it requires automated software tools. Once completed, this calculation provides the information on the equivalent magnetic pressure which is acting upon surfaces that separate iron from nonmagnetic domains. With the equivalent magnetic pressure readily available, the forces and torque can be calculated by performing surface integration along all the relevant surfaces. The outcome of such calculations is the torque value equal to the value obtained by assuming that the force LIB acts on each conductor and that magnetic field lines pass equally through slots as they do through teeth. The last assumption is equivalent to considering the machine where the rotor surface is smooth cylinder with no slots, while the rotor conductors reside in the air gap, attached to the rotor surface. In such hypothetical case, the expression $F = LIB$ is more obvious.

---

[5] Magnetic field creates forces acting on surfaces delimiting different domains. These forces can be described by introducing equivalent pressure $p(\text{N/m}^2)$. The force acting on surface S is equal to $F = pS$. The energy density of the magnetic field in the first domain, next to the boundary, is $w_1 = \mu_1 H_1^2/2$. Across the boundary, in the second domain, the energy density is $w_2 = \mu_2 H_2^2/2$. Equivalent pressure is equal to $w_1 - w_2$.

Chapter 18
Synchronous Machines

The following chapters study principles of operation, construction, mathematical model, and basic characteristics of synchronous machines. Along with induction machines, synchronous machines belong to the group of AC machines. Their operating principles are different. The rotor of induction machines revolves at the speed slightly lower than the synchronous speed, thus their name *asynchronous* machines. The rotor of synchronous machines revolves at the synchronous speed.

In both induction and synchronous machines, revolving magnetic field is created by AC currents in the three-phase windings of the stator. The stator magnetic circuits and the stator windings of induction and synchronous machines are very much the same. In both cases, three-phase system of stator currents creates rotating magnetomotive force and rotating field of magnetic inductance. The field revolves at the speed which is determined by the angular frequency of the stator currents, also called power supply frequency. Synchronous machines and induction machines have different construction of their rotors. The rotor winding in most induction machines is a short-circuited cage made of aluminum bars which are placed in the rotor slots. Rotor in synchronous machines may have excitation winding or permanent magnets. The rotor with excitation winding is supplied with DC currents that create the rotor magnetomotive force and the rotor flux. Instead of excitation windings, rotor of synchronous machines may have permanent magnets built into the rotor magnetic circuit. In this case, the rotor does not have any windings.

The principles of operation of induction machines have been described in Chap. 14. The rotor windings of an induction machine are short-circuited. When the rotor of an induction machine falls behind the revolving field by the amount of slip, the electromotive forces are induced in short-circuited rotor windings, and the rotor currents appear as a consequence. By joint action of the induced rotor currents and magnetic field, induction machine generates the electromagnetic torque, proportional to the slip. The torque generation process requires the rotor to revolve somewhat slower than the field, so that the revolving field advances with respect to the rotor by the amount of slip. Certain amount of slip is required in order to change the rotor flux and create electromotive forces and electrical currents in rotor bars.

S.N. Vukosavic, *Electrical Machines*, Power Electronics and Power Systems, 521
DOI 10.1007/978-1-4614-0400-2_18, © Springer Science+Business Media New York 2013

It has been shown in Chap. 14 that electromagnetic torque of an induction machine depends on the angular frequency of rotor currents $\omega_{slip} = \omega_e\text{-}p\Omega_m$, which is determined by the rotor lagging with respect to revolving field. In rated operating conditions, the rotor of an induction machine does not rotate synchronously with the field. Therefore, induction machines are also called *asynchronous machines*.

Rotor in synchronous machines is either an electromagnet or a permanent magnet. Position of the rotor flux is uniquely defined by the position of the rotor. In rated operating conditions, the rotor revolves synchronously with the stator field. Electromagnetic torque is proportional to the vector product of the stator and rotor flux vectors. The synchronous rotation of the rotor is the reason for this type of electrical machines to be called *synchronous machines*.

In this chapter, basic operating principles of synchronous machines are introduced and explained. The torque generation is discussed for the machines with an excitation winding on the rotor and for the machines with permanent magnets. The construction of stator windings and stator magnetic circuit is rather similar to that of the induction machine. It is reinstated briefly, along with generation of the revolving magnetic field of the stator winding. The rotor construction is explained in more detail. The available methods of supplying the DC excitation current to rotors with an excitation winding are introduced, explained, and discussed, including the sliding rings with brushes and the transformer with revolving secondary and rectifier circuit on the rotor. This chapter reviews most significant characteristics of permanent magnet materials. Two different ways of inserting permanent magnets into the rotor magnetic circuits are explained and discussed. Magnetic and electrical properties of synchronous machines with buried magnets and surface-mounted magnets are studied. In particular, the difference in self-inductance of the stator windings is discussed and explained for buried magnet machines and surface-mounted magnet machines.

18.1 Principle of Operation

Like induction machines, synchronous machines have AC currents in stator windings. Stator currents create the stator magnetomotive force F_S which revolves at the speed $\Omega_e = \omega_e/p$, also called synchronous speed. The synchronous speed is determined by the angular frequency ω_e of stator currents and by the number of pole pairs p. The stator magnetomotive force creates the stator flux $\Phi_S = F_S/R_\mu$, which depends on magnetic resistance R_μ. The flux Φ_S rotates at the same speed Ω_e as the magnetomotive force. Rotor in synchronous machines may have permanent magnets built into magnetic circuit or excitation windings supplied by DC current. In both cases, the rotor flux Φ_R has the course and direction determined by the rotor position. The flux vector Φ_R rotates at the rotor speed Ω_m. With $\Omega_m = \Omega_e$, both the stator flux vector Φ_S and the rotor flux vector Φ_R revolve at the same speed. They do not change their relative position and maintain the angle between the two vectors constant. The torque and power of electromechanical conversion are dependent on the vector product of the two flux vectors. The vector product depends on the angle between the two vectors. With $\Omega_m = \Omega_e$, the two flux vectors do not change their relative position. The electromagnetic torque is proportional to the product of the two flux amplitudes and to the sine of the angle between the two flux vectors. With constant flux amplitudes and constant relative position between the two flux vectors, the electromagnetic torque remains constant. The operation of synchronous machines and the torque development require synchronous rotation ($\Omega_m = \Omega_e$) of the stator field, created by the stator currents, and the rotor field, created by excitation winding or permanent magnets.

18.2 Stator Windings

Stator of a synchronous machine is very much the same as the stator of an induction machine. Current circuits on the stator side have three separate parts called stator phase windings or stator phases. Each phase is obtained by connecting a number of turns in series. Relevant conductors are distributed along the machine circumference and placed into slots. The stator slots are carved into the inner side of the stator magnetic circuit, facing the air gap. Distribution of stator conductors along the machine circumference is quasisinusoidal. That is, an attempt is made to achieve sinusoidal change of the conductor density along the circumference. Ideal sinusoidal distribution cannot be achieved due to a finite number of slots. Conductors cannot be placed in an arbitrary position. They have to be placed in one of the slots. Hence, there are a limited number of discrete locations for the placement of stator conductors. For this reason, sinusoidal distribution of stator conductors along the machine circumference cannot be achieved in full. Yet, the windings are made in such way that the distribution of conductors gets as close to sinusoidal as possible. Each of the three-phase windings has two terminals. One end of each phase winding is connected to the three-phase voltage source which supplies the stator windings. Stator terminals of synchronous generators may be connected to three-phase electrical loads. Remaining ends of the three-phase windings are wired together into the node called *star point*. This way of connecting the stator phases is called *star connection*. With the sum of the three-phase currents equal to zero, $i_a(t) + i_b(t) + i_c(t) = 0$, there is no need to connect the star point to the source; thus, no return line is required. Most machines[1] have their star points disconnected from the power

---

[1] Electrical machines supplied the mains with line voltages $U < 1$ kV are also called low-voltage machines. Most low-voltage machines have star connected stator windings with floating star point. Namely, the star connection of the three phases is not connected to any other node. In most cases, the star point is even inaccessible, namely, it is not made available to the user. Machines with stator voltages in excess to 1 kV may have their star point connected to the neutral or to the ground by means of a series impedance. This connection reduces the overvoltage stress. In most cases, the impedance used for grounding the star point has a minor effect on the equation $i_a + i_b + i_c = 0$.

supply (floating). In some cases, the three phases of the stator winding are
connected in triangle, and this way of connecting the phase windings is called
delta connection. There is no star point in a delta connection.

Each phase winding creates magnetomotive force which is determined by the
number of turns of the phase winding and by the electrical current of the phase
winding. Each phase winding with AC currents creates variable magnetomotive
force. Currents i_a, i_b и i_c in individual phases of the stator winding create
magnetomotive forces F_a, F_b, and F_c. Corresponding magnetomotive force vectors
are positioned along magnetic axis of each phase winding. Magnetic axes of the
phases that make a two-pole winding system are displaced by $2\pi/3$. Resulting stator
magnetomotive force F_S is obtained by vector summation of the three
magnetomotive forces, F_a, F_b, and F_c, created by the three-phase windings.

18.3 Revolving Field

Magnetic axes of the stator phases as well as magnetomotive forces F_a, F_b, and F_c
are displaced in space by $2\pi/(3p)$, where p is the number of pairs of magnetic poles.
The stator winding can be arranged so as to produce magnetic field with more than
two magnetic poles. Example of four-pole stator winding has been explained in
Chap. 17.

Revolving magnetic field with two magnetic poles is established by AC currents
in two-pole stator windings. With phase currents of the same amplitude I_m and
frequency ω_e, and with their initial phases displaced by $2\pi/3$, a two-pole stator
winding creates two-pole stator field which revolves at the speed $\Omega_e = \omega_e/p = \omega_e$,
also called synchronous speed. With stator windings that create magnetic field with
more than one pair of magnetic poles ($p > 1$), the stator field revolves at the speed
$\Omega_e = \omega_e/p$.

Magnetomotive forces of the three-phase windings are given in (Fig. 18.1),
assuming that the currents have the same amplitude I_m, the same angular frequency
ω_e, and the initial phases mutually shifted by $2\pi/3$. It is assumed that each phase
winding has N_S turns:

$$F_a = N_s I_m \cos \omega_e t,$$

$$F_b = N_s I_m \cos\left(\omega_e t - \frac{2\pi}{3}\right),$$

$$F_c = N_s I_m \cos\left(\omega_e t - \frac{4\pi}{3}\right), \tag{18.1}$$

By summing the magnetomotive forces of individual phases, one obtains the
resultant magnetomotive force of the stator winding, represented by the vector F_S in
Fig. 18.2. This vector revolves at synchronous speed. Calculation of the amplitude

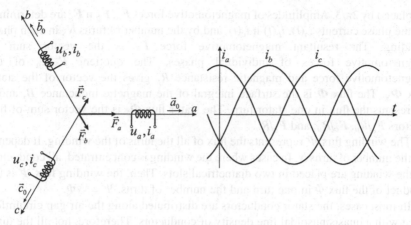

Fig. 18.1 Three-phase stator winding of synchronous machine

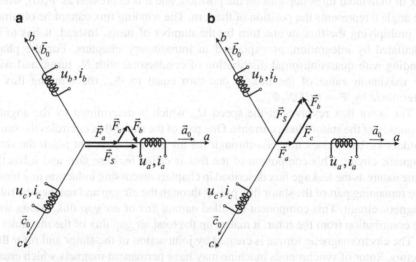

Fig. 18.2 Spatial orientation of the stator magnetomotive force

and spatial orientation of the magnetomotive force created by three-phase stator windings is explained in Chap. 15. Considering (18.1), magnetic field revolves at the speed $\Omega_e = \omega_e/p$ and maintains a constant amplitude.

Initial phase of stator currents determines the spatial orientation of the stator magnetomotive force F_S at instant $t = 0$. At steady state, the vector F_S in synchronous machines has to revolve in synchronism with the rotor. The torque depends on the sine of the angle between the rotor flux vector and the vector F_S of the stator magnetomotive force. When the stator field F_S/R_μ revolves in synchronism with the rotor, synchronous machine develops a constant torque and constant power.

Magnetomotive force vectors of individual phases are oriented along magnetic axes of respective phase windings. With two-pole stator winding, the axes are displaced by $2\pi/3$. Amplitudes of magnetomotive forces F_a, F_b и F_c are determined by the phase currents $i_a(t)$, $i_b(t)$ и $i_c(t)$, and by the number of turns N_S in each phase winding. The resultant magnetomotive force F_S is the vector sum of magnetomotive forces of individual phases. The quotient F_S/R_μ of the magnetomotive force and magnetic resistance R_μ gives the vector of the stator flux Φ_S. The flux Φ is the surface integral of the magnetic inductance B, and it represents the flux in one stator turn. The stator flux Φ_S is the vector sum of flux vectors F_a/R_μ, F_b/R_μ, and F_c/R_μ.

The winding flux Ψ represents the flux of all the turns of the winding. It depends on the number of turns N. In cases where the winding is concentrated, all conductors of the winding are placed in two diametrical slots. Then, the winding flux Ψ is the product of the flux Φ in one turn and the number of turns, $\Psi = N\Phi$.

In most cases, the stator conductors are distributed along the air gap circumference with a quasisinusoidal line density of conductors. Therefore, not all the turns of the phase winding are in the same place, and they do not have the same flux. The flux in individual turns depends on the position, and it is expressed as $\Phi_S(\theta)$, where the angle θ represents the position of the turn. The winding flux cannot be obtained by multiplying the flux in one turn by the number of turns. Instead, it has to be calculated by integration, as explained in introductory chapters. For the phase winding with quasisinusoidal distribution of conductors, with N_S turns, and with the maximum value of the flux in one turn equal to Φ_m, the winding flux is determined by $\Psi = (\pi/4)N_S\Phi_{Sm}$.

The stator flux revolves at the speed Ω_e, which is determined by the angular frequency of the stator phase currents. One part of the stator flux encircles the stator conductors, but it does not pass through the air gap and does not reach the rotor magnetic circuit. This component of the flux is called *leakage flux*, and it has the same nature as the leakage flux described in chapters discussing induction machines. The remaining part of the stator flux passes through the air gap and reaches the rotor magnetic circuit. This component is called *mutual flux* or *air gap* flux. Along with the contribution from the rotor, it makes up the total air gap flux of the machine.

The electromagnetic torque is created by joint action of the stator and rotor flux vectors. Rotor of synchronous machine may have permanent magnets which create the rotor flux. There are also synchronous machines with an *excitation winding* on the rotor. DC currents in excitation winding create the rotor magnetomotive force and the rotor flux.

The flux of the stator winding system comprising three-phase windings is defined in (18.2), where R_μ is magnetic resistance and N_S is the number of turns per phase. The stator flux vector is denoted by Ψ_S. It is obtained by dividing the magnetomotive force F_S by magnetic resistance, and it is shown in Fig. 18.3. The flux linkages of individual phases are denoted by Ψ_a, Ψ_b, and Ψ_c, and they depend on the amplitude of the flux vector Ψ_S and its relative position with respect to magnetic axis of relevant phase winding:

$$\vec{\Psi}_s = N_S \frac{\vec{F}_s}{R_\mu}. \tag{18.2}$$

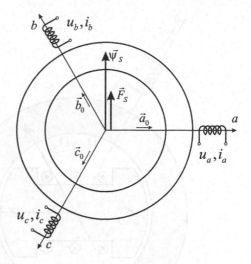

Fig. 18.3 Vectors of the stator magnetomotive force and flux

18.4 Torque Generation

With permanent magnets built into the rotor magnetic circuit, the rotor generates the flux which moves along with the rotor. The same way, the rotor with an excitation winding creates the rotor magnetomotive force and the rotor flux which moves with the rotor. Hence, the rotor flux vector has the position which is equal to position of the rotor. At the same time, the stator currents create the magnetomotive force F_S and the stator flux Φ_S. For both stator and rotor flux vectors, corresponding magnetic poles are identified as the regions where the lines of magnetic field enter or exit the iron parts of magnetic circuit. Regarding the stator flux, position of corresponding magnetic poles is defined by direction of the revolving magnetomotive force F_S, created by the stator currents. Magnetic poles of the rotor flux are defined by the rotor position. Electromagnetic forces tend to bring the opposite stator and rotor magnetic poles in close vicinity (Fig. 5.3). When the stator field rotates in such way that the north stator pole leads and remains ahead of the south rotor pole, there is a constant electromagnetic torque which tends to increase the rotor speed. Considering vectors shown in Fig. 18.4, the torque acts toward bringing the rotor flux Ψ_R closer to the vector F_S. The torque can be expressed as the vector product between the stator magnetomotive force and the rotor flux:

$$\vec{T}_{em} = k_t \vec{\Psi}_R \times \vec{F}_S.$$

Developing constant torque and performing continuous electromechanical conversion require that relative position of the stator and rotor vectors remains unchanged. This condition is met when the rotor revolves at synchronous speed Ω_e, which is the angular speed of the stator field F_S.

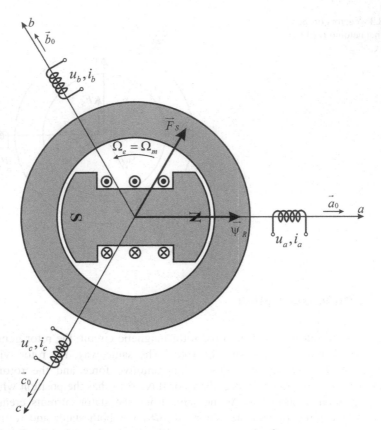

Fig. 18.4 Position of rotor flux vector and stator magnetomotive force

Dividing the vector F_S by the number of turns N_S, one obtains the vector with course and direction of F_S, and with the amplitude equal to the stator current. In further considerations, the vector F_S/N_S is called *stator current vector*. The torque can be expressed as the vector product of the rotor flux $\boldsymbol{\Phi}_R$ and the vector of the stator current, $T_{em} = k_t N_S\, \boldsymbol{\Phi}_R \times i_S = k_t N_S\, |\boldsymbol{\Phi}_R||i_S|\,\sin(\xi)$, where ξ denotes the angle between the rotor flux vector and the stator current vector, while $|i_S|$ denotes the amplitude of the stator current.

For the given rotor flux, required torque can be obtained with different pairs of values ($|i_S|$, ξ). The amplitude of the stator current required to obtain desired torque is lower when $\sin(\xi)$ is higher, and it reaches the minimum in cases where the relevant vectors are orthogonal. With $\xi = \pm\pi/2$ and with the smallest possible $|i_S|$, corresponding copper losses in the stator winding reach their minimum. Condition $\xi = \pm\pi/2$ provides the possibility to obtain the maximum torque for the given current amplitude $|i_S|$. In other words, it maximizes the *torque-per-Ampere* ratio $T_{em}/|i_S|$.

In order to maximize the torque-per-Ampere ratio in synchronous machines, it is necessary to establish the phase currents i_a, i_b, and i_c so as to obtain the stator magnetomotive force vector F_S which is perpendicular to the rotor flux. Hence, whatever the rotor speed or position, the vector F_S is to be locked to the rotor position, advancing with respect to the rotor flux by $\pi/2$. In such cases, the electromagnetic torque is determined by the amplitude of the stator current, $T_{em} = k_t N_S |\boldsymbol{\Phi}_R| |i_S|$. Negative values of the torque are obtained when the vector F_S falls behind the rotor flux by $\pi/2$. This approach is used in controlling the torque of permanent magnet synchronous machines used in motion control applications. These synchronous machines are also called synchronous servomotors, and they are distinguished by low inertia of the rotor and high ratio of the peak torque and the rated torque. In motion control applications, each synchronous servomotor is supplied from a separate three-phase inverter which supplies the motors with the phase currents required for generating the set torque. By using the current regulator with pulse width modulation control, the stator voltages are adjusted so as to obtain desired phase currents i_a, i_b, and i_c. Three-phase inverters with pulse with modulation and with current regulator are also called *current regulated pulse width modulated inverters* or CRPWM inverters.

Question (18.1): Consider two induction machines with short-circuited rotor cage and with the same dimensions of stator and rotor magnetic circuits. One of these is a two-pole induction machine while the other is a four-pole machine. The rotors of both machines are taken out of their stators. A new machine is made by inserting the rotor of the second machine into the stator of the first machine. Is it possible for the new machine to develop any torque?

Answer (18.1): The torque development is based on interaction of the stator and rotor magnetic fields. In order to obtain the electromagnetic torque, it is necessary that the rotor and stator fields have the same number of magnetic poles. Magnetic field of the stator of an induction machine is created by magnetomotive forces caused by electrical currents in the stator windings. The stator windings of the first machine generate a two-pole magnetic field, while the stator windings of the second machine create a four-pole magnetic field. The latter has two north magnetic poles and two south magnetic poles. In both machines, magnetic field of the rotor depends on the currents induced in the rotor conductors. The rotor currents appear as a consequence of electromotive forces induced in the rotor cage. Electromotive forces depend on magnetic induction in the air gap, and they also depend on the slip speed. The change in amplitude and direction of the rotor electromotive forces and currents reflects the change of the air gap inductance B. Therefore, the number of magnetic poles of consequential rotor flux is the same as the number of poles of the stator magnetic field. Hence, one and the same rotor creates a two-pole field while operating within a two-pole stator and a four-pole field when operating in a four-pole stator. Therefore, the machine made by combining the stator of the first machine with the rotor of the second machine will be capable of running as a proper induction machine, and it will develop the torque.

Question (18.2): Consider two synchronous machines, each with permanent magnet rotor. Both machines have equal dimensions of their rotor and stator magnetic circuits. The first machine is a two-pole machine while the second is a four-pole machine. A new machine is made by inserting the rotor of the second machine into the stator of the first machine. Would this new machine be able to develop any torque?

Answer (18.2): Unlike induction machines, synchronous machines have the rotor flux created by an excitation winding on the rotor or by permanent magnets build into the rotor magnetic circuit. With permanent magnet excitation, the number of magnetic poles depends on configuration of the permanent magnets. With an excitation winding, the excitation current is not induced from the stator side. Instead, it is provided from a separate source of DC current. The way of making the excitation winding determines the number of magnetic poles of the rotor field. Hence, with both permanent magnet excitation and with excitation winding, the number of magnetic poles of the rotor field of synchronous machines depends on the rotor design. In other words, the number of magnetic poles of the rotor field does not depend on the number of poles of the stator field, which was the case with induction machines. Synchronous machine operates properly and develops electromagnetic torque only in the case when the stator and rotor have the same number of magnetic poles. A combination of a two-pole stator with a four-pole rotor would not develop any electromagnetic torque.

18.5   Construction of Synchronous Machines

Synchronous machines have stator with three-phase windings and rotor with either excitation winding or with permanent magnets. The stator terminals are connected to a symmetrical system of three-phase voltages and currents. The stator currents create revolving magnetic field in the air gap of the machine. For the proper operation of synchronous machine, the stator field has to revolve at the same speed as the rotor. Electromagnetic torque is created from interaction of the two magnetic fields. Synchronous machines can be made in the form of discs or cylinders; they can have hollow rotor, and there are also linear synchronous machines which perform translation rather than rotation. Synchronous machines are mostly cylindrical machines (Fig. 18.5).

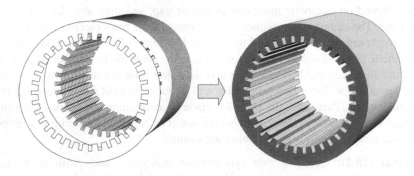

Fig. 18.5 Stator magnetic circuit is made by stacking iron sheets

18.6 Stator Magnetic Circuit

Stator of cylindrical synchronous machines is a hollow cylinder which accommodates the rotor. The main parts of the stator are magnetic circuit and current circuits, also called phase windings. Magnetic circuit is made of ferromagnetic materials, usually iron alloys, while the current circuits consist of insulated copper conductors.

Stator of synchronous machines is entirely the same as the stator of an induction machine. Magnetic induction B in the stator magnetic circuit changes due to rotation of the magnetic field with respect to the stator core. Induction B varies with the frequency ω_e, which is the angular frequency of the stator currents. Variation of magnetic induction causes eddy current losses and hysteresis losses in conductive ferromagnetics. The losses due to eddy currents are proportional to the square of the frequency, while the losses due to hysteresis are proportional to the frequency. In synchronous machines supplied from the mains, magnetic induction in the stator magnetic circuit pulsates at the frequency of $f_e = 50$ Hz (60 Hz). In order to reduce losses in the stator magnetic circuit, it is made by stacking the iron sheets. The sheets are separated by a thin layer of electric insulation.

The iron sheets are actually made of iron alloys comprising small quantities of manganese and other elements. Lamination of magnetic circuit does not reduce the hysteresis losses, but it suppresses the eddy currents and reduces the overall iron losses in magnetic circuit caused by the pulsation of magnetic induction. Since the magnetic field revolves with respect to the stator, magnetic induction in stator magnetic circuit changes direction. For this reason, it is essential for the iron sheets to have the same magnetic properties in all directions. Such sheets are obtained by *hot rolling* of steel, and they are called *hot rolled sheets*. On the other hand, magnetic circuits used in power transformers have the field which is always directed along the same path. Therefore, it is important to have improved magnetic properties along the flux path, while the properties in directions perpendicular to the path are of lesser importance. In such cases, it is beneficial to use anisotropic[2] material, optimized to provide a low magnetic resistance along the flux path. The sheets with such properties are obtained by cold rolling. Hence, the iron sheets used in power transformers are *cold rolled sheets*. Cold rolling results in reduced magnetic resistance in direction of rolling, which should correspond to the flux path within power transformer.

By stacking the sheets, one obtains a hollow cylinder with slots on the inner surface which faces the air gap. The slots are intended for placing conductors of the stator winding. Parts of the stator magnetic circuit between the slots are called teeth. The flux passing toward the air gap is directed through the teeth, made of iron of high-permeability and low magnetic resistance. The flux does not pass through the slots, where permeability is μ_0 and where magnetic resistance is high. Therefore, magnetic induction inside teeth is higher compared to the rest of the stator magnetic circuit, which results in increased teeth iron losses.

---

[2] Anisotropic – having different properties in different directions. For example, anisotropic ferromagnetic may have different permeability in direction of axes x, y, and z.

18.7 Construction of the Rotor

The rotor of synchronous machines with permanent magnets built into magnetic circuit is shown in the part (A) of Fig. 18.6. Permanent magnets are ferromagnetic materials capable of providing the rotor flux without an excitation winding. Magnetizing characteristic $B(H)$ of permanent magnets has a relatively high remanent induction B_r. In absence of rotor windings, there is no rotor magnetomotive force $F = NI$; thus, there are no external means to create and control the rotor flux but the permanent magnets. With sufficient remanent induction of permanent magnets, it is possible to achieve significant values of the rotor flux. By inserting permanent magnets into the rotor magnetic circuit, the rotor flux is obtained in a lossless manner, without a need of making the rotor windings. This simplifies the machine construction, reduces the losses, and increases efficiency. One shortcoming of this solution is the lack of possibility to change the rotor flux. Without the flux control, the field-weakening operation of permanent magnet synchronous machines is virtually impossible.

The part (b) of Fig. 18.6 shows the rotor with electromagnetic excitation. This rotor has an excitation winding with N_R turns carrying DC current i_R. The magnetomotive force of the excitation winding $F_R = N_R i_R$ determines the excitation flux $\Phi_R = N_R i_R / R_\mu$ and the flux of the excitation winding $\Psi_R = N_R \Phi_R = (N_R^2/R_\mu)i_R = L_R i_R$. There is a small amount of the rotor flux which encircles the excitation winding, but it does not

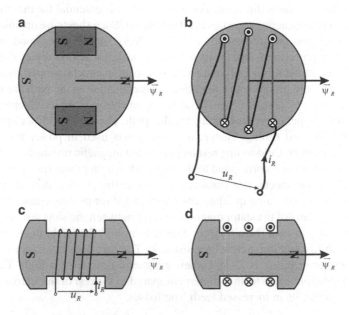

Fig. 18.6 (**a**) Rotor with permanent magnets. (**b**) Rotor with excitation winding. (**c**) Rotor with excitation winding and salient poles. (**d**) Common symbol for denoting the rotor in figures and diagrams

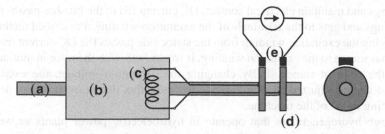

Fig. 18.7 Passing the excitation current by the system with slip rings and brushes. (**a**) Shaft. (**b**) Magnetic circuit of the rotor. (**c**) Excitation winding. (**d**) Slip rings. (**e**) Brushes

pass through the air gap and does not reach the stator magnetic circuit and the stator windings. This flux is called excitation leakage flux. Major part of the excitation flux encircles both the rotor and stator winding; it contributes to the mutual flux Ψ_m and it is denoted by Ψ_{mR}. This flux is the bases for the process of electromechanical energy conversion and for creation of the electromagnetic torque.

Magnetic core of the rotor with excitation winding may have *salient poles*, and one such case is shown in Fig. 18.6c. Salient pole rotors have a small air gap and a low magnetic resistance along the path of the excitation flux, while the air gap in direction perpendicular to the flux is larger, causing a larger magnetic resistance.

Commonly used symbol that represents the rotor in figures and diagrams is shown in Fig. 18.6d. Although it resembles a salient pole rotor, it is also used as a simplified representation of cylindrical rotors, such as the one in Fig. 18.6b, which have the same air gap and the same magnetic resistance along the circumference.

Advantage of electromagnetic excitation over permanent magnet excitation is that the latter allows variation of the excitation flux by varying the excitation current i_R. Shortcomings of this solution are increased losses and more complex construction of the machine, owing to the need to feed DC current i_R to the winding which is mounted on the rotor. The excitation voltage u_R supplied to the terminals of the excitation winding needs to be wired to an external DC source. This source is placed on the stator side, the side that does not move, while the excitation winding resides on the rotor which revolves with respect to the stator. This brings up the problem of passing the excitation current to the moving part of the machine (Fig. 18.7).

18.8 Supplying the Excitation Winding

Supplying the excitation winding can be performed by means of the slip rings fastened to the rotor shaft. The two terminals of the excitation winding can be wired to a pair of slip rings, both of them with mutual electrical insulation and insulated from the shaft. Two conductive brushes can be fastened to the stator and pressed against respective slip ring, providing in this way an electrical access to the excitation winding. While the rotor is in motion, the brushes slide against the rings and maintain electrical contact. DC current fed to the brushes passes to the slip rings and gets to the terminals of the excitation winding. Prescribed method of supplying the excitation winding from the stator side passes the DC current from an external source to the excitation winding. It works both with the rotor in motion and with the rotor at standstill. By changing the excitation voltage, the excitation current can be adjusted so as to provide the rotor flux that corresponds to desired operating mode of the machine.

Large hydrogenerators that operate in hydroelectric power plants as well as turbogenerators that operate in thermal power plants have electromagnetic excitation that includes the excitation winding on the rotor. Electrical current in excitation winding is adjusted so as to obtain desired voltage across the stator terminals. The stator electromotive force determines the stator voltage. At the same time, the electromotive force is proportional to the rotor flux. Hence, the stator voltage of synchronous generators can be increased or decreased by changing the excitation current. Variation of generator voltages is required in order to compensate for variable voltage drops between the power plants and electrical consumers. As the consumption of electrical power at the consumer side increases, the generator load increases as well as the stator current. An increased current produces larger voltage drops in transmission lines, power transformers, and distribution cables. In order to keep the consumer voltages constant, it is necessary to increase the voltage of synchronous generators. This increase is required to compensate for the increased voltage drops in transmission and distribution. The goal is achieved by increasing the excitation current of synchronous generators.

One shortcoming of the excitation system with slip rings is that the excitation current passes through the contact between the brushes and slip rings. The rotor motion makes the contact surfaces slide against each other. With unsteady electrical contact, electric arc may app7ear in the course of rotation. Sporadic arcing presents the fire risk and increases electromagnetic interference. Besides, arcing and friction contribute to ware of the sliding rings and brushes. This in turn requires repairs and maintenance. Besides, the presence of slip rings increases the axial length of the machine.

18.9   Excitation with Rotating Transformer

The transfer of the excitation power to the rotor without mechanical contact can be accomplished by using a rotating transformer, the principles of which are depicted in Fig. 18.8. Primary winding of this transformer is on the stator side of the machine, along with one half of the transformer magnetic circuit. Both transformer parts are attached to the stator and do not move. The stator side of the transformer magnetic circuit has the shape of a ring. Along the circumference, a slot is carved into the inner surface of the ring, and the primary winding is placed into this slot. The other part of the transformer magnetic circuit is attached to the rotor shaft, and it moves with the rotor. It has the shape of a ring of a smaller diameter so that it fits into the stator part of the transformer magnetic circuit. A slot

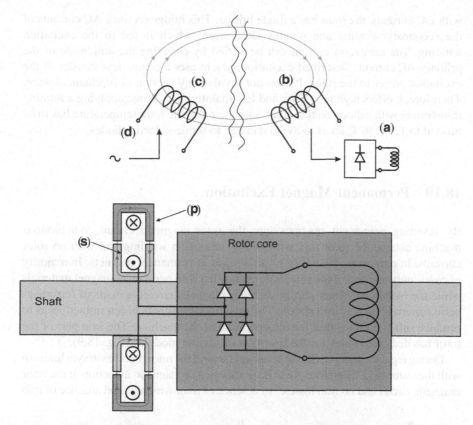

Fig. 18.8 Contactless excitation system with rotating transformer. (a) Diode rectifier on the rotor side. (b) Secondary winding. (c) Primary winding. (d) Terminals of the primary fed from the stator side. (P) Stator part of the magnetic circuit. (S) Rotor part of the magnetic circuit

is carved into the outer side of this ring, and it houses the secondary winding of the rotating transformer. It is observed in Fig. 18.8 that conductors of primary and secondary windings of the rotating transformer have tangential direction. Namely, they are wound around the shaft. The circumstance that one part of the magnetic circuit rotates with respect to the other part does not preclude establishing the mutual flux of the transformer. This flux couples the primary and secondary windings and provides for customary transformer function of passing the electrical energy from the primary to the secondary side. The field lines that represent the mutual flux are shown in Fig. 18.8.

Primary winding of the rotating transformer is supplied by an external source of AC current with the frequency ranging from several hundred Hz to several kHz. Due to magnetic coupling between the primary and secondary windings, the AC currents are transferred from the primary to the secondary side of the transformer. In this way, the secondary winding provides the source of AC currents that are made available at the rotor side of the machine. In order to supply the excitation winding with DC currents, the rotor has a diode bridge. This bridge rectifies AC currents of the secondary winding and obtains DC current which is fed to the excitation winding. The excitation current can be varied by changing the amplitude of the primary AC current. Described excitation system uses a contactless transfer of the excitation power to the rotor. It does not involve any friction or mechanical ware. Therefore, it offers high reliability and low maintenance. When applying a rotating transformer with a diode rectifier built into the rotor, the rotor temperature has to be limited to 125–150°C so as to avoid damage to semiconductor diodes.

18.10 Permanent Magnet Excitation

By inserting permanent magnets into the rotor magnetic circuit, synchronous machine obtains the rotor flux without any excitation winding and with no rotor currents. In rare cases, the whole rotor is made of permanent magnets. In majority of cases, only one part of the rotor volume is filled with permanent magnet materials while the remaining, larger part of the rotor magnetic circuit is made of ferromagnetic materials such as iron sheets. The quantity of magnets is determined so as to produce sufficient rotor flux for the operation of the machine. The iron part of the rotor has dedicated holes for the insertion of magnet modules (Fig. 18.9).

During regular operation of synchronous machine, the rotor revolves in synchronism with the stator field. Therefore, there is no variation of magnetic induction in the rotor magnetic circuit and no iron losses.[3] In absence of rotor windings and absence of iron

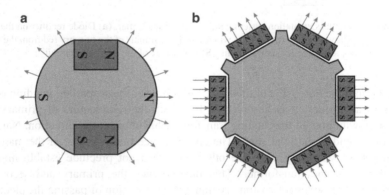

Fig. 18.9 (a) Rotor with interior magnets. (b) Surface-mounted magnets

---

[3] In synchronous motors supplied from three-phase inverters with pulse width modulation, there is a certain current ripple, a high-frequency component of the stator current with an amplitude of $0.02 \ldots 0.03 \, I_n$ and with the frequency which is equal to the switching frequency of power transistors. Due to this nonsinusoidal supply, there are some small, high-frequency variations of the magnetic induction within the rotor circuit, notwithstanding the fact that the fundamental flux component revolves in synchronism with the rotor.

losses in rotor magnetic circuit, there is no heat generated within the rotor. As a consequence, there is no need to devise any particular measures for cooling of the rotor. Without the rotor heat, cooling of the stator winding and the stator magnetic circuit is easier to achieve in synchronous machines than in induction machines. For this reason, there is a certain margin to increase the current and flux densities in synchronous machine. With increased current density and increased magnetic induction, synchronous machines can deliver more torque and more power from the same volume. Alternatively, for the given torque and given power, synchronous machines can be designed with lower size and lower weight than equivalent induction machines. Hence, specific torque[4] and specific power of synchronous permanent magnet machines are higher compared to induction machines.

The problem of synchronous machines with permanent magnets is the absence of possibility of changing the rotor flux. For this reason, permanent magnet machines have difficulties operating in field-weakening mode. Permanent magnets can be built on the surface of the rotor magnetic circuit (*surface mount*) or within the interior part of the magnetic circuit (*interior magnet* or *buried magnet*). The way of inserting the magnets has significant impact on the machine parameters and characteristics. It mostly affects the stator inductance L_S. The winding inductance is inversely proportional to the magnetic resistance, and the latter is greatly affected by the method of inserting the magnets. In synchronous machines with interior magnets, magnetic resistance is relatively small. The winding inductance has relative value that ranges from 0.1 up to 0.5. In machines with surface mount magnets, relative value of the winding inductance ranges from 0.01 up to 0.1.

Synchronous machines with permanent magnets have virtually no rotor losses, and their efficiency is higher compared to other kinds of electrical machines. Comparing the power balance of permanently excited synchronous machines to the power balance of induction machines, it has to be noted that the synchronous machine does not have the rotor losses sP_δ. With the rotor revolving at synchronous speed, relative slip s of synchronous machines is equal to zero. The absence of rotor losses contributes to a significant increase in the efficiency of synchronous machines.

Question (18.3): Efficiency of a two-pole induction motor running at the rated speed of $n_n = 2,850$ rpm is 90%. It is known that the iron losses are relatively small compared to copper losses. Make an estimate of efficiency of a permanent magnet synchronous motor of the same rated speed, same power, voltage, current, and dimensions.

Answer (18.3): Based on the problem formulation, the iron losses can be neglected in both machines. Mechanical losses are relatively small and similar in both cases. Therefore, they can be neglected in making an estimate of the efficiency. Required estimate should be made considering copper losses alone.

---

[4] Specific torque is the ratio T_{em}/m, the quotient of the torque, and mass of the machine. Alternatively, specific torque can be defined as T_{em}/V, the quotient of the torque and volume.

In synchronous machines, there are copper losses in the stator winding only. Compared to synchronous machines, induction machines have the rotor losses as well, and their amount is $sP_\delta = P_{Cu2}$. The stator and rotor currents in induction machines have approximately the same amplitude. At the same time, the cross section of the stator and rotor slots is also similar, as well as the current density. Therefore, it is reasonable to assume that the copper losses in stator are comparable to the copper losses in rotor of induction machines. The rotor losses $s_n P_{\delta n}$ are close to $(3,000–2,850)/3,000 = 5\%$ of the rated power. Hence, a rough estimate of the stator losses of synchronous machine is 5%. Efficiency of an equivalent synchronous machine is close to 95%.

18.11   Characteristics of Permanent Magnets

Magnetizing characteristic of permanent magnet material is shown in Fig. 18.10. The abscissa represents the external field H, that is, the field brought into the considered domain by means of the factors outside the magnet. In most cases, external field is created by electrical currents in coils or windings placed in close vicinity of the magnet. Remanent induction B_R exists with no external field, with $H = 0$, and it may exceed 1 T. Smaller values of remanent induction such as 0.3 T are encountered with ferrite magnets. Majority of ferromagnetic materials such as iron have a small level of remanent inductance, which appears due to tendency of oriented magnetic dipoles to retain their direction after removing the external field (Fig. 18.11). A certain remanent induction exists even in iron sheets which are used in making magnetic circuits of electrical machines. This value has rather small remanent induction of $B_R < 50$ mT, which is readily reduced to zero by an external magnetic field H of opposite direction or by exposing magnetic material to elevated temperatures.

Relation between the magnetic induction within permanent magnet and the external field H is given in Fig. 18.10. By introducing an external magnetic field H of moderate intensity $(H > -H_c/2)$ and direction opposite to remanent induction,

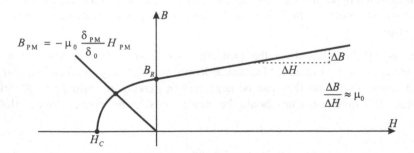

Fig. 18.10 Magnetizing characteristic of permanent magnet

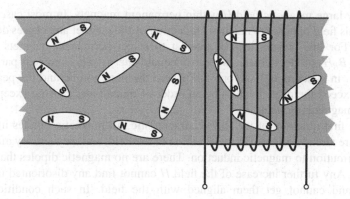

Fig. 18.11 Ferromagnetic material viewed as a set of magnetic dipoles

the operating point moves from $(0, B_R)$ toward the left, which results in some decrease of magnetic induction. Upon removing demagnetizing field H, the operating point returns to the initial position, to the point $(0, B_R)$ in Fig. 18.10.

In cases where the external field H reaches the value of *coercitive field* $-H_c$, magnetic induction drops to zero. This may cause permanent damage to the magnet. Namely, starting from the point $(-H_c, 0)$ and removing the field $-H_c$ does not necessarily bring the operating point back to $(0, B_R)$. Instead, the operating point may return to the point $(0, B_{R1})$ with much lower remanent induction $B_{R1} < B_R$. In such cases, the original remanent induction cannot be restored, and the damage to the magnet remains permanent. The remanent induction can be reduced by a factor of two or three. Damage to the magnet can be evaluated after removing the demagnetizing field $H < 0$.

In majority of permanent magnet materials, characteristic $B(H)$ has a point of inflection in the second quadrant called the knee point, where the external field $H < 0$ makes the magnetic induction reduce at an increased rate. The field strength at the knee point is approximately $H = -H_c/2$. In most cases, permanent damage to the magnet occurs when the operating point passes the knee point and proceeds the left, reaching the zones with $H < -H_c/2$.

Larger values of remanent induction B_R have positive impact on characteristics of synchronous machines with permanent magnets, since the rotor flux is proportional to B_R and the torque is proportional to the rotor flu. It is also beneficial to have a large coercitive field H_c, which delineates the magnet capability to withstand demagnetizing field. Quality of permanent magnets is measured by the product $B_R H_c$, which has dimensions of energy density. Magnets with *higher energies* have larger values of remanent induction and larger values of coercitive fields. They are more robust with respect to external fields, that is, they could operate with larger demagnetizing fields H without suffering any damage. This property helps the operation of permanent magnets built into the rotor magnetic circuit, where they get exposed to demagnetizing fields. The stator currents of synchronous machines produce considerable magnetomotive forces. These magnetomotive forces produce relatively large magnetic field H within permanent magnets. In most cases, direction of this field opposes to magnetic induction of the magnet and causes demagnetization. For this reason, it is of interest to select permanent magnets with an adequate $B_R H_c$ product, that is, with an adequate *energy*. This choice is particularly important in synchronous servomotors where the peak torque and the peak stator current exceed their rated values by an order of magnitude, causing exceptionally large demagnetizing fields.

In the first quadrant of the $B(H)$ characteristic, all magnetic dipoles inside the magnet are already oriented in the direction of the field, and they already make their full contribution to magnetic induction. There are no magnetic dipoles that are not oriented. Any further increase of the field H cannot find any disoriented magnetic dipoles and cannot get them aligned with the field. In such conditions, any further increase ΔB of magnetic induction is equal to $\mu_0 \Delta H$. Therefore, differential permeability $\Delta B/\Delta H$ of permanent magnets in the first quadrant is close to μ_0. In other words, the permanent magnet response ΔB to an external field ΔH is equal to that of the air or vacuum, $\Delta B = \mu_0 \Delta H$. Considering magnetic resistance to external field, permanent magnet behaves as the air. Magnetic resistance R_μ to an external magnetomotive force does not change by inserting permanent magnet into the air-filled space, nor does it change by extracting permanent magnet from the magnetic circuit.

18.12 Magnetic Circuits with Permanent Magnets

It is of interest to analyze magnetic circuits that comprise permanent magnets. As an example, magnetic circuit in Fig. 18.12 is studied in order to obtain the values of the field H and magnetic induction B in different parts of the circuit. The circuit has permanent magnet of the length δ_{PM}, an air gap δ_0, and an iron part of magnetic

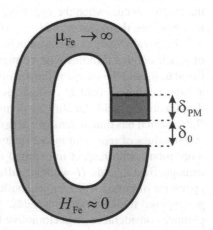

Fig. 18.12 Magnetic circuit comprising a permanent magnet and an air gap

circuit of permeability μ_{Fe} which is very large. Therefore, magnetic field H_{Fe} in iron is neglected.

Considered structure does not have any windings; thus, $Ni = H_{PM}\delta_{PM} + H_0\delta_0 = 0$. All parts of the circuit have the same cross section S. Considering the flux conservation law $\Phi_{PM}=\Phi_0=\Phi_{Fe}$, one obtains that $B_{PM} = B_0 = B_{Fe}$. Since $H_0 = B_0/\mu_0 = B_{PM}/\mu_0$, relation between magnetic induction B_{PM} and field H_{PM} is given by expression $B_{PM} = - \mu_0(\delta_{PM}/\delta_0)\, H_{PM}$. The intersection of nonlinear magnetizing characteristic $B(H)$ of permanent magnet material and the line $B_{PM} = -\mu_0(\delta_{PM}/\delta_0)H_{PM}$ gives the operating point of considered magnetic circuit. Coordinates of this intersection represent the values of magnetic induction and magnetic field within the magnet. In B-H diagram, the intersection is located in the second quadrant, where magnetic induction is positive while magnetic field H is negative.

In cases where magnetic circuit does not contain any air gap, the field $H_{PM} = B_{PM}(\delta_0/\delta_{PM})/\mu_0$ inside the magnet is equal to zero, while the magnetic induction is equal to remanent induction B_R of permanent magnet material. By increasing the air gap, the slope of the straight line $B_{PM} = -\mu_0(\delta_{PM}/\delta_0)H_{PM}$ decreases. The intersection with the curve $B(H)$ assumes smaller value of magnetic induction, while the field H_{PM} assumes negative value. Hence, when the magnet is placed in magnetic circuit with an air gap, magnetic induction B_{PM} and field H_{PM} within the magnet are of opposite direction.

18.13 Surface Mount and Buried Magnets

Permanent magnets can be mounted on the rotor surface or buried within the rotor magnetic circuit. Example of permanent magnets mounted on the rotor surface is given in Fig. 18.13.

The lines of magnetic field pass from the stator teeth into the air gap δ_0; they proceed and enter the permanent magnet of thickness δ_{PM} and then pass to the iron parts of the rotor magnetic circuit. Passing through the rotor core and getting to the opposite magnetic pole of the rotor, the field lines return from the rotor into the stator passing through the iron parts of the rotor, then through the permanent magnet, proceeding through the air gap, and reaching the stator teeth.

Resultant flux of the stator winding has the component Ψ_{mR} that comes from permanent magnets and the component $L_S i_S$ caused by the stator currents, the latter proportional to the stator self-inductance L_S. Stator self-inductance L_S depends on magnetic resistance R_μ encountered along the path of the stator flux. Permeability in iron is very large ($\mu_{Fe} \sim \infty$), making magnetic field H_{Fe} in iron parts of magnetic circuit negligible. Therefore, magnetic resistance on the path of the stator flux reduces to the air gap resistance and the resistance of permanent magnets:

$$R_\mu = \frac{2\delta_0}{\mu_0 S} + \frac{2\delta_{PM}}{\mu_{PM}S}. \tag{18.3}$$

Fig. 18.13 Surface-mounted permanent magnets. (a) Air gap. (b) Permanent magnet

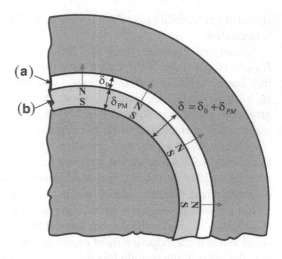

Fig. 18.14 Permanent magnets buried into the rotor magnetic circuit

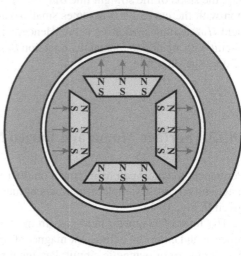

In the first quadrant of B–H plane, differential permeability of permanent magnet materials $\mu_{PM} = \Delta B/\Delta H$ is equal to μ_0. In the presence of the stator magnetomotive force, behavior of permanent magnets of thickness δ_{PM} is the same as behavior of an additional air gap δ_{PM}. Therefore, the equivalent magnetic gap δ is equal to the sum of the two,

$$\delta = \delta_0 + \delta_{PM}, \tag{18.4}$$

where δ_0 represents the actual mechanical air gap, that is, the distance between the stator and rotor, while δ_{PM} represents the thickness of the magnet. For the purposes of L_S calculation, the magnet can be replaced by an additional air gap of the thickness δ_{PM}. Thickness of the magnet is many times larger than the air gap. For this reason, magnetic resistance on the path of the stator flux is significantly larger in synchronous machines with surface-mounted magnets. Hence, the stator self-inductance L_S is very low. Synchronous machines with surface-mounted magnets have the stator inductance of the order 1%. The stator inductance of synchronous machines with interior magnets has the values ranging from 10% up to 70%.

A small inductance L_S reduces the reactance of the machine $X_S = \omega_e L_S$. Synchronous machines are usually supplied from three-phase transistor inverters with pulse width modulation, where reduced inductance of the stator winding results in an increased current ripple ΔI. In inverter supplied machines, pulse width-modulated pulses of the amplitude E repeat across the stator terminals with commutation frequency $f_{PWM} = 1/T \approx 10$–20 kHz, and they create oscillations of the stator current around the fundamental component. These oscillations are called *ripple*. The amplitude of current ripple is roughly $\Delta I \approx ET/(4L_S)$. With low stator inductance L_S of synchronous machines with surface-mounted magnets, the ripple ΔI is increased. Therefore, in order to contain the stator current ripple in synchronous servomotors, it is necessary to increase the commutation frequency of three-phase transistor inverters and to operate with $f_{PWM} > 10$ kHz.

Rather than being mounted on the surface, permanent magnets can be inserted into dedicated openings made within inner iron parts of the rotor magnetic circuit (Fig. 18.14). These openings can be placed further away from the rotor surface, in deeper areas of the rotor core and closer to the shaft. Hence, the magnets are *buried* into the rotor core. With buried magnets, the stator teeth do not face the permanent magnets across the air gap. Instead, the stator flux passes from the stator teeth into the iron parts of the rotor magnetic circuit. The lines of magnetic field caused by the stator currents pass through the air gap and enter the surface parts of the rotor magnetic circuit which are made of iron having a very high permeability. Because of this, the equivalent magnetic gap $\delta = \delta_0 + \delta_{PM}$ is equal to the air gap δ_0, and it is much smaller than the equivalent magnetic gap with surface-mounted magnets. Therefore, magnetic resistance along the stator flux path is considerably smaller, and the stator inductance L_S is considerably higher. With buried magnets, relative value of stator inductance ranges from 10% up to 70%.

18.14 Characteristics of Permanent Magnet Machines

Surface-mounted magnets result in a very small stator inductance L_S. The rate of change of the stator current $di_a/dt = (u_a - e_a)/L_S$ is very high and approaches the value of $di_a/dt \approx 7{,}000$ $[I_n/s]$,[5] allowing a very high rate of change of the electromagnetic torque T_{em}, which depends on the rotor flux and the stator current, $T_{em} = k|\Psi_{Rm} \times i_S|$. With surface-mounted magnets, the torque rise time from zero up to the rated value can be achieved in 100–200 μs. For this reason,

---

[5] Starting from zero, the rated current I_n can be reached in $1/7{,}000$ s.

synchronous motors with surface-mounted magnets are used in motion control applications such as the industry automation and robotics, where the closed loop speed and position control depend on the ability to effectuate very fast torque changes.

One shortcoming of synchronous machines with surface-mounted magnets is their limited ability to operate in the field-weakening region, where the rotor speed exceeds the rated speed, increasing as well the electromotive force $j\omega_e\Psi_S$ induced in the stator windings. Above the rated speed, the stator flux Ψ_S has to be reduced in order to maintain the electromotive force $j\omega_e\Psi_S$ within the rated limits, $|j\omega_e\Psi_S| \leq U_n = \Psi_n\omega_n$. To this aim, the flux $\Psi_S = \Psi_{Rm} + L_S i_S$ should be decreased according to the law $\Psi_S(\omega_e) = \Psi_n(\omega_n/\omega_e)$. The flux Ψ_{Rm} of permanent magnets cannot be altered, and the flux reduction requires a demagnetizing component of the stator current i_S. The amount of demagnetizing stator current depends on the stator inductance. For machines with very small inductance $L_S \sim 1\%$, very high stator currents are required in order to reduce the flux. Therefore, the operation of synchronous machines with surface-mounted magnets in the zone of flux weakening is not feasible. Hence, synchronous machines with surface-mounted magnets are used in motion control applications that require high peak torque capability, fast torque changes, and low inertia.

There are applications of synchronous machines that do not require fast torque changes, but they do require the field-weakening operation. In such cases, synchronous machines are used with permanent magnets built into interior[6] parts of the rotor magnetic circuit. With stator teeth facing the iron parts of the rotor magnetic circuit, magnetic resistance is decreased while the stator self-inductance is increased. With elevated L_S, reduction of the stator flux $\Psi_S = \Psi_{Rm} + L_S i_S$ can be performed with relatively low stator currents. In absence of losses in the rotor, synchronous machines with buried magnets are applied in all cases where the efficiency is of particular importance. Some of these examples are renewable energy sources and autonomous electrical vehicles.

---

[6] Synchronous machines with permanent magnets that are not placed on the rotor surface, and do not face the air gap, but reside instead in dedicated holes and chambers carved within inner regions of the rotor magnetic circuits, located further away from the surface and closer to the shaft are called *buried magnet* or *interior magnet* machines.

Chapter 19
Mathematical Model of Synchronous Machine

This chapter introduces and explains mathematical model of synchronous machines. The model considers three-phase synchronous machines with excitation windings or permanent magnets on the rotor. This model does not include damper windings, which are introduced and explained in Chap. 21. The model represents transient and steady state behavior in electrical and mechanical subsystems of synchronous machines. Analysis and discussion introduce and explain Clarke and Park coordinate transforms. The model includes differential equations that express the voltage balance in stator and rotor windings, inductance matrix which relates flux linkages and currents, Newton differential equation of motion, expression for the air gap power, and expression for the electromagnetic torque. The model development process is very similar to that of the induction machine, which is detailed in Chap. 15. Therefore, some considerations are shortened or removed. The model obtained in this chapter is suitable for both isotropic (cylindrical) and anisotropic (salient pole) machines. This chapter closes with some basic considerations on the reluctant torque and synchronous reluctance machines.

19.1 Modeling Synchronous Machines

Dealing with synchronous generators and synchronous motors requires some basic knowledge on their electrical and mechanical properties. Electrical properties of machines involve steady state and transient relations between the machine voltages, currents, and flux linkages. Mechanical properties have to do with the rotor speed, electromagnetic torque, and motion resistances. For the purpose of designing the power supply and controls, it is required to know the equations and expressions that relate voltages and currents of synchronous machine. Steady state analysis relies on equivalent circuits that involve the machine parameters and phasors representing the relevant currents, voltages, and fluxes. Mechanical properties of synchronous machines are required to design and specify the interface to mechanical loads or driving turbines.

High-power synchronous machines are used as generators in electrical power plants. They are connected to three-phase network of constant frequency of $f = 50$ Hz or $f = 60$ Hz. Mechanical work is obtained via shaft from a steam or water turbine; it is converted into electrical energy and supplied to the network. Most synchronous generators have wound rotors. The excitation winding offers the possibility of changing the excitation flux by varying the excitation current. Changes in excitation flux affect the electromotive forces and stator voltages. This opens the possibility of controlling the voltage of the network and to change the reactive power. Synchronous machines of lower power are used in applications such as motion control, vehicle propulsion, industrial robots, and automated production machines. In these applications, synchronous machines act as motors. Their tasks include overcoming the motion resistances, driving the work machines, and controlling the speed and position of tools and work pieces. There is a tendency of designing and manufacturing medium- and low-power synchronous machines with constant excitation based on permanent magnets.

During transients, processes that take place in synchronous motors and generators, voltages, and currents are related by differential equations describing the voltage balance in the windings. The voltage balance equations describe the *electrical subsystem* of the machine. The *mechanical subsystem* is described by Newton differential equation of motion. The set of differential equations and expressions describing behavior of a synchronous machine is called *mathematical model* or *dynamic model*. In what follows, the mathematical model of synchronous machine is determined, with the aim of establishing the mechanical characteristic and the steady state equivalent circuit:

- Mechanical characteristic of synchronous machine is dependence of the electromagnetic torque on the rotor speed in steady state for given frequency and amplitude of the stator voltage.
- Steady state equivalent circuit is a network of resistances and inductances which serves for calculation of electrical currents and flux linkages of synchronous machine.

In the course of modeling, several approximations are taken, and certain phenomena of secondary importance are neglected. The iron losses are considered as negligible, and all ferromagnetic materials are assumed to be linear and free from saturation phenomena. Moreover, all the parasitic capacitances are neglected as distributed parameter effects. The model of synchronous machine will have:

- N differential equations expressing the voltage balance in windings
- Inductance matrix
- Expression for the torque
- Newton equation of motion

19.2 Magnetomotive Force

Most synchronous machines have the stator winding system with three-phase windings. There are also winding systems with 5-, 7-, 9-, 17-, or even more phase windings, but they are seldom used. The operation of synchronous machines with $N_{ph} > 3$ phases can be represented by an equivalent three-phase machine. In a three-phase synchronous machine with symmetrical power supply, the phase currents are

$$i_a = I_m \cos \omega_e t,$$
$$i_b = I_m \cos(\omega_e t - 2\pi/3),$$
$$i_c = I_m \cos(\omega_e t - 4\pi/3). \tag{19.1}$$

They have the same angular frequency and the same amplitude. Their initial phases are displaced by $2\pi/3$. At the same time, magnetic axes of the phase windings are spatially displaced by $2\pi/(3p)$, where p is the number of pole pairs. Construction of three-phase stator windings is discussed in Chap. 14, while the aspects of machines with multiple-pole pairs are discussed in Chap. 17.

The stator phase currents create the stator magnetomotive force represented by the vector $\mathbf{F}_S = \mathbf{F}_a + \mathbf{F}_b + \mathbf{F}_c$. The phase currents given in (19.1) produce magnetomotive force which rotates at the speed $\Omega_e = \omega_e/p$, maintaining a constant amplitude of $F_S = 3/2\, NI_m$. In order to represent the vector $\mathbf{F}_S$, which resides in $\alpha\beta$ plane in Fig. 19.1, it is split in two components, corresponding to projections of the vector on the axes α- and β- of the orthogonal coordinate system. The $\alpha\beta$ coordinate

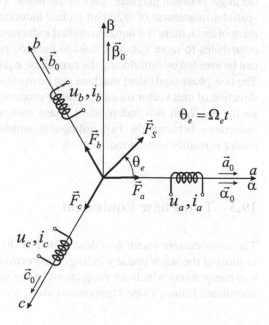

Fig. 19.1 Revolving vector of the stator magnetomotive force

system is positioned in such a way that the axis α coincides with the axis of the phase winding a. Directions of α– and β– axes are defined by unit vectors $\boldsymbol{\alpha}_0$ and $\boldsymbol{\beta}_0$. Direction of axes of the phase windings a, b, and c can be expressed in terms of these unit vectors, $a_0 = \boldsymbol{\alpha}_0$, $b_0 = -\frac{1}{2}\boldsymbol{\alpha}_0 + \boldsymbol{\beta}_0$ sqrt(3)/2, and $c_0 = -\frac{1}{2}\boldsymbol{\alpha}_0 - \boldsymbol{\beta}_0$ sqrt (3)/2. Magnetomotive forces of individual phases are

$$\vec{F}_a = Ni_a\vec{\alpha}_0,$$

$$\vec{F}_b = Ni_b\left(-\frac{1}{2}\vec{\alpha}_0 + \frac{\sqrt{3}}{2}\vec{\beta}_0\right),$$

$$\vec{F}_c = Ni_c\left(-\frac{1}{2}\vec{\alpha}_0 - \frac{\sqrt{3}}{2}\vec{\beta}_0\right), \tag{19.2}$$

and the resultant magnetomotive force is equal to

$$\vec{F}_s = \vec{F}_a + \vec{F}_b + \vec{F}_c = \frac{3}{2}NI_m\left[\vec{\alpha}_0\cos\omega_e t + \vec{\beta}_0\sin\omega_e t\right] \tag{19.3}$$

The phase currents in a three-phase winding are not independent variables, and this makes the modeling more difficult. For star-connected phase windings, the phase currents are related by $i_a + i_b + i_c = 0$. Similar difficulty exist in delta-connected phase windings, where $u_a + u_b + u_c = 0$. With $i_c = -i_a - i_b$, there are only two independent currents in voltage balance equations. Therefore, it is necessary to replace i_c by $(-i_a - i_b)$ in voltage balance equations for phase c. In addition, the angle between magnetic axes of the phase windings is $2\pi/3$. For windings with spatial displacement of $\pi/2$, their mutual inductance is equal to zero. With displacement of $2\pi/3$, there is a nonzero mutual inductances between all the phases, and this contributes to more complex voltage balance equations. The above shortcomings can be avoided by introducing the two-phase equivalent of the three-phase machine. The two-phase equivalent machine can be made with one-phase winding oriented in direction of unit vector $\boldsymbol{\alpha}_0$ and the other-phase winding oriented in direction of unit vector $\boldsymbol{\beta}_0$. With two independent phase currents i_α and i_β and with the mutual inductance between the two orthogonal windings equal to zero, the two-phase model is readily understood.

19.3   Two-Phase Equivalent

The stator current vector i_S is determined by dividing the vector F_S by the number of turns of the stator phase winding. The vector i_S can be represented as the sum of two components which are projections of the vector on the axes of $\alpha\beta$ orthogonal coordinate frame. These components are i_α and i_β, and they can be considered as

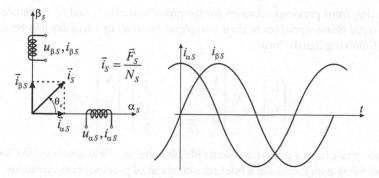

Fig. 19.2 Two-phase representation of the stator winding

the phase currents of the equivalent two-phase winding, with the phase windings aligned with $\alpha-$ and $\beta-$axes. Introduction of the two-phase equivalent makes the mathematical model more concise and intuitive. The number of phase currents is reduced to two, and the mutual inductance between the phase windings is equal to zero. The two-phase winding system with phase currents i_α and i_β produces the magnetomotive force with projections F_α and F_β. These projections are proportional to corresponding currents, namely, $F_\alpha = Ni_\alpha$ and $F_\beta = Ni_\beta$. At the same time, the flux $\Psi_{S\alpha}$ of the phase winding α is projection of the stator flux vector $\boldsymbol{\Psi}_S$ on the axis α. The absence of mutual inductance and the circumstance that the currents and flux linkages in two phases correspond to projections of relevant vectors facilitate understanding and using the two-phase mathematical model.

While introducing the two-phase equivalent, it is of interest to maintain the same flux, the same magnetic field energy, and the same torque as in original, three-phase machine. This is achieved if the amplitude and spatial orientation of the magnetomotive force $\boldsymbol{F}_S$ remain unaltered. With the same magnetomotive force, the same flux and the same energy of magnetic field are obtained. There is no unique way to maintain the same vector $\boldsymbol{F}_S$ with the two-phase equivalent. It can be obtained with a set of windings with lower currents and with more turns per phase or with another set of windings with larger currents and lower number of turns per phase.

Removal of the three-phase winding and its replacement with the two-phase winding can be carried out in such way that the number of turns remains unchanged, $N_{abc} = N_{\alpha\beta}$. Then, vector of the stator magnetomotive force $\boldsymbol{F}_S$ remains the same provided that $i_\alpha(t) = i_a(t) - i_b(t)/2 - i_c(t)/2 = 3/2\, I_m \cos\omega_e t$ and $i_\beta(t) = (\text{sqrt}(3)/2) \cdot (i_b(t) - i_c(t)) = 3/2\, I_m \sin\omega_e t$. It should be noted that in the considered case, the peak and rms values of the phase currents in the two-phase equivalent are 50% larger than corresponding currents in the three-phase winding (Fig. 19.2).

Starting from previous relations for the phase currents i_α and i_β, transformation of the three-phase variables to their two-phase equivalent variables can be written in the following matrix form:

$$
\begin{bmatrix} i_\alpha \\ i_\beta \end{bmatrix} = K_I \begin{bmatrix} 1 & \dfrac{1}{2} & -\dfrac{1}{2} \\ 0 & \dfrac{\sqrt{3}}{2} & -\dfrac{\sqrt{3}}{2} \end{bmatrix} \begin{bmatrix} i_a \\ i_b \\ i_c \end{bmatrix}
\tag{19.4}
$$

Three-phase to two-phase transform ($3\Phi/2\Phi$ transform) was discussed in Chap. 15. The following paragraphs are a brief reinstatement of previous considerations.

19.4 Clarke $3\Phi/2\Phi$ Transform

Three-phase to two-phase transform is known as Clarke transform, named after the author who proposed the first formulation of the transform. The matrix given in (19.4) is called *transformation matrix*. Presence of the adjustable coefficient K_I allows for a certain degree of freedom in defining the two-phase equivalent for the original three-phase machine. The number of turns of the equivalent machine does not have to be the same to that of the original machine, provided that the magnetomotive force remains unaltered. Namely, the condition of invariability of the magnetomotive force F_S can be accomplished by choosing a two-phase equivalent with $N_{\alpha\beta} = mN_{abc}$ turns. Then, the currents in the phase windings $i_\alpha(t)$ and $i_\beta(t)$ should be m times lower in order to maintain the same value of F_S. For that reason, the coefficient K_I in (19.4) has to be $K_I = 1/m$. The ratio of the peak currents of the two-phase equivalent and the three-phase original is equal to $(2/3)/m$.

In order to obtain the model of the two-phase machine, it is also necessary to transform the voltages and fluxes of the original machine to α-β coordinate system. Clarke transform for the voltages and fluxes is given by the following expressions:

$$
\begin{bmatrix} u_\alpha \\ u_\beta \end{bmatrix} = K_U \begin{bmatrix} 1 & \dfrac{1}{2} & -\dfrac{1}{2} \\ 0 & \dfrac{\sqrt{3}}{2} & -\dfrac{\sqrt{3}}{2} \end{bmatrix} \begin{bmatrix} u_a \\ u_b \\ u_c \end{bmatrix}
\tag{19.5}
$$

$$
\begin{bmatrix} \Psi_\alpha \\ \Psi_\beta \end{bmatrix} = K_\Psi \begin{bmatrix} 1 & \dfrac{1}{2} & -\dfrac{1}{2} \\ 0 & \dfrac{\sqrt{3}}{2} & -\dfrac{\sqrt{3}}{2} \end{bmatrix} \begin{bmatrix} \Psi_a \\ \Psi_b \\ \Psi_c \end{bmatrix}
\tag{19.6}
$$

There is possibility to assign different values to coefficients K_U, K_I, and K_Ψ. For practical reasons, $3\Phi/2\Phi$ transform is adopted where all the three coefficients are equal, $K_U = K_I = K_\Psi = K$. In this way, the two-phase equivalent maintains the same ratio of currents, voltages, and fluxes as the original machine, and the 3Φ/2Φ transform is invariant in terms of impedances and inductances.[1]

Question (19.1): Is it possible to actually make a two-phase machine with $N_{\alpha\beta} = mN_{abc}$ turns which produces the same magnetomotive force F_S of the stator winding as the original three-phase synchronous machine and which has the voltages and currents $(u_\alpha, u_\beta, i_\alpha, i_\beta)$ that are equal to those obtained by applying the Clarke transform to the voltages and currents of the original machine? It is assumed that transformation matrices used for the flux, voltage, and current are equal, namely, that they have the same leading coefficient ($K_U = K_I = K_\Psi$).

Answer (19.1): The same magnetomotive force F_S results in the same flux $\Phi_S = F_S/R_\mu$. At the same time, the angular speed of the revolving magnetic field has to be the same in both 2-phase and 3-phase machines. For that reason, both machines have the same electromotive force induced in one turn. With that in mind, the phase voltages of the two-phase equivalent are equal to $u_{\alpha\beta} = m \, u_{abc}$. Maintaining the amplitude of F_S requires that the phase currents of the two-phase equivalent are $i_{\alpha\beta} = (3/2)\cdot(i_{abc}/m)$. Finally, maintaining the ratio of voltages and currents requires that the ratio u_{abc}/i_{abc} is equal to the ratio $u_{\alpha\beta}/i_{\alpha\beta}$. Summarizing the above considerations,

$$\frac{u_{\alpha\beta}}{i_{\alpha\beta}} = \frac{m \, u_{abc}}{(3/2)(i_{abc}/m)} = \frac{2m^2}{3}\frac{u_{abc}}{i_{abc}} = \frac{u_{abc}}{i_{abc}}$$

$$\Rightarrow \quad m = \sqrt{\frac{3}{2}} \quad \Rightarrow \quad K_U = K_I = K_\Psi = \sqrt{\frac{2}{3}}.$$

* * *

Solution to Question 19.1 proves the possibility to actually make the two-phase equivalent machine with the same flux, torque, and magnetomotive force and with the same impedances and inductances as the original machine. The voltages and currents of the actual two-phase machine are obtained by applying the Clarke transform to the original variables. Considered two-phase equivalent must have sqrt(3/2) times more turns than the original machine. Voltages, currents, and fluxes of the two-phase equivalent are obtained by Clarke transform of the original variables, that is, voltages, currents, and fluxes of the three-phase winding. The 3Φ/2Φ transform to be applied must have the coefficients $K_U = K_I = K_\Psi =$ sqrt(3/2). The actual two-phase machine has the same torque and power as the original machine. Therefore, besides being impedance-invariant, considered Clarke transform is also power-invariant.[2]

---

[1] Invariability of impedances and inductances is discussed in Chap. 15. Impedance-invariant transform results in an equivalent machine where all the impedances are the same as relevant impedances of the original machine.

[2] Coordinate transforms do not have to be power-invariant. Generally, coordinate transforms such as Clarke transform provide another mathematical formulation of the considered dynamic system. The new model does not have to correspond to any real system. An example to that is the model of a resistor, $u_1 = Ri_1$, obtained by applying the "transform" $u_1 = 2u$, $i_1 = 2i$ to the original resistor, described by $u = Ri$. The model $u_1 = Ri_1$ is impedance-invariant, but it is not power-invariant.

If the three-phase winding is removed from the magnetic circuit of the original machine and replaced by the two-phase winding with sqrt(3/2) times more turns, one obtains a two-phase synchronous machine which replaces the three-phase machine. The voltages and currents obtained by using $3\Phi/2\Phi$ transform with $K = $ sqrt(2/3) are equal to voltages and currents that can be measured on the actual two-phase machine windings. Hence, considered $3\Phi/2\Phi$ transform with $K = $ sqrt (2/3) results in the two-phase equivalent machine that can be actually made. Notice at this point that the possibility of actually making the machine that comes out of the $3\Phi/2\Phi$ transform is not of particular importance.

Clarke transform with $K = $ sqrt(2/3) is invariant in terms of the magnetomotive force, power, impedance, and inductance. In addition, it can be practically applied by making a machine with voltages and currents equal to those obtained by the transform. In spite of that, transform with coefficient $K = $ sqrt(2/3) is seldom used. The presence of irrational number such as sqrt(2/3) in calculations is the reason for this particular transform to be rarely used.

In selecting the coordinate transform, there is freedom to choose the coefficient K so as to facilitate the use of the model. In practice, transforms with $K_U = K_I = K_\Psi = 2/3$ and with $K_U = K_I = K_\Psi = 1$ are often used. With $3\Phi/2\Phi$ transform of currents which uses $K_I = 1$, one obtains the currents i_α and i_β with peak values and rms values 50% larger than those of the original variables. Transform which results in currents i_α and i_β with peak and rms values equal to those of the phase current i_a, i_b, and i_c, it is required to apply the coefficient $K = 2/3$. This choice has two shortcomings.

Clarke transform that uses $K = 2/3$ is not power-invariant. Namely, the power calculated in $\alpha\beta$ domain is not the same as the power of the original machine. Notice at this point that the rms values of voltages and currents are the same for $\alpha\beta$ and abc variables. The original machine has three-phase windings, while the equivalent machine in $\alpha\beta$ domain has only two-phase windings. Therefore, calculated power is not the same since $P_{\alpha\beta} = (2/3)P_{abc}$. As a consequence, whenever using the $3\Phi/2\Phi$ transform with $K = 2/3$, the actual power of the original machine has to be calculated on the basis of the expression $P_{abc} = (3/2)\, P_{\alpha\beta}$.

It is not possible to make an actual two-phase machine that would have the same voltages and currents as those calculated by the transform. In other words, it is not possible to rewind the existing three-phase machine and make an equivalent two-phase machine with voltages and currents that correspond to the values obtained by the considered $3\Phi/2\Phi$ transform. On the other hand, the choice $K = 2/3$ results in transformed voltages and currents in $\alpha\beta$ domain that have the same amplitude and the same rms value as those of voltages and currents of the original three-phase machine. With $K = 2/3$, the virtual two-phase equivalent must have a larger number of turns. In order to maintain the amplitude of the vector F_S, it is necessary to have $N_{\alpha\beta} = (3/2)N_{abc}$. With the same flux per turn $\Phi = F_S/R_\mu$, the turn electromotive forces are the same in both the original and the equivalent machine. Hence, in an attempt to make an actual two-phase equivalent, the two-phase machine has to be made with the same electromotive force per turn and with 50% more turns than the original machine. With $N_{\alpha\beta} = (3/2)N_{abc}$, the phase voltages of such a machine are $u_{\alpha\beta} = (3/2)\,u_{abc}$, 50% larger than the original phase voltage. At the same time, $3\Phi/2\Phi$ transform with $K = 2/3$ results in $u_{\alpha\beta} = u_{abc}$. This shows that the transform that uses $K = 2/3$ results in a two-phase machine that does not have a real, actual counterpart.

Favorable properties of the $3\Phi/2\Phi$ transform with coefficient $K = 2/3$ are invariability in terms of impedance and inductance, as well as the circumstance that the peak and rms values of the original machine are equal to those of the two-phase equivalent. The absence of invariability in terms of power is corrected by applying the formula $P_{abc} = (3/2)P_{\alpha\beta}$.

In further analysis, it will be assumed that the three-phase stator winding is replaced by the two-phase equivalent by applying Clarke $3\Phi/2\Phi$ transform which uses the coefficient $K = 2/3$. Unless otherwise stated, the analysis starts with an assumption that considered synchronous machine has one pair of magnetic poles ($p = 1$) and that the electrical frequency ω corresponds to mechanical speed Ω.

19.5   Inductance Matrix and Voltage Balance Equations

A synchronous machine with two-phase stator winding and wound rotor is shown in Fig. 19.3. The voltage balance equations of the considered windings are

$$u_{\alpha s} = R_s i_{\alpha s} + \frac{\mathrm{d}}{\mathrm{d}t}\,\Psi_{\alpha s}, \quad u_{\beta s} = R_s i_{\beta s} + \frac{\mathrm{d}}{\mathrm{d}t}\,\Psi_{\beta s}, \quad u_R = R_R i_R + \frac{\mathrm{d}}{\mathrm{d}t}\,\Psi_R. \qquad (19.7)$$

Variables $u_{\alpha s}$ and $u_{\beta s}$ represent the phase voltages supplied to the two-phase stator winding, while u_p represents the voltage supplied to the excitation winding. The fluxes $\Psi_{\alpha s}$, $\Psi_{\beta s}$, and Ψ_R represent total flux linkages of the windings α_s and

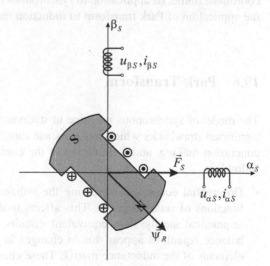

Fig. 19.3 Synchronous machine with the two-phase stator winding and the excitation winding on the rotor

β_s and of the excitation winding. Relation between currents $i_{\alpha s}$, $i_{\beta s}$, and i_R and flux linkages is given by the inductance matrix. Since the rotor is in motion, the angular displacement of the excitation winding magnetic axis with respect to α_s axis of the stator is $\theta_m = \Omega_m t$, assuming that the rotor speed Ω_m does not change. With variable θ_m, mutual inductances of the inductance matrix are variable, and the inductance matrix is a nonstationary matrix:

$$
\begin{bmatrix} \Psi_{\alpha s} \\ \Psi_{\beta s} \\ \Psi_R \end{bmatrix} = \begin{bmatrix} L_s & 0 & L_m \cos \theta_m \\ 0 & L_s & L_m \sin \theta_m \\ L_m \cos \theta_m & L_m \sin \theta_m & L_R \end{bmatrix} \cdot \begin{bmatrix} i_{\alpha s} \\ i_{\beta s} \\ i_R \end{bmatrix} \tag{19.8}
$$

The flux in phase α of the stator winding is equal to $\Psi_{\alpha s} = L_s i_{\alpha s} + L_m \cos(\theta_m) i_R$. Variable mutual inductances result in the voltage balance equations that have trigonometric functions such as $\cos(\theta_m)$, where the state variable θ_m appears as the function argument. This causes difficulties in deriving a legible steady state equivalent scheme:

$$
u_{\alpha s} = R_s i_{\alpha s} + \frac{\mathrm{d}}{\mathrm{d}t} (L_s i_{\alpha s} + L_m \cos(\theta_m) i_R)
$$
$$
= R_s i_{\alpha s} + L_s \frac{\mathrm{d}}{\mathrm{d}t} i_{\alpha s} + L_m \cos(\theta_m) \frac{\mathrm{d}}{\mathrm{d}t} i_R - \omega_m L_m i_R \sin(\theta_m).
$$

In addition, considered model has the state variables such as the phase currents $i_{\alpha s}$ and $i_{\beta s}$ which have sinusoidal change even in the steady state. Therefore, their time derivatives assume nonzero values in steady state conditions, and this creates difficulties in the steady state analysis and hinders formulation of controls of the machine. The above mentioned shortcomings are removed by applying Park coordinate transform. This transform represents the machine vectors in revolving dq coordinate frame. Its application to synchronous machines is very much the same as the application of Park transform to induction machines, discussed in Chap. 15.

19.6   Park Transform

The model of synchronous machine in stationary α_S–β_S coordinate frame has two significant drawbacks which make its use more difficult and hinder the analysis, conclusion making, and formulation of the control law. These drawbacks are the following:

- Differential equations expressing the voltage balance comprise trigonometric functions of state variables. This affects usability of the model and results in impractical steady state equivalent circuits. Variable coefficients in voltage balance equations appear due to changes in mutual inductances, which make elements of the inductance matrix. These changes take place due to variation in relative position between the rotor, which carries the excitation winding, and the stator, which carries the stator phase windings.

- First-time derivatives of state variables have nonzero values even in steady state conditions. State variables of the machine model which is formulated in stationary α_S–β_S coordinate frame are projections of relevant vectors on α_S- and β_S-axes. At steady state, the rotor speed, the flux amplitude, the electromagnetic torque, and the conversion power are all constant. Yet, the stator phase currents, flux linkages of individual phases, their magnetomotive forces, and electromotive forces are all AC quantities. They exhibit sinusoidal change even in steady state, and their first-time derivatives are not equal to zero. This circumstance makes the analysis of steady state operation more difficult.

The model deficiencies mentioned above are removed by applying transformation of all the state variables to rotating coordinate system with orthogonal d- and q-axes. This dq coordinate frame revolves at the rotor speed, and it is called *synchronous coordinate frame*. In most cases, d-axis is made collinear with magnetic axis of the excitation winding. In permanent magnet machines, d-axis is made collinear with the rotor flux vector which is produced by permanent magnets. Transformation of currents, voltages, flux linkages, and magnetomotive forces into dq coordinate frame is called *Park transform*, and it has been explained in detail in Chap. 15.

Park transform can be conceived as a replacement of the existing stator phases α_S and β_S by imaginary, *virtual windings* residing in synchronously rotating dq coordinate frame, positioned in such way that the d-axis coincides with magnetic axis of the excitation winding. This new coordinate system rotates at the same speed as the rotor, therefore the name *synchronous coordinate frame*. The voltages, currents, and flux linkages of the excitation winding do not need to be transformed, as they reside already within the target dq coordinate frame. Transformation procedure and notation used hereafter are the same as the ones used in Chap. 15. The subsequent paragraphs are reduced to a brief reinstatement of the Park transform, already detailed in Chap. 15.

With $\theta_{dq} = \theta_m$, the new dq coordinate system revolves at the same speed as the rotor flux Ψ_R. In steady state conditions, relative position between vectors Ψ_R and F_S does not change. Hence, dq frame revolves in synchronism with the vector of the stator magnetomotive force F_S as well. Therefore, in steady state conditions, projections of both vectors on d- and q-axes are constant. The same conclusion applies to voltage, current, and stator flux vectors. Hence, Park transform provides a set of variables in synchronously rotating dq frame that all have constant values in steady state conditions.

Advantage of placing the d-axis along magnetic axis of the excitation winding is the circumstance that all of the excitation flux extends along the d-axis. Namely, q component of the excitation flux Ψ_R is equal to zero. For the setup in Fig. 19.4, which includes virtual windings d and q and the excitation winding, mutual inductance between the virtual phase q and the excitation winding is equal to zero, which simplifies further considerations. In synchronous machines with

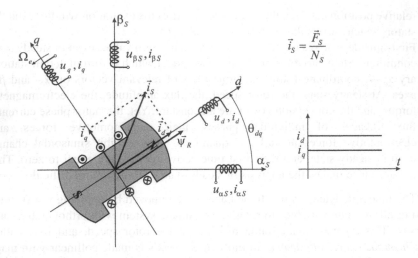

Fig. 19.4 Transformation of stator variables to a synchronously rotating coordinate system. The angle θ_{dq} is equal to the rotor angle θ_m

permanent magnets, the d-axis is positioned so as to coincide with direction of the rotor flux Ψ_R, the flux which is caused by permanent magnets.

By adopting the synchronously rotating coordinate system with d-axis aligned with the rotor magnetic axis, one obtains the system of windings shown in Fig. 19.4, with constant mutual inductances between virtual stator phases and the excitation winding and with steady state variables that assume constant values in steady state conditions. As an example, projections i_d and i_q of the stator current vector $i_S = F_S/N_S$ on axes of the dq frame are constant in the steady state, when the vector i_S maintains both the amplitude and relative position with respect to dq-axes. The same can be proved for all relevant state variables. By transforming stator variables from stationary $\alpha_S - \beta_S$ frame to synchronously rotating dq frame, one obtains the model with all the relevant state variables constant in steady state conditions. This facilitates analysis of steady state operating regimes.

The virtual stator windings d and q cannot be actually made. Instead, they serve as graphical means of grasping the effects of Park transform. One can make a thought experiment of *removing* the $\alpha_S - \beta_S$ stator windings and *mounting* new, virtual windings, with their magnetic axes collinear with the axes of dq coordinate frame (Fig. 19.4).

19.7 Inductance Matrix in *dq* Frame

Virtual dq windings do not change relative positions with respect to the excitation winding. Virtual stator phase winding of d-axis coincides with magnetic axis of the excitation winding. The mutual inductance $L_{d\text{-}R}$ between the two windings is equal to L_m. At the same time, mutual inductance between virtual stator phase winding of q-axis and the excitation winding is equal to zero. In addition, mutual inductance between mutually orthogonal windings d and q is also zero. Hence, the inductance matrix with $3 \times 3 = 9$ elements has only five nonzero elements. Their values L_S, L_R, and L_m are constant:

$$
\begin{bmatrix} \Psi_d \\ \Psi_q \\ \Psi_R \end{bmatrix} = \begin{bmatrix} L_s & 0 & L_m \\ 0 & L_s & 0 \\ L_m & 0 & L_R \end{bmatrix} \cdot \begin{bmatrix} i_d \\ i_q \\ i_R \end{bmatrix}. \tag{19.9}
$$

Currents i_d and i_q in these new, virtual windings must produce the same vector of the stator magnetomotive force F_S that was created previously with phase windings α_S and β_S. To meet this condition, currents i_d and i_q in virtual windings must assume the values $i_d = i_{\alpha s}\cos(\theta_m) + i_{\beta s}\sin(\theta_m)$ and $i_q = -i_{\alpha s}\sin(\theta_m) + i_{\beta s}\cos(\theta_m)$. Matrix form of these relations is given in (19.11). All the remaining variables,[3] such as the voltages and flux linkages, have to be transformed from $\alpha_S - \beta_S$ frame into synchronously rotating dq frame. Applying Park transform to all the relevant variables, one obtains voltages, currents, flux linkages, and magnetomotive forces in synchronously rotating dq frame. Each variable such as i_d and i_q is equal to projection of relevant vector on the axes of the dq coordinate frame. At this point, it is necessary to derive the voltage balance equations in dq frame.

The angle $\theta_{dq} = \theta_m$ between the d-axis and the α_S-axis is

$$
\theta_{dq} = \theta_{dq}(0) + \int_0^t \omega_m d\tau = \theta_{dq}(0) + \int_0^t p\Omega_m d\tau \tag{19.10}
$$

where p is the number of magnetic pole pairs. Park transform is given in expression

$$
\begin{bmatrix} i_d \\ i_q \end{bmatrix} = [T] \cdot \begin{bmatrix} i_{\alpha s} \\ i_{\beta s} \end{bmatrix} = \begin{bmatrix} \cos\theta_m & \sin\theta_m \\ -\sin\theta_m & \cos\theta_m \end{bmatrix} \cdot \begin{bmatrix} i_{\alpha s} \\ i_{\beta s} \end{bmatrix} \tag{19.11}
$$

For a two-pole machine, electrical representation of the rotor speed $\omega_m = p\Omega_m$ is equal to the mechanical speed Ω_m of the rotor. With multiple-pole pair machines, where $p > 1$, spatial displacement between the north and south magnetic poles becomes π/p, while the actual distance between orthogonal axes becomes $\pi/(2p)$. Therefore, Figs. 19.4 and 19.5 do not represent anymore an adequate spatial

---

[3] Winding currents can be treated as the state variables. In this case, flux linkages cannot be the state variables, as they are calculated from currents in (19.9). On the other hand, one can promote the flux linkages into state variables. In the latter case, winding currents are calculated from flux linkages and therefore do not represent the state variables. The voltages across the windings are external driving forces and do not represent the state variables. In mechanical subsystem, the state variables are the speed and position of the rotor.

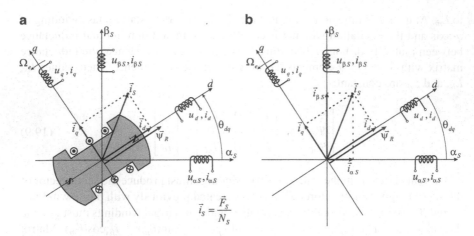

Fig. 19.5 Transformation of stator variables to a synchronously rotating coordinate system. The angle θ_{dq} is equal to the rotor angle θ_m

disposition of relevant axes and windings. In both figures, an angular displacement of 2π corresponds to the actual spatial displacement of $2\pi/p$ in synchronous machines with $2p$ magnetic poles. Equations 19.10 and 19.11 can be used to determine the angle θ_{dq} and to perform Park rotational transform in both two-pole and multipole synchronous machines.

19.8   Vectors as Complex Numbers

Vectors of the current, voltage, and flux can be represented by using complex notation, whereby projections of each vector on orthogonal axes $\alpha\beta$ or dq are represented by real and imaginary parts of a complex number, such as $\underline{i}_{dq} = i_d + ji_q$, $\underline{i}_{\alpha\beta} = i_{\alpha s} + ji_{\beta s}$. Taking into account that $e^{j\theta} = \cos(\theta) + j\sin(\theta)$, Park transform of stator currents from $\alpha\beta$ to dq coordinate frame can be represented by expression $\underline{i}_{dq} = e^{-j\theta}\,\underline{i}_{\alpha\beta}$. Using complex numbers to represent vectors simplifies the process of deriving the voltage balance equations in dq coordinate frame. By introducing complex notation in representing the stator voltage $\underline{u}_{\alpha\beta s} = u_{\alpha s} + ju_{\beta s}$, the voltage balance equations for the stator winding can be represented by a single equations (19.12). This equation employs complex numbers $\underline{u}_{\alpha\beta s}$, $\underline{i}_{\alpha\beta s}$, and $\underline{\Psi}_{\alpha\beta s}$:

$$u_{\alpha s} = R_s i_{\alpha s} + \frac{\mathrm{d}}{\mathrm{d}t}\Psi_{\alpha s}, \quad u_{\beta s} = R_s i_{\beta s} + \frac{\mathrm{d}}{\mathrm{d}t}\Psi_{\beta s},$$

$$\Rightarrow \quad \underline{u}_{\alpha\beta s} = R_s \underline{i}_{\alpha\beta s} + \frac{\mathrm{d}}{\mathrm{d}t}\underline{\Psi}_{\alpha\beta s}. \tag{19.12}$$

Voltages, flux linkages, and currents in synchronous, dq coordinate frame can be represented by complex numbers. Real part of a complex number corresponds to projection of relevant vector on d-axis, while imaginary part corresponds to projection of relevant vector on q-axis. By doing so, dq plane is interpreted as a complex plane with real d-axis and imaginary q-axis:

$$\underline{i}_{dq} = i_d + j i_q$$

It is also possible to define another complex plane, with real α-axis and imaginary β-axis. The voltages, flux linkages, and currents in stationary $\alpha\beta$ coordinate frame can be represented by complex numbers defined in $\alpha\beta$ plane. Real and imaginary parts of a complex number correspond to projection of relevant vector on $\alpha-$ and $\beta-$axes:

$$\underline{i}_{\alpha\beta} = i_{\alpha s} + j i_{\beta s}$$

Park transform of stator currents from $\alpha\beta$ frame into dq frame is written as

$$\underline{i}_{dq} = i_d + j i_q = e^{-j\theta_m}(i_{\alpha s} + j i_{\beta s}) \qquad (19.13)$$

Relations between complex representations of voltages and flux linkages in dq and $\alpha\beta$ coordinate frames are

$$\underline{u}_{dq} = e^{-j\theta_m} \cdot \underline{u}_{\alpha\beta s},$$
$$\underline{\Psi}_{dq} = e^{-j\theta_m} \cdot \underline{\Psi}_{\alpha\beta s}. \qquad (19.14)$$

19.9 Voltage Balance Equations

Voltages u_d and u_q across the virtual stator phases d and q in synchronously rotating coordinate frame can be obtained by applying Park transform to $\alpha\beta$ voltages:

$$\underline{u}_{dq} = u_d + j u_q = \underline{u}_{\alpha\beta s} e^{-j\theta_m} = \left(R_s \underline{i}_{\alpha\beta s} + d\underline{\Psi}_{\alpha\beta s}/dt\right) e^{-j\theta_m}. \qquad (19.15)$$

Variables $\underline{i}_{\alpha\beta s}$ и $\underline{\Psi}_{\alpha\beta s}$ of the stationary coordinate frame can be expressed in terms of dq variables by applying the inverse Park transform, $\underline{i}_{\alpha\beta s} = \underline{i}_{dq} \exp(-j\theta_m)$:

$$\underline{u}_{dq} = \left(R_s \underline{i}_{dq} e^{+j\theta_m} + d(\underline{\Psi}_{dq} e^{+j\theta_m})/dt\right) e^{-j\theta_m}$$
$$= R_s \underline{i}_{dq} + d\underline{\Psi}_{dq}/dt + j\omega_m \underline{\Psi}_{dq} \qquad (19.16)$$

where $\omega_m = p\Omega_m$. Therefore, the voltage balance equations in dq frame do not have the usual form of $u = Ri + \mathrm{d}\Psi/\mathrm{d}t$. They comprise an additional factor which appears as a consequence of applying rotational transform. Angle θ_m denotes the position of magnetic axis of the rotor with respect to the stator phase a. The angular frequency $p\Omega_m = \omega_m$ represents the rotor speed. In cases where synchronous machine has more than one pair of magnetic poles ($p > 1$), it is necessary to take into account the circumstance that $p\Omega_m = \omega_m$. The angle θ_m defines the transformation matrix of Park rotational transform. With $p > 1$, this angle is p times larger than the mechanical displacement of the rotor. Starting from the voltage balance equations in stationary $\alpha\beta$ coordinate frame

$$\underline{u}_{\alpha\beta s} = R_s \underline{i}_{\alpha\beta s} + \frac{\mathrm{d}}{\mathrm{d}t} \underline{\Psi}_{\alpha\beta s},$$

one obtains

$$\underline{u}_{dq} = e^{-\mathrm{j}\theta_m} \left(R_s \underline{i}_{\alpha\beta s} + \frac{\mathrm{d}}{\mathrm{d}t} \underline{\Psi}_{\alpha\beta s} \right) = e^{-\mathrm{j}\theta_m} \left(R_s e^{\mathrm{j}\theta_m} \underline{i}_{dq} + \frac{\mathrm{d}}{\mathrm{d}t} \left(e^{\mathrm{j}\theta_m} \underline{\Psi}_{dq} \right) \right)$$

Equation which expresses the voltage balance of the stator windings in synchronously rotating dq coordinate system is

$$\underline{u}_{dq} = R_s \underline{i}_{dq} + \frac{\mathrm{d}}{\mathrm{d}t} \underline{\Psi}_{dq} + \mathrm{j}\omega_m \underline{\Psi}_{dq} \qquad (19.17)$$

Equation 19.17 can be split into real and imaginary parts. Each of them represents a scalar equation

$$\mathrm{Re}\{\underline{u}_{dq}\} \rightarrow u_d = R_s i_d + \frac{\mathrm{d}}{\mathrm{d}t} \Psi_d - \omega_m \Psi_q,$$

$$\mathrm{Im}\{\underline{u}_{dq}\} \rightarrow u_q = R_s i_q + \frac{\mathrm{d}}{\mathrm{d}t} \Psi_q + \omega_m \Psi_d.$$

Therefore, the voltage balance equation with complex variables can be split in two scalar equations:

$$u_d = R_s i_d + \frac{\mathrm{d}}{\mathrm{d}t} \Psi_d - \omega_m \Psi_q,$$

$$u_q = R_s i_q + \frac{\mathrm{d}}{\mathrm{d}t} \Psi_q + \omega_m \Psi_d. \qquad (19.18)$$

The voltage balance equations of the stator windings get affected by Park transform, and they obtain additional factors such as $\omega_m \Psi_d$. At the same time, the voltage balance equation in excitation winding remains unchanged. This equation does not get affected by Park transform. Actually, the excitation current, voltage, and flux linkage do not get transformed by Park transform. Namely, the excitation winding is fastened to the rotor, and its magnetic axis coincides with d-axis of synchronously rotating coordinate frame. Thus, transient phenomena of the excitation winding get modeled by the following equation:

$$u_R = R_R i_R + \frac{\mathrm{d}}{\mathrm{d}t}\Psi_R \tag{19.19}$$

19.10   Electrical Subsystem of Isotropic Machines

Mathematical model of synchronous machine describes *electrical* and *mechanical* subsystem. The former is described by the voltage balance equations and the latter by Newton equation of motion. With two equivalent phase windings on the stator and one excitation winding on the rotor, synchronous machine has three windings and, hence, three differential equations expressing the voltage balance in these windings. Besides, the model includes the inductance matrix, which provides relations between flux linkages and currents, as well as the expression for the electromagnetic torque.

Rotor magnetic circuit of a synchronous machine may be of cylindrical form, and in this case, magnetic resistance R_μ is the same in all directions. Cylindrical rotor with which maintains the same magnetic resistance in all directions is called *isotropic*. Since the self-inductance of each winding depends on the ratio N^2/R_μ, a constant magnetic resistance results in constant self-inductances. The stator phase windings have the same number of turns, and therefore, $L_d = L_q = L_S$. In cases where magnetic resistances in d- and q-axes are different, the rotor is called *anisotropic*. In this case, self-inductances are not the same, $L_d \neq L_q$.

Voltage balance equations in (19.8) are written for the windings shown in Fig. 19.6, and they are applicable to both isotropic and anisotropic synchronous machines. The difference between the former and the latter appears in the inductance matrix.

Voltage balance equations in virtual stator phase windings in dq coordinate frame and in the excitation winding are given by

$$u_d = R_s i_d + \frac{\mathrm{d}\Psi_d}{\mathrm{d}t} - \omega_m \Psi_q \,,$$

$$u_q = R_s i_q + \frac{\mathrm{d}\Psi_q}{\mathrm{d}t} + \omega_m \Psi_d \,,$$

$$u_R = R_R i_R + \frac{\mathrm{d}\Psi_R}{\mathrm{d}t} \,, \tag{19.20}$$

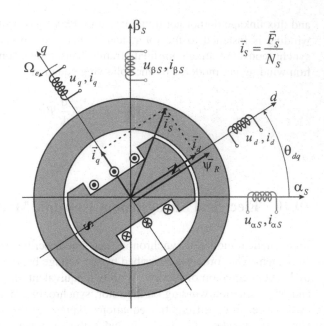

Fig. 19.6 Model of a synchronous machine in the *dq* coordinate system

while the relations between fluxes and currents are determined by the inductance matrix, given in (19.9). This matrix equation can be split into three scalar expressions:

$$\Psi_d = L_s^* i_d + L_m i_R, \tag{19.21}$$

$$\Psi_q = L_s^* i_q, \tag{19.22}$$

$$\Psi_R = L_R i_R + L_m i_d \tag{19.23}$$

In isotropic machines, $L_d = L_q = L_s^*$. With anisotropic machines where $L_d \neq L_q$, parameter L_s^* in (19.21) is replaced by L_d and parameter L_s^* in (19.22) is replaced by L_q. Flux of the excitation winding is equal to $\Psi_R = L_R i_R + L_m i_d$. Coefficient L_R represents self-inductance of the excitation winding, while L_m is mutual inductance of the excitation winding and virtual stator phase winding in *d*-axis. Mutual inductance L_m can be determined from measurement of the largest mutual inductance between the excitation winding and one of the stator phase windings, *a*, *b*, and *c*. The mutual inductance assumes the largest value when the rotor position and the position of magnetic axis of the excitation winding coincide with magnetic axis of the stator phase winding. For synchronous machines with isotropic cylindrical rotor, where the magnetic circuit has the same magnetic resistance in all directions, the flux linkage of the virtual stator phase winding in *d*-axis is equal to $\Psi_d = L_s i_d + L_m i_R$, while the flux of the virtual stator phase winding in *q*-axis is equal to $\Psi_q = L_s i_q$.

19.11 Torque in Isotropic Machines

Electromagnetic torque of an isotropic synchronous machine can be derived by analyzing the balance of power. It is necessary to consider the electrical power, delivered to synchronous machine from a three-phase voltage source, mechanical power on the rotor shaft, and the power of losses in iron, copper, and mechanical subsystem. Starting from expression for electrical power delivered by the source, $p_e = (3/2) (u_d i_d + u_q i_q)$, and by using the voltage balance equations for the stator windings, one obtains the following relation:

$$
\begin{aligned}
p_e &= \frac{3}{2}\left(u_d i_d + u_q i_q\right) \\
&= \frac{3}{2} R_s \left(i_d^2 + i_q^2\right) + \frac{3}{2}\left(i_d \frac{\mathrm{d}\Psi_d}{\mathrm{d}t} + i_q \frac{\mathrm{d}\Psi_q}{\mathrm{d}t}\right) + \frac{3}{2}\omega_m\left(\Psi_d i_q - \Psi_q i_d\right) \\
&= p_{Cu1} + \frac{\mathrm{d}W_m}{\mathrm{d}t} + p_{em}.
\end{aligned}
\tag{19.24}
$$

In the above expression, the first factor on the right represents the power of losses in the stator windings, also called copper losses. There are also iron losses in stator magnetic circuit, where magnetic inductance B changes with an angular frequency of ω_m. One of the four approximations taken in modeling electrical machines is that the iron losses are rather small and therefore negligible (see Chap. 6, Sect. 6.2). Therefore, considered balance of power does not include the iron losses. The second factor in the above equation is $\mathrm{d}W_m/\mathrm{d}t$, and it defines the rate of change of the energy accumulated in magnetic coupling field. If the machine operates with a constant flux, the energy of the magnetic field is constant as well, and the factor $\mathrm{d}W_m/\mathrm{d}t$ is equal to zero. The third and the last factor of the above equation is $p_{em} = (3/2)\omega_m(\Psi_d i_q - \Psi_q i_d)$. It is obtained by subtracting the losses from the input electrical power. Therefore, it represents the rate of change of the electrical energy into mechanical work. It is transferred through the air gap to the rotor by means of electromagnetic interaction of the stator and rotor fields in the air gap. The power p_{em} is called *power of electromechanical conversion*.

Synchronous machines have the stator magnetic field revolving in synchronism with the rotor. In steady state, the angular frequency of the source ω_e is equal to the angular frequency of rotation $p\Omega_m = \omega_m$. The question arises whether $\omega_e = \omega_m$ during transients. Namely, during transient processes, where the electromagnetic torque T_{em} changes, the angle between the stator field $\boldsymbol{F}_S$ and the rotor flux $\boldsymbol{\Psi}_R$ may change as well. The vector $\boldsymbol{F}_S$ revolves at the speed ω_e/p, while the rotor flux $\boldsymbol{\Psi}_R$ revolves at the rotor speed Ω_m. When the angle between the two vectors changes, there is a temporary difference between ω_e and $p\Omega_m = \omega_m$. When the machine enters a new steady state, the vectors assume and maintain a new relative position, and the equation $\omega_e = p\Omega_m = \omega_m$ is reinstated.

The change between ω_e and ω_m allows the angle between vectors $\boldsymbol{F}_S$ and $\boldsymbol{\Psi}_R$ to change during transients. Yet, they do remain in synchronism, namely, they revolve at the same speed in steady state conditions. If the dq coordinate system is introduced by aligning the axes d with the rotor magnetic axis, then the speed of rotation of this dq coordinate frame remains equal to ω_m in both steady state and transient conditions. Therefore, the voltage balance equation (19.20) comprises the angular frequency ω_m in both steady state and transient conditions.

It is of interest to calculate the amount of p_{em} that is passed to the output shaft as mechanical power. With rotor excitation based on permanent magnets, there is no excitation winding and no losses in the rotor. The power p_{em} passed to the rotor through the air gap is converted into mechanical power. This mechanical power may not be equal to the output mechanical power of the machine, due to losses in mechanical subsystem such as friction and air resistance to rotor motion. Therefore, it is called *internal mechanical power*, and it is denoted by $p_{mR} = p_{em}$. When the rotor has an excitation winding, this winding has copper losses $R_R i_R{}^2$. These losses are supplied from an external power source that provides the voltage u_R and the current i_R to the excitation winding. In steady state, this power source supplies the power $u_R i_R = R_R i_R{}^2$ to the excitation winding. Therefore, equation $p_{mR} = p_{em}$ applies as well to synchronous machines with electromagnetic excitation.

Internal mechanical power is equal to the product of the internal electromagnetic torque T_{em} and the rotor speed $\Omega_m = \omega_m/p$. This torque is a mechanical interaction between the stator and the rotor, caused by electromagnetic forces generated by the coupling field. The electromagnetic torque is calculated from expression

$$T_{em} = \frac{p_{mR}}{\Omega_m} = p\frac{p_{em}}{\omega_m} = \frac{3p}{2}\left(\Psi_d i_q - \Psi_q i_d\right) \tag{19.25}$$

The above expression is further simplified for isotropic machines, where the magnetic circuit has the same magnetic resistance in all directions and where equation $L_d = L_q = L_s$ applies. By introducing relations $\Psi_d = L_s i_d + L_m i_R = L_s i_d + \Psi_{Rm}$ and $\Psi_q = L_s i_q$ in the above torque expression, one obtains

$$T_{em} = \frac{3p}{2}\left(\Psi_d i_q - \Psi_q i_d\right) = \frac{3p}{2}L_m i_R i_q = \frac{3p}{2}(L_m i_R)i_q = \frac{3p}{2}\Psi_{Rm} i_q \tag{19.26}$$

The flux component $\Psi_{Rm} = L_m i_R$ represents the part of the excitation flux which encircles the stator winding. It is slightly smaller than the flux of the excitation winding due to a finite amount of magnetic leakage flux. In machines with permanent magnet excitation, the flux Ψ_{Rm} represents the part of the rotor flux, caused by permanent magnets, that encircles the stator windings. A small amount of the flux of permanent magnets does not reach the stator core and does not contribute to the process of electromechanical conversion.

19.12 Anisotropic Rotor

The rotor magnetic circuit may have a noncylindrical form which introduces the difference in magnetic resistances along d- and q-axes. This results in different self-inductances L_d and L_q of virtual stator phase windings in dq frame. Cylindrical structures, where magnetic resistance is not dependent on direction of the field, are called *isotropic*, which means the ones that are having the same properties in all directions. In isotropic machines, inductances L_d and L_q are equal. When magnetic resistance changes with direction of the field, then, the machine is *anisotropic*, and inductances L_d and L_q are different. Salient features of anisotropic machines will be presented in the following section, along with the corresponding expression for the torque.

The flux of the virtual stator phase winding in d-axis is equal to $\Psi_d = L_d i_d + L_m i_R$, while the flux in the virtual stator phase winding q is equal to $\Psi_q = L_q i_q$. Excitation flux does not contribute to the stator flux in q-axis. With cylindrical rotor, inductances L_d and L_q have the same value, $L_d = L_q = L_s$. Construction of an anisotropic rotor is shown in Fig. 19.7, with different magnetic resistances in d- and q-axes. In the left side of the figure, there is a cross section of magnetic circuit of the rotor with electromagnetic excitation. This magnetic circuit is shaped to achieve a low magnetic resistance in d and to facilitate establishing the excitation flux. Conductors of the excitation winding are placed on the sides of the magnetic circuit, directed along q-axis. For this reason, magnetic resistance to the flux directed along q-axis is relatively high because the path of the q-axis flux includes relatively large air-filled segments. A higher magnetic resistance results in a smaller inductance $L \sim N^2/R_\mu$; thus, the circumstance that $R_{\mu d} < R_{\mu q}$ results in $L_d > L_q$.

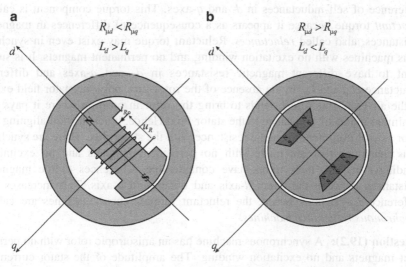

a $R_{\mu d} < R_{\mu q}$ $L_d > L_q$

b $R_{\mu d} > R_{\mu q}$ $L_d < L_q$

Fig. 19.7 (a) Anisotropic rotor with excitation winding and with different magnetic resistances along d- and q-axes. (b) Anisotropic rotor with permanent magnets

The second example in Fig. 19.7 shows a rotor magnetic circuit with permanent magnets built in the interior of the rotor. The magnets are inserted along d-axis. Differential permeability of permanent magnets is close to μ_0, and their presence on the flux path in d-axis increases magnetic resistance $R_{\mu d}$. With $R_{\mu d} > R_{\mu q}$, the phase winding inductances are $L_d < L_q$.

Flux linkages of an anisotropic machine are

$$
\begin{aligned}
\Psi_d &= L_d i_d + L_m i_R, \\
\Psi_q &= L_q i_q, \\
\Psi_R &= L_R i_R + L_m i_d.
\end{aligned}
\tag{19.27}
$$

19.13   Reluctant Torque

Differences between self-inductances L_d and L_q of virtual stator phases d and q affect the expression for the electromagnetic torque:

$$
\begin{aligned}
T_{em} &= \frac{3p}{2} \left(\Psi_d i_q - \Psi_q i_d \right) \\
&= \frac{3p}{2} \Psi_{Rm} i_q + \frac{3p}{2} \left(L_d - L_q \right) i_d i_q.
\end{aligned}
\tag{19.28}
$$

The torque expression contains an additional component, proportional to the difference of self-inductances in d- and q-axes. This torque component is called *reluctant* torque because it appears as a consequence of differences in magnetic resistances, also called *reluctances*. Reluctant torque may exist even in synchronous machines with no excitation winding and no permanent magnets. It is sufficient to have different magnetic resistances in d- and q-axes and different inductances L_d and L_q. In the absence of the rotor flux, only the stator field exists in the air gap. This torque tends to bring the rotor in position where it pays the minimum magnetic resistance to the stator flux. Hence, it acts toward aligning the rotor axis of minimum magnetic resistance with the stator flux. There are synchronous machines that are made with no permanent magnets and no excitation windings. Instead, their rotors have considerable differences of the magnetic resistances between the direct d-axis and quadrature q-axis. This increases the difference $L_d - L_q$ and, hence, the reluctant torque. These machines are called *synchronous reluctance machines*.

Question (19.2): A synchronous machine has an anisotropic rotor with no permanent magnets and no excitation winding. The amplitude of the stator current is limited to I_m. Determine the largest possible value of the reluctant torque. All parameters affecting the reluctant torque are known.

Answer (19.2): Reluctant torque is proportional to the product of currents i_d and i_q. When the amplitude of the stator current is limited, the current components can be expressed as $i_d = I_m\cos(\xi)$ and $i_q = I_m\sin(\xi)$, where ξ is the angle between the stator current vector and d-axis. The reluctant torque is then

$$T_{em} = \frac{3p}{2}\left(L_d - L_q\right)I_m^2 \cos\xi\,\sin\xi = \frac{3p}{4}\left(L_d - L_q\right)I_m^2 \sin 2\xi.$$

The highest reluctant torque is obtained for $\xi = \pi/4$, when $i_d = i_q = I_m/\mathrm{sqrt}(2)$, and it is equal to $(3p/4)(L_d - L_q)I_m^2$.

19.14 Reluctance Motor

Synchronous reluctance machines are used in applications where the size and weight have no particular importance and where the prevailing goal is to have a construction which is robust, simple, and low cost. Reluctance machines have no active parts on the rotor. Rotor has only magnetic circuit made to have different magnetic resistances in direct and quadrature axes. The rotor magnetic circuit can be obtained by stacking the iron sheets in the way shown in Fig. 19.8. By stacking the iron sheets, one obtains a small magnetic resistance along the sheets and a large magnetic resistance in direction perpendicular to the sheets. The flux that passes in direction perpendicular to the sheets passes a number of times through the insulation gaps between the sheets, which increase the equivalent magnetic resistance. In the prescribed way, one obtains an anisotropic rotor, the rotor with different magnetic resistances in d- and q-axes. At the same time, the external appearance of the rotor is cylindrical, with no salient poles and with a low air drag. Therefore, the rotor may reach high speeds without significant motion resistances and without jeopardizing mechanical integrity of the machine.

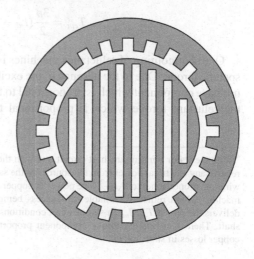

Fig. 19.8 Rotor of a reluctant synchronous machine

Advantage of synchronous reluctance machines, compared to wound rotor machines, is that the former have considerably simpler construction and they do not need to have external supply to the excitation winding through the slip rings. Compared to permanent magnet machines, reluctance machines are easier and cheaper to manufacture. One advantage of reluctance machines is the fact that the air gap flux depends on the stator current only, which opens the possibility to adjust the flux to different operating conditions, such as the field weakening.

With stator currents of reluctance machine equal to zero, the flux is equal to zero as well. At the same time, the electromotive force $\omega_m \Psi$ is equal to zero as well. In cases where the stator winding of synchronous reluctance machine is disconnected from the source and opened, the voltages across the terminals are equal to zero, and they do not pose electrical hazard. At the same time, the fact that there are no induced voltages between the open winding terminals of a reluctance motor means that short-circuiting the terminals would not result in any stator currents. This means that reluctance motor with short-circuited stator winding and with rotor in motion does not have any stator current and does not generate any torque. Synchronous machines with permanent magnet excitation do not have the same behavior. The rotor flux of a permanent magnet machine is present even in cases where the machine is disconnected from the source. The voltage across the terminals of the opened stator winding is equal to the electromotive force $E_0 = \omega_m \Psi_{Rm}$. When the stator terminals are short-circuited, a relatively high short-circuit current is established, approximately equal to $I_{ks} \approx E_0/X_S = \Psi_{Rm}/L_S$. With low stator inductances L_S of synchronous machines with surface mounted magnets, the short-circuit currents are very large, and they result in the stator magnetomotive forces that may demagnetize and destroy the magnets. Besides, the short-circuit current I_{ks} produces braking torque.[4] Reluctance machines do not have the short-circuit current, they have no braking torque in short-circuit conditions, and their voltage between the stator terminals is equal to zero in cases where the machine is disconnected from the source:

$$T_{em} = \frac{3p}{2} \left(L_d - L_q \right) i_d i_q .$$ (19.29)

One shortcoming of reluctance machines is their lower specific power and lower specific torque. Due to the absence of the excitation flux, the torque expression does not have component which is proportional to the product $\Psi_{Rm} i_q$. Instead, it has only the reluctant torque which is proportional to the product of currents i_d and i_q.

---

[4] With the stator winding in short circuit and with the short-circuit current I_{KS}, machine does not receive any electrical power from the source. At the same time, there are copper losses in the stator winding, proportional to $R_S I_{KS}^2$. The power of copper losses is supplied from the only access to the machine which remains available, this access being the shaft, where the machine receives or delivers the power $T_{em} \Omega_m$. In short-circuit conditions, mechanical power is absorbed through the shaft. There is a braking torque component proportional to $R_S I_{KS}^2/\Omega_m$ which accounts for the copper losses in short circuit.

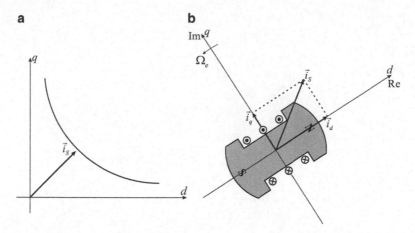

Fig. 19.9 (a) Constant torque hyperbola in the $i_d - i_q$ diagram. (b) Positions of the rotor, dq coordinate system, and complex plane

A thought experiment of inserting permanent magnets into existing reluctance machine would add the component $\Psi_{Rm}i_q$ to the torque, hence increasing the specific torque and power of the machine. It can be concluded that the ratio T_{em}/I_S between the electromagnetic torque and the stator current is considerably lower in reluctance machines than in synchronous machines with permanent magnets or with excitation winding. For developing the same torque, the stator current of reluctance machine will be significantly larger compared to the required stator current of an excited synchronous machine. For this reason, the power and efficiency of reluctance machines are lower than those of excited synchronous machines.

In order to mitigate the unfavorable ratio T_{em}/I_S of reluctance machines, their control is conceived to maximize the torque-per-amp ratio. It is of interest to obtain the maximum possible torque for the given amplitude of the stator current. The left side of Fig. 19.9 shows a dq coordinate frame with currents i_d and i_q on the abscissa and ordinate. The root locus denoting constant torque $T_{em} \sim i_d i_q$ is represented by a hyperbola. Various pairs of i_d and i_q provide the same product $i_d i_q$. Thus, there is a certain degree of freedom in controlling the machine, and it is to be used to minimize the losses and maximize the torque-per-amp ratio. In Fig. 19.9, the amplitude of the stator current is proportional to the radius vector that starts from the origin and ends at the selected operating point on hyperbola $T_{em} \sim i_d i_q$. The smallest amplitude is achieved with radius vector at an angle of $\pi/4$ with respect to d- and q-axes, that is, with $i_d = i_q$. Wherever possible, reluctance machines are controlled with $i_d = i_q$.

Chapter 20
Steady-State Operation

Mathematical model of electrical machine contains differential and algebraic equations describ,ing the machine operation in given supply conditions and given load. Using the model, it is possible to derive the changes of the rotor speed, electromagnetic torque, the air-gap flux, and phase currents during transients and in steady-state conditions. The model is needed to design the power supply of the machine and to devise control algorithm. At the same time, the model is used to predict performance of the machine in operating conditions of interest and to evaluate whether the machine is suitable for given application.

The study of characteristics of synchronous machines and their performance begins with steady-state analysis. It deals with steady-state operating conditions, where the machine operates with constant speed, torque, and flux amplitude.

In this chapter, mathematical model is used to derive the steady-state equivalent circuit of synchronous machine and to obtain the torque and power expressions in steady-state conditions. Operation at steady state is analyzed and illustrated by means of phasor diagrams. Some basic notions on phasors are reinstated and exercised for isotropic and anisotropic machines. The *power angle* is introduced and defined, and the steady-state torque is expressed in terms of the power angle. This chapter discusses and explains electrical and mechanical properties of synchronous machines supplied from stiff three-phase network. Steady-state operation of synchronous motors and generators is analyzed on the basis of corresponding phasor diagrams, and the expressions for the torque, active power, and reactive power are derived in terms of the power angle.

20.1 Voltage Balance Equations at Steady State

Park transform of the state variables has been carried out with the aim of obtaining the model of a synchronous machine where all the state variables are constant at steady state. Projections of vectors of the stator current, voltage, and flux on *dq* axes of synchronously rotating coordinate frame do not change in steady-state conditions. Therefore, their first-time derivatives are equal to zero. This facilitates obtaining the steady-state equivalent scheme. At steady state, the complex numbers $\underline{u}_{dq}$, $\underline{i}_{dq}$, and $\underline{\Psi}_{dq}$, which represent the vectors of the stator voltage, current, and flux, become complex constants. Their constant, steady-state values are denoted by $\underline{u}_s$, $\underline{i}_s$, and Ψ_s. These complex constants have their amplitude and angle and therefore may be treated as phasors. It is important to notice that complex constants $\underline{u}_s$ and $\underline{i}_s$ represent the steady-state voltages and currents of in three-phase stator winding. Hence, unlike common phasors, the numbers $\underline{u}_s$ and $\underline{i}_s$ represent alternating voltages and currents in a three-phase system.

The absolute value of a common phasor corresponds to the rms value of the AC quantity represented by the phasor. On the other hand, the absolute values of complex numbers $\underline{u}_s$ and $\underline{i}_s$ depend on the leading coefficient of Clarke $3\Phi/2\Phi$ transform. Namely, relation between the rms value of the stator phase voltages and the absolute value $|\underline{u}_s| = \mathrm{sqrt}(u_d^2 + u_q^2)$ of the complex constant $\underline{u}_s$ is determined by coefficient K of $3\Phi/2\Phi$ transform which is used in deriving the two-phase equivalent of the three-phase machine. With $K = 2/3$, the absolute value $|\underline{u}_s|$ is equal to the peak value of the stator phase voltages. To facilitate the analysis of synchronous machines, it is desirable to represent the voltage balance equations by the steady-state equivalent scheme. At steady state, there is no change of the excitation current and the excitation flux; thus,

$$i_R = \mathrm{const.} = I_R$$

$$L_m I_R = \Psi_{Rm}.$$

In synchronous machines with permanent magnets on the rotor, the rotor flux that encircles the stator winding is constant and equal to Ψ_{Rm}. In steady state, the rotor speed is equal to the synchronous speed; thus, $\Omega_m = \Omega_e$. With angular frequency ω_e of the stator voltages and currents, synchronous speed is equal to $\Omega_e = \omega_e/p$, and it determines the speed of rotation of the stator magnetic field. Therefore, $\Omega_m = \Omega_e = \omega_e/p = \omega_m/p$. Synchronously rotating dq coordinate frame is selected so as to have the d-axis collinear with the excitation flux, that is, with the flux of permanent magnets. Therefore, dq frame revolves at the same speed as the rotor. Relation between the electrical frequency ω and the mechanical speed of rotation Ω in multipole machines is $\omega = p\Omega$. Hence, the frequency which appears in voltage balance equations is equal to $\omega_m = p\Omega_m$. At steady state, equation $\omega_e = \omega_m$ holds.

For isotropic synchronous machine that operates in the steady state, the voltage balance equations of the stator windings are given below:

$$u_d = R_s i_d + L_s \frac{\mathrm{d}i_d}{\mathrm{d}t} - \omega_e L_s i_q = R_s i_d - \omega_e L_s i_q$$

$$u_q = R_s i_q + L_s \frac{\mathrm{d}i_q}{\mathrm{d}t} + \omega_e L_s i_d + \omega_e \Psi_{Rm} = R_s i_q + \omega_e L_s i_d + \omega_e \Psi_{Rm}.$$

On the basis of the two previous equations, it is possible to write

$$u_d = R_s i_d - \omega_e L_s i_q = R_s i_d - p\Omega_m L_s i_q,$$
$$ju_q = jR_s i_q + jp\Omega_m L_s i_d + jp\Omega_m \Psi_{Rm}.$$

By adding the two previous equations, (20.1) is obtained, which represents the steady-state voltage balance in the stator windings and which employs the phasors $\underline{u}_s$ and $\underline{i}_s$:

$$\underline{u}_s = R_s \underline{i}_s + j\omega_e \underline{\Psi}_s = R_s \underline{i}_s + j\omega_e L_s \underline{i}_s + j\omega_e \Psi_{Rm}$$
$$= R_s \underline{i}_s + jp\Omega_m L_s \underline{i}_s + jp\Omega_m \Psi_{Rm} \qquad (20.1)$$

where

$$\underline{u}_s = u_d + ju_q, \quad \underline{i}_s = i_d + ji_q.$$

Flux $\Psi_{Rm} = L_m i_R$ is the part of the excitation flux which encircles the stator windings. With permanent magnet excitation, the flux Ψ_{Rm} represents the part of the flux created by permanent magnets which passes through the air gap and encircles the stator windings. Since direct d-axis is collinear with the rotor flux, projection of the flux Ψ_{Rm} on quadrature q-axis is equal to zero. Therefore, in complex notation, the flux $\underline{\Psi}_{Rm}$ is a real number, that is, $\underline{\Psi}_{Rm} = \Psi_{Rm} + j0$.

20.2 Equivalent Circuit

At steady state, (20.1) takes the form

$$\underline{U}_s = R_s \underline{I}_s + j\omega_e \underline{\Psi}_s = R_s \underline{I}_s + j\omega_e L_s \underline{I}_s + j\omega_e \Psi_{Rm}. \qquad (20.2)$$

where the voltage and current are denoted in uppercase letters which designate the steady-state values. Resistance R_S and inductance L_S are the parameters of the stator phase windings, while the product $\omega_e \Psi_{Rm}$ represents the electromotive force. In cases where $I_S = 0$, the stator voltage is equal to $\underline{E}_0 = j\omega_e \Psi_{Rm}$.

Product $\underline{E}_0 = j\omega_e \Psi_{Rm}$ represents the electromotive force and also the no load stator voltage. This voltage ($\underline{E}_0$) appears across the stator terminals when the stator current is equal to zero. On the basis of the (20.2), the steady-state equivalent circuit can be represented as a series connection of electromotive force $\underline{E}_0$, reactance $X_S = \omega_e L_S$, and resistance R_S. It should be noted that the equivalent circuit shown in Fig. 20.1 corresponds to an isotropic synchronous machine where $L_d = L_q = L_S$.

The voltage $\underline{U}_S$ is shown on the left side of the equivalent circuit, and it represents the stator voltage $\underline{U}_S = U_d + jU_q$. When using the equivalent circuit

Fig. 20.1 Steady-state
equivalent circuit

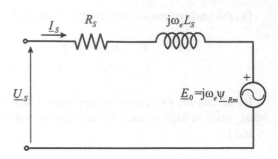

in Fig. 20.1, where $\underline{U}_S = U_d + jU_q$, it has to be noted that the amplitude of
phasors corresponds to peak values of corresponding phase variables. The same
equivalent circuit can be used with phasor amplitudes that correspond to rms
values.

When synchronous machine operates as motor, electrical power is drawn from
the source and brought into machine. The active (real) component of the phasor I_S
that represents the stator current is directed from the left to the right in Fig. 20.1.
When synchronous machine operates as a generator, the phasor $\underline{U}_S$ represents the
voltage across the electrical load that receives the electrical energy obtained from
the generator. The current supplied to the load is $-\underline{I}_S$. The active component of the
stator current in this operating mode is directed from the right to the left, opposite to
the reference direction denoted in Fig. 20.1.

20.3   Peak and rms Values of Currents and Voltages

In many cases where the equivalent circuit in Fig. 20.1 is used, it is assumed that the
amplitude of relevant phasors corresponds to the peak value of related phase
variables. This assumption relies on Clarke transform performed with the leading
coefficient $K = 2/3$. Starting with the phase voltages $u_a(t)$, $u_b(t)$, and $u_c(t)$ and
applying Clarke and Park transforms, one obtains the stator voltage representation
in synchronously rotating coordinate system. The voltage $U_d + jU_q$ has an ampli-
tude $|\underline{U}_S|$ which is equal to the peak value of the phase voltage, $U_{eff} \cdot \text{sqrt}(2)$.
Decision to assign the peak values of the phase quantities to phasor amplitudes
has to be applied uniquely to all the phasors in the equivalent scheme, namely, to all
the voltage, currents, electromotive forces, and flux linkages. Hence, the amplitude
$|E_0|$ of the electromotive force corresponds to the peak value of the no load phase
voltage. The phasor $\underline{\Psi}_{Rm}$ represents the rotor flux that encircles the stator windings.
Its amplitude $|\underline{\Psi}_{Rm}|$ corresponds to the amplitude of the vector $\boldsymbol{\Psi}_{Rm}$, namely, to the
peak value the flux reaches in the stator phase windings. The phasor $\underline{\Psi}_S = L_S\underline{I}_S +
\Psi_{Rm}$ has the amplitude $|\underline{\Psi}_S|$ which corresponds to the amplitude of the stator flux
vector. With the above considerations in mind, the input power to the machine is

equal to $P_e = (3/2) \text{Re}(\underline{U}_S \underline{I}_S^*)$.[1] The power of losses in the stator windings is calculated as $P_{Cu1} = (3/2)R_S I_S^2$. The power of electromechanical conversion is equal to $P_{em} = P_e - P_{Cu1} = (3/2) \text{Re}(\underline{E}_0 \underline{I}_S^*)$. Since flux vector Ψ_{Rm} coincides with d-axis, which is assigned to be the real axis of the complex $d + jq$ plane, the electromotive force phasor is equal to $\underline{E}_0 = j\omega_e \Psi_{Rm}$ and it is collinear with quadrature axis (q), which is at the same time the imaginary axis of the complex $d + jq$ plane. For this reason, the product $\underline{E}_0 \underline{I}_S^*$ assumes the value $\omega_e \Psi_{Rm} I_q$. The electromagnetic torque is equal to the ratio of the power P_{em} and the synchronous speed $\Omega_e = \omega_e/p$:

$$T_{em} = \frac{P_{em}}{\Omega_e} = \frac{3p}{2\omega_e} \omega_e \Psi_{Rm} I_q = \frac{3p}{2} \Psi_{Rm} I_q. \tag{20.3}$$

It is also possible to interpret the phasors of the equivalent circuit as complex numbers with amplitudes that represent rms values of relevant phase variables. In this case, the equivalent circuit operates with rms voltages and currents. The phasor $\underline{U}_S$ on the left side of the equivalent circuit represents the stator voltage, and it has an amplitude $|\underline{U}_S|$ that corresponds to the rms value U_{rms} of the stator phase voltages. Now, the electromotive force $|\underline{E}_0|$ corresponds to the rms value of the no load phase voltage. For many, working with rms values of voltages and currents is more handy. When using the phasors that correspond to rms values, the flux phasors have to be treated in the same way. Therefore, the phasor Ψ_{Rm}, as calculated from the equivalent circuit, obtains an amplitude which is sqrt(2) times smaller than the amplitude of the flux vector Ψ_{Rm}. At the same time, the value $|\Psi_S| = |L_S I_S + \Psi_{Rm}|$ is equal to the amplitude of the stator flux vector divided by sqrt(2). Adopting the phasors that correspond to rms values, the input power to the machine is calculated as $P_e = 3 \text{Re}(\underline{U}_S \underline{I}_S^*)$; the power of losses in the stator winding is obtained as $P_{Cu1} = 3R_S I_S^2$, while the power of the electromechanical conversion is equal to $P_{em} = P_e - P_{Cu1} = 3 \text{Re}(\underline{E}_0 \underline{I}_S^*) = 3\omega_e \Psi_{Rm} I_q$. Notice at this point that both Ψ_{Rm} and I_q in the previous expression assume the values that are sqrt(2) times smaller than the values in (20.3). When using phasors that correspond to rms values, the electromagnetic torque is calculated as

$$T_{em} = \frac{P_{em}}{\Omega_e} = \frac{3p}{\omega_e} \omega_e \Psi_{Rm} I_q = 3p \Psi_{Rm} I_q. \tag{20.4}$$

Question (20.1): A two-pole synchronous machine has the stator self-inductance L_S, while the stator resistance R_S is small and it can be neglected. Machine has permanent magnets on the rotor, and they create the flux Ψ_{Rm} in the stator windings. Machine operates at steady state, connected to a three-phase network of frequency $f_e = 50$ Hz, wherein the rms value of phase voltages is U_{Sn}. The stator voltages

---

[1] $\text{Re}(\underline{z}) = z_r$ is the real pert of the complex number $\underline{z} = z_r + jz_i$. The value $\underline{z}^*$ is equal to $\underline{z}^* = z_r - jz_i$.

have a phase advance δ with respect to corresponding no load electromotive forces. Calculate the power delivered by the network to the machine.

Answer (20.1): The equivalent circuit is analyzed by assuming that the involved phasors represent the rms values. The stator current of the machine is equal to $\underline{I}_S = (\underline{U}_S - \underline{E}_0)/jX_S$, where $\underline{E}_0 = j\omega_e \Psi_{Rm}$. The flux $\underline{\Psi}_{Rm} = \Psi_{Rm}$ remains on the real axis. Since the stator voltages lead with respect to electromotive forces by δ, while the electromotive force $\underline{E}_0 = j\omega_e \Psi_{Rm}$ resides on the imaginary axis, the stator voltages lead with respect to the real axis by $\delta + \pi/2$. Components of the stator voltage in dq frame are $U_d = -U_{Sn}\sin(\delta)$ and $U_q = U_{Sn}\cos(\delta)$. Therefore,

$$\underline{U}_S = -U_{Sn}\sin\delta + jU_{Sn}\cos\delta,$$

$$\underline{I}_S = \frac{\underline{U}_S - \underline{E}_0}{jX_S} = \frac{-U_{Sn}\sin\delta + jU_{Sn}\cos\delta - j\omega_e \Psi_{Rm}}{jX_S}$$

$$= \frac{U_{Sn}\cos\delta - \omega_e \Psi_{Rm}}{X_S} + j\frac{U_{Sn}\sin\delta}{X_S}.$$

Electrical power received from the network is calculated from the following expression:

$$\underline{S} = 3\underline{U}_S\underline{I}_S^* = 3\frac{U_{Sn}(\omega_e \Psi_{Rm})}{X_S}\sin(\delta) + j3\frac{U_{Sn}(U_{Sn} - \omega_e \Psi_{Rm}\cos(\delta))}{X_S} = P_e + jQ_e.$$

The active power received by the machine is $P_e = 3U_{Sn}E_0\sin(\delta)/X_S$.

20.4 Phasor Diagram of Isotropic Machine

Analysis of steady states based on the equivalent circuit relies on complex notation where the vectors of the stator voltages, currents, and flux linkages are represented by corresponding phasors. The voltage balance in the steady-state equivalent circuit can be represented by phasor diagram. The phasor diagram in Fig. 20.2 represents the balance of voltages in isotropic synchronous machine that operates in motoring mode. d-Axis of synchronously rotating coordinate frame is assigned as the real axis of the complex plane.[2] Thus, the q-axis becomes imaginary axis. Park rotational transform is

---

[2] Steady-state voltages and currents can be represented as phasors in an arbitrary complex plane. The angle between the phasor and the real axis determines the initial phase of considered AC voltages and currents. Apparently, this imposes a constraint on the choice of the position of the real axis. On the other hand, the initial phase is the value of the phase at the instant $t = 0$. Therefore, by choosing the instant $t = 0$, it is possible to select complex planes with different position of their real and imaginary axes. Phasor diagrams of synchronous machines are mostly drawn in the complex plane where the real axis coincides with d-axis of synchronously rotating coordinate system. It is also of interest to notice that other choices are legitimate as well. When solving some problems, the calculations are more simple when the real axis is selected to be aligned with stator voltage or the stator current.

Fig. 20.2 Phasor diagram of
an isotropic machine in
motoring mode

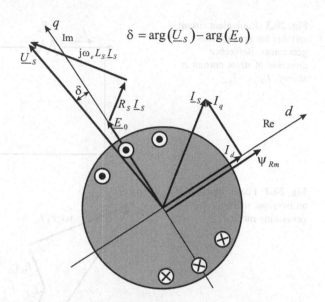

$$\delta = \arg(\underline{U}_S) - \arg(\underline{E}_0)$$

introduced with d-axis being collinear with the excitation flux. Therefore, the phasor $\underline{\Psi}_{Rm}$ resides on the real axis, and it is equal to $\underline{\Psi}_{Rm} = \Psi_{Rm} + j0$.

The phasor $\underline{\Psi}_{Rm}$ represents this part of the excitation flux that encircles the stator windings. The no load electromotive force $\underline{E}_0 = j\omega_e \Psi_{Rm}$ has a phase advance of $\pi/2$, and it resides on imaginary axis. The stator voltage $\underline{U}_S$ is obtained by adding the voltage drop across the stator impedance $\underline{Z}_S = R_S + j\omega_e L_S$ to the no load electromotive force. The voltage drop across the stator resistance is in phase with the stator current $\underline{I}_S$, while the voltage drop across reactance X_S leads by $\pi/2$. The angle δ which determines the phase delay of the electromotive force $\underline{E}_0$ behind the voltage $\underline{U}_S$ is called *power angle*. The apparent power of the machine is equal to

$$
\begin{aligned}
\underline{S} &= 3\underline{U}_S \underline{I}_S^* \\
&= 3\frac{U_{Sn}(\omega_e \Psi_{Rm})}{X_S}\sin(\delta) + j3\frac{U_{Sn}(U_{Sn} - \omega_e \Psi_{Rm}\cos(\delta))}{X_S} \\
&= P_e + jQ_e.
\end{aligned}
\tag{20.5}
$$

The active power P_e delivered to the machine by the electrical source is determined by the power angle δ:

$$
P_e = 3\frac{U_{Sn}E_0}{X_S}\sin(\delta).
\tag{20.6}
$$

A positive value of power angle results in a positive power and positive torque. Therefore, in cases where the voltage phasor has a phase advance with respect to the electromotive force, the machine operates as a motor and develops a positive torque.

Fig. 20.3 Equivalent circuit suitable for synchronous generators. Reference direction of stator current is altered, $I_G = -I_S$

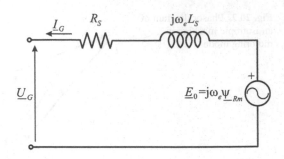

Fig. 20.4 Phasor diagram of an isotropic machine in generating mode

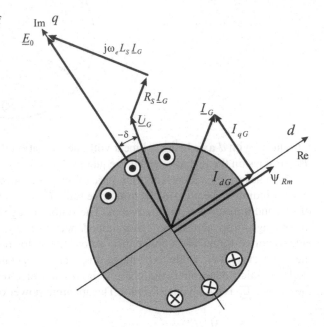

With $\delta < 0$, the voltage lags behind the electromotive force, the power and torque are negative, and thus, the machine operates as a generator.

When an isotropic synchronous machine operates as a generator, the active component of the stator current is negative. Steady-state analysis of synchronous generators is more straightforward if the reference direction of the stator current changes. In Fig. 20.3, the steady-state equivalent circuit is redrawn with altered reference direction for stator current. The new circuit is more intuitive, as it represents the generator which supplies the electrical consumers U_G with generator current I_G. The phasor diagram of an isotropic synchronous generator is given in Fig. 20.4.

At steady state, rotor of synchronous machine revolves at synchronous speed, which is determined by the frequency of the power supply ω_e, this $\omega_e = \omega_m = p\Omega_m$. Whenever a change in the supply voltages or the change in the mechanical subsystem occurs, the machine enters in transient state. An approximate insight of transient behavior of the machine can be obtained from the phasor diagram in Fig. 20.2. In a thought experiment with synchronous motor, a sudden increase in the load torque T_m results in a decrease of the rotor speed. The stator voltage vectors revolve at the speed Ω_e, determined by the supply frequency, while the motion of the electromotive force is determined by the rotor speed. The speed difference $\Omega_e - \Omega_m$ affects the power angle. The electromotive force E_0 revolves at the same speed as the rotor flux. Therefore, it starts revolving slower than the phasor U_S, which revolves at constant, synchronous speed $\Omega_e = \omega_e/p$. For that reason, the power angle δ starts increasing. According to (20.6), the input power p_e increases, which results in an increased electromagnetic torque T_{em}. An increased T_{em} counteracts the torque T_m and brings the synchronous machine into a new equilibrium, a new steady-state operating conditions.

Question (20.2): Make a phasor diagram for synchronous machine starting from the example given in Fig. 21.2 and assuming that stator current I_S lags behind the electromotive force by $3\pi/2$. The stator resistance R_S is negligible.

Answer (20.2): The electromotive force E_0 and the stator voltage U_S reside on imaginary axis of the diagram. The voltage amplitude is smaller than the electromotive force by the amount of $X_S I_S$.

Question (20.3): A two-pole synchronous machine operates at steady state with power angle of $\delta = 0$. The stator voltage amplitude is equal to the no load electromotive force. With $U_S = E_0$, the stator current is equal to zero. At instant $t = 0$, the shaft is loaded by the torque T_m in direction opposite to motion. Discuss the changes in the rotor speed. Assume that the number of pole pairs is $p = 1$, resulting in $\Omega_m = \omega_m$ and $\Omega_e = \omega_e$.

Answer (20.3): At steady state, the rotor revolves at synchronous speed. Therefore, relative position of the stator voltage U_S and the electromotive force E_0 does not change. With $\delta = 0$ and $U_S = E_0$, there is no stator current and no electromagnetic torque. The change of the rotor speed is determined by $J \, d\Omega_m/dt = T_{em} - T_m$. Following the increase of the load torque, the rotor speed decreases. The rotor starts lagging behind the voltage U_S and it falls behind the stator magnetic field. The voltage phasor U_S leads with respect to E_0; thus, the angle δ increases. This increase affects the input electrical power P_e and the electromagnetic torque T_{em}. As the torque T_{em} increases with δ, it compensates the increase in the load torque T_m and prevents further decrease of the rotor speed. For this transient to decay, it is necessary to restore the rotor speed $\Omega_m < \Omega_e$ to the synchronous speed Ω_e. Therefore, the torque T_{em} must exceed the motion resistance torque T_m for a brief interval of time, so as to achieve a positive value of acceleration $d\Omega_m/dt$. This short interval of acceleration is required to restore the rotor speed to the original value, to the synchronous speed Ω_e. Derivative $d\delta/dt$ of power angle is determined by the difference between the synchronous speed and the rotor speed, $d\delta/dt = \omega_e - \omega_m = p(\Omega_e - \Omega_m)$. New state of equilibrium is reached when $\omega_e = \omega_m$, resulting in $d\delta/dt = 0$.

According to (20.6), the power of synchronous machine depends on no load electromotive force, on stator voltage, and on power angle. Electromotive force E_0 is determined by the excitation current. Different pairs of values (E_0, δ) produce the same power and the same torque, provided that the product $E_0\sin(\delta)$ remains unchanged. Hence, the machine can maintain the same power with different values of the excitation current and different values of the excitation flux Ψ_{Rm}. This degree of freedom can be used for to adjust reactive power Q_e (20.6) exchanged between the machine and the supply network.

Considering the sign of reactive power Q_e, there is convention to consider positive the reactive power taken from the network and delivered to electrical loads of inductive character, such as coils, where the load current lags behind the supply voltage. Reactive power taken from the network by receivers such as capacitors is considered negative. Capacitor current leads with respect to the supply voltage. All the loads of this nature can be considered as *generators* of reactive power. Majority of loads and devices connected to distribution networks are of inductive nature, including electrical motors, transformers, and all the devices that include a series inductance. Parallel capacitors are often connected and used as reactive power compensators that make up for the reactive power generated by other loads.

On the basis of (20.5), reactive power taken from the network by an isotropic synchronous machine is equal to

$$Q_e = 3\frac{U_{Sn}(U_{Sn} - E_0\cos(\delta))}{X_S}. \tag{20.7}$$

By reducing the excitation current I_R, no load electromotive force $E_0 = \omega_m L_m I_R$ reduces as well, and it may become smaller than the voltage of the network. With $U_{Sn} > E_0$, reactive power Q_e is positive; therefore, the machine acts as an inductive load. With sufficient increase in excitation current and electromotive force E_0, expression (20.7) becomes negative. Thus, the increase in excitation current results in negative values of reactive power. In such case, machine acts as a capacitive load. Hence, the change in excitation current changes the electromotive force and makes the machine absorb or produce reactive power Q_e.

Synchronous generators in hydroelectric and thermal power plants supply active power consumed by all the electrical loads that are connected to the power grid. Most of electrical loads have inductive nature, and they absorb reactive power. Therefore, besides generating the active power, most generators provide reactive power as well. The amount of reactive power produced by generators is controlled by the excitation current. In a symmetrical three-phase system with sinusoidal voltages and currents, relation $S^2 = P^2 + Q^2$ connects apparent power S, active power P, and reactive power Q. Apparent power S in continuous service is limited by rated voltages and currents. Therefore, an increase in reactive power reduces the available active power that the machine can deliver in continuous service.

An increase in reactive power Q_e increases apparent power. Therefore, it increases the stator current as well. The stator current which is sustainable in continuous service is limited due to the copper losses in stator windings. Excessive current results in overheating. Therefore, the steady-state current cannot exceed the rated current I_n. With rated voltages, the rated current results in the rated apparent power S_n. Starting from the steady-state conditions where $P = P_n$ and $Q = Q_n$, an increase in reactive power increases the apparent power above the rated level S_n. In order to avoid overheating, the active power obtained from the generator has to be reduced.

In cases where the reactive power of power consumers is compensated by using parallel capacitors distributed across transmission and distribution networks, reactive power request imposed on synchronous generators is waived, and their active power does not have to be reduced due to reactive power generation.

20.5 Phasor Diagram of Anisotropic Machine

Anisotropic machine has different inductances in virtual stator phases that reside in d- and q-axes of synchronously rotating dq frame. Therefore, phasor diagram of an anisotropic machine is more complex than phasor diagram of isotropic machine. With $L_d \neq L_q$, the voltage balance in stator winding cannot be expressed by relation $\underline{U}_S = \underline{E}_0 + \underline{Z}_S\underline{I}_S = \underline{E}_0 + (R_S + j\omega_m L_S)\underline{I}_S$ because the voltage drops $L_d\omega_m I_d$ and $L_q\omega_m I_q$ across the stator inductances have different values of self-inductances. Calculation of real and imaginary components of the stator voltage must be calculated separately as they cannot be expressed by $j\omega_m L_S\underline{I}_S$. Thus, $U_d = R_S I_d - \omega_m L_q I_q$, while $U_q = E_0 + RI_q + \omega_m L_d I_d$. Reactances $L_d\omega_m$ and $L_q\omega_m$ are denoted by X_d and X_q, respectively.

The phasor diagram is drawn for steady-state operation where the electrical representation of the rotor speed $\omega_m = p\Omega_m$ gets equal to the angular frequency of the supply ω_e. The voltage balance equations along the d- and q-axes are

$$U_d = -U_s \sin\delta = R_s I_d - \omega_e L_q I_q,$$
$$U_q = +U_s \cos\delta = R_s I_q + \omega_e L_d I_d + \omega_e \Psi_{Rm}. \tag{20.8}$$

In these equations, variable $\Psi_{Rm} = L_m I_R$ represents the part of the rotor flux which passes through the air gap and encircles the stator windings. This definition is also used in permanent magnet machine, where variable Ψ_{Rm} represents the part of the flux of permanent magnets which encircles the stator windings. By solving (20.8), one obtains the stator currents I_d and I_q (20.9). Relation between phasors of voltages, currents, electromotive force, and flux in an anisotropic machine are presented in Fig. 20.5:

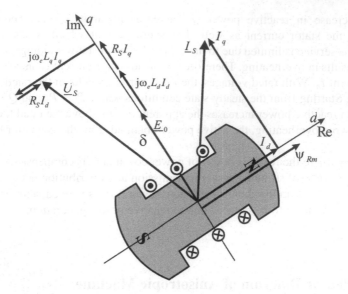

Fig. 20.5 Phasor diagram of an anisotropic machine ($\omega_m = \omega_e$)

$$I_q = +\frac{U_s \sin \delta}{\omega_e L_q} = \frac{U_s \sin \delta}{X_q},$$

$$I_d = \frac{U_s \cos \delta - \omega_e \Psi_{Rm}}{\omega_e L_d} = \frac{U_s \cos \delta - E_0}{X_d}. \tag{20.9}$$

20.6 Torque in Anisotropic Machine

By selecting the complex plane with the real axis collinear with the rotor flux, as shown in Fig. 20.5, the stator voltage phasor $\underline{U}_S = U_d + jU_q$ is equal to $-U_S \sin (\delta) + jU_S \cos(\delta)$, where $U_S = |\underline{U}_S|$, while δ represents the power angle, the angle between $\underline{U}_S$ and $\underline{E}_0$. On the basis of (20.8), which gives voltage balance in the windings, one can calculate currents I_d and I_q. In large synchronous machines, the voltage drop across the stator resistance R_S can be neglected. From (20.9),

$$I_q = \frac{U_s \sin \delta}{X_q}, \quad I_d = \frac{U_s \cos \delta - E_0}{X_d}.$$

Since the notation used above denotes the rms values of voltages and currents, the power absorbed by the machine from the supply network is equal to $P_e = \mathrm{Re}\,(3\underline{U}_S\underline{I}_S^*) = 3(U_d I_d + U_q I_q)$. With $R_S \approx 0$, there are no significant copper losses in the winding. Moreover, the iron losses in the stator magnetic circuit have been neglected as well. Therefore, the input power P_e gets passed through the air gap to the rotor and converter into mechanical power. Hence, P_e is equal to the power of electromechanical conversion. The torque can be determined by dividing the power P_e by the synchronous speed $\Omega_e = \omega_e/p$:

$$
\begin{aligned}
T_{em} &= \frac{P_e}{\Omega_e} \\
&= \frac{3p}{\omega_e}\left[-U_S\sin(\delta)\left(\frac{U_S\cos(\delta)-E_0}{X_d}\right)+U_S\cos(\delta)\left(\frac{U_S\sin(\delta)}{X_q}\right)\right] \\
&= \frac{3p}{\omega_e}\frac{U_S E_0 \sin(\delta)}{X_d}+\frac{3p}{\omega_e}\frac{U_S^2}{2}\left(\frac{1}{X_q}-\frac{1}{X_d}\right)\sin(2\delta)=T_{EXC}+T_{REL}. \quad (20.10)
\end{aligned}
$$

The torque T_{em} has component T_{EXC} which is the product of the excitation flux and the stator currents, and it depends on the electromotive force and the stator voltage. It is created by electromagnetic forces that result from interaction of the rotor field and the stator currents. This torque component is equal to

$$
T_{EXC} = \frac{3p}{\omega_e}\frac{U_S E_0 \sin(\delta)}{X_d}. \quad (20.11)
$$

The torque component T_{REL} is called *reluctant torque*. It does not depend upon excitation flux Ψ_{Rm}, and it exists even in machines where the excitation flux is equal to zero. Reluctant torque is thrusting the rotor toward position of minimum magnetic resistance. Namely, the rotor is driven in position where the magnetic resistance along the path of the stator flux assumes the smallest magnetic resistance. In cases where $L_d > L_q$, reluctance torque acts toward moving the d-axis in position aligned with the stator flux. In terms of the phasor diagram, it acts toward aligning the q-axis and the stator flux $\underline{U}_S$. The torque component T_{REL} depends on the square of the stator voltage:

$$
T_{REL} = \frac{3p}{\omega_e}\frac{U_S^2}{2}\left(\frac{1}{X_q}-\frac{1}{X_d}\right)\sin(2\delta). \quad (20.12)
$$

20.7 Torque Change with Power Angle

Diagram in Fig. 20.6. shows the torque change of an anisotropic machine in terms of the power angle. The maximum value of the torque is reached for an angle smaller than $\pi/2$. The maximum torque is limited by the machine reactances. Likewise the maximum torque of induction machines, the maximum torque of synchronous machines is larger with smaller reactances (inductances).

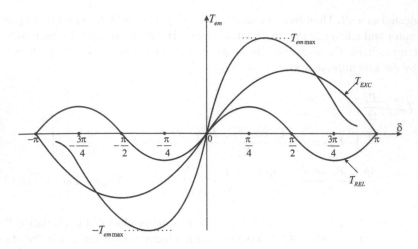

Fig. 20.6 Torque change in anisotropic machine in terms of power angle δ

In the case of an isotropic machine, $X_d = X_q = X_S$. The maximum value of the torque which is available from an isotropic machine is obtained with the power angle $\delta = \pi/2$, and it is equal to

$$T_{em\ max} = \frac{3p}{\omega_e} \frac{U_S E_0}{X_d}. \tag{20.13}$$

20.8   Mechanical Characteristic

Mechanical characteristic $T_{em}(\Omega_m)$ of an electrical machine is the steady-state dependence of electromagnetic torque and the rotor speed. It depends on the power supply conditions, namely, on the amplitude and frequency of supply voltages. Mechanical characteristic obtained with the rated power supply conditions is called *natural characteristic*. Mechanical characteristic of a synchronous machine is shown in Fig. 20.7. Previous diagram (Fig. 20.6) is not a mechanical characteristic. Instead, it shows dependence of the electromagnetic torque and the power angle, $T_{em}(\delta)$, while mechanical characteristic $T_{em}(\Omega_m)$ gives dependence of the electromagnetic torque and the rotor speed. Both $T_{em}(\delta)$ and $T_{em}(\Omega_m)$ dependences are defined in steady-state conditions. Synchronous machine operates in steady state only in cases where the rotor speed Ω_m corresponds to the speed of rotation of the stator magnetic field, denoted by Ω_e and called synchronous speed. Therefore, mechanical characteristic $T_{em}(\Omega_m)$ of synchronous machine is a straight line defined by $\Omega_m = \Omega_e$. The peak values of the torque are reached at the ends of the straight line $\Omega_m = \Omega_e$ given in Fig. 20.7. The peak torque is limited by the machine reactances, and it is equal to $T_{em\ max} = 3pU_S E_0/(\omega_e X_S)$.

Fig. 20.7 Mechanical characteristic

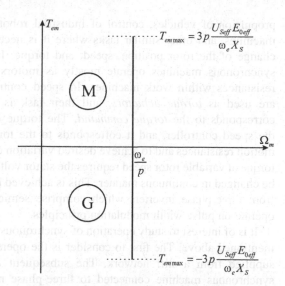

$$T_{em\,max} = 3p \frac{U_{Seff} E_{0eff}}{\omega_e X_s}$$

$$T_{em\,max} = -3p \frac{U_{Seff} E_{0eff}}{\omega_e X_s}$$

20.9 Synchronous Machine Supplied from Stiff Network

Most synchronous machines are connected to three-phase network with AC voltages having the line frequency of $f_e = 50$ or 60 Hz. Other machines are supplied from static power converters which provide a three-phase system of voltages of variable frequency and amplitude. In the former case, the stator frequency ω_e and the voltage amplitude are determined by external factors and cannot be changed. The network where the voltage frequency and amplitude do not change and remain constant is called *stiff network*. In synchronous machines supplied from a stiff network, the synchronous speed cannot be changed, and the steady-state value of the rotor speed remains constant. In the latter case, the machine is supplied from a separate source, from static power converter which produces a three-phase system of voltages of adjustable amplitude and frequency. In most such cases, static power converter supplies only one machine. Therefore, the frequency and amplitude of the stator voltages can be varied and adjusted to achieve desired flux and desired variation of the rotor speed.

In hydropower plants and thermal power plants, large power synchronous machines are used as generators, and they are connected to three-phase network of constant frequency. Mechanical power is obtained via shaft from steam turbines or waterwheels and turbines. This power is converted into electrical and delivered to the network. The rated voltage of large synchronous generators ranges from 6 up to 25 kV. They are connected to transmission networks with rated voltages from 110 up to 700 kV. Each synchronous generator has a dedicated transformer that connects the stator terminal to high-voltage transmission network.

Synchronous machines of lower power are used in motion control applications, where each machine has a dedicated static power converter that provides the supply voltages of variable amplitude and frequency. Motion control applications include propulsion of vehicles, control of industrial robots, motion tasks in production machines, and other similar tasks where it is necessary to provide a continuous change of the rotor position, speed, and torque. In motion control applications, synchronous machines operate mostly as motors which overcome the motion resistances within work machines. In speed control loops, synchronous motors are used as *torque actuators*, and their task is to provide the torque which corresponds to the *torque command*. The torque command is calculated within the speed controller, and it corresponds to the torque required to overcome the motion resistances and to achieve desired variation of the speed. Providing variable torque at variable rotor speed requires the stator voltage frequency and amplitude to be changed in continuous manner. This is achieved by supplying the stator winding from three-phase inverters which comprise semiconductor power switches and operate on pulse width modulation principles.

It is of interest to study operation of synchronous machines in both cases that are mentioned above. The first to consider is the operation of synchronous machines supplied from a stiff network. The subsequent analysis considers three-phase synchronous machine connected to three-phase network with symmetrical AC voltages, wherein the voltages have constant line frequency and constant amplitude. Without the lack of generality, it is assumed that the shaft of synchronous machine is connected to driving turbine which provides the source of mechanical work and that the synchronous machine operates as a generator which converts mechanical work into electrical energy. The steady-state performance of synchronous machine is considered, described by the steady-state equivalent circuit in Fig. 20.1, phasor diagram in Fig. 20.2. In further discussion, it is of interest to investigate the impact of changes in the power angle δ on the electromagnetic torque (20.3), active and reactive power (20.5). In steady-state conditions, gradual changes in the torque of the driving turbine result in changes of the power angle and the active power, while the changes in excitation current affect the reactive power.

20.10   Operation of Synchronous Generators

Synchronous generators in thermal and hydropower plants have the rated voltages that range from 6 to 25 kV and rated power that ranges from several tens of MW up to several hundreds of MW. Generator shaft is coupled to a steam turbine or a water turbine which provides the driving torque $T_T > 0$. This torque supports the motion of the rotor; thus, Newton equation of motion has the form $Jd\Omega_m/dt = T_{em} + T_T$. At steady state, $d\Omega_m/dt = 0$; thus, the electromagnetic torque of the generator is negative and equal to $T_{em} = -T_T$. The power of electromechanical conversion $P_e = T_m\Omega_m$ is also negative, illustrating the fact that the electromechanical energy conversion within the machine has opposite direction, since the mechanical work gets converter into electrical energy. In the phase diagram of generator (Fig. 20.4), the electromotive force $\underline{E}_0$ leads with respect to the voltage $\underline{U}_S$; thus, the power angle δ is negative. Power P_e delivered to the machine from the network is negative in generator mode, and it is equal to

$$P_e = 3\frac{U_{Sn}E_0}{X_S}\sin(\delta) = -P_G, \tag{20.14}$$

where P_G denotes the active power delivered from the generator to the network. Assuming that the network has a constant line frequency, the change in power angle is described by $d\delta/dt = \omega_e - p\Omega_m$. With constant ω_e, an increase in the rotor speed causes the power angle to reduce. An equilibrium point is reached when $\Omega_m = \omega_e/p$.

It is of interest to study behavior of synchronous generators in cases where the driving turbine torque T_T exhibits small changes and also in cases where the line frequencies ω_e changes by a very small amount.

20.10.1 Increase of Turbine Power

The power of steam turbines or water turbines can be increased or decreased according to requirements. Variation of the steam turbine power is achieved by means of opening or closing the valves that feed the steam from the boiler to the steam turbine and also by operating the valves that change the air and coal dust supply to the boiler. The power of a water turbine changes in a similar manner. An increased turbine power results in a larger turbine torque T_T which is passed to the synchronous generator, where it tends to increase the rotor speed. Starting from the state of equilibrium where $T_{em0} = -T_{T0}$, an increase of the turbine torque to a new value of $T_{T1} = T_{T0} + \Delta T$ leads to an increase of the rotor speed according to equation $Jd\Omega_m/dt = T_{em} + T_T = +\Delta T$.

An increase of the rotor speed changes the power angle δ according to $d\delta/dt = \omega_e - p\Omega_m$. The motion of the electromotive force $\underline{E}_0 = jp\Omega_m\Psi_{Rm}$ is determined by the rotor speed, while the voltage $\underline{U}_S$ rotates at synchronous speed $\Omega_e = \omega_e/p$, determined by the network frequency. With an increase in the rotor speed, the power angle delta decreases, and the phase lead $-\delta$ of electromotive force with respect to the stator voltage increases. The power angle assumes negative values in generator mode, and it goes further in negative direction. The generator power $P_G = -P_e = -3(U_{Sn}E_0/X_S)\sin(\delta)$ is increased. Along with that, the electromagnetic torque $T_{em} = P_e/\Omega_e < 0$ increases in magnitude. With $Jd\Omega_m/dt = T_{em} + T_T$, a new equilibrium with $Jd\Omega_m/dt = 0$ can be reached when the electromagnetic torque reaches the value of $T_{em} = -T_{T1}$. The nature of transient phenomena that take place while reaching the new equilibrium is discussed later. In this new steady state, electrical power delivered by the generator to the transmission network is increased. It is assumed that the excess power in the network gets counteracted by an increase in electrical power consumption of electrical loads that are connected to the network. If the assumption that the network is stiff holds, the end of the excess power does not put in question the above considerations. In an actual network, however large, the line frequency may exhibit small changes in consequence to variation in electrical power consumption or variation in mechanical power delivered by steam and water turbines.

In a thought experiment with an electrical network that has several generators and a number of constant power electrical consumers, it is of interest to consider the events that take place when all the steam and water turbines increase their mechanical power at the same time. If all the electrical consumers retain the same power, then the sum of power generated by all the generators in the network must remain constant as well. Namely, unless otherwise stated, the network does not have means to accumulate or store the excess of electrical energy. Therefore, the sum of power generated by all the synchronous generators has to remain constant. With $P_G = -P_e = $ const., the electromagnetic torque T_{em} has to remain constant as well. According to Newton equation of motion $J d\Omega_m/dt = T_{em} + T_T$, an increase in T_T in conditions where T_{em} remains constant results in an increase of the rotor speed. Hence, in the considered thought experiment, the rotor speed in all generators will increase, increasing in such way the line frequency of the considered network. Excess energy that comes from the steam and water turbines is not converted into electrical energy. Instead, it is stored as kinetic energy of revolving masses, $W_{kin} = \sum \frac{1}{2} J \Omega_m^2$. If the situation with excess turbine power persists, the line frequency of the network is continuously increased due to continuous increase in the rotor speed of synchronous machines.

In practical power networks, an increase in the line frequency indicates the excess power received from the steam and water turbines, while a decrease in the line frequency indicates the lack of mechanical power delivered to generators and/ or an excessive consumption.

Question (20.4):   The problem statement relies on the above thought experiment, where the power of steam and water turbines increases, where the power of electrical consumers remains constant, and where the rotor speed and the line frequency increase. Assume that the power system considered above has one high-voltage transmission line connected to a much larger power system. This second, larger power system can be treated as a stiff network. Series reactance X_S of the transmission line is known, as well as the voltage amplitudes U_1 and U_2 of the two power systems. Both systems have symmetrical systems of three-phase voltages. Consider the effects of such connection on the behavior of the system with the excess power.

Answer (20.4):   The two electrical power systems with voltages U_1 and U_2, connected by means of the transmission line with the series reactance X, can be represented as two voltage sources connected by a series impedance jX. This representation is similar to the equivalent circuit in Fig. 20.1, where the voltage U_S gets connected to the voltage E_0 across the series impedance.

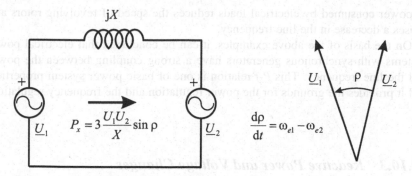

Phasors $\underline{U}_1$ and $\underline{U}_2$ have the phase difference ρ. This phase difference changes according to equation $d\rho/dt = \omega_{e1} - \omega_{e2}$, where ω_{e1} is the line frequency of the first power system while ω_{e2} is the angular frequency of the second power system. According to (20.6), the power exchanged between the two power systems can be expressed as $P_x = 3(U_1 U_2/X) \sin \rho$. As the mechanical power of turbines in the first power system increases, the line frequency ω_{e1} will increase as well due to an increase in rotor speed of synchronous generators within the first power system. At the same time, the line frequency ω_{e2} in the second, larger power system remains the same. The angle ρ will increase due to $d\rho/dt > 0$. Therefore, the phasor $\underline{U}_1$ increases the phase advance with respect to the phasor $\underline{U}_2$, and the power $P_x \sim \sin \rho$ increases. Hence, the excess power of the first power system gets passed to the second power system. On the long run, a new steady-state condition appears with $\omega_{e1} = \omega_{e2}$.

Notice at this point that the energy and power exchange between the two power systems takes place automatically, without any need for the system operator to commutate any switches or to issue any commands. At the same time, connection between the two power systems helps in resolving problems of temporary excess of power in one of the systems as well as problems of temporary increase in power consumption.

20.10.2 Increase in Line Frequency

Line frequency in power network determines the phase angle of the stator voltage $\underline{U}_S$. When the network frequency increases, the power angle δ changes. Considering synchronous generator with $\delta < 0$ and with the power angle time derivative $d\delta/dt = \omega_e - p\Omega_m$, negative value of the power angle becomes closer to $\delta = 0$. Generator power $P_G = -3(U_{Sn}E_0/X_S)\sin(\delta)$ decreases, as well as the magnitude of the electromagnetic torque. These considerations can be applied to any and all synchronous generators connected to the electrical power network. Therefore, whenever the line frequency $\omega_e = 2\pi f_e$ of the power system increases, the power received from synchronous generators reduces. At the same time, a sudden increase of power consumed by electrical loads reduces the speed of revolving rotors and causes a decrease in the line frequency.

On the basis of the above examples, it can be concluded that electrical power systems with synchronous generators have a strong coupling between the power and the line frequency. This P-f relation is one of basic power system properties, and it provides the grounds for the power regulation and the frequency regulation.

20.10.3 Reactive Power and Voltage Changes

A synchronous generator delivers the power $P_G = -3(U_{Sn}E_0/X_S)\sin(\delta)$ with different values of the electromotive force and different values of power angle. In order to maintain a constant power, it is of interest to keep the product $E_0\sin(\delta)$ constant. The electromotive force E_0 can be varied by changing the excitation current of the generator. It is possible to change the excitation current, to change the flux Ψ_{Rm} and the electromotive force $E_0 = \omega_m\Psi_{Rm}$, and yet to maintain the same power, provided that the product $E_0\sin(\delta)$ remains the same. This degree of freedom can be used to change the reactive power Q_e absorbed by the machine from the three-phase network. Reactive power of an isotropic synchronous machine is proportional to the voltage difference between the stator voltage and the electromotive force E_0:

$$Q_e = 3\frac{U_{Sn}(U_{Sn} - E_0\cos(\delta))}{X_S}. \tag{20.15}$$

By increasing the excitation current I_R, electromotive force $E_0 = \omega_m L_m I_R$ becomes larger than the stator voltage; thus, the reactive power Q_e becomes negative. In this way, synchronous generator becomes a source of reactive power, and its equivalent impedance has capacitive nature. Majority of electrical consumers has an inductive power factor and consumes reactive power, behaving as a coil. Therefore, electrical power system must comprise adequate sources of reactive power.

Transmission of reactive power across transmission and distribution line results in significant voltage drops. Most transmission lines have their equivalent series reactance X considerably larger than equivalent series resistance R. In cases where a three-phase electrical load gets connected at the end of the transmission line, where it absorbs reactive power Q and has the voltage $\underline{U}_{END} = U_{END}$, the current of the transmission line is equal to $\underline{I} = -jQ/(3U_{END})$. The voltage at the beginning of the transmission line is equal to $\underline{U}_{BEG} = \underline{U}_{END} + jX\underline{I} = U_{END} + XQ/(3U_{END})$. The voltage drop $jX\underline{I}$ is collinear (in phase) with the voltages. Hence, the value XI is directly subtracted from U_{BEG} in order to obtain U_{END}.

Notice at this point that replacing reactive power consumer by resistive load which absorbs active power $P = Q$ results in a significant reduction of the voltage drop. With resistive load, the current of the same amplitude results in much smaller difference in amplitudes of $\underline{U}_{BEG}$ and $\underline{U}_{END}$. With $\underline{I} = P/(3U_{END})$, $\underline{U}_{BEG} = \underline{U}_{END} + jXP/(3U_{END})$. The voltage drop $jX\underline{I}$ is perpendicular (phase shifted by $\pi/2$) with respect to U_{END}, and this circumstance results in $|\underline{U}_{BEG}| - |\underline{U}_{END}| << XI$.

Whenever electrical loads absorb reactive power, there are considerable voltage drops across the transmission lines. As a consequence, it is necessary to increase the stator voltage of synchronous generators in order to maintain constant voltage at electrical loads. For this to achieve, it is necessary to increase the excitation current I_R of synchronous generators, which leads to increased flux Ψ_{Rm} and increased electromotive force E_0. At the same time, an increase in E_0 contributes to increased reactive power delivered from generators to the network.

From the above considerations, it is concluded that the voltage across the network and the reactive power flow are strongly related. This U-Q relation constitutes the bases for the voltage control in electric power systems.

20.10.4 Changes in Power Angle

Changes in mechanical power received from steam or water turbines result in transient response of synchronous machine which ends in a new steady-state condition with a different value of power angle. A sudden increase in electrical load of the generator produces the same effect. It is of interest to investigate the transient response of the machine torque, power, and power angle during such transients.

With constant power supply frequency, the stator voltage vector rotates at constant synchronous speed $\Omega_e = \omega_e/p$, determined by the supply frequency ω_e. The rotor flux is created by the current I_R in the excitation winding, which revolves with the rotor. Therefore, the vector of the rotor flux revolves at the rotor speed Ω_m. The excitation flux $\underline{\Psi}_{Rm}$ creates the electromotive force $\underline{E}_0 = j\underline{\Psi}_{Rm}\Omega_m$ in the stator windings. In steady state, the stator voltage and electromotive force are represented by phasors $\underline{U}_S$ and $\underline{E}_0$. Power angle δ represents the difference in initial phases of the stator voltage and the electromotive force. It changes according to the law[3] $d\delta/dt = \omega_e - \omega_m = \omega_e - p\Omega_m$. The phase of the stator voltage depends on the supply frequency ω_e, while the phase of the electromotive force depends on the rotor speed $p\Omega_m$. Therefore, changes in power angle δ are defined by the speed difference between the synchronous speed and the rotor speed. At the same time, the electromagnetic torque of synchronous machine operating in the steady state depends on the product of U_S, E_0, and sine of the power angle.

---

[3] In phasor diagrams, phasors $\underline{U}_S$ and $\underline{E}_0$ represent an AC stator voltage and an AC electromotive force. The power angle δ represents the phase difference or the *electrical* shift between the stator voltage and the electromotive force. Therefore, the first-time derivative of power angle is equal to $d\delta/dt = p(\Omega_e - \Omega_m) = \omega_e - p\Omega_m = \omega_e - \omega_m$, where p is the number of pole pairs.

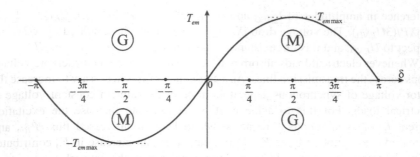

Fig. 20.8 Torque change in isotropic machine in terms of power angle

Variation of the torque in terms of the power angle δ is presented in Fig. 20.8. If the machine operates with no load, the power angle is equal to zero, as well as the electromagnetic torque. If the motion resistance torque T_m appears, directed toward reducing the rotor speed, the power angle increases because its derivative $d\delta/dt = \omega_e - p\Omega_m$ becomes positive. The increase in the power angle increases the electromagnetic torque. A new equilibrium is reached when $T_{em} = T_m$.

Assuming that the machine is connected to steam or water turbine which provides the torque T_T, directed toward increasing the rotor speed, the rotor speed Ω_m exceeds the synchronous speed Ω_e, and the power angle $d\delta/dt = \omega_e - p\Omega_m$ obtains a negative first derivative and a negative value. A negative value of δ means that the electromotive force phasor $\underline{E}_0$ leads with respect to the voltage $\underline{U}_S$. Machine develops electromagnetic torque $T_{em} < 0$ which is opposite to the rotor motion; thus, the machine operates as a generator, converting mechanical work into electrical energy.

In cases where the power angle is relatively small, it is justified to adopt the approximation $\sin(\delta) \approx \delta$ and to represent the torque T_{em} by an approximate expression $T_{em} \approx k\delta$. At steady state, the electromagnetic torque is in equilibrium with the torque components resulting from the mechanical subsystem. When the machine operates as motor, electromagnetic torque is equal to motion resistances T_m. When the machine operates as generator, electromagnetic torque is equal to the torque provided by steam or water turbines. In both cases, the rotor speed does not change since $Jd\Omega_m/dt = T_{em} - T_m = 0$ in motoring mode and $Jd\Omega_m/dt = T_{em} + T_T = 0$ in generator mode. At steady state, the rotor revolves at synchronous speed $\Omega_m = \Omega_e$. It is of interest to determine the character of transients which appear due to disturbances such as the load torque changes or changes in the line frequency ω_e. In the subsequent considerations, it is assumed that the number of pairs of magnetic poles is $p = 1$; thus, the angular frequencies of electrical variables ω_m and ω_e are equal to corresponding mechanical speeds Ω_m and Ω_e. Therefore, Newton equation of motion is written as $J\,d\omega_m/dt = T_{em} - T_m$, where T_m represents the motion resistances. Alternatively, $-T_m = T_T$ represents the driving torque of the steam or water turbine. With that in mind, the change in the power angle is determined by

$$\frac{d\delta}{dt} = \omega_e - p\Omega_m = \omega_e - \omega_m. \tag{20.16}$$

Newton equation for a two-pole machine takes the form

$$J\frac{d\Omega_m}{dt} = J\frac{d\omega_m}{dt} = T_{em} - T_m, \qquad (20.17)$$

while the change in the power angle depends on differential equation

$$J\frac{d^2\delta}{dt^2} = -T_{em}(\delta) + T_m. \qquad (20.18)$$

Staring from no load conditions where $\delta = 0$, $T_{em} = 0$, and $\omega_m = \omega_e$, and introducing the step change in the load torque T_m (or the turbine torque T_T), a transient interval follows where the rotor speed, electromagnetic torque, and power angle change and fluctuate before entering another steady-state condition. During these transients, the rotor speed is not equal to the synchronous speed. Taking into account that position of the dq coordinate system is determined by the rotor position, while the vector of the stator voltage depends on the power supply frequency, it is concluded that the stator voltage vector will move with respect to selected dq coordinate frame. For this reason, projections U_d and U_q of the stator voltage vector U_S on the axes of dq frame will change during transients. Electrical subsystem of synchronous machine is represented by voltage balance equations in virtual stator phases that are placed in d-axis and q-axis of dq frame. Changes in corresponding voltages U_d and U_q introduce nonzero derivatives of flux linkages Ψ_d and Ψ_q, thus bringing the electrical subsystem in transient state. Hence, during transients in mechanical subsystem, where the electromagnetic torque, the rotor speed, and the power angle exhibit transient changes, the electrical subsystem of synchronous machine does not remain in steady-state condition, and it enters transient state of its own.

While in transient state, electrical subsystem of synchronous machines cannot be represented by the steady-state equivalent circuit. At the same time, the transient torque cannot be represented by simplified expression such as $T_{em} \approx k\delta$, which is derived from the steady-state equivalent circuit.

Within the next chapter, it will be shown that the time constants of mechanical subsystems are considerably larger than the time constants of electrical subsystem. This circumstance will be used to simplify transient analysis of synchronous machines.

Newton equation for a two-pole machine takes the form

$$J \frac{d\Omega}{dt} = \ldots$$

(20.17)

while the change in the power angle depends on quite a just equation

$$\frac{d\delta}{dt} = \ldots$$

(20.18)

Chapter 21
Transients in Sychronous Machines

In this chapter, transient response of synchronous machines connected to stiff network is analyzed and discussed. Analysis of transients in electrical and mechanical subsystems of synchronous machines is relatively complex due to a relatively large number of state variables, such as the rotor position and speed, and the winding currents and flux linkages. Complexity of mathematical model does not help the process of understanding the nature of transients and hinders deriving corresponding conclusions. The analysis can be simplified by introducing the assumption that transients in electrical subsystem decay considerably faster than those of mechanical subsystem. In this way, analysis of transients in mechanical subsystem can be performed by using steady state model of electrical subsystem. In this way, results are made more legible and intuitive.

This chapter begins with introducing and explaining the time constants that characterize transient response in electrical subsystem and mechanical subsystem of synchronous machines. For synchronous machine supplied from stiff network, electromagnetic torque is expressed in terms of the power angle. Transient response of the rotor speed, power angle, and the electromagnetic torque is obtained from Newton differential equation of motion, and from the torque-power angle relation. Damper winding is introduced and explained as the means of suppressing oscillations of the torque, power, and speed caused by sudden changes at electrical or mechanical accesses to the machine. Some typical realizations of damper winding are introduced and discussed.

In the second half of this chapter, analysis is focused on the short-circuit transient in synchronous generators. Short-circuit analysis is simplified by the assumption that transient processes in damper windings are the first to decay during the subtransient interval. Within the subsequent interval called transient interval, transient processes in excitation winding cease. Eventually, transient phenomena disappear, and the stator current reduces to steady state short-circuit current. A series of reasonable assumptions and engineering considerations results in deriving subtransient and transient time constants, and in subtransient, transient, and steady state reactances and short-circuit currents. This chapter ends by indicating most typical values of relevant time constants and reactances.

S.N. Vukosavic, *Electrical Machines*, Power Electronics and Power Systems,   595
DOI 10.1007/978-1-4614-0400-2_21, © Springer Science+Business Media New York 2013

21.1   Electrical and Mechanical Time Constants

Transient phenomena in electrical subsystem have to do with winding currents and flux linkages in the air gap and in corresponding magnetic circuits. Transient response in machine windings is characterized by electrical time constants. Electrical current in a winding of resistance R, with self-inductance L and with the voltage $u = 0$, decays exponentially, according to the law $i(t) = i(0) \cdot e^{-t/\tau}$, where $\tau = L/R$ is the electrical time constant of the winding. The change $i(t) = i(0) \cdot e^{-t/\tau}$, the winding current takes place in cases where the winding does not have any coupling with other windings. In synchronous machines, the stator phase windings and the excitation winding are coupled, and they interact and affect the change of currents and flux linkages. Therefore, transient response may not be exponential, and the time constant can be different than $\tau = L/R$. Nonetheless, the time constant $\tau = L/R$ serves as a rough description of dynamic processes in electrical subsystem. In smaller machines, time constants of the windings are of the order of several tens of milliseconds, while high-power machines have time constants $\tau = L/R$ in excess to 1 s. On the other hand, transient phenomena in mechanical subsystem are considerably slower. Synchronous generators have massive rotors with significant inertia. They are coupled to steam or water turbines with certain inertia J as well. Due to large inertia, rotor speed changes, and transient phenomena in mechanical subsystem are considerably longer than the time constants of electrical subsystem. With that in mind, analysis of transients in mechanical subsystem can be simplified. While introducing disturbances into mechanical subsystem, such as the torque changes, electrical subsystem of the machine gets disturbed as well. Due to much shorter electrical time constants, transient processes in electrical subsystem would quickly decay. Following that, electrical subsystem can be modeled by steady state equivalent circuit. For the largest part of the transient process in mechanical subsystem, electromagnetic torque can be modeled by (20.11) and considered proportional to $\sin(\delta)$.

Neglecting transient phenomena in the windings and considering them much quicker than transient processes in mechanical subsystem, the changes in electromagnetic torque can be considered proportional to $\sin(\delta)$. Analysis of relatively small torque changes in close vicinity of no load operating point where $T_{em} = 0$ and $\delta = 0$, electromagnetic torque can be approximated by $T_{em} \approx k\delta$.

21.2   Hunting of Synchronous Machines

During transient phenomena, the power angle changes according to $d\delta/dt = \omega_e - \omega_m$. The second derivative of the power angle can be determined as $d^2\delta/dt^2 = d\omega_e/dt - d\omega_m/dt$. Assuming that the supply frequency ω_e does not change, it follows that $d\omega_m/dt = -d^2\delta/dt^2$. For a two-pole machine where $p = 1$ and $\Omega_m = \omega_m$, Newton equation of motion takes the form

$$J\frac{d^2\delta}{dt^2} = -k\delta + T_m. \tag{21.1}$$

Differential equation (21.1) can be transferred into algebraic equation by applying Laplace transform. Instead of the time function $\delta(t)$, algebraic equation has the complex image $L(\delta(t)) = \delta(s)$, where s denotes Laplace operator. It is assumed that the load torque disturbance $T_m(t)$ changes at instant $t = 0$ from $T_m(0^-) = 0$ to $T_m(0^+) = T_M$. Prior to the load torque step, the system was in steady state with $T_{em} = 0$, $\delta(0) = 0$, and $d\delta/dt(0) = 0$. With that in mind, Laplace transform is applied to differential equation (21.1), and the latter is converted into the following algebraic equation:

$$J\frac{d^2\delta(t)}{dt^2} + k\delta(t) = T_m(t) \quad \Rightarrow \quad Js^2\delta(s) + k\delta(s) = T_m(s) = \frac{T_M}{s}.$$

Complex image $T_m(s) = T_M/s$ represents Heaviside function $h(t)\cdot T_M$ which describes the step increase of the torque at the initial $t = 0$. Based upon the above algebraic equation, the power angle $\delta(s)$ can be expressed in terms of the torque:

$$\delta(s) = \frac{1}{Js^2 + k}\frac{T_M}{s} = W(s)\frac{T_M}{s}.$$

The function $W(s)$ is called transfer function because it relates the input and the output of the system. Complex image $T_m(s)$ of the load torque $T_m(t)$ is considered to be the input to the system since the torque changes originate transient phenomena in the system. Complex image $\delta(s)$ of the power angle $\delta(t)$ is the system response, and it is considered to be the output of the system. Polynomial $f(s) = Js^2 + k$ in denominator of the transfer function is called *characteristic polynomial*.

Zeros of characteristic polynomial are the roots of equation $f(s) - 0$, and they are called *poles of the transfer function*. These poles determine character of the system response, namely, whether response is aperiodic or with oscillations. The poles also determine the response speed and decay time of eventual oscillations. With characteristic polynomial of the form $f(s) = (s - s_1)(s - s_2)$, which has two finite zeros s_1 and s_2, the system response $\delta(t)$ to the input step change $T_m(t)$ comprises factors exp (s_1t) and $\exp(s_2t)$. With $s_1 = -1/\tau$, the system response comprises factors such as $\exp(-t/\tau)$. With $s_1 = \pm j\omega_0$, the system response comprises factors such as exp $(\pm j\omega_0t) = \cos(\omega_0t) \pm j\sin(\omega_0t)$. With $s_1 = -1/\tau \pm j\omega_0$, the system response comprises factors such as $\exp(-t/\tau \pm j\omega_0t) = \exp(-t/\tau)\cdot\cos(\omega_0t) \pm j\cdot\exp(-t/\tau)$ $\cdot\sin(\omega_0t)$.

Characteristic polynomial $f(s) = Js^2 + k$ has two finite zeros $s_{1/2} = \pm j\cdot\text{sqrt}(k/J)$ $= \pm j\omega_0$. Therefore, the load step response of considered synchronous machine is oscillatory. According to (21.1), characteristic polynomial has two poles on imaginary axis of s-plane, $\pm j\omega_0$. This means that the response does not decay.

Namely, the oscillations caused by the load step would persist indefinitely. In practice, the oscillations gradually decay due to friction and secondary effects and losses that were not modeled in (21.1). Though, decay may be extremely slow.

Conclusion concerning oscillatory character of the response can be derived without preceding discussion that involves characteristic polynomial and response character. Instead, one can rely on similarity between the considered mechanical subsystem and a series LC circuit. It is necessary to consider LC circuit which stays at rest for $t < 0$, with $u_C = 0$ and $i_L = 0$, and then gets connected to voltage source $+E$ at instant $t = 0$. Complex image of the voltage step $E\,h(t)$ is E/s. Impedances of capacitor and inductor are $Z_C = 1/(sC)$ and $Z_L = sL$, respectively. Complex image of the capacitor voltage is $u_C(s) = (E/s){\cdot}Z_C/(Z_C + Z_L) = (E/s)/(1 + LCs^2)$. It is well known that series LC circuit exhibits oscillations following the step change of the input voltage. The oscillations of the capacitor voltage and inductor current of LC circuit have angular frequency $\omega_0 = 1/\mathrm{sqrt}(LC)$, determined by the roots of equation $1 + LCs^2 = 0$, $s_{1/2} = \pm j/\mathrm{sqrt}(LC)$. In LC circuit where $u_C = 0$ and $i_L = 0$ for $t < 0$, and where the input voltage $E\,h(t)$ exhibits a step change at $t = 0$, the voltage across capacitor changes as

$$u_C(t) = E\left[1 - \cos\left(\frac{t}{\sqrt{LC}}\right)\right].$$

By analogy with LC circuit, synchronous machine with characteristic polynomial $f(s) = Js^2 + k$ in denominator of the transfer function $W(s)$ oscillates with angular frequency $\omega_0 = \mathrm{sqrt}(k/J)$. Whatever the inertia J and coefficient k, complex image $\delta(s)$ of the power angle is obtained as a solution of algebraic equation

$$Js^2\delta(s) + k\delta(s) = T_m(s),$$

while the response $\delta(t)$ and its character depend on zeros of characteristic equation

$$f(s) = s^2 + \frac{k}{J} = 0.$$

Time change of the electromagnetic torque during transients that follow the step change of the load torque is shown in Fig. 21.1.

Oscillatory responses of the speed, torque, and power angle are not acceptable, in particular, in large synchronous machines.[1] On the basis of Fig. 21.1, a step change of the load torque from $T_m = 0$ to $T_m = T_M$ leads to sustained, undamped

---

[1] Sustained oscillations in torque and current increase the rms value of the stator currents, increase the power of copper losses, contribute to mechanical stress and ware of shaft and transmission elements, increase the acoustic noise, and reduce the peak torque and peak power capability of the machine.

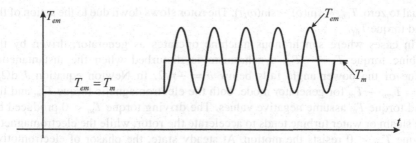

Fig. 21.1 Torque response of synchronous machine following the load step

oscillations in electromagnetic torque T_{em} which changes between $T_{em} = 0$ and $T_{em} = 2T_M$. The oscillations caused by the step change in mechanical subsystem of synchronous machines supplied from a stiff network are called *hunting of synchronous machine*. Variation of the torque and power angle following the step can be described by the following expression:

$$T_{em}(t) = T_M(1 - \cos \omega_0 t), \quad \delta(t) = \frac{T_{em}(t)}{k} = \frac{T_M}{k}(1 - \cos \omega_0 t).$$

In the considered example, the first time derivative of the power angle is equal to $d\delta/dt = \omega_e - \omega_m = (T_M/k)\omega_0 \sin\omega_0 t$, and it has the average value equal to zero. Hence, during oscillations of synchronous machine, the rotor speed Ω_m oscillates around synchronous speed Ω_e. During transients, the instantaneous value of the rotor speed $\Omega_m(t)$ is derived from $\Omega_e(t)$ according to $\omega_e - \omega_m = (T_M/k)\omega_0 \sin\omega_0 t$. Yet, the average value of the rotor speed remains equal to the synchronous speed even during transients. In an attempt to visualize the transient response in phasor diagram shown in Fig. 20.2, the phasor $\underline{E}_0$ oscillates around $\underline{U}_S$, but they do remain in close vicinity.

During oscillations, power angle δ may approach the value of $\pi/2$. It reaches $\delta = \pi/2$ in cases where the electromagnetic torque $T_{em}(t)$ reaches the peak electromagnetic torque $T_{em \, max}$ (20.13). With the load torque $T_M = T_{em \, max}/2$, oscillations of electromagnetic torque reach twice the value of the initial step, and, consequently, the power angle δ reaches the value of $\pi/2$. Notice in the region $\delta > \pi/2$ in Fig. 20.6 that electromagnetic torque does not increase with the power angle. Instead, for $\delta > \pi/2$, the electromagnetic torque decreases, and the value $dT_{em}(\delta)/d\delta$ becomes negative. Coefficient k in (21.1) represents the slope of the torque-power angle curve, and it is equal to $dT_{em}(\delta)/d\delta$. With $k < 0$, $T_M > 0$, and with $\delta > \pi/2$, the second time derivative of the power angle $d^2\delta/dt^2 = d\omega_e/dt - d\omega_m/dt$ becomes positive. Therefore, the difference $\omega_e - \omega_m$ does not decrease. Instead, it continues to rise, and the machine falls out of synchronism. This actually means that the rotor speed reduces to such an extent that the return into synchronism is no longer possible. With the speed difference $\omega_e - \omega_m$ which has a nonzero average value ω_k, the power angle progressively increases according to $\delta(t) \approx \omega_k t \approx (\omega_e - \omega_m)t$. Therefore, the average value of electromagnetic torque becomes equal to zero, $T_{em} \sim \sin(\delta) \sim \sin(\omega_k t)$. The rotor slows down due to the action of the load torque T_M.

In cases where synchronous machine operates as generator, driven by the turbine torque $T_T = -T_m$, synchronism is disturbed when the instantaneous value of the power angle falls below $\delta = -\pi/2$. In Newton equation $J \, d\Omega_m/dt = T_{em} - T_m$ for generator mode, both the electromagnetic torque T_{em} and the load torque T_m assume negative values. The driving torque $T_m < 0$ produced by the steam or water turbine tends to accelerate the rotor, while the electromagnetic torque $T_{em} < 0$ resists the motion. At steady state, the phasor of electromotive force $\underline{E}_0$ leads with respect to the phasor $\underline{U}_S$, and the power angle becomes negative, $0 > \delta > -\pi/2$, which results in negative electromagnetic torque and power. Falling out of synchronism occurs when the power angle drops below $-\pi/2$. Following that, power angle progressively reduces, producing electromagnetic torque $T_{em}(t)$ which oscillates and which has zero average value. The driving torque of the turbine continues to deliver a positive torque which accelerates the rotor and increases the rotor speed above the synchronous speed. In order to prevent the synchronous machine from falling out of synchronism, it is necessary to provide damping of oscillations.

21.3 Damped LC Circuit

In order to get an insight into possible ways of damping oscillations of synchronous machine, it is of interest to recall the analogous phenomena in an LC circuit. Figure 21.2 shows an LC circuit with series resistance R, with the voltage step E applied at instant $t = 0$.

Following the voltage step, the voltages and currents in this RLC circuit oscillate at frequency $\omega_0 = \omega_n \sqrt{1 - \xi^2}$, where ξ is the damping coefficient of the RLC circuit. In the case when resistance R is equal to zero, damping coefficient ξ is equal to zero as well, and the frequency of oscillations is equal to $\omega_0 = \omega_n = 1/\sqrt{LC}$; thus, variation of the voltage across the capacitor C can be represented by

$$u_C(t) = E(1 - \cos(\omega_n t)).$$

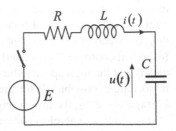

Fig. 21.2 Damped
oscillations of an LC circuit

In the case when $R > 0$, the roots of the equation $f(s) = 0$ are complex numbers $s_{1/2} = -\xi\omega_n \pm j\omega_0$. They have negative real part $-\xi\omega_n = -1/\tau = -R/(2\,L)$. Since the response $u_C(t)$ contains factor of the form $\exp(s_1 t)$, the amplitude of oscillations decays according to the law $\exp(-\xi\omega_n t) = \exp(-t/\tau)$. Characteristic polynomial which determines response of the RLC circuit is

$$f(s) = s^2 + 2\xi\omega_n s + \omega_n^2,$$

and its zeros are

$$s_{1/2} = -\xi\omega_n \pm j\omega_n\sqrt{1 - \xi^2} = -\xi\omega_n \pm j\omega_0.$$

Responses of the current and voltage contain factors

$$e^{-s_1 t} = e^{-\xi\omega_n t}\,e^{+j\omega_0 t}.$$

The voltage across the capacitor can be determined by applying the inverse Laplace transform to the complex image $U_C(s) = (E/s)/(1 + RCs + LCs^2)$, obtaining in this way

$$u_C(t) \approx E - E\,e^{-\xi\omega_n t}\cos(\omega_0 t) - \xi\,E\,e^{-\xi\omega_n t}\sin(\omega_0 t).$$

According to the previous equation, the voltage across the capacitor exhibits damped oscillations at the frequency ω_0. These oscillations decay exponentially. Time constant $\tau = 2L/R$ determines the time required for the oscillations amplitude to decrease by the factor of $e = 2.71$. In cases where the time constant τ exceeds the period of oscillations $T = 2\pi\cdot\text{sqrt}(LC)$ several times, the oscillations are weakly damped. Higher values of resistance R result in shorter time constants τ, and they produce higher values of the damping coefficient ξ, resulting in a better damping of oscillations. Damping coefficient ξ is equal to the ratio of real part and natural frequency ω_n of the poles $s_{1/2}$. It can be calculated as cosine of the angle between negative part of the real axis of s-plane and the radius that starts from the origin and ends at the pole s_1. Time response having damping coefficient of $\xi < 1$ is shown in Fig. 21.3. The left side of the figure is denoted by (a), and it shows position of the poles $s_{1/2}$ in s-plane, while the right side of the figure, denoted by (b), shows the time response, that is, the voltage across the capacitor of the RLC circuit.

In cases where damping coefficient exceeds one, the response does not contain any oscillations. Instead, it has exponential change,

$$u_C(t) \approx K_1 + K_2\,e^{-t/\tau 1} + K_3\,e^{-t/\tau 2},$$

where time constants $\tau_1 = -1/s_1$ and $\tau_2 = -1/s_2$ are reciprocal values of the poles $s_{1/2}$. With $\xi \geq 1$, the poles are negative real numbers that do not have an imaginary part (Fig. 21.4).

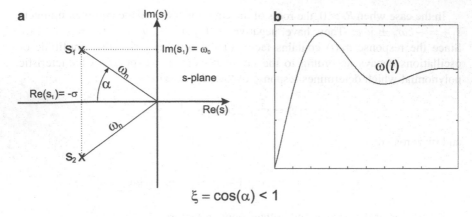

Fig. 21.3 Response with conjugate complex zeros of characteristic polynomial

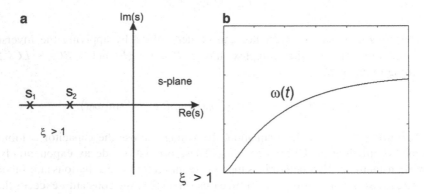

Fig. 21.4 Response with real zeros of the characteristic polynomial

21.4   Damping of Synchronous Machines

In order to introduce damping into transient response of synchronous machine, it is necessary to provide measures and actions that increase coefficient a which multiplies Laplace operator s of characteristic polynomial $f(s) = s^2 + as + b = s^2 + 2\zeta\omega_n s + \omega_n^2$. Characteristic polynomial $f(s) = Js^2 + k$ obtained from Newton differential equation (21.1) does not have any damping, and it results in undamped, sustained oscillations of the power angle δ. In order to design damping actions, it is of interest to consider the following equation:

$$J\frac{d^2\delta(t)}{dt^2} + k\delta(t) = T_m \quad \Rightarrow \quad f(s) = Js^2 + k,$$

where $Jd\Omega_m/dt$ represents the inertial torque while $T_{em} = k\delta$ represents the electromagnetic torque. In order to damp oscillations of the synchronous machine connected to a stiff network, it is necessary to extend the above differential equation by adding the torque component proportional to the first derivative of the power angle, $k_P \cdot d\delta/dt$. This adds the factor $k_P \cdot s$ into characteristic polynomial $f(s)$:

$$J\frac{d^2\delta(t)}{dt^2} + k_P\frac{d\delta(t)}{dt} + k\delta(t) = T_m \quad \Rightarrow \quad f(s) = Js^2 + k_P s + k.$$

Having in mind that characteristic polynomial with two conjugate complex zero has the form

$$f(s) = s^2 + 2\xi\omega_n s + \omega_n^2,$$

the torque component $k_P \cdot d\delta/dt$ contributes to damping $\xi = k_P/(2J\omega_n)$. The required torque component can be achieved by introducing small variations of the load torque. With $T_m(t) = T_{m0} - k_P \, d\delta/dt$, it is possible to obtain a stable, well-damped response of the power angle δ and the electromagnetic torque of the synchronous machine. Yet, it is very difficult to make such changes of the torque T_m. In synchronous generators, the torque $T_T = -T_m$ is provided by the steam or water turbines, which cannot be controlled with desired dynamics. Synchronous motors are loaded by the torque T_m of respective mechanical loads and work machines. Neither this torque can include the component $k_P \, d\delta/dt$.

Damping torque $k_P \, d\delta/dt$ can be obtained from the very synchronous machine, provided that the electromagnetic torque T_{em} included two components, where the first is proportional to $\sin(\delta) \sim \delta$ while the second is proportional to the first time derivative of the power angle, $d\delta/dt$. Desired variation of the torque is determined by expression $T_{em} = k\delta + k_P \, d\delta/dt$.

Electromagnetic torque of an AC machine depends on the flux and also on the stator current. On the other hand, the stator current can be changed by altering the stator voltages. Therefore, generally speaking, desired variation of electromagnetic torque can be achieved by changing the stator voltages. With changes in stator voltages determined by $d\delta/dt$, the stator currents and the electromagnetic torque would exhibit changes that depend on $d\delta/dt$. However, the subject of this analysis is a synchronous machine connected to a stiff network, where the voltages cannot change according to $d\delta/dt$. Instead, the stator voltages have amplitude and frequency that do not change. Therefore, the damping torque $\Delta T_{em} = k_P \, d\delta/dt$ requires the change of the machine construction and introduction of a new set of windings.

21.5 Damper Winding

Considering that $d\delta/dt = \omega_e - \omega_m$, the required damping torque $\Delta T_{em} = k_P \, d\delta/dt$ is proportional to the slip $\Omega_e - \Omega_m$. At steady state, synchronous speed Ω_e is equal to the rotor speed Ω_m. During transients caused by the load torque disturbances, the

Fig. 21.5 Damper winding
built into heads of the rotor
poles. Conductive rotor bars
are short-circuited at both
sides by conductive plates

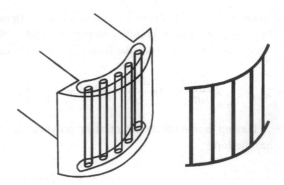

rotor speed oscillates around the synchronous speed, which gives rise to a nonzero slip $\omega_e - \omega_m$ frequency. By adding a short-circuited cage in the rotor of a synchronous machine, it is possible to obtain performance similar to that of an induction machine. Squirrel cage rotor of induction machines provides electromagnetic torque which is proportional to the slip. The same way, a short-circuited cage winding mounted on the rotor of synchronous machine contributes to the torque component which is proportional to the first derivative of the power angle $d\delta/dt = \omega_e - \omega_m = \omega_k$.

In a synchronous machine with a cylindrical, laminated magnetic circuit of the rotor, short-circuited cage winding is inserted in the same way as in induction machines. Conductive cage bars are inserted within the rotor magnetic circuit in axial direction, next to the rotor surface, in close vicinity of the air gap. At both ends of the rotor cylinder, the bars are short-circuited by conductive plates or rings. In cases where the rotor magnetic circuit has salient poles, as shown in Fig. 21.5, the bars are built into the pole heads. In high-power synchronous generators used in thermal power plants, designed to operate with high peripheral speeds, the rotor magnetic circuit has elements made of nonlaminated steel, so as to achieve mechanical robustness in the presence of large centrifugal forces. In such cases, it is not necessary to use short-circuited cage on the rotor. During transient response, where the difference between the synchronous speed Ω_e and the rotor speed Ω_m contributes to the slip frequency $\omega_e - \omega_m = \omega_k$, magnetic induction within the rotor magnetic circuit changes and produces eddy currents within the rotor parts made of nonlaminated conductive steel. Eddy currents within pieces of homogeneous steel create the effect which is the same as the effect made by a short-circuited cage.

The above considerations propose insertion of a new set of windings on the rotor of synchronous machines, called damper winding. This new winding is used to suppress the oscillations in synchronous machines supplied from a stiff network. The question arises whether this new winding changes the steady state behavior of the machine.

A short-circuited cage in the rotor of synchronous machine does not have any impact on the machine operation at steady states. With the stator field which revolves in synchronism with the rotor, there is no relative motion between the rotor and the field. Therefore, there is no change of magnetic induction within the rotor magnetic circuit. For this reason, the flux of short-circuited cage winding does not change, and it produces no electromotive forces. Consequently, there are no electrical currents within the cage windings, and no change in the torque and flux. Hence, the machine operates in the same way as it would perform without the cage winding. During transient processes, where the rotor moves relative to the stator field, the first derivative of the power angle is equal to $\omega_e - \omega_m = \omega_k \neq 0$. For that reason, the flux within the cage winding changes, and it produces electromotive forces which depend on the slip ω_k. This electromotive force produces electrical currents in short-circuited cage. The angular frequency of these currents is equal to $d\delta/dt = \omega_k$. Through the interaction with the magnetic field, the cage currents contribute to electromagnetic torque which is proportional to the slip, $\Delta T_{em} = k_P \, d\delta/dt$. This torque does not exist in steady state conditions. The torque ΔT_{em} has stabilizing effects, and it contributes to decay of oscillations in synchronous machines supplied from stiff networks.

21.6 Short Circuit of Synchronous Machines

Large power synchronous generators used in hydro power plants or thermal power plants have the rated power from several tens to several hundreds of MW. The line voltages across the stator windings range from 6 up to 25 kV. In most cases, each generator has its own power transformer, called *block transformer*. The stator terminals are connected to the primary winding of this power transformer. The transformer has secondary voltages ranging from 110 up to 700 kV. The secondary winding is connected to the three-phase high-voltage transmission network. The network extends and reaches large cities and industrial areas, where the high voltage is supplied to a set of power transformers which reduce the voltage. Following that, distribution networks feed the electrical energy to individual consumers.

Electrical power system comprises electrical generators, transmission lines, power transformers, distribution lines, and also various consumers. It has large size and spreads throughout whole countries. Consequently, there are quite frequent faults such as short circuit, caused by component failures, weather conditions, human error, and other reasons. Electrical power systems include a sophisticated protection system which detects the short-circuit faults, interrupts the short-circuit currents, and isolates the faulty part of the network. Proper functionality of said protection system relies on synchronous generators supplying the short-circuit current, so as to enable distinguishing the short-circuit event, its type, and location. For that reason, it is of interest to study behavior of synchronous machines brought into short circuit.

When a short circuit occurs on the high-voltage transmission line, next to the power plant, the short-circuit current which is fed from synchronous generator is directly proportional to the stator electromotive force and inversely proportional to the equivalent short-circuit impedance. The latter depends on the series impedance of the high-voltage transmission line, on the series impedance of the block transformer, and on the internal impedance of the generator windings.

In a synchronous generator with constant excitation current, and with negligible resistance R_S, the short-circuit current can be determined from the steady state equivalent circuit in Fig. 20.1. It is obtained by dividing the electromotive force E_0 and the series reactance $X_S = \omega_e L_S$. Excitation winding of large synchronous generators is supplied from controllable sources of DC voltage which have a finite output impedance. Therefore, during the short-circuit transients, the assumption $i_R = I_R = $ const. does not hold. Moreover, large synchronous generators also have sets of damper windings. The presence of damper windings in the rotor of synchronous generator reduces its equivalent impedances and increases the short-circuit current. More exact calculation of the short-circuit current requires the model of synchronous machine to be extended by including short-circuited damper windings.

In Chap. 15, it is shown that the short-circuited cage winding can be modeled by a system of two orthogonal short-circuited rotor windings. Complete model of synchronous machine with damper windings comprises five coupled windings. These five windings are (1) the excitation winding, (2) the stator winding of d-axis, (3) the stator winding of q-axis, (4) short-circuited damper winding in d-axis, and (5) short-circuited damper winding in q-axis. The use of complete model with five coupled windings allows more precise prediction of short-circuit currents in the stator winding. This model is relatively complex, and it is not developed nor used in this chapter. Instead, an approximate calculation of the short-circuit current is proposed which takes into account the impact of the excitation winding and the impact of the damper winding.

When a short circuit occurs at terminals of the stator winding, the stator voltage drops to $\underline{U}_S = 0$. In an actual short-circuit condition across the high-voltage transmission line, the stator voltage is not equal to zero. There is a certain series impedance $\underline{Z}_1$ between the stator terminals and the short-circuit location. The analysis of this condition can be simplified by adding the impedance $\underline{Z}_1$ to the stator impedance and considering that $\underline{U}_S = 0$. This means that the reactance and resistance of the stator winding are to be increased by the amount corresponding to the equivalent series reactance and series resistance between the machine terminals and the short-circuit site. In further considerations, it will be assumed that the external inductance and resistance which separate the machine from the short-circuit site are included in the inductance and resistance of the stator winding. With that in mind, further analysis is carried out assuming that the stator winding is short-circuited ($\underline{U}_S = 0$), even in cases where the short circuit occurs along the high-voltage transmission line.

The subsequent analysis of short circuits is carried out under assumption that the stator current before the fault is equal to zero. In addition, stator resistance R_S in high-power machines can be neglected; thus, the voltage balance equations take the form

$$\underline{u}_{dq} = R_S \underline{i}_{dq} + d\underline{\Psi}_{dq}/dt + j\omega_m \underline{\Psi}_{dq},$$
$$\Rightarrow \quad \underline{u}_s = 0 = d\underline{\Psi}_{dq}/dt + j\omega_m \underline{\Psi}_{dq}. \tag{21.2}$$

Transient processes in a short-circuited synchronous machine include variations of the flux components in d- and q-axes. Therefore, it is not justified to assume that the machine operates in the steady state, where the first derivatives of the flux components in dq coordinate frame are equal to zero. On the other hand, complete mathematical model of synchronous machine with stator windings, excitation winding, and damper windings is rather complex. In order to obtain an estimate of shortcircuit currents, it is of interest to simplify the analysis by introducing certain approximations. These approximations are based on the fact that the time constants that characterize transient processes in damping cage windings decay quickly, while transient processes in excitation winding last longer. Therefore, the short-circuit process is split in three intervals:

- *Subtransient* interval, where transient processes in damping cage have not ceased and where the damping cage current contributes to the short-circuit current.
- *Transient* interval that follows subtransient interval and starts after the damping cage currents has decayed to zero.
- *Steady state* interval, which follows after the transient phenomena in both damping and excitation winding have ceased and where the short-circuit current can be determined from the steady state equivalent circuit in Fig. 20.1.

21.6.1   DC Component

The steady state operation of electrical circuits with AC currents and voltages can be disturbed by stepwise changes of voltage or current sources, by changing the circuit impedance, or by introducing short circuits or open circuits. Whenever such transients occur, the process of passing from the previous steady state condition into a new steady state condition may involve a certain amount of DC current that exists during transients and that decays as the circuit enters the new steady state condition. Well-known example involves connecting the coil of the self-inductance L to a voltage source $u_L(t) = U_m \sin(\omega t)$ at instant $t = 0$, where $i_L(0) = 0$. The coil current for $t > 0$ is equal to $i_L(t) = (U_m/L/\omega)\cdot(1 - \cos(\omega t))$, and it contains a DC component. In coils with a finite resistance R, this DC current decays with the time constant of $\tau = L/R$. Similar processes take place in short-circuited synchronous machines.

During short circuit of the stator winding, the phase voltages are equal to zero. Therefore, the voltage balance equation in phase a is $u_a = 0 = R_S i_a + d\Psi_a/dt$. In large synchronous machine, the voltage drop $R_S i_a$ is very small. Therefore, equation reduces to $d\Psi_a/dt = 0$, meaning that the flux in each phase tends to retain the value $\Psi_a(0^-)$ which corresponds to the flux at the instant of making the short circuit. Therefore, it is assumed that the flux in short-circuited windings remains constant during the short-circuit fault. Even large power synchronous machines have a finite value of the stator resistance $R_S > 0$. Although very small, the voltage $R_S i_a$ does influence the winding flux, in particular, over longer intervals of time.

Considering that the short circuits are of limited duration, the assumption $d\Psi_a/dt = 0$ can be successfully used to facilitate calculation of short-circuit currents.

During steady state operation before the short circuit, the flux in phase a varies as sinusoidal function of the frequency $\omega_m = \omega_e$. The average value of the flux Ψ_a prior to the short circuit is equal to zero. At the instant of the short circuit, the value of flux is $\Psi_a = \Psi_a(0^-)$. If the voltage drop $R_S i_a$ is negligible, the flux in phase a retains the value $\Psi_a(0^-)$ during the short-circuit conditions. Therefore, flux Ψ_a comprises a DC component. Due to the presence of a small but finite voltage drop $R_S i_a$, the DC component of the flux exponentially decays. The time constant $\tau = L_{\gamma eS}/R_S$ determines the rate of change of the DC component, and it depends on the equivalent inductance $L_{\gamma eS}$ and the stator resistance R_S. Larger values of the winding resistance result in quicker decay of DC component. Detailed analysis of changes in the machine torque, speed, and $\alpha\beta$ and dq variables that are caused by the DC component of the stator flux requires evaluation of the complete mathematical model. This model includes the damping cage windings, and it is rather involved and difficult to evaluate. Therefore, further considerations are based on the assumption that the time constant $\tau = L_{\gamma eS}/R_S$ is relatively small and that the DC components of the stator flux decay rather quickly. With DC component decay time significantly shorter than the duration of the short-circuit phenomena, it is possible to neglect the DC component and to assume that the initial value of the flux is equal to zero. With that in mind, the initial value of the phase flux linkages at the short-circuit instant is given by equations $\Psi_a(0^+) = \Psi_b(0^+) = \Psi_c(0^+) = 0$, which results in $\Psi_d(0^+) = \Psi_q(0^+) = 0$.

The purpose of this discussion is to obtain an approximate change of the short-circuit current. The analysis is simplified by the assumption that $\Psi_d(0^+) = \Psi_q(0^+) = 0$. The assumption will be used in calculating the short-circuit current I_{SC}.

The short-circuit current is affected by the value of the excitation current I_R at the instant of the short circuit and on the values of the stator currents at the same instant. The short-circuit current of synchronous machine that was running with no load prior to the fault is different than the short-circuit current of loaded machine. In order to keep the analysis simple, the subsequent steps consider the situation where the short circuit occurs while the machine is not loaded. Hence, the stator currents are equal to zero in the wake of the short circuit, resulting in $i_d(0^-) = 0$ and $i_q(0^-) = 0$. Notwithstanding the abovementioned assumptions, the subsequent analysis helps the reader understand the basic intervals of short-circuit transients, and it also provides the means to obtain a rough estimate of the short-circuit currents in each interval:

- During *subtransient* interval, both the damping cage and the excitation winding contribute to short-circuit current I_{SC3}.
- In the following *transient* interval, the cage currents have ceased, and the short-circuit current I_{SC2} is aided by the excitation winding.
- *Steady state* short-circuit current I_{SC1} is calculated assuming that transient processes in the excitation winding have ended and that the excitation current I_R is constant.

Within the subsequent considerations, the short-circuit current I_{SC1} is calculated first, assuming that the excitation current $i_R(t) = I_R$ does not change. In calculation of I_{SC1}, it is justifiable to consider that the excitation winding is supplied from a current source and that the machine does not have damping cage.

Then, the short-circuit current I_{SC2} is calculated, considering transient phenomena in the excitation winding which is supplied from the voltage source.

Finally, the short-circuit current I_{SC3} is calculated by taking into account the impact of transient processes in damping cage and transient processes in excitation winding on the current in short-circuited stator windings.

21.6.2   Calculation of I_{SC1}

It is of interest to calculate the short-circuit current I_{SC1} which exists in the stator windings when the transient processes in other windings have decayed and when the excitation current $i_R(t) = I_R$ does not change. In this calculation, it is justifiable to assume that the damper winding does not exist and that the excitation winding is supplied from the current source.[2] Considering condition $\Psi_d = \Psi_q = 0$, relations $L_q i_q = 0$ and $L_d i_d + L_m I_P = 0$ are obtained for a synchronous machine which does not have damper windings.[3] Components of the current I_{SC1} are given by (21.3), where E_0 represents the no load electromotive force obtained with the excitation current I_R:

$$i_q = 0,$$

$$i_d = -\frac{L_m I_R}{L_d} = -\frac{\omega_m L_m I_R}{\omega_m L_d} = -\frac{\omega_m L_m I_R}{\omega_m L_d} = -\frac{E_0}{X_d},$$

$$I_{SC1} = \frac{E_0}{X_d} \tag{21.3}$$

Therefore, the rms value of the short-circuit current is $0.707 \cdot I_{SC1} = 0.707 E_0 / X_d$. When the short-circuit fault persists, the current I_{SC1} does not decay in time, and it is retained in the stator winding until the short circuit is disconnected. Instantaneous values of the corresponding phase currents can be obtained by applying the inverse Park and inverse $3\Phi/2\Phi$ transform to the components obtained in (21.3).

---

[2] Excitation windings of large synchronous machines are supplied from adjustable sources of DC voltage. The source voltage is used as the driving force which is used to control the excitation current. At the wake of the short circuit event, the changes in the excitation voltage can be neglected, and it is justifiable to assume that the excitation winding is supplied from the voltage source that provides a constant voltage. Later on, as the transient phenomena decay while the short circuit persists, it is justifiable to assume that the excitation current is constant, namely, that the excitation winding is supplied from a source of constant current.

[3] In the considered case, the machine is equipped by damper windings, but the transient processes in these windings have ceased, and the electrical currents in damper windings are equal to zero.

21.6.3 Calculation of I_{SC2}

The excitation winding of large synchronous machines is supplied either from an auxiliary DC generator, which provides adjustable excitation voltage and which is called *exciter*, or from a static power converter with large thyristors, which converts the three-phase AC voltages into adjustable DC voltage. Both the exciter machine and the static power converter can be modeled as a DC voltage source which provides adjustable voltage, and which has a finite, relatively low internal resistance. The excitation voltage is adjusted by the excitation controller. As the shortcircuit takes place all of a sudden, the excitation voltage remains constant for a while, and it retains the value $U_R = R_R I_R$ that existed before the short circuit. Notice that I_R denotes the value of the excitation current prior to the short circuit, while $i_R(t)$ denotes the change of the excitation current during the transients.

The excitation winding is magnetically coupled to the stator winding, which is short-circuited. Due to a finite mutual inductance between the stator windings and the excitation winding, the short-circuit stator currents affect the mutual flux. With the excitation winding supplied from a voltage source, these changes introduce variation in the excitation current. Assuming that the voltage drop $U_R = R_R I_R$ across the resistance of the excitation winding is relatively small, the voltage balance equation for the excitation winding reduces to the expression $d\Psi_R/dt = 0$. Therefore, with excitation winding supplied from the voltage source, and with $R_R I_R \approx 0$, the flux $\Psi_R = L_R i_R + L_m i_d$ retains the value $L_R I_R$ which existed prior to the short-circuit fault.

On the bases of the previous considerations, the stator flux components maintain the values $\Psi_d(0^+) = \Psi_q(0^+) = 0$ throughout the short-circuit transients. One part of the stator flux encircles the excitation winding. The axis d of synchronous dq frame coincides with magnetic axis of the excitation winding. Therefore, condition $\Psi_d(0^+) = 0$ reduces the flux in the excitation winding. Variation of the flux in the excitation winding causes induction of an electromotive force which acts toward increasing the excitation current $i_R(t)$, in an attempt to suppress the reduction in the excitation flux and to maintain the flux linkage that existed prior to the short circuit. Neglecting the voltage drop $R_R i_R$ in the voltage balance equation for the excitation winding, it is reasonable to assume that the relation $d\Psi_R/dt \approx 0$ holds for relatively short intervals of time. With that in mind, the excitation flux Ψ_R tends to retain the value Ψ_{R0} that existed prior to the short-circuit fault. Therefore,

$$L_q i_q = 0, \quad \Rightarrow \quad i_q = 0,$$

$$\Psi_R = L_R i_R + L_m i_d = \Psi_{R0} = L_R I_R, \quad \Rightarrow \quad i_R = I_{R0} - \frac{L_m}{L_R} i_d, \tag{21.4}$$

$$\Psi_d = L_d i_d + L_m i_R = 0 = L_d i_d + L_m I_{R0} - \frac{L_m^2}{L_R} i_d.$$

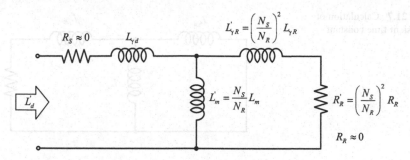

Fig. 21.6 Simplified equivalent scheme of short-circuited synchronous machine with no damper winding and with the excitation winding supplied from voltage source

Current i_d during transient interval is given by (21.5):

$$i_d = -\frac{L_m I_{R0}}{L_d - \frac{L_m^2}{L_R}} = -\frac{L_m I_{R0}}{\left(\frac{L_d L_R - L_m^2}{L_R}\right)} = -\frac{L_m I_{R0}}{L_d'}. \qquad (21.5)$$

Inductance denoted by L_d' in the previous equation resembles the equivalent primary leakage inductance of the transformer whose secondary winding is short-circuited. In the considered case, the stator winding in d-axis corresponds to the primary winding, and the short-circuited excitation winding corresponds to the secondary winding.[4] The equivalent scheme of this transformer is given in Fig. 21.6, where $L_{\gamma R}' = (N_S/N_R)^2 L_{\gamma R}$ is the leakage inductance of the excitation winding referred to the stator side, while $L_m' = (N_S/N_R)L_m$ is the mutual inductance referred to the stator side. By referring the inductances to the stator side, the self-inductance of the stator winding in d-axis can be represented as $L_d = L_{\gamma d} + L_m'$, where $L_{\gamma d}$ is the stator leakage inductance. Self-inductance of the excitation winding referred to the stator side can be determined as $L_R' = L_{\gamma R}' + L_m'$.

Since leakage inductances are much lower than mutual inductances, parameter L_d' is approximately equal to the sum of the leakage inductance of the stator winding $L_{\gamma d}$ and the leakage inductance of the excitation winding $(N_S/N_R)^2 L_{\gamma R}$ referred to the stator side,

$$L_d' = \frac{X_d'}{\omega_m} = \frac{L_d L_R - L_m^2}{L_R} \approx L_{\gamma d} + L_{\gamma R}'. \qquad (21.6)$$

Inductance L_d' is dependent on the coefficient of magnetic coupling k between the stator windings and the excitation winding. Practical values of inductance L_d' are considerably smaller than the self-inductance L_d. The stator current amplitude

---

[4] The assumption $R_R I_R \approx 0$ reduces the voltage across the excitation winding to zero, which means that it behaves as a short-circuited winding.

Fig. 21.7 Calculation of
transient time constant

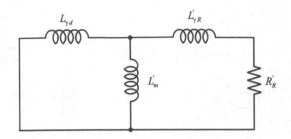

I_{SC2} which exists in the stator phase windings during the transient interval of the short circuit is given by (21.7). It should be noted that $I_{SC1} < I_{SC2}$:

$$i_d = -\frac{L_m I_{R0}}{L'_d}, \quad I_{SC2} = \frac{L_m I_{R0}}{L'_d} = \frac{E_0}{X'_d} \tag{21.7}$$

The short-circuit current I_{SC2} during transient interval is larger than the steady state short-circuit current I_{SC1}, due to the fact that transient processes in the excitation winding contribute to the short-circuit current. If the short circuit persists as the transient processes in the excitation winding decay, the transient current I_{SC2} reduces toward the steady state value I_{SC1}.

Negative value of the stator current component i_d contributes to an increase in the excitation current i_R, which tends to maintain the excitation flux at the value which existed before the short-circuit event. Since the excitation voltage does not change, the voltage balance equation in the excitation winding has the form $U_R = R_R I_R = R_R i_R + d\Psi_R/dt$. During transient processes, the first derivative of the excitation flux is negative, $d\Psi_R/dt = R_R(I_R - i_R) < 0$, and therefore, the excitation flux reduces. Reduction of the excitation flux reduces the electromotive force induced in the stator winding. As a consequence, the short-circuit current in the stator windings is reduced as well.

The change of the excitation flux and the excitation current during transient interval can be analyzed on the bases of equivalent circuit obtained from Fig. 21.6 and shown in Fig. 21.7, where the stator resistances are removed while the stator winding is short-circuited.

The above RL circuit is characterized by the time constant $\tau' = L'_e/R'_R$ which determines the exponential change of electrical currents in the circuit. Parameter L'_e is the equivalent inductance connected in series with the resistance R_R'. Therefore, $L'_e = L_{\gamma R}' + L_{\gamma d} L_m'/(L_{\gamma d} + L_m')$. The transient time constant τ' is given in (21.8):

$$\tau' = \frac{1}{R'_R}\left(L'_{\gamma R} + \frac{L_{\gamma d}L'_m}{L_{\gamma d} + L'_m}\right). \tag{21.8}$$

In most synchronous machines, the transient time constant τ' is considerably larger than one period of the stator current $T = 2\pi/\omega_m \approx 2\pi/\omega_e = 20$ ms. The change of the flux amplitude during one period T is very small. With small values of

T/τ', exponential decay within one period of stator currents can be approximated by the expression $\exp(-T/\tau') \approx 1 - T/\tau' \approx 1$. The stator currents during short circuit can be regarded as sinusoidal, with an amplitude that decays exponentially.

From the previous analysis, it is possible to obtain the change of the short-circuit current in a synchronous machine with no damping cage winding and with voltage-supplied excitation winding. In absence of the damper windings, the short-circuit phenomena do not have subtransient interval. During the transient interval, immediately upon the short circuits is established, the amplitude of the stator phase current is I_{SC2} (21.7). It decays exponentially and reaches the steady state value I_{SC1}. The change of the current amplitude is determined by the time constant τ'. The actual change of the phase current $i_a(t)$ depends on the initial phase of the stator voltages prior to the short circuit, on the instant of establishing the short circuit, and also on the circuit resistances that have been neglected. As an example, the phase current $i_a(t)$ may change as

$$i_a(t) \approx \left\{ I_{SC1} + (I_{SC2} - I_{SC1}) \cdot \exp\left(-\frac{t}{\tau'}\right) \right\} \sin(\omega_m t + \varphi_a).$$

The analysis performed above takes into account dynamic phenomena in d-axis. In the considered case, the currents and fluxes in q-axis were equal to zero prior to the short circuit. The accuracy in calculating the short-circuit currents can be further enhanced by taking into account the coupling of transient processes d-axis and q-axis. The coupling between transient phenomena in orthogonal axes can be understood from differential equations expressing the voltage balance in virtual d-winding and q-winding. It has to be noted at this point that the q-axis does not have magnetic coupling with the excitation winding. The rotor of synchronous machines comprises only one excitation winding, and this winding has magnetic axis aligned with d-axis of the dq coordinate frame. For this reason, (21.3) and (21.7) provide an adequate approximation of the short-circuit current during transient and steady state intervals.

21.6.4 Calculation of I_{SC3}

The presence of short-circuited damping cage increases the amplitude of the short-circuit current. Short-circuited windings exhibit the tendency to maintain the flux. Any flux change results in induced electromotive forces and electrical currents in short-circuited windings that oppose to the flux change and act toward retaining the flux at the previous level. While the short-circuit event tends to drive the stator flux to zero, the damper windings act toward maintaining the flux. With larger flux, the stator electromotive forces are larger, which results in an increased amplitude of the short-circuit currents. This effect dies out as the transient processes in the damper windings cease.

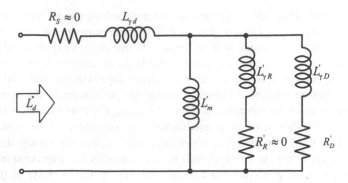

Fig. 21.8 Simplified equivalent scheme of short-circuited machine with damper winding and voltage-supplied excitation winding

The damper windings can be represented by two orthogonal short-circuited windings, one aligned with d-axis and the other aligned with q-axis. Adding the short-circuited rotor cage to the machine described in the previous example, the machine model in synchronously rotating dq coordinate frame obtains three windings that are aligned with d-axis. One of them is virtual stator phase in d-axis, the second is the excitation winding, while the third is the damper winding aligned with d-axis. The flux of the stator winding in d-axis is denoted by Ψ_d, the excitation flux by Ψ_R, while the flux of the short-circuited damper winding of d-axis is denoted by Ψ_D.

The three coupled windings residing in d-axis resemble a transformer with primary, secondary, and tertiary winding. The primary is the stator winding, while the secondary and tertiary windings are the excitation winding and the damper winding of d-axis. Considerations and assumptions adopted for transient interval of the short circuit are valid for subtransient interval as well. Namely, I_{SC3} calculation can be based on the assumption that the excitation winding and the damper winding are short-circuited throughout the subtransient interval. The equivalent transformer is shown in Fig. 21.8, and its secondary and tertiary windings are in short circuit.

The short-circuit current I_{SC3} during the first, subtransient interval can be determined from the inductance L''_d called *subtransient inductance*. The subtransient inductance corresponds to the equivalent primary inductance of the short-circuited transformer in Fig. 21.8. As a matter of fact, it is the equivalent leakage inductance of the three-winding transformer that is observed from the primary side. Inductance L''_d determines the subtransient reactance $X''_d = \omega_m L''_d$. The short-circuit current I_{SC3} is inversely proportional to the subtransient reactance X''_d. The subtransient inductance L''_d is given in (21.9):

$$L''_d = L_{\gamma d} + \left(\frac{1}{L'_m} + \frac{1}{L'_{\gamma R}} + \frac{1}{L'_{\gamma D}} \right) \approx L_{\gamma d} + \frac{L'_{\gamma R} L'_{\gamma D}}{L'_{\gamma R} + L'_{\gamma D}}. \qquad (21.9)$$

The flux linkages and electrical currents of the three-winding system shown in Fig. 21.8 are related by the inductance matrix which is given in (21.10). Subtransient inductance L''_d can be determined by using the inductance matrix and comparing the flux values before the short-circuit event and after the short-circuit event. In order to determine the initial value of the short-circuit current I_{SC3}, it is necessary to define the flux values $\Psi_d(0^-)$, $\Psi_R(0^-)$, and $\Psi_D(0^-)$ before the short circuit as well as the flux values $\Psi_d(0^+)$, $\Psi_R(0^+)$, and $\Psi_D(0^+)$ immediately after the short circuit.

Preliminary assumptions in this short-circuit analysis are that the initial values of the stator currents are equal to zero ($i_d(0^-) = i_q(0^-) = 0$) and that the excitation current is equal to the steady state value I_{R0} ($i_R(0^-) = I_{R0}$). The machines enter the short-circuit transient while operating at steady state. Therefore, the rotor speed is equal to the synchronous speed ($\omega_m = \omega_e$), while the electrical current of the d-axis damper winding is equal to zero ($i_D(0^-) = 0$). Relation between currents and flux linkages of the three d-axis windings is given by the following inductance matrix:

$$[\Psi] = \begin{bmatrix} \Psi_d \\ \Psi_D \\ \Psi_R \end{bmatrix} = \begin{bmatrix} L_d & L_{dD} & L_m \\ L_{dD} & L_D & L_{RD} \\ L_m & L_{RD} & L_R \end{bmatrix} \begin{bmatrix} i_d \\ i_D \\ i_R \end{bmatrix} = [L_{dRD}][i_{dRD}]. \tag{21.10}$$

The element L_{dD} of the matrix denotes the mutual inductance between the stator d-axis winding and the damper winding of d-axis. The element L_{RD} denotes the mutual inductance between the excitation winding and the damping of d-axis, while the element L_m denotes the mutual inductance between the stator d-axis winding and the excitation winding. Coefficients L_D, L_d, and L_R represent the self-inductances of the relevant windings. Since the values of electrical currents in considered windings were $i_d(0^-) = i_D(0^-) = 0$, and $i_R(0^-) = I_{R0}$, corresponding values of the flux linkages prior to the short-circuit event are $\Psi_D(0^-) = L_{RD}I_{R0}$, $\Psi_R(0^-) = L_R I_{R0}$, and $\Psi_d(0^-) = L_m I_{R0}$. The damper winding is short-circuited, while the excitation winding is connected to a voltage source of a relatively low internal resistance. Neglecting the resistance R_R of the excitation winding, it is justifiable to assume that both the excitation voltage $R_R I_{R0}$ and the voltage drop $R_R i_R$ are negligible and that the excitation winding is short-circuited during the subtransient interval, as well as the damper winding.

Electromotive forces induced in short-circuited windings tend to maintain the winding flux. Any flux changes results in induced electromotive forces and consequential currents that oppose to flux changes and suppress any rapid change of the flux. Therefore, the flux linkages at the wake of the short circuit are $\Psi_D(0^+) = \Psi_D(0^-)$ and $\Psi_R(0^+) = \Psi_R(0^-)$.

The initial flux of the stator winding was discussed while considering transient interval and steady state short circuit, and it was pointed out that the initial value of the stator flux is to be considered equal to zero. In this way, calculation of the short-circuit current is simplified while still providing an appropriate estimate of the current amplitude. The same discussion will be repeated hereafter.

With negligible voltage drop $R_S i_S$, the voltage balance in the stator phase a following the short-circuit event reduces to $d\Psi_a/dt = 0$. Therefore, the stator flux retains the value $\Psi_a(0^-)$; hence, $\Psi_a(0^+) = \Psi_a(0^-)$. In a hypothetical case where $R_S = 0$ and where the stator winding is short-circuited, the stator flux Ψ_a retains the initial value indefinitely. Since the stator resistance has a small but finite value, the voltage drop across the stator resistance acts toward reducing the flux in short-circuited stator winding. Depending on the short-circuit instant, the stator flux and current retain a certain initial value that can be regarded as a DC component. This DC component decays exponentially. The time constant of this exponential decay is shorter for larger values of the stator resistance. Eventually, the DC components of the stator currents and flux linkages cease, while the AC component of the short circuit remains at an amplitude determined by the electromotive forces and reactances of the machine. The AC component of the short-circuit current is considered of primary interest. Therefore, the impact of the initial DC component on the rms and peak values of the short-circuit stator currents are neglected in this analysis. For this reason, it is justifiable to assume that the initial values of the stator flux linkages at the instant of the short circuit are equal to zero. The subsequent analysis starts with the assumption that the initial flux of the stator phases reduces quickly from the initial values $\Psi_a(0^-)$, $\Psi_b(0^-)$, and $\Psi_c(0^-)$ down to zero. On that grounds, it is assumed that $\Psi_a(0^+) = \Psi_b(0^+) = \Psi_c(0^+) = 0$, which results in $\Psi_d(0^+) = \Psi_q(0^+) = 0$. Now the d component of the stator current $i_d(0^+)$ is calculated from (21.11):

$$\begin{bmatrix} i_d(0^+) \\ i_D(0^+) \\ i_R(0^+) \end{bmatrix} = [L_{dRD}]^{-1} \begin{bmatrix} \Psi_d(0^+) \\ \Psi_D(0^+) \\ \Psi_R(0^+) \end{bmatrix} = \begin{bmatrix} L_d & L_{dD} & L_m \\ L_{dD} & L_D & L_{RD} \\ L_m & L_{RD} & L_R \end{bmatrix}^{-1} \begin{bmatrix} 0 \\ L_{RD}I_{R0} \\ L_R I_{R0} \end{bmatrix},$$

$$i_d(0^+) = -\frac{L_m\left(L_R L_D - L_{RD}^2\right)}{L_d L_D L_R - L_d L_{RD}^2 - L_R L_{dD}^2 + 2L_{dD}L_m L_{RD} - L_D L_m^2} I_{R0}. \qquad (21.11)$$

The above expression is relatively complex and needs to be elaborated. Without the lack of generality, it is possible to refer all the inductances to the stator side, by multiplying them with the squared transformation ratio, the number determined by the number of turns in relevant windings.

Mutual flux of the three windings passes through the same magnetic circuit of magnetic resistance R_μ. Transforming the excitation winding and the damper winding to the stator side actually means representing these windings by equivalents which have the same number of turns as the stator winding. Therefore, it is reasonable to assume that all the mutual inductances transformed to the stator side are equal, $L'_m = L'_{RD} = L'_{dD}$. At the same time, all the self-inductances can be represented as sums of corresponding leakage inductances and mutual inductances; hence, $L_d = L'_m + L_{\gamma S}$, $L'_D = L'_m + L'_{\gamma D}$, and $L'_R = L'_m + L'_{\gamma R}$. The values L'_{RD}, L'_{dD}, L'_D, $L'_{\gamma D}$, L'_R, and $L'_{\gamma R}$ correspond to relevant inductances transformed to the stator side. Assuming that each of the leakage inductances is considerably smaller

Fig. 21.9 Calculating the subtransient time constant

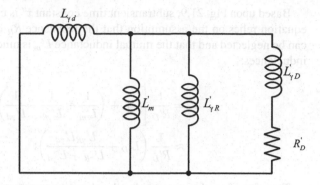

than the mutual inductance, expression for the current $i_d(0^+)$ can be reduced to the following form:

$$i_d(0^+) = -\frac{L_m(L'_{\gamma R} + L'_{\gamma D})}{L'_{\gamma R}L'_{\gamma D} + L_{\gamma d}L'_{\gamma D} + L_{\gamma d}L'_{\gamma R}}I_{R0} = -\frac{L_m I_{R0}}{\frac{L'_{\gamma R}L'_{\gamma D}}{L'_{\gamma R}+L'_{\gamma D}} + L_{\gamma d}} = -\frac{L_m I_{R0}}{L''_d}.$$

Inductance L''_d in the previous expression is subtransient inductance, and it has been given in (21.9). The amplitude of subtransient short-circuit current I_{SC3} is given in (21.12). This amplitude depends on the electromotive force E_0, and it is inversely proportional to subtransient reactance. The current I_{SC3} exists in the stator phase windings at the very beginning of the short-circuit fault, during subtransient interval. It is larger than transient current I_{SC2}, which determines the short-circuit current amplitude during the second, transient interval of the short-circuit event:

$$I_{SC3} = \frac{L_m I_{R0}}{L''_d} = \frac{E_0}{\omega_m L''_d}, \qquad (21.12)$$

$$I_{SC3} = \frac{L_m I_{R0}}{\frac{L'_{\gamma R}L'_{\gamma D}}{L'_{\gamma R}+L'_{\gamma D}} + L_{\gamma d}} > \frac{L_m I_{R0}}{L'_{\gamma R} + L_{\gamma d}} = I_{SC2}. \qquad (21.13)$$

The previous analysis shows that electrical currents induced in the damper windings contribute to the amplitude of the stator currents in short circuit. The presence of the damper winding reduces the equivalent leakage inductance of the machine from the value of $L'_{\gamma e} = L_{\gamma d} + L'_{\gamma R}$ to the value of $L''_{\gamma e} = L_{\gamma d} + L'_{\gamma R} L'_{\gamma D}/(L'_{\gamma R} + L'_{\gamma D})$, thus increasing the short-circuit current. Electrical current $i_D(t)$ exists in the bars of short-circuited damping cage. Due to a finite resistance of these bars, the current decays exponentially. The time constant τ'' which defines the exponential decay of damping currents depends on the resistance R_D' of the damper windings and on the equivalent inductance that exists in the circuit with the current $i_D(t)$. According to Fig. 21.9, this inductance includes parallel connection of $L'_{\gamma d}, L'_m$, and $L'_{\gamma R}$ connected in series with $L'_{\gamma D}$.

Based upon Fig. 21.9, subtransient time constant τ'' is calculated in (21.14). This equation relies on the assumption that the resistance R_R of the excitation winding can be neglected and that the mutual inductance L'_m is much larger than the leakage inductances:

$$\tau'' = \frac{1}{R'_D}\left[L_{\gamma D} + \left(\frac{1}{L'_m} + \frac{1}{L'_{\gamma R}} + \frac{1}{L'_{\gamma d}}\right)^{-1}\right]$$

$$\approx \frac{1}{R'_D}\left(L_{\gamma D} + \frac{L'_{\gamma R}L'_{\gamma d}}{L'_{\gamma R} + L'_{\gamma d}}\right). \tag{21.14}$$

Transferred to the stator side, the damping resistance R'_D is larger than the resistance of the excitation winding. Therefore, the subtransient time constant τ'' is shorter than the transient time constant τ'. Therefore, the short-circuit event in synchronous machines has three intervals. It begins with a brief subtransient interval, proceeds with somewhat longer transient interval, and, unless interrupted, enters the steady state short-circuit interval. The three intervals are discussed below.

21.7   Transient and Subtransient Phenomena

21.7.1   Interval 1

The first interval of the short circuit is subtransient interval, which starts immediately after the instant of establishing the short-circuit conditions. The current of the amplitude I_{SC3} is established in the stator windings. This current is directly proportional to the excitation current and inversely proportional to the subtransient inductance of the machine, denoted by L''_d. Subtransient current decays exponentially with the subtransient time constant τ''. After several intervals of τ'', the current amplitude reduces to I_{SC2}. At the end of subtransient interval, transient processes in damper winding have decayed, and damping currents have reduced to zero. At this point, the damper windings cease to make any contribution to short-circuit currents.

21.7.2   Interval 2

During the second interval, called transient interval, the amplitude of short-circuit currents in the stator windings is I_{SC2}. This current is directly proportional to the excitation current and inversely proportional to transient inductance of the machine L'_d. The current amplitude decays exponentially with the time constant τ'. After several transient time constant τ', the amplitude of the stator currents reduces to I_{SC1}.

At the end of transient interval, transient processes in the excitation winding have decayed, and the excitation current settles to the steady state value $I_{R0} = U_R/R_R$. At this point, the excitation winding does not have any transient currents which contribute to short-circuit currents in the stator windings.

21.7.3   Interval 3

After subtransient and transient interval, transient processes in the excitation winding and the damper windings decay. Damping currents reduce to zero, while the excitation current settles on the steady state value. At this point, the short-circuit current can be determined from the steady state equivalent circuit in Fig. 20.1. The steady state short-circuit current in the stator winding has amplitude I_{SC1}, which is directly proportional to the excitation current and inversely proportional to the stator inductance $L_d \approx L_m$. This amplitude does not change until protection mechanisms are activated to disconnect the stator winding from the short circuit.

Interval 1 is commonly called *subtransient interval*, and the reactance $X''_d = \omega_m L''_d$ is called *subtransient reactance*. Interval 2 is called *transient interval*, while reactance $X'_d = \omega_m L'_d$ is called *transient reactance*. Expressions for the short-circuit currents in transient and subtransient intervals are obtained by adopting a series of approximations aimed to simplify calculations and to make the basic short-circuit behavior more obvious. Expressions such as (21.12, 21.13, and 21.14) are relatively simple, and they give an insight into the impact of the machine parameters on the amplitude and dynamic behavior of the short-circuit current. More precise calculations require solving the complete mathematical model of synchronous machine which comprises five magnetically coupled windings.

In synchronous generators, practical relative values of synchronous reactance x_d range from 0.8 up to 2. Relative values of transient reactance x'_d range from 0.2 up to 0.5, while relative values of subtransient reactance range from 0.1 up to 0.3. Transient time constant τ' has values from 400 ms up to 2 s, while subtransient interval τ'' runs from 30 to 150 ms.

Chapter 22
Variable Frequency Synchronous Machines

This chapter studies the operation and characteristics of three-phase synchronous machines connected to three-phase inverters, static power converters capable of adjusting the stator voltages by means of changing the width of the voltage pulses supplied to the stator terminals. The average voltage of the pulse train is adjusted to suit the machine needs. Variable speed operation of synchronous machine is achieved with variable frequency and variable amplitude of stator voltages. This chapter introduces and explains some basic torque and speed control principles. The need of controlling the stator currents is discussed and explained. Fundamental principles of stator current control are introduced, relying on PWM-controlled three-phase inverter as the voltage actuator. Field-weakening performance of inverter-supplied synchronous machines with buried magnets and surface-mounted magnets is analyzed and explained. The limits of constant power operation in field-weakening mode are determined, explained, and expressed in terms of the stator self-inductance. Based upon the study of operating limits of the machine and operating limits of associated three-phase inverter, steady-state operating area and transient operating area are derived in $T - \Omega$ plane and studied for inverter-supplied synchronous machines.

22.1 Inverter-Supplied Synchronous Machines

Synchronous machines of low and medium power are used in applications such as motion control, vehicle propulsion, industrial robots, or production machines. In these applications, synchronous machines have task to overcome motion resistance and provide the required acceleration and deceleration of moving parts. Synchronous motor is used as an actuator which develops the torque required to control the speed or position and to overcome the motion resistances while driving

S.N. Vukosavic, *Electrical Machines*, Power Electronics and Power Systems,   621
DOI 10.1007/978-1-4614-0400-2_22, © Springer Science+Business Media New York 2013

the controlled object along predefined trajectories. Synchronous motors are better suited to these tasks than the other electrical motors. Advantages of synchronous machines over the other types of electrical machines include their high specific power, high specific torque, a low inertia, and relatively low losses. Synchronous machines with permanent magnet excitation have no rotor windings and no rotor losses. Therefore, energy efficiency of these motors is considerably improved over other types of motors. With permanent magnets, the rotor flux is obtained without power losses such as $U_R I_R = R_R I_R^2$, encountered in machines which have the excitation winding. Moreover, there are no iron losses in the rotor magnetic circuit of synchronous machine. Due to synchronous rotation of the rotor and the stator field, there are no pulsations of magnetic induction B in magnetic circuit of the rotor.

With no rotor losses and no heat generated within the rotor, the cooling of permanent magnet synchronous machines is greatly simplified, and it is possible to reach larger current densities in the stator winding and larger magnetic induction, which results in higher specific power.[1] Compared to an induction machine of the same rated power, synchronous machine with permanent magnets has 20–30% lower mass and volume. The power balance charts drawn for induction machines and synchronous machines have the same losses in the stator winding, in the stator magnetic circuit, and in the mechanical subsystem. Induction machines have power losses in the rotor which are equal to sP_δ, where s is relative slip while P_δ is the air-gap power. Synchronous machines do not have the losses sP_δ, and this greatly increases their energy efficiency.

In applications such as motion control, vehicle propulsion, and industry automation, it is required to provide continuous change of the rotor speed. In synchronous machines, the rotor speed corresponds to the synchronous speed $\Omega_e = p\omega_e$, which is determined by the supply frequency ω_e. For this reason, synchronous motors in motion control applications have to be supplied from separate power sources that produce three-phase voltages of variable frequency and variable amplitude, in accordance with the motor needs. Two synchronous motors within the same industrial robot or electrical vehicle most often rotate at different rotor speeds. Therefore, each motor has its own supply frequency, and it requires a separate power source that provides the stator voltages and currents of desired frequency. Such power sources are usually three-phase inverters with transistor power switches. Commutation of power transistors produces a train of variable width voltage pulses. The pulse width of these pulses affects the average voltage within one commutation period. By sinusoidal change of the pulse width, the average voltage exhibits sinusoidal change with an adjustable amplitude and frequency. Pulse-width-modulation techniques enable generation of pulse trains of an average value that corresponds to the motor needs.

---

[1] Specific power is the quotient of the rated power of the machine and the machine volume or weight. Similar definition holds for specific torque.

22.2 Torque Control Principles

In applications where the speed or position of the moving object is controlled and enforced to track some predefined trajectory, synchronous permanent magnet motors are used as *torque actuators*, *executive organs* that provide the driving torque that overcomes the motion resistances and forces the speed or position to maintain desired reference values. Speed regulators and position regulators calculate the difference between the controlled variable Ω_m (θ_m) and its reference value Ω^* (θ^*), and they calculate the electromagnetic torque T_{em}^* required to remove the error $\Delta\Omega = \Omega^* - \Omega_m$ and drive the controlled variable back into the reference track. The speed change is determined by Newton equation of motion, $J\, d\Omega_m/dt = T_{em} - T_m$, where T_m denotes the load torque disturbance, while T_{em} denotes the electromagnetic torque provided by the synchronous motor. The motor torque T_{em} has to overcome the load torque T_m and to provide the acceleration $d\Omega_m/dt$ which is needed to achieve desired speed change. While the speed (position) regulator calculates the torque reference signal T_{em}^*, the synchronous permanent magnet motor has the task of providing the actual shaft torque T_{em} which corresponds to this reference signal. The speed and accuracy of motion control rely on the torque actuator capability of providing the torque which corresponds to the reference. In other words, the electromagnetic torque has to track the reference T_{em}^* accurately and with negligible delay, resulting in $T_{em}^* = T_{em}$.

In Chap. 19, it has been shown that the electromagnetic torque of synchronous machines depends on the vector product of the stator flux and the stator current. Hence,

$$T_{em} = \frac{3p}{2}\left(\Psi_d i_q - \Psi_q i_d\right).$$

In permanent magnet motors, the stator flux components are $\Psi_d = \Psi_{Rm} + L_d i_d$ and $\Psi_q = L_q i_q$, where Ψ_{Rm} is the flux of permanent magnets which passes through the air gap and encircles the stator windings. Introducing the flux Ψ_{Rm}, the torque expression becomes

$$T_{em} = \frac{3p}{2}\Psi_{Rm} i_q + \frac{3p}{2}\left(L_d - L_q\right) i_d i_q.$$

Most synchronous permanent magnet motors have cylindrical form of the rotor magnetic circuit and a very small difference between magnetic resistances in d-axis and q-axis. Therefore, it is justifiable to consider that $L_d = L_q$. With that in mind,

$$T_{em} = \frac{3p}{2}\Psi_{Rm} i_q.$$

Hence, the electromagnetic torque of synchronous permanent magnet motor is determined by the q-component of the stator current. In order to reduce the copper losses in the stator windings and to reduce the rating of the power converter that supplies the stator winding, it is beneficial to deliver desired torque T_{em} with the smallest possible stator current. The ratio T_{em}/I_S between the electromagnetic torque and the stator current I_S is equal to $T_{em}/\mathrm{sqrt}(i_d^2 + i_q^2)$, and it has the minimum value when $i_d = 0$, that is, when the stator current vector is aligned with q-axis. In this case, the rotor flux vector and the stator current vector are displaced by $\pi/2$.

Since the flux Ψ_{Rm} of permanent magnets does not change, the torque of permanent magnet synchronous motor is controlled by changing the stator current i_q. Given the torque reference T_{em}^*, it is necessary to establish the stator currents with their dq components $i_d = 0$ and $i_q = (2/(3p)) \cdot T_{em}^*/\Psi_{Rm} = K \cdot T_{em}^*$. The torque control of permanent magnet synchronous motor is accomplished by regulating the stator current and delivering the stator current component i_q so that it corresponds to the desired torque. The speed and accuracy in delivering desired torque are uniquely defined by the speed and accuracy in regulation of the stator current. Current regulation relies on supplying the stator winding from the three-phase inverter, static power converter which employs switching power transistors in order to adjust the stator voltage and achieve desired phase currents.

Figure 22.1 shows a three-phase inverter with transistor power switches. The inverter supplies the stator winding of a synchronous motor. The mains voltages u_R, u_S, and u_T are rectified within the diode rectifier that makes use of six power diodes. The rectifier provides the DC voltage E which exists in intermediate circuit also called DC link circuit. The elements L_{DC} and C_{DC} serve as the filter that removes AC components from the rectifier output. Braking unit with transistor Q_K and resistor R_K serves to reduce the DC voltage E in braking intervals, when the motor operates as generator which passes the braking power back into the DC link, charging the DC link capacitor and increasing the DC link voltage. The braking energy cannot be returned to the mains due to the nature of the three-phase diode rectifier. Therefore, the breaking power gets dissipated in the resistor R_K. Six power transistors Q_1–Q_6 are commutated in order to obtain pulse-shaped phase voltages. The state of the power transistor switches is determined so as to drive the three-phase currents toward their reference values $i_a^*(t)$, $i_b^*(t)$, and $i_c^*(t)$.

The problem of current control in three-phase motors requires more detailed study, and it is beyond the scope of this chapter. For this reason, only some basic information and principles are presented in further text.

It is of interest to determine the reference values for the phase currents of a synchronous permanent magnet motor with flux Ψ_{Rm}, with the rotor position equal to θ_m, with $p = 1$ pole pairs, and with the torque reference T_{em}^*. Assuming that the current regulator manages to maintain the phase currents on their reference values, the phase current references

$$i_a^*(t) = \left(\frac{2}{3}\frac{T_{em}^*}{\Psi_{Rm}}\right) \cos\left(\theta_m + \frac{\pi}{2}\right),$$

$$i_b^*(t) = \left(\frac{2}{3}\frac{T_{em}^*}{\Psi_{Rm}}\right) \cos\left(\theta_m + \frac{\pi}{2} - \frac{2\pi}{3}\right),$$

$$i_c^*(t) = \left(\frac{2}{3}\frac{T_{em}^*}{\Psi_{Rm}}\right) \cos\left(\theta_m + \frac{\pi}{2} - \frac{4\pi}{3}\right), \tag{22.1}$$

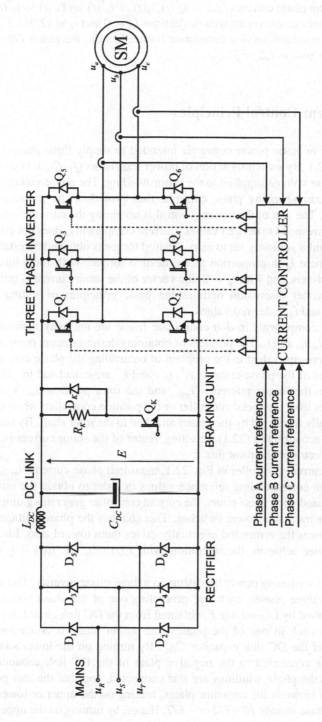

Fig. 22.1 Power converter topology intended to supply synchronous permanent magnet motor and the associated current controller

result in the stator phase currents $i_a(t) = i_a{}^*(t)$, $i_b(t) = i_b{}^*(t)$, and $i_c(t) = i_c{}^*(t)$. They produce the stator current vector with projections $i_d = 0$ and $i_q = (2/3) \cdot T_{em}{}^* / \Psi_{Rm}$ on the axes of synchronous d–q coordinate frame. In turn, the motor delivers the electromagnetic torque $T_{em} = T_{em}{}^*$.

22.3   Current Control Principles

Basic topology of static power converter intended to supply three-phase motor is given in Fig. 22.1. By switching action of power transistors Q_1–Q_6, it is possible to change the phase voltage supplied to the stator winding. The phase voltages u_a, u_b, and u_c affect corresponding phase currents, thus providing the grounds for the current control. The aim of the current control is obtaining the current components i_d and i_q that correspond to desired values. Current component i_d does not contribute to the torque, and it is mostly set to zero. Desired torque is obtained by establishing current component i_q in proportion to the desired torque. With rotor flux vector collinear with d-axis, and with $i_d = 0$, the vector of the stator current is orthogonal to the flux. Further discussion outlines the basic principles of setting current components i_d and i_q to desired values.

The current components in d–q coordinate frame are uniquely defined by the phase currents i_a, i_b, and i_c. The problem of obtaining desired current components i_d and i_q is therefore equivalent to the problem of controlling the phase currents. The reference values of the phase currents $i_a{}^*$, $i_b{}^*$, and $i_c{}^*$ are calculated in (22.1), and they depend on the torque reference $T_{em}{}^*$ and the rotor position θ_m. The torque reference comes from the speed controller or the position controller, while the rotor position is usually measured by the sensor attached to the rotor shaft. By setting the phase currents according to (22.1), resulting vector of the stator current is perpendicular to the vector of the rotor flux.

Within the current controller in Fig. 22.1, measured phase currents i_a, i_b, and i_c are compared to corresponding reference values in order to obtain the errors Δi_a, Δi_b, and Δi_c. Based upon these errors, the current controller generates gating signals that commutate transistor power switches. This changes the phase voltages in the way which reduces the errors and eventually drives them toward zero. Ideally, the current controller achieves the operation with $i_a(t) = i_a{}^*(t)$, $i_b(t) = i_b{}^*(t)$, and $i_c(t) = i_c{}^*(t)$.

There are six switching power transistors in a three-phase inverter. The switches are grouped in three phases, each one generating one of the phase voltages. The inverter is supplied by DC voltage E, obtained from the DC link circuit. By turning on the upper switch in one of the phases, the output phase is connected to the positive plate of the DC link capacitor C_{DC}. By turning on the lower switch, the output phase is connected to the negative plate of the DC link capacitor C_{DC}. Assuming that the phase windings are star connected, and that the star point has the potential in between the capacitor plates, turning on the upper or lower switch results in the phase voltage of $+E/2$ or $-E/2$. Hence, by turning on the upper switch

Q_1 of phase a, the phase voltage $u_a(t)$ becomes $+E/2$, while turning on the lower switch Q_2 results in $u_a(t) = -E/2$.

During one commutation period of $T \approx 100$ µs, the upper switch in phase a is closed (on) during time interval t_{ON}, while in the remaining part of the commutation period, $T - t_{ON}$, the upper switch is opened (off) while the lower switch is closed (on). The phase voltage $u_a(t)$ is equal to $+E/2$ during time interval t_{ON}, and then it remains $-E/2$ for the rest of the period $(T - t_{ON})$. The width t_{ON} of the positive pulse can be varied continuously within the range $0 < t_{ON} < T$. The average value of the phase voltage during one period T is equal to $u_{a(av)} = E \cdot (t_{ON}/T - 1/2)$. Continuous change of the pulse width t_{ON} results in continuous change in the average value of the phase voltage which ranges between $-E/2$ and $+E/2$. Variation of the instantaneous value of the phase current $i_a(t)$ depends on the difference between the instantaneous value of the phase voltage and the electromotive force induced in the winding. Variation of the phase current during one commutation period T depends on the average phase voltage $u_{a(av)}$ during one commutation period. The change in the phase current is based on the voltage balance equation $u_a = R_S i_a + \mathrm{d}\Psi_a/\mathrm{d}t$, where $R_S i_a$ is the voltage drop across the stator resistance while $\mathrm{d}\Psi_a/\mathrm{d}t$ is the electromotive force induced in one phase.

Distribution of magnetic field along the air gap depends on characteristics of permanent magnet and the way of their mounting. The air-gap field is also affected by the stator currents. The flux in the stator phase winding a is $\Psi_a = L_S i_a + \Psi_{Rm(a)}$, and it has two components. Component $\Psi_{Rm(a)}$ is produced by permanent magnets, while the flux $L_S i_a$ depends on the stator currents.[2] Component $\Psi_{Rm(a)}$ depends on the amplitude and spatial orientation of the flux Ψ_{Rm}, which is generated by permanent magnets, which passes through the air gap, and which encircles the stator windings. Amplitude Ψ_{Rm} is determined by characteristics of permanent magnets, by the air gap δ, and by properties of magnetic circuits. The spatial orientation of the permanent magnet flux is determined by the rotor position θ_m. Therefore, the flux $\Psi_{Rm(a)}$ depends on the amplitude Ψ_{Rm} and on the rotor position θ_m. With $\theta_m = 0$, the rotor is aligned with magnetic axis of the phase a, resulting in $\Psi_{Rm(a)} = \Psi_{Rm}$.

The changes in the stator flux Ψ_a produce electromotive force $\mathrm{d}\Psi_a/\mathrm{d}t$. The later has two components, $L_S \, \mathrm{d}i_a/\mathrm{d}t$ and $\mathrm{d}\Psi_{Rm(a)}/\mathrm{d}t$. The second component $e_a = \mathrm{d}\Psi_{Rm(a)}/\mathrm{d}t$ is caused by the flux of permanent magnets, and it depends on the rotor speed $\Omega_m = \mathrm{d}\theta_m/\mathrm{d}t$.

With constant rotor speed, the orientation of the permanent magnet flux is determined by $\theta_m = \theta_0 + p\Omega_m t$. The part of this flux which encircles the phase winding a is equal to $\Psi_{Rm}\cos(\theta_m)$. Besides, the phase winding has also the flux $L_S i_a$; thus, the total flux of this phase is $\Psi_a = \Psi_m\cos(\theta_m) + L_S i_a$. The voltage balance equation becomes $u_a = R_S i_a + L_S \, \mathrm{d}i_a/\mathrm{d}t - \omega_m \Psi_m \sin(\theta_m)$. By neglecting the

---

[2] Notice at this point that the flux Ψ_a in phase a depends on the phase current i_a but also on the phase currents i_b and i_c, due to a finite mutual inductance between the phase windings. Due to $i_a = -i_b - i_c$ and with reasonable assumption that $L_{ab} = L_{ac}$, the flux component $L_a i_a + L_{ab}i_b + L_{ac}i_c = L_a i_a + L_{ab}(i_b + i_c)$ can be rewritten as $L_S i_a$.

voltage drop across the stator resistance and by adopting notation $e_a = \mathrm{d}\Psi_{Rm(a)}/\mathrm{d}t = -\omega_m \Psi_m \sin(\theta_m)$, the voltage balance equation becomes $u_a = L_S\, \mathrm{d}i_a/\mathrm{d}t + e_a$. With that in mind, the rate of change of the stator current i_a can be expressed as

$$\frac{\mathrm{d}i_a}{\mathrm{d}t} = \frac{1}{L_S}(u_a - e_a) = \frac{1}{L_S}(\pm E - e_a). \qquad (22.2)$$

In operating conditions where $E \geq |e_a|$, the phase voltage u_a prevails in the above expression. For $u_a = +E/2$, the rate of change of the stator current is positive, notwithstanding the electromotive force e_a. For $u_a = -E/2$, the rate of change of the stator current is negative. In other words, it is possible to obtain an increase of i_a by turning on the switch Q_1 and a decrease of i_a by turning on the switch Q_2. These are the basic prerequisites for the current control. If the phase current i_a does not correspond to the reference value, the current controller detects the error $\Delta i_a = i_a^*$ $- i_a$. Then, it generates command signals for the inverter commutation so that the error Δi_a is reduced and eventually removed. In cases where the error is positive ($\Delta i_a > 0$), it is necessary to turn the upper switch Q_1 on and to obtain $\mathrm{d}i_a/\mathrm{d}t > 0$. This would reduce the error $\Delta i_a = i_a^* - i_a$. In cases where the error is negative ($\Delta i_a < 0$), it is necessary to turn on the lower switch Q_2 and to obtain reduction in the phase current ($\mathrm{d}i_a/\mathrm{d}t < 0$). This would drive the error $\Delta i_a = i_a^* - i_a$ toward zero. A simple control law that proves efficient in controlling the stator currents of three-phase AC motors is expressed by the following equation:

$$u_a = \frac{E}{2} \cdot \mathrm{sign}(i_a^* - i_a), \quad u_b = \frac{E}{2} \cdot \mathrm{sign}(i_b^* - i_b), \quad u_c = \frac{E}{2} \cdot \mathrm{sign}(i_c^* - i_c).$$

With the above control law, transistor power switches Q_1–Q_6 are commutated according to the sign of corresponding current error Δi. Whenever the error becomes positive, the upper switch is turned on, which increases the phase current and reduces the error. As soon as the error becomes negative, the lower switch is turned on to reduce the phase current and drive the error in positive direction. Hence, the straight application of the above control law may result in elevated commutation frequencies of the power transistors and increased commutation losses. In order to contain the commutation frequency, it is possible to introduce some hysteresis[3] into comparators which determine the sign of the current error. In this case, the commutation frequency is restrained by the hysteresis of comparators.

Practical current controllers operate in d–q coordinate frame. This means that the errors $\Delta i_d = i_d^* - i_d = 0 - i_d$ and $\Delta i_q = i_q^* - i_q$ are calculated from references and feedback signals in d–q coordinate frame. The current controller calculates the voltages u_d and u_q conceived to eliminate detected errors and drive the currents

---

[3] Comparator with hysteresis H generates the output on the basis of the previous comparison and on the new input. With previous output equal to $+1$, the input signal has to fall below $-H$ in order to produce the new output of -1. With previous output equal to -1, the input signal has to climb above $+H$ in order to produce the new output of $+1$.

i_d and i_q to their references. The voltages u_d and u_q are transformed to the stationary α–β coordinate system by inverse Park transform, and the obtained voltages u_α and u_β are passed to the inverse Clarke transform to obtain the voltage references for the phase voltages. Pulse-width-modulation-controlled three-phase inverter is used to supply the desired voltages to the three-phase motor.

The algorithm which calculates the control variable $u_{dq} = u_d + j u_q$ from the current error $\Delta i_{dq} = \Delta i_d + j \Delta i_q$ is called *control algorithm*, while the *regulator* or *controller* is the device which implements the algorithm and provides the output. Control algorithms are usually applied in digital form, by means of a program which is executed by digital signal processor. Regulator can often be described by its transfer function. The speed and quality of dynamic response depend on the transfer function of the controller.

The power converter with six switching power transistors supplies the three-phase stator winding by adjustable voltages. Therefore, it represents a controllable voltage supply. By closing the feedback loop which controls the stator phase currents, an appropriate current controller achieves that $i_a = i_a^*$, $i_b = i_b^*$, and $i_c = i_c^*$. This turns the current-controlled three-phase inverter into a current source. Hence, the three-phase machine behaves as if it were supplied from a current source. The operation of synchronous permanent magnet motor and controlled stator currents is similar to the operation of DC machine with constant excitation with controller armature current. The speed and accuracy in delivering the torque are fully dependent on characteristics of the current controller.

22.4 Field Weakening

Current control can be accomplished in conditions where the electromotive force e induced in the stator phase windings does not exceed the available phase voltage $E/2$, where E is the DC voltage across the intermediate circuit called DC link circuit of the three-phase inverter. Condition $E/2 \geq |e_a| = |\omega_m \Psi_m \sin(\theta_m)|$ must be fulfilled at every instant. In operation with $i_d = 0$, the stator flux in d-axis of the machine is equal to

$$\Psi_d = \Psi_{Rm} + L_d i_d = \Psi_{Rm}, \qquad (22.3)$$

while the stator flux in q-axis is $\Psi_q = L_q i_q$. In synchronous machines with surface-mounted magnets, the stator inductances L_d and L_q are very low, and they range from 0.01 to 0.05 relative units. Therefore, the flux along q-axis is considerably lower than the flux Ψ_{Rm}. For this reason, it is justifiable to assume that the air-gap flux amplitude equals $\Psi_m \approx \Psi_{Rm} + L_m i_d = \Psi_{Rm} = \Psi_n$, where Ψ_n is the rated flux. At the rated rotor speed $\Omega_n = \omega_n/p \approx E/\Psi_{Rm}/p = E/\Psi_n/p$, induced electromotive force is equal to the rated voltage, which is at the same time the maximum voltage available from the three-phase switching inverter. With $i_d = 0$, permanent magnet synchronous motor cannot exceed the rated speed.

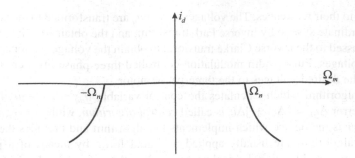

Fig. 22.2 Variation of defluxing current i_d in the field-weakening region

When approaching the rated speed, the electromotive force approaches the maximum available voltage. There is no more voltage margin and no possibility to control the stator current. In other words, there is no possibility to establish the stator currents required to obtaining the desired torque.

Operation at speeds above the rated speed requires the electromotive force to remain below the rated voltage. For that to achieve, it is necessary to reduce the flux, namely, to achieve the flux weakening. In order to keep the electromotive force $\omega_m \Psi_m$ within the boundaries of the available voltage, the air-gap flux Ψ_m has to be changed in terms of the rotor speed. With flux that changes as $\Psi_m(\Omega_m)$ $= \Psi_n \cdot \Omega_n/\Omega_m = \Psi_{Rm} \cdot \omega_n/\omega_m$, induced electromotive force at speeds above the rated speed retains its rated value $\Psi_{Rm} \cdot \omega_n$. It should be noted that the q-axis flux $L_q i_q$ is considerably smaller than the flux of permanent magnets. The later assumption holds due to very small inductances L_q and L_d in synchronous machines with surface-mounted magnets. With $\Psi_{Rm} \gg \Psi_q$ and $\Psi_m \approx \Psi_d$, desired change of the flux Ψ_m determines desired change of the flux Ψ_d, which changes according to

$$\Psi_d(\Omega_m)|_{\Omega_m > \Omega_n} = \Psi_{Rm} \frac{\Omega_n}{\Omega_m} = \Psi_{Rm} \frac{\omega_n}{\omega_m}. \tag{22.4}$$

With $\Psi_m \approx \Psi_{Rm} + L_d i_d$, it is concluded that the field weakening requires a certain negative current in d-axis, $i_d < 0$, which performs defluxing or demagnetization. The change of this defluxing current with the rotor speed is given in (22.5) and in Fig. 22.2. The rated flux of the machine Ψ_n is equal to Ψ_{Rm}:

$$i_d(\omega_m)|_{|\omega_m| \geq \omega_n} = -\frac{\Psi_{Rm}}{L_m}\left(1 - \frac{\Omega_n}{\Omega_m}\right) \approx -\frac{\Psi_n}{L_d}\left(1 - \frac{\omega_n}{\omega_m}\right). \tag{22.5}$$

Question (22.1): An isotropic synchronous machine with permanent magnets on the rotor has the stator inductance $L_S = 0.05$. Determine the maximum rotor speed in steady state, with no mechanical load attached to the shaft.

Answer (22.1): In operation with high speeds that exceed the rated speed, it is necessary to provide defluxing current in d-axis which has a negative value, $i_d < 0$. This current reduces d-axis flux and maintains the electromotive force within the limits of the available supply voltage. In the absence of the load torque, the electromagnetic torque and q-axis current are equal to zero. With $i_q = 0$, all of the available stator current can be used for weakening the field in d-axis. At steady state, the current in the stator windings must not exceed the rated current. Therefore, relative value of the defluxing current that can be maintained permanently is $i_d = -1$. By using the expression for the defluxing current $i_d(\omega_m)$ in the region of field weakening, one obtains $(1 - \omega_n/\omega_m) = L_S/\Psi_{Rm} = L_S/\Psi_n$. Relative value of the rated flux is equal to one, thus resulting in $\omega_n/\omega_m = 1 - 0.05 = 0.95$, which gives $\omega_m = 1.0526\,\omega_n$.

***

Previous example shows that synchronous motors with surface-mounted magnets and with very low stator inductance cannot perform any significant increase of the rotor speed above the rated value. Therefore, they cannot be used in the field-weakening operation. On the other hand, induction servomotor can operate in field-weakening mode and achieve the rotor speeds that exceed the rated speed by several times. This is due to the fact that the flux in induction motors gets produced by the stator current and not by permanent magnets. This shortcoming of synchronous servomotors is the principal reason that induction servomotors are still in use. Namely, other characteristics of these two types of servomotors go in favor of synchronous motors. Induction machines have the power of losses significantly larger than synchronous machines. While the rotor losses in synchronous machines are virtually zero, there are considerable rotor losses $P_{Cu2} = sP_n$ in induction machines, proportional to the slip. The absence of the rotor losses facilitates cooling of synchronous motors and allows larger current and flux densities. Therefore, synchronous servomotors have superior specific power and torque, and they are smaller than the equivalent induction servomotors. Moreover, their rotors have much lower inertia J, which is beneficial in most motion control applications. There are, however, applications of electrical actuators where it is very important to provide the operation in the field-weakening region and to provide the rotor speed that goes well beyond the rated speed. In these applications, induction servomotors are advantageously used.

Synchronous motors with surface-mounted permanent magnets have a very low stator inductance. This proves as an advantage in motion control applications such as industrial robots and manipulators. With a low inductance L_S, the rate of change of the stator current $di_a/dt = (u_a - e_a)/L_S$ is very large and it reaches $di_a/dt \approx 10^4$ I_n/s, which allows the electromagnetic torque $T_{em} = k\,|\Psi_{Rm} \times i_S|$ to make the change from zero to the rated value in some 100 μs. Fast torque response contributes to an increased bandwidth and improved closed loop performance in speed control and position control applications. For this reason, synchronous motors with surface-mounted magnets are applied in industry automation, robotics, and many other motion control applications that employ speed and position control loops. The absence of the field-weakening operations in these applications is not considered a significant shortcoming.

22.5   Transient and Steady-State Operating Area

Due to a very low inductance of the stator winding of $L_S = 0.01 \ldots 0.05$ relative units, the operation of synchronous motors with permanent magnets mounted on the rotor surface is limited to the range of $\Omega_m \in [-\Omega_n \ldots + \Omega_n]$. The maximum no load speed at steady state is $\Omega_{max(ss)} = \Omega_n/(1 - L_S)$ which exceeds the rated speed by several percents. The crossing of the steady-state operating limits and the abscissa in $T_{em} - \Omega_m$ diagram is at the speed $\Omega_n/(1 - L_S)$, which is just slightly above the rated speed. Regarding the torque steady-state operating limits, continuous torque is determined by the rated stator current. The steady-state torque has to remain within the limits of the rated torque, $-T_n \le T_{em} \le +T_n$. Steady-state operating area and transient operating area are shown in Fig. 22.3.

The transient operating limits depend on the maximum available instantaneous value of the stator current. The stator current may exceed the rated current for a brief interval of time. This does not have to result in excessive motor temperature, provided that in between the current pulses there is sufficient time with reduced current, so that the motor can release the heat and cool down. All the motion control tasks are usually performed in cycles. Within each cycle, the torque of relatively large value is delivered within relatively short interval of time, in order to perform desired acceleration or deceleration. These short intervals are mostly followed by prolonged intervals with reduced torque and reduced stator current. For this reason, the ratio between the peak torque and the rms value of the torque within one motion cycle is very large. The same conclusion holds for the peak and rms values of the stator current. Therefore, the transient operating area is limited by the peak torque which exceeds the rated torque by several times. Synchronous permanent magnet motors are designed and manufactured to sustain overloads of I_{max}/I_n in excess to five.

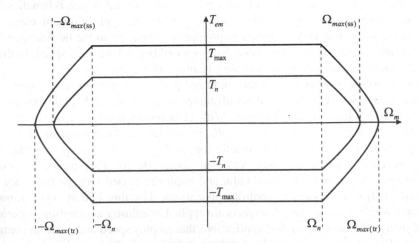

Fig. 22.3 The transient and steady-state operating limits of synchronous motors with permanent magnet excitation

The peak current capability is limited by the motor construction but also by the characteristics of the three-phase inverter which supplies the stator windings. Semiconductor power switches within the inverter comprise tiny silicon crystals with very low thermal capacity and with limited density of electrical current. In cases where the current density within semiconductor exceeds certain limit, there is an abrupt increase in temperature which changes the structure of the crystal and causes permanent damage to the semiconductor power switch. Semiconductor power switches can withstand electrical currents that exceed declared limits but only for relatively short intervals of time, measured in milliseconds. Large peak currents can also damage the synchronous servomotor. Large stator current may damage the permanent magnets. The stator currents produce the stator magnetomotive force, which depends on the current amplitude and on the number of turns. With large currents, the stator magnetomotive force produces large demagnetization field within the magnets. In *B-H* characteristic of permanent magnets, the operating point *(B,H)* of the magnets is moved closer to coercive field $H_C < 0$, where the induction *B* of permanent magnets reduces to zero. For most permanent magnet materials, reaching the coercive field H_C damages the magnet. Having reached this operating point, their remanent induction cannot return to the initial value. Instead, the remanent induction is decreased by two or three times. Therefore, the maximum permissible stator current is set to the value that does not bring the risk of damaging the magnets. At the same time, it has to be compatible with peak current capability of semiconductor power switches that are used within the three-phase inverter attached to the stator windings.

The maximum torque T_{max} which defines the limits of the transient operating area is determined by the peak current I_{max}. The maximum no load speed $\Omega_{max(tr)}$ that can be reached over short time intervals is determined by expression $\Omega_{max(tr)} = \Omega_n/(1 - I_{max}L_S)$, where the stator inductance and the peak current are both expressed in relative units.

Question (22.2): An isotropic machine with permanent magnets on the rotor has stator self-inductance of $L_S = 0.05$. Find the maximum rotor speed that can be reached for a short interval of time. The peak current capability is $I_{max} = 5$, while the mechanical load attached to the shaft is equal to zero.

Answer (22.2): By using expression for demagnetizing current $i_d(\omega_m)$ in the field-weakening region, one obtains that $(1 - \omega_n/\omega_m) = I_{max}L_S/\Psi_{Rm} = I_{max}L_S/\Psi_n$. Relative value of the rated flux is equal to one; thus, the ratio of the rated speed and the maximum speed is determined by $\omega_n/\omega_m = 1 - 5 \cdot 0.05 = 0.75$, resulting in $\Omega_m = 1.33 \cdot \Omega_n$.

***

Synchronous permanent magnet machines are also used in applications that require high efficiency, low losses, and high specific power, but where it is not necessary to effectuate quick changes of the electromagnetic torque. These applications do not include motion control tasks, and they do not use synchronous machines in speed control and position control loops. Some of the examples are the motors used in blowers, household appliances, and HVAC systems but also the generators in renewable sources of electrical energy, such as the wind turbines, the motors in electrical vehicle propulsion, auxiliary drives in automotive field, and similar. Superior efficiency, lower weight, and lower inertia of synchronous permanent magnet motors are the reasons for their use instead of corresponding induction machines. In order to improve the field-weakening performance of synchronous permanent magnet machines, it is possible to remove the permanent magnets from the rotor surface and to burry them deep into the rotor magnetic circuit. This results in larger values of the stator self-inductance. According to (22.5), larger L_S improves capability of synchronous machines to operate above the rated speed. Yet, the field-weakening performance of synchronous permanent magnet machine is inferior to that of induction machines. Synchronous permanent magnet machines that operate in the field-weakening range require considerable demagnetization current $i_d < 0$ which does not contribute to the torque, while increasing the amplitude of the stator current $I_S = \text{sqrt}(i_d^2 + i_q^2)$ and increasing the copper losses. For that reason, the applications requiring prolonged operation in field-weakening mode with considerable ratio Ω_m/Ω_n call for an induction machine.

Bibliography

1. Kloeffler RC, Kerchner RM, Brenneman JL (1949) Direct current machinery. McMillan, New York
2. Kimbark EW (1956) Power system stability. Synchronous machines, vol III. Wiley, New York
3. White DC, Woodson HH (1959) Electromechanical energy conversion. Wiley, New York
4. Shortley GH, Williams DE (1961) Elements of physics. Prentice-Hall, Englewood Cliffs
5. Edwards JD (1973) Electrical machines: introduction to principles and characteristics. Intertext Books International Textbook Company Limited, Aylesbury
6. Slemon GR, Straughen A (1980) Electrical machines. Addison Wesley, Reading
7. Petrović M (1987) Ispitivanje električnih mašina. Naučna knjiga, Beograd
8. Vukosavić SN (2007) Digital control of electrical drives. Springer, New York
9. Wildi T (2006) Electrical machines, drives and power systems. Pearson Prentice Hall, Upper Saddle River
10. Seeley S (1958) Electromechanical energy conversion. McGraw-Hill, New York
11. Brown D, Hamilton EP (1984) Electromechanical energy conversion. McMillan, New York
12. Fitzgerald AE, Kingsley C (1961) Electrical machinery. McGraw-Hill, New York
13. Alger PL (1965) Induction machines. Gordon and Breach, New York
14. Chalmers BJ (1988) Electric motor handbook. Butterworth, London
15. Krause PC, Wasynczuk O, Sudhoff S (1995) Analysis of electric machinery. IEEE Press, Piscataway

S.N. Vukosavic, *Electrical Machines*, Power Electronics and Power Systems,
DOI 10.1007/978-1-4614-0400-2, © Springer Science+Business Media New York 2013

Bibliography

Index